The Dynamics and Structure of the Liquid–Liquid Interface

Fitzwilliam College, Cambridge
September 1–3, 2004

FARADAY DISCUSSIONS
Volume 129, 2005

RS•C

The Dynamics and Structure of the Liquid–Liquid Interface

A General Discussion on The Dynamics and Structure of the Liquid–Liquid Interface was held at Fitzwilliam College, Cambridge, UK on 1st, 2nd and 3rd September, 2004

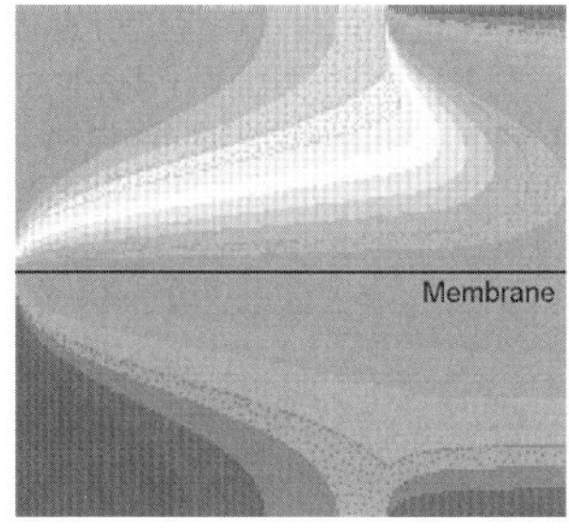

Cover
The image shows concentration distributions of a reagent introduced within a flowing solvent stream from the upper phase. The solute can then partition into the immiscible (lower) solution phase: both phases are flowing, under laminar conditions. Finite element methods were used to simulate the concentration distributions within each phase as a function of flow rate. This approach was thus used to optimise the electrode sensor position for the detection of reagents transferring across the liquid/liquid interface for experimentally feasible transport rates.

Image kindly supplied by Dr R. Dryfe, University of Manchester (UMIST), UK and Dr A. Fisher, University of Cambridge, UK.

Chemical biology articles published in this journal also appear in the *Chemical Biology Virtual Journal*:
www.rsc.org/chembiol

contents

INTRODUCTORY LECTURE

PAPERS AND DISCUSSIONS

This journal is © The Royal Society of Chemistry 2004

Spiers Memorial Lecture

Recent experimental advances in studies of liquid/liquid interfaces

Megan A. Leich and Geraldine L. Richmond

Department of Chemistry, University of Oregon, Eugene OR 97403-1253, USA

Received 11th October 2004, Accepted 11th October 2004
First published as an Advance Article on the web 22nd November 2004

Liquid/liquid interfaces play a key role in many important processes. Studying the molecular structure and interactions that occur at these interfaces can aid in our understanding of more complicated processes such as molecular transport across cell membranes. A variety of techniques have been applied to this pursuit. Here we present selected examples of exciting recent studies using different techniques to examine liquid/liquid interfaces.

Introduction

Liquid surfaces and interfaces have a central part in many chemical, physical, and biological processes. Many important processes occur at the interface between water and a hydrophobic liquid. Separation techniques are possible due to the hydrophobic/hydrophilic properties of liquid/ liquid interfaces. In biological systems, protein folding and membrane formation rely on the interaction of a hydrophobic surface with water, and ion and solute transport across these and other liquid/liquid interfaces are dependent on the interaction between the interfacial molecules in the hydrophobic fluid and the hydrophilic fluid with the ion or solute being studied. Reactions that proceed at interfaces are also highly dependent on the interactions between the interfacial solvent and solute molecules.

The interfacial structure and properties of molecules at these interfaces are generally very different from those in the bulk. Therefore, an understanding of the chemical and physical properties of these systems is dependent on an understanding of the interfacial molecular structure. However, these properties have traditionally been difficult to study at liquid/liquid interfaces, due to the buried nature of the system. For example, most spectroscopic techniques analyze the molecules in the bulk as well as those at the interface. Since there are considerably more bulk molecules than interfacial molecules, the spectral features arising from molecules at the interface are often overwhelmed by those from the bulk molecules. It is therefore necessary to use a surface-sensitive technique to study these interfacial molecules.

Recent developments in experimental surface techniques have facilitated the study of liquid/liquid interfaces. Some of these methods specifically probe interfacial molecules, while others are traditionally bulk techniques that have been adapted to study the interface. Second-harmonic generation and vibrational sum-frequency spectroscopy both provide molecular information that is inherently surface-specific. Ellipsometry and Brewster angle microscopy are also interfacial techniques. Ellipsometry has been used to provide information about adsorption and changes in

DOI: 10.1039/b415753m

Faraday Discuss., 2005, **129**, 1–21   **1**

molecular orientation at interfaces, while Brewster angle microscopy has recently been adapted to yield images of the adsorbate domain formation at liquid/liquid interfaces. Fluorescence spectroscopy has been used in several unique ways to provide insight into the environment of interfacial molecules. X-ray and neutron scattering have also recently been applied to liquid/liquid interfaces with impressive results. Here, we will present a few examples of recent applications of experimental techniques to the study of liquid/liquid interfaces. This is by no means meant to be a comprehensive review of the field. Instead, snapshots of recent progress in selected areas are provided, and more detailed information is available in the literature.

Surface second harmonic generation

Since the technique was pioneered more than twenty years ago, second harmonic generation (SHG) has proven to be a useful tool for studying interfacial molecular properties such as symmetry, orientation and number density of adsorbed molecules, and interfacial dielectric properties.[1–4] The first application of this technique to the study of liquid/liquid interfaces occurred in 1988.[5] Since then, a variety of liquid/liquid interfaces have been studied with SHG.[6,7]

Second harmonic generation is a second-order nonlinear spectroscopy and, under the electric dipole approximation, is forbidden in isotropic media, making it inherently surface-specific.[8] In SHG, a laser beam is directed onto an interface and generates another beam at twice the incident frequency. The intensity of the SHG beam is related to the resonant and the nonresonant contributions to the nonlinear susceptibility

$$I(2\omega) \propto |\chi_{\mathrm{NR}}^{(2)} + \chi_{\mathrm{Re},s}^{(2)}|^2 \, I^2(\omega) \tag{1}$$

where χ_{NR} and $\chi_{\mathrm{Re},s}$ are the nonresonant and resonant nonlinear susceptibility, respectively, and $I(\omega)$ is the intensity of the incident beam. The nonresonant contribution is generally much smaller than the resonant, which can be described by

$$\chi_{\mathrm{Re},s}^{(2)} = N \sum_{k,e} \frac{\langle A_{k,e} \rangle}{(\omega_{gk} - \omega - i\Gamma)(\omega_{eg} - 2\omega + i\Gamma)} \tag{2}$$

where N is the number of probed molecules, ω_{ij} is the transition energy between the ground state and the states k and e, $\langle A_{k,e} \rangle$ is the orientational average of the molecular hyperpolarizability, and Γ is the transition line width. When 2ω is resonant with ω_{eg}, $\chi_{\mathrm{Re},s}^{(2)}$ becomes large, enhancing the SHG intensity. By measuring the scaled intensity $I(2\omega)/I^2(\omega)$ as a function of 2ω, the excitation spectrum of a solute at an interface can be obtained. Since the effect relies on the resonant enhancement of $\chi_{\mathrm{Re},s}^{(2)}$, it is necessary to use a probe molecule that has a chromophore sensitive at an appropriate wavelength. Two selected examples of how SHG has been used to probe liquid/liquid interfaces are given below.

Knowledge of the changes in polarity at liquid/liquid interfaces is key to understanding the processes that occur at these interfaces. Polarity scales have helped to provide insight into processes in bulk solvents, but such a scale did not exist for liquid/liquid interfaces until recently. In 1998, Wang *et al.* developed an interfacial polarity scale based on their SHG measurements of the polarities of the water/chlorobenzene and water/1,2-dichloroethane interfaces.[9] Using the probe molecule *N*,*N*-diethyl-*p*-nitroaniline (DEPNA), shown in Fig. 1a, the positions of intramolecular π–π* charge transfer (CT) absorption bands were measured for DEPNA at these interfaces and used as an indicator of the interfacial solvent polarity. In order to develop a polarity scale that is independent of the probe molecule, the polarity of the air/water interface was measured by both DEPNA and another probe molecule, 4-(2,4,6-triphenylpyridinium)-2,6-diphenylphenoxide ($E_T(30)$), shown in Fig. 1b. The bulk polarity scale of $E_T(30)$, known as the $E_T(30)$ scale, is well established, and is one of the most comprehensive polarity scales. From the measured excitation wavelengths for both probe molecules, similar polarities of the air/water interface were calculated, indicating that the measured value is a good representation of the polarity of the air/water interface.

The interfacial CT wavelengths were also compared with the transition wavelengths in bulk liquids (Fig. 2). From these results, it was determined that the interfacial polarity has a very simple relationship to the polarities of the two liquids forming the interface: the interfacial polarity is the

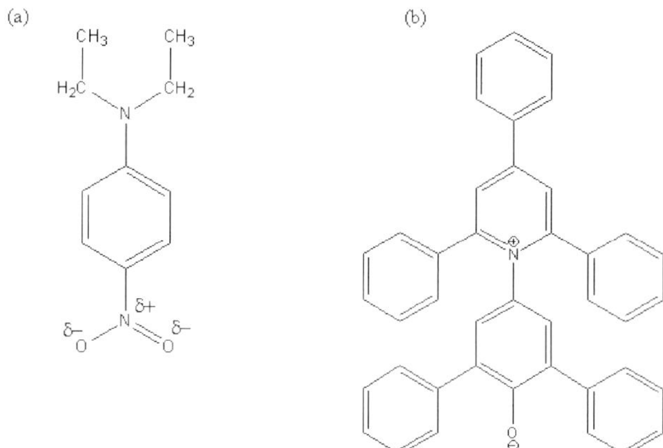

Fig. 1 (a) *N,N*-diethyl-*p*-nitroaniline (DEPNA). (b) 4-(2,4,6-Triphenylpyridinium)-2,6-diphenylphenoxide ($E_T(30)$).

average of the polarities of the two bulk phases

$$P_{A/B} = \frac{P_A + P_B}{2} \tag{3}$$

where $P_{A/B}$ is the polarity of the interface between liquids A and B, and P_A and P_B are the polarities of the bulk liquids. This relationship suggests that the difference in the ground and excited state interface solvation energies may be largely due to long-range, or solute–bulk solvent, interactions rather than local, or solute–interfacial solvent, interactions.

A second recent example of the application of SHG to liquid/liquid interfaces focuses on measuring the width of the interfacial region. Steel and Walker used a class of molecules known as "molecular rulers" to measure the width of liquid/liquid interfaces.[10] These molecules consist of a hydrophobic, solvatochromic probe based on *p*-nitroanisole with an anionic sulfate group attached by different length alkyl spacers (Fig. 3). At the liquid/liquid interface, the anionic sulfate group remained in the aqueous phase, while the hydrophobic spacer and chromophore extended into the organic phase. Using different length alkyl spacers allowed the distance the probe penetrated into the organic phase to be adjusted. By monitoring the excitation wavelength of the probe molecule as a function of chain length, they were able to determine the distance required to change the solvent polarity from the aqueous to the organic limit.

Fig. 4 shows SHG spectra of molecular rulers at the weakly associating cyclohexane/water interface. In bulk cyclohexane, the excitation maxima of the chromophores are centered around 295 ± 2 nm, and in bulk aqueous solution the maxima are at 318 ± 2 nm. When *p*-nitroanisole without an alkyl chain is adsorbed at the cyclohexane/water interface, the intensity maximum in the SHG spectrum occurs at 308 nm, which is consistent with the above assertion that interfacial polarity is the average of the polarities of the two bulk liquids.[9] As the chain length of the alkyl spacer increases from two to four to six carbons, the SHG maximum exhibits a shift towards the cyclohexane limit, indicating that the probe molecule is in an increasingly polar environment. Since the six-carbon chain molecular ruler samples a nonpolar environment, the interfacial environment changes from polar to nonpolar in less than 9 Å. These results suggest that the dipolar width, which is the distance for a dielectric environment to change from one phase to another, is molecularly sharp.

In contrast with the cyclohexane/water interface, the 1-octanol/water interface undergoes strong hydrogen bonding interactions between the two layers. Fig. 5 shows SHG spectra of the 1-octanol/water interface with C_2, C_4, C_6, and C_8 molecular rulers adsorbed to the interface. The window between the bulk excitation wavelengths for 1-octanol and water extends from 303 nm (bulk 1-octanol) to 318 nm (bulk water), but the SHG maximum for the adsorbed C_2 molecular rulers adsorbed at the 1-octanol/water interface is at 285 ± 2 nm, which is well outside the window. As the molecular ruler length is increased, the SHG maximum shifts to longer wavelengths. From these

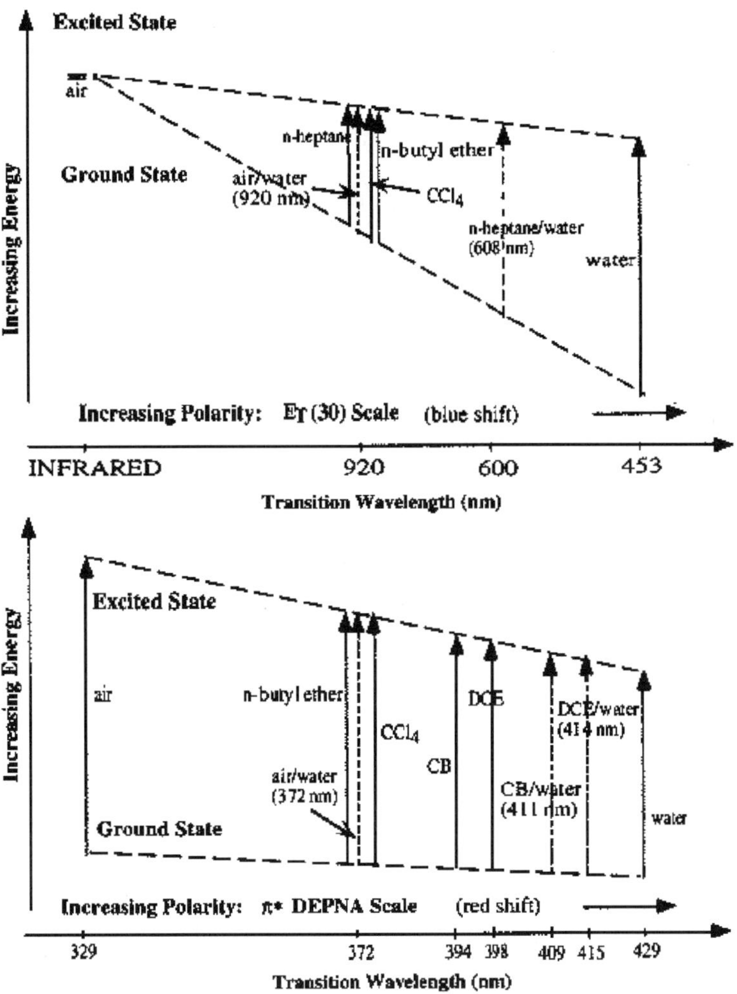

Fig. 2 $E_T(30)$ and DEPNA polarity scales. Observed transition wavelengths for $E_T(30)$ and DEPNA in bulk liquids and interfaces. CB is chlorobenzene and DCE is 1,2-dichloroethane. Solid arrows represent transitions in bulk liquids, and dotted arrows represent transitions at interfaces. The bracketed wavelengths are the predicted interface transition wavelengths. Reprinted with permission from ref. 9. © American Chemical Society 1998.

results, it was concluded that the adsorbed C_2 molecular ruler experienced an interfacial polarity that was less than that of either bulk phase. As the ruler length increased, the dielectric environment approached the polarity of the bulk solution. These spectra suggest that the 1-octanol undergoes surface-induced ordering, causing the interfacial octanol molecules to orient with the OH group toward the water layer and the eight-carbon chain pointing into the bulk octanol, creating a hydrophobic region between the polar layers of 1-octanol and water.

Vibrational sum-frequency spectroscopy

Vibrational sum-frequency spectroscopy (VSFS) is another second-order nonlinear optical method that is being increasingly applied to liquid/liquid interfaces. Like SHG, VSFS is inherently surface-specific. However, VSFS has the advantage over SHG of measuring the vibrational spectrum of the interfacial molecules. VSFS relies on the resonance between the infrared vibrational modes of the interfacial molecules and the tunable infrared beam for enhancements to the signal intensity. In

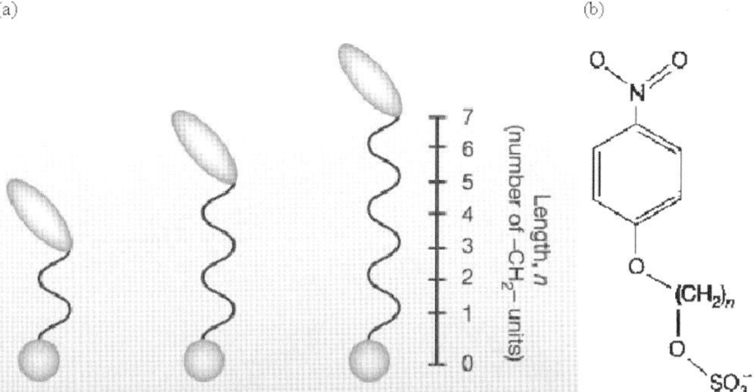

Fig. 3 Schematic representation of adsorbed molecular ruler surfactants, and their general structure. (a) Schematic representation of molecular ruler surfactants adsorbed to a liquid/liquid interface. As the alkyl spacer between the headgroup (circle) and solvatochromic chromophore (ellipse) lengthens, the hydrophobic probe can extend further into the organic phase. Correlating ruler length (in $-CH_2-$ units) with chromophore excitation wavelength enables experiments to determine the distance required for interfacial solvent polarity to converge to the organic limit. (b) General structure of molecular ruler surfactants. Surfactants are referred to as C_n rulers, where n corresponds to the number of methylene ($-CH_2-$) groups in the alkyl spacer. Reprinted with permission from ref. 10. © Nature Publishing Group 2003.

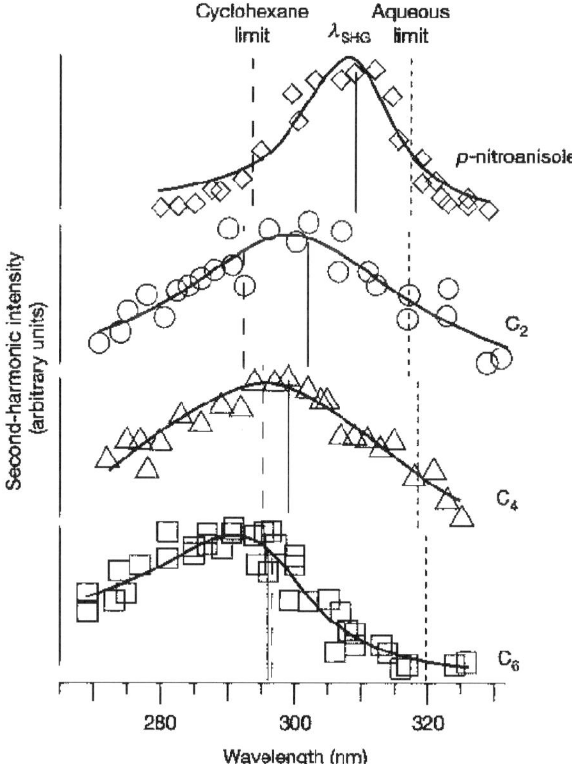

Fig. 4 Resonance-enhanced SHG spectra of (top to bottom) p-nitroanisole, C_2 rulers, C_4 rulers, and C_6 rulers adsorbed to a cyclohexane/water interface. Dashed and dotted lines denote excitation maxima in bulk cyclohexane and water, respectively. Solid vertical lines correspond to SHG maxima (λ_{SHG}) as determined by fitting the data. Note that SHG maxima do not always correspond to the wavelengths with highest SHG intensity, owing to the nonresonant contribution to $\chi^{(2)}$. Reprinted with permission from ref. 10. © Nature Publishing Group 2003.

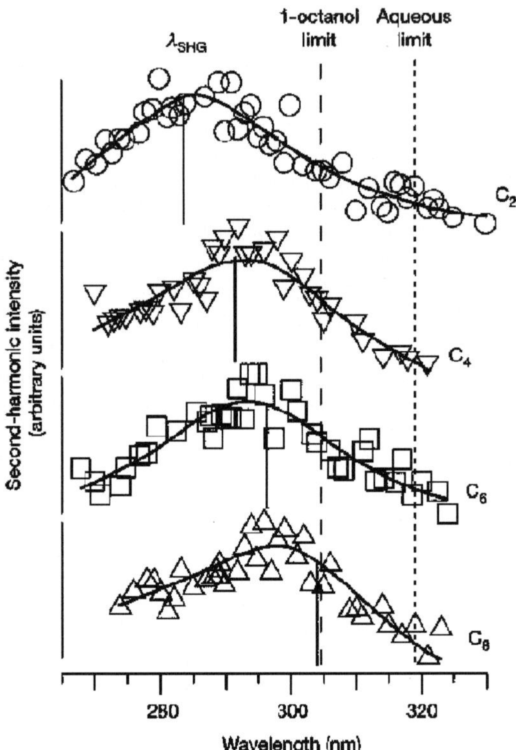

Fig. 5 Resonance-enhanced SHG spectra of (top to bottom) C_2 rulers, C_4 rulers, C_6 rulers, and C_8 rulers adsorbed to a 1-octanol/water interface. Dashed and dotted lines denote excitation maxima in bulk octanol and water, respectively. Solid vertical lines correspond to SHG maxima (λ_{SHG}) as determined by fitting the data. Reprinted with permission from ref. 10. © Nature Publishing Group 2003.

VSFS, two laser beams, one visible and one tunable infrared (IR), are overlapped on the interface, both in time and in space. These two beams then generate a third beam at the sum of the frequencies of the two incident beams. The intensity of the generated sum-frequency beam is given by

$$I(\omega_{SF}) \propto \left| \chi_{NR}^{(2)} + \sum_{v} \chi_{v}^{(2)} \right|^2 I(\omega_{vis})I(\omega_{IR}) \qquad (4)$$

where χ_{NR} and χ_{v} are the nonresonant and resonant nonlinear susceptibility, respectively. $I(\omega_{vis})$ is the intensity of the visible beam, and $I(\omega_{IR})$ is the intensity of the infrared beam. Additionally, χ_{v} can be written as

$$\chi_{v}^{(2)} \propto \frac{A_K M_{IJ}}{(\omega_v - \omega_{IR} - i\Gamma_v)} \qquad (5)$$

where A_K is the IR transition moment, M_{IJ} is the Raman transition polarizability, ω_v is the vibrational transition frequency, ω_{IR} is the frequency of the IR beam, and Γ_v is the transition line width. Tuning the IR beam over a vibrational transition therefore results in an enhancement of the generated sum-frequency signal, giving a spectrum of the interfacial molecules. Therefore, VSFS can be used to study molecules that have vibrations that lie within the range of tunability of the IR source, making it applicable to a wider range of molecules. The polarizations of all three beams can be selected in order to obtain orientational information about the interfacial molecules. The most commonly employed polarization scheme is ssp, where the sum-frequency is s-polarized, the visible is s-polarized, and the infrared is p-polarized.

The first application of VSFS to the study of a liquid/liquid interface was by Richmond and coworkers, who used it to study the adsorption of sodium dodecyl sulfate (SDS) at the CCl_4/D_2O interface.[11] These initial measurements demonstrated that VSFS was a powerful tool that could be applied to obtain information about these buried interfaces. D_2O was used in these studies instead of H_2O to eliminate the overlap of the intensity of the O–H stretching modes with the C–H modes of interest. The success of this experiment was largely due to the use of a total internal reflection (TIR) geometry, which enhanced the weak signal to a detectable level.

Since VSFS is a nonlinear optical technique, the generated signal is weak, especially with nanosecond lasers, which have lower peak intensities relative to picosecond and femtosecond lasers. In addition to the resonant enhancement, the sum-frequency intensity is also dependent on the linear Fresnel factors for the input beams and the nonlinear Fresnel factor for the generated sum-frequency beam. In the TIR geometry, the visible beam is brought to the interface through the liquid with the higher index of refraction, and with the incident angle at or just past the critical angle such that it is totally internally reflected. By using this configuration, the sum-frequency signal is enhanced by several orders of magnitude.[12]

In addition to proving that VSFS is a viable technique for studying liquid/liquid interfaces, the work by Messmer *et al.* also provided evidence for the interfacial ordering of the adsorbed surfactant.[11,12] By using the ssp polarization scheme and comparing the relative intensities of the symmetric stretches of the methyl group on the end of the chain (2866 cm^{-1}) and the methylenes in the chain (2844 cm^{-1}), they were able to determine that the chain ordering increased as the interfacial surfactant concentration increased. The methyl symmetric stretch is largest for alkyl chains in an entirely *trans* configuration. As the number of *gauche* defects in the chains increases, the intensity of the methyl peak decreases, and the intensity of the methylene peak increases. The methylene intensity remained relatively constant as the concentration was increased, but the methyl intensity increased significantly, indicating that the chains were reorienting such that a larger number of chains were oriented with the dipole transition moments of the methyl groups having a component normal to the interfacial plane.

More recent surfactant studies at liquid/liquid interfaces detailed the adsorption of other surfactants at the CCl_4/D_2O interface[13] and at the d_{34}-hexadecane/D_2O interface.[14] The work by Watry and Richmond[13] examined the adsorption of linear alkylsulfonate and linear alkylbenzene sulfonate surfactants at the CCl_4/D_2O interface. These two classes of molecules make up a significant percentage of the surfactants used in the detergent industry, making it important to understand their properties at interfaces. It was found that the alkyl chains of dodecylbenzene-sulfonate are much more disordered at the interface than those of dodecanesulfonate. The increased degree of disorder is attributed to the presence of the benzene ring, which exists at the interface in a staggered arrangement.

Bain and coworkers have developed a new experimental arrangement to acquire VSFS spectra of hexadecyltrimethylammonium bromide (CTAB) at the hexadecane/water interface (Fig. 6).[14] By using a thin film of d_{34}-hexadecane and D_2O with the protonated surfactant, they were able to minimize the absorption of the infrared beam by the C–H stretches of the hexadecane. At low concentrations, it was found that CTAB forms conformationally disordered monolayers. As the concentration of the CTAB approached the critical micelle concentration (c.m.c.), the spectra show a change to a more ordered, upright chain conformation, as might be expected from the previous surfactant studies already discussed.

Another focus of VSFS studies of liquid/liquid interfaces has been on the interfacial structure and orientation of water at the water/organic liquid interface. The CCl_4/water interface provides an excellent starting point, since the CCl_4 is transparent in the C–H and O–H stretching regions, in contrast to many hydrophobic liquids. The VSFS spectrum of the neat CCl_4 interface is presented in Fig. 7, along with the individual peaks that were used to fit the spectrum.[15] The free OH feature arises from water molecules that straddle the interface, with one O–H pointing into the CCl_4 and the other pointing into the water, as depicted in Fig. 8. The O–H pointing into the CCl_4 cannot hydrogen bond, and is therefore at a higher energy than the other O–H bond, or donor OH, which is hydrogen-bonded to other water molecules. The frequency of the free OH band is slightly red-shifted from what is measured for the air/water interface. This demonstrates the presence of a weak CCl_4/H_2O interaction at this interface that results in the molecular orientation of the water molecules. Also observed are donor OH modes, tetrahedrally coordinated water molecules within

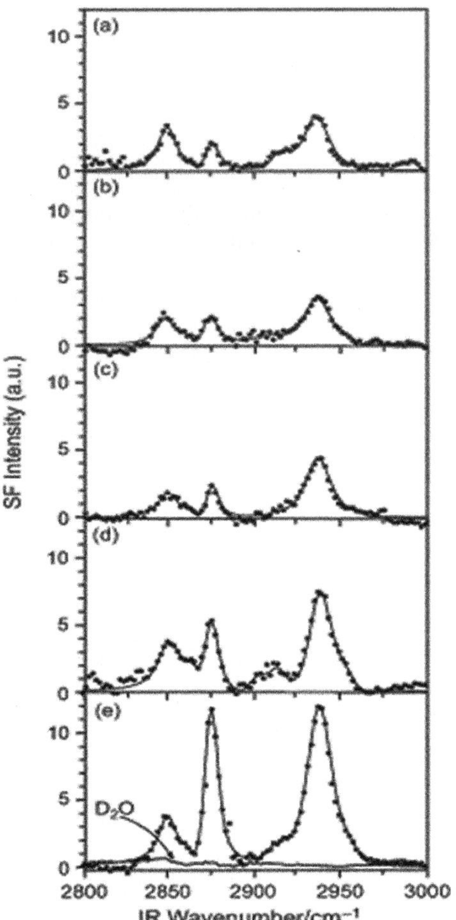

Fig. 6 SF spectra of CTAB at the oil/water interface (ssp-polarization). From top to bottom: (a) 0.05 mM CTAB, (b) 0.1 mM CTAB, (c) 0.3 mM CTAB, (d) 0.5 mM CTAB, (e) 0.6 mM CTAB compared with pure D_2O. Reprinted with permission from ref. 14. © American Chemical Society 2003.

the interfacial region, and the symmetric and antisymmetric stretching modes of monomeric water molecules.

The interfacial structure and orientation of alkane/water interfaces have also been studied, and subsequently compared with the CCl_4/water results.[16] Fig. 9 shows the VSFS spectra of three different alkane/water interfaces using the ssp polarization combination. The spectra of the interfaces of water with hexane, heptane and octane are all very similar to that of the CCl_4/water spectrum. However, the free OH is slightly blue-shifted from that of the CCl_4/water interface, indicating that the interaction between the water and the hexane, heptane and octane is slightly weaker. The free OH intensities are also lower than for the CCl_4/water interface, which is consistent with the weaker interaction between water and the alkanes. The weaker interaction could give the water molecules more rotational freedom, producing more possible orientations of the interfacial water molecules, and thus lower VSF intensities. To obtain a more complete picture of the interfacial water structure, Richmond and coworkers have complimented their VSFS studies with molecular dynamics simulations that allow the calculation of the number density of various interfacial water species.[16] This combination of theory and experiment allows the development of other interfacial molecular models while also providing information about isotropically oriented molecules that is difficult to obtain with VSFS.

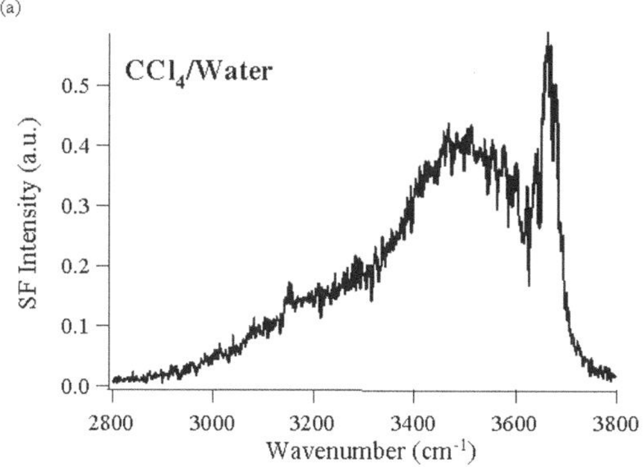

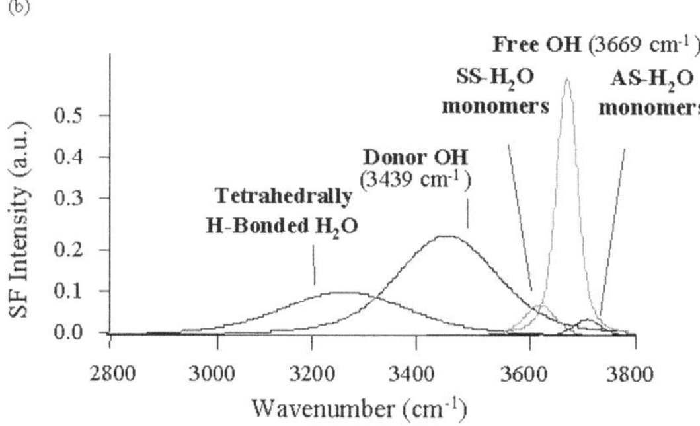

Fig. 7 (a) VSFS spectrum of the neat CCl_4/water interface (ssp polarization). (b) Individual peaks that contribute to the spectrum, derived from a nonlinear least squares fit to the data.

The 1,2-dichloroethane (DCE)/water interface has drawn considerable attention due to its electrochemical relevance as an interface between two immiscible electrolyte solutions.[6,17,18] Of particular issue is the question of whether the interfacial region consists of two separate and distinct phases or whether there is a significant degree of mixing between the two layers. To help further the understanding of this system, VSFS spectra were taken of the $(CCl_4 + DCE)$/water interface with varying mole fractions of DCE, from pure CCl_4 to pure DCE, and are presented in Fig. 10.[19] The loss of the free OH peak in the DCE/water spectrum and the overall decrease in the sum-frequency

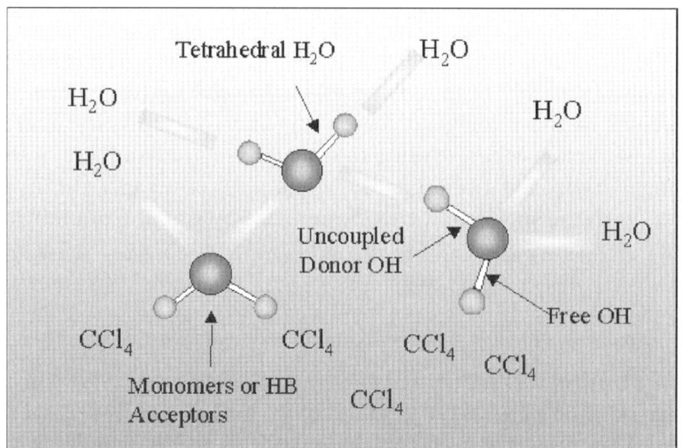

Fig. 8 Schematic of the interfacial water species contributing to the VSF spectrum of the neat CCl_4/water interface. Labeled species correspond to individual peaks from the spectra fit.

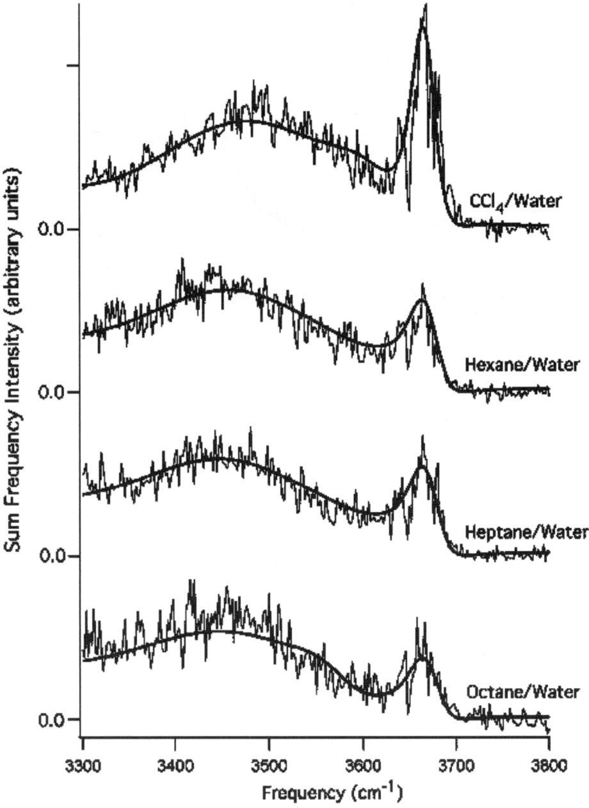

Fig. 9 VSF spectra of the CCl_4/water, hexane/water, heptane/water, and octane/water interfaces (ssp polarization). The solid lines are fits to the experimental data. Reprinted with permission from ref. 16. © American Chemical Society 2003.

intensity as the mole fraction of DCE increases indicate that the interfacial water molecules are more randomly oriented than at the CCl_4/water interface, and are hydrogen bonding with other water molecules.

The solvation of charge at a liquid/liquid interface has been the focus of a series of VSFS studies by Scatena and Richmond.[20] These studies examine the structure and bonding of water around the charged head group of isolated surfactants adsorbed to the CCl_4/water interface. As shown in Fig. 11, trace amounts of SDS have a dramatic effect on the interfacial water spectrum, making it possible to measure the solvation shell spectrum at very low surfactant concentrations. Water in the solvation shell is found to be highly oriented and displays weak hydrogen bonding interactions with adjacent water molecules in the interfacial region. As the concentration increases, the hydration shells begin to interact with each other, leading to the observed red-shifting of the water peak, which is indicative of stronger hydrogen bonding. By monolayer coverage, the hydrogen bonding has become even stronger, with the spectrum being dominated by signal from tetrahedrally coordinated water molecules.

Ellipsometry

Ellipsometry is another optical technique that can be applied to the study of liquid/liquid interfaces. By comparing the polarization states of the beam before and after it is reflected off of the interface, many optical material properties can be determined, such as the interfacial thickness and refractive index. As with many other properties, the surface refractive index is often different than that of the bulk for sensitive studies. Using the settings for the waveplates and polarizers in the ellipsometer,

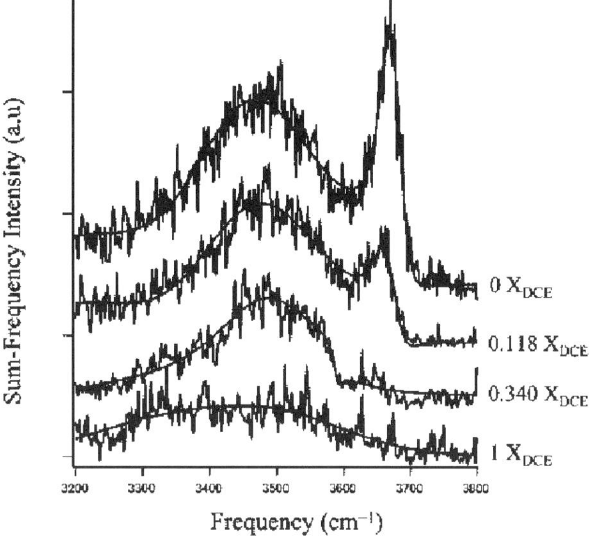

Fig. 10 VSF spectra of water at (DCE + CCl$_4$)/H$_2$O interfaces (ssp polarization). The top graph (X$_{DCE}$ = 0) corresponds to the pure CCl$_4$/H$_2$O interface, and the bottom graph (X$_{DCE}$ = 1) corresponds to the pure DCE/H$_2$O interface; the remaining graphs correspond to indicated mole fractions of DCE at the (DCE + CCl$_4$)/H$_2$O interface. Spectra are offset vertically for visual clarity. Reprinted with permission from ref. 19. © American Chemical Society 2004.

the ellipsometric parameters Ψ and Δ can be determined. These parameters, which characterize the polarization of the reflected light, can be used to calculate the complex ratio of the Fresnel coefficients.

$$\frac{r_p}{r_s} = \rho = \tan \psi e^{i\Delta} \tag{6}$$

where r_p is the reflection coefficient for p-polarized light and r_s is the reflection coefficient for s-polarized.

While surface freezing had been observed for long-chain alkanes at the air/alkane interface,[21] it had not been observed at the alkane/water interface. Recently, however, Lei and Bain have shown that the addition of a simple cationic surfactant can induce surface freezing at the tetradecane/water interface.[22] Using ellipsometry, they have shown that the presence of hexadecyltrimethylammonium bromide (CTAB) can induce surface freezing at the tetradecane/water interface. The coefficient of ellipticity, $\bar{\rho}$, was measured as a function of temperature (Fig. 12). $\bar{\rho}$ is defined as Im(r_p/r_s) at the Brewster angle θ_B, where θ_B is the angle where Re(r_p/r_s). The coefficient of ellipticity, which is related to the interfacial thickness, shows a large change indicative of a first order phase transition at 12 °C, which is 6 °C above the bulk melting point of tetradecane. The effect can be observed even for mole fractions of CTAB as low as 0.1. By modeling the ellipsometric data, it was determined that the frozen layer has a thickness of one monolayer, and the chains are densely packed and oriented nearly normal to the interface. This study is the first known observation of surfactant-induced surface freezing at an alkane/water interface.

Another recent study using ellipsometry to study liquid interfaces is that of Binks *et al.*, in which monodisperse silica particles (25 nm diameter) were studied at the toluene/water interface.[23] Ellipsometric studies were performed to determine the contact angles of the hydrophobized particles at the interface. However, the agreement between experiment and theory is not sufficient to calculate the contact angle. Although the results showed that ellipsometry alone is not sufficient for determining the contact angles in these systems, other interesting information was obtained. It was determined that the surface is not smooth, and is about three layers thick, as shown in Fig. 13. They also observed a large change in the ellipsometric parameter Δ, but little change in Ψ with

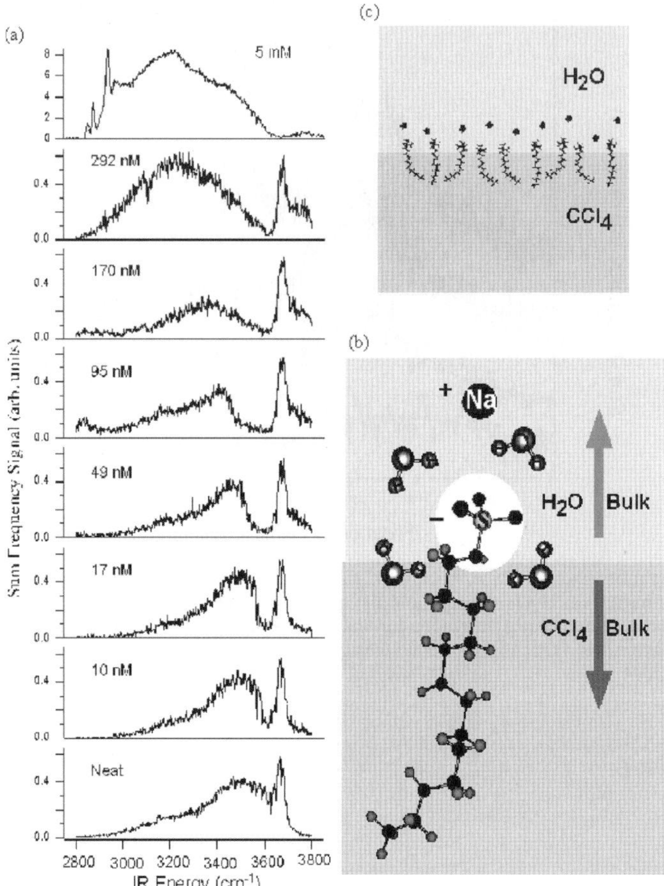

Fig. 11 (a) VSF spectra of the CCl$_4$/water interface with increasing concentrations of SDS, from the neat interface to 5 mM SDS (monolayer coverage) (ssp polarization). (b) Schematic of an SDS molecule adsorbed to the interface, along with the water molecules solvating the head group. (c) Schematic of a monolayer of SDS molecules ordering at the CCl$_4$/water interface.

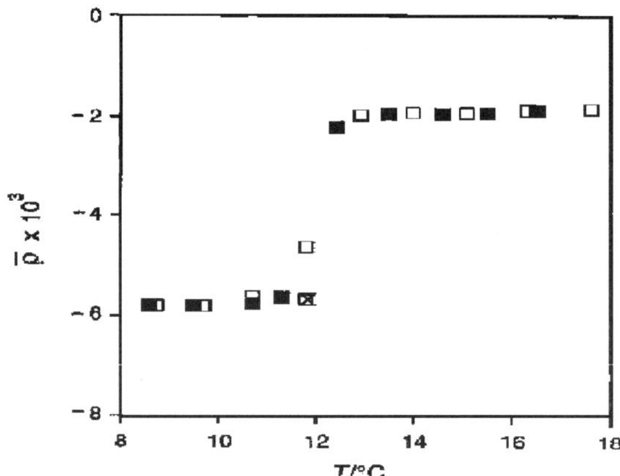

Fig. 12 Coefficient of ellipticity $\bar{\rho}$ as a function of temperature for the interface between a 0.6 mM solution of CTAB and tetradecane. Reprinted with permission from ref. 22. © American Physical Society 2004.

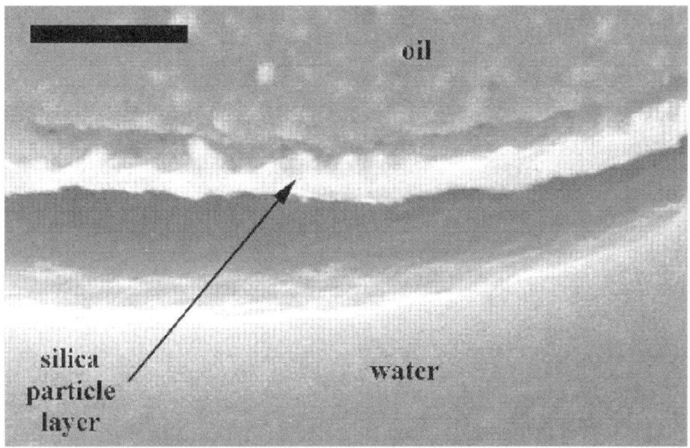

Fig. 13 Freeze–fracture SEM image of an oil-in-water emulsion drop stabilized with 25 nm diameter hydrophobized silica particles, showing a section through the oil/water interface. The bar is 500 nm. Reprinted with permission from ref. 23. © American Chemical Society 2003.

increasing amounts of particles spread at the interface, which can be seen in Fig. 14. At low concentrations up to monolayer coverage, theoretical calculations of Δ agree well with experimental results. At high concentrations, the results are consistent with a surface coverage greater than a close-packed monolayer or a corrugated monolayer with an amplitude smaller than the particle radius.

Brewster angle microscopy

Brewster angle microscopy (BAM) has been used for a number of years to study air/liquid interfaces, but has only recently been applied to liquid/liquid interfaces.[24] BAM is a space-resolved imaging ellipsometric technique in which a beam is brought to the neat interface at Brewster's angle. The subsequent adsorption of surfactants to the interface then changes the interfacial refractive

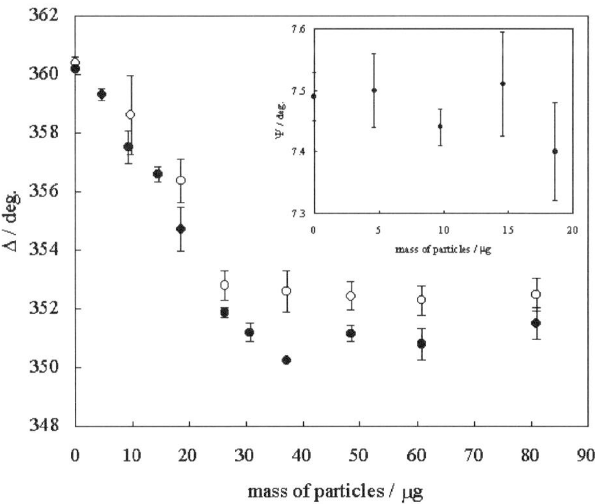

Fig. 14 Ellipsometric parameter Δ for hydrophobized silica particles, with a diameter of 25 nm, at the toluene/water interface, plotted against the mass of particles spread. ○, particles spread from methanol; ●, particles spread from 2-propanol. The inset shows the ellipsometric parameter Ψ plotted against the mass of the particles. Reprinted with permission from ref. 23. © American Chemical Society 2003.

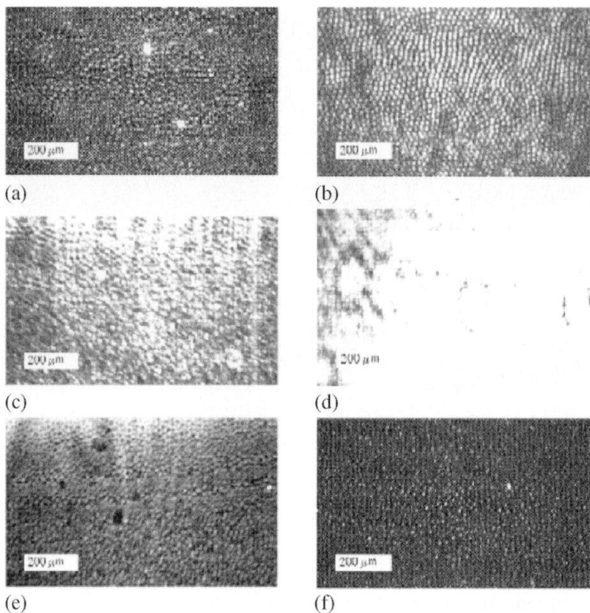

Fig. 15 BAM images of an adsorbed film of $FC_{12}OH$ at the hexane/water interface at different temperatures: (a) 21 °C, (b) 20 °C, (c) 15 °C, (d) 12 °C, (e) 15 °C, (f) 23 °C; (a–d) stepwise decrease of T; (e, f) stepwise increase of T. Reprinted with permission from ref. 24. © American Chemical Society 1999.

index and therefore the magnitude of the reflected portion of the beam, allowing the imaging of domain patterns of adsorbates.

The first application of BAM to the study of liquid/liquid interfaces was by Uredat *et al.*, in which BAM was used to study temperature-induced phase transitions in Gibbs monolayers of octadecanol ($C_{18}OH$) and 1,1,2,2-tetrahydroperfluorododecanol ($FC_{12}OH$) at the hexane/water interface.[24] The BAM images of $FC_{12}OH$ are shown in Fig. 15. At 21 °C, small domains of the condensed film are visible at the interface. As the temperature is decreased, the size and interfacial coverage of these domains increase. At the lowest temperature, 12 °C, the condensed phase seems to have entirely covered the interface. The final two frames of Fig. 15 show the films as the temperature is increased. Increasing the temperature from 12 °C to 15 °C produces dark patches in the films, which are attributed to "holes" in the film caused by a gas-like film within the condensed phase, even though the temperature is still well below the phase transition temperature ($T_t = 24 \pm 1$ °C). Finally, when the temperature is increased to the phase transition temperature, small domains of the condensed phase can be seen in coexistence with the gas-like phase.

Fig. 16 presents BAM images showing the domain patterning of $C_{18}OH$ at the hexane/water interface at 15 °C, 12 °C and 10 °C. The condensed phase domains are generally larger for $C_{18}OH$ than for $FC_{12}OH$. All of these temperatures are below the phase transition temperature of 24 °C. At 15 °C and 12 °C, the images are dominated by regions that are roughly circular, but at 12 °C these circular regions coexist with continuous regions of the condensed phase, as seen in Fig. 16b. When the temperature is decreased to 10 °C, the image shows bands of the continuous regions and oval, deformed domain regions oriented along the direction of flow within the interface. When the temperature is held constant and the system is monitored over time (Fig. 16d–f), the domains become polyhedral in shape and eventually are composed almost entirely of the continuous, condensed phase. The final image shows the crystallization of the interface after 111 min. Surprisingly, the coexistence of the condensed and expanded phases of the $C_{18}OH$ is not limited to a single temperature, as is expected for a first-order phase transition, but instead extends over ~ 15 K below the phase transition temperature. This behavior can be attributed to trace surface-active impurities.

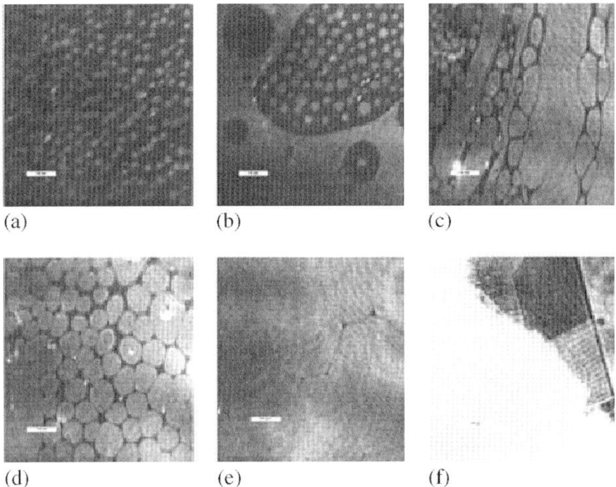

Fig. 16 BAM images of $C_{18}OH$ at the hexane/water interface, showing domain structures along a descending scan below the phase transition temperature ($T_t = 24$ °C): (a) 15 °C, (b) 12 °C, (c–f) 10 °C, and illustrating the evolution of domain patterns as a function of time after the temperature step from 12 to 10 °C for (c) 45 min, (d) 100 min, (e) 108 min, and (f) 111 min. The white bar represents a length of 100 μm. Reprinted with permission from ref. 24. © American Chemical Society 1999.

Total internal reflection fluorescence spectroscopy (TIRFS)

Through the use of a fluorescent dye as a probe, total internal reflection fluorescence spectroscopy (TIRFS) provides a method for studying adsorption at liquid/liquid interfaces. In TIRFS, a laser beam is directed onto a liquid/liquid interface in a total internal reflection geometry. The incident beam is polarized perpendicular to the incident plane (s-polarized), and the incident angle is chosen to be sufficiently greater than the critical angle for total internal reflection. When the beam hits the interface in a TIR geometry, it induces an evanescent wave that decays exponentially with vertical distance from the interface, allowing molecules at or near the interface to be probed. Since fluorescence is not inherently surface-specific, signals can arise from molecules in the bulk liquid. To remove any unwanted bulk signal, the fluorescence is measured both from the interface and from the bulk incident medium to allow subtraction of the background fluorescence signal.[25]

In a recent study, Yamashita *et al.* used TIRFS to study the solvation dynamics of two fluorophores (Fig. 17) at the heptane/water interface.[26] The fluorophores, 12-(9-anthroyloxy) stearic acid (12-AS) and 4-(9-anthroyloxy) butanoic acid (4-ABA), both have an anthroyloxy group as their fluorophore, and the emission wavelength is known to be a good indicator of the polarity of the solvent. Both probe molecules have emission wavelengths of ~ 480 nm in methanol and ~ 460 nm in heptane, which shows that the emission wavelength depends on the fluorophore

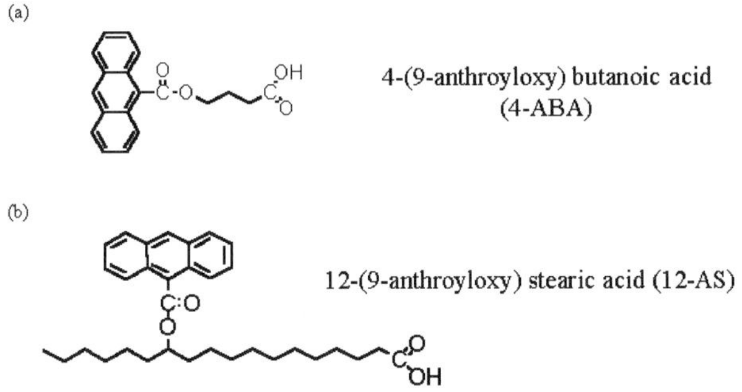

Fig. 17 Structures of (a) 4-(9-anthroyloxy) butanoic acid (4-ABA) and (b) 12-(9-anthroyloxy) stearic acid (12-AS).

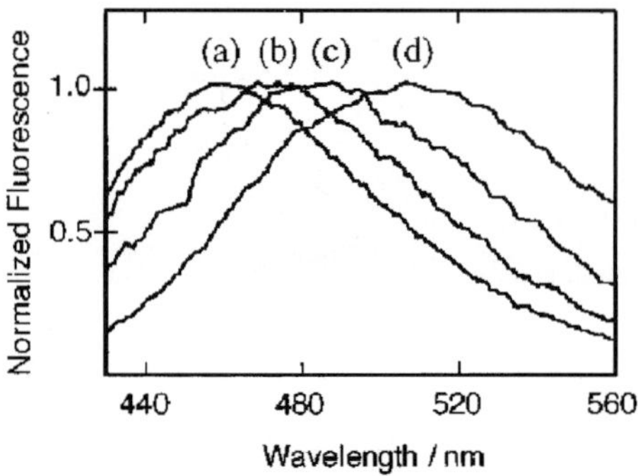

Fig. 18 Fluorescence spectra of 12-AS and 4-ABA. (a) 1×10^{-5} M 12-AS in heptane. (b) 12-AS at the heptane/water interface. [12-AS] $= 1 \times 10^{-8}$ M in heptane. (c) 4-ABA at the heptane/water interface, [4-ABA] $= 5 \times 10^{-6}$ M in phosphate buffer (pH 7.0). (d) 1×10^{-5} M 4-ABA in aqueous phosphate buffer (pH 7.0). Reprinted with permission from ref. 26. © American Chemical Society 2003.

and not on the structure of the rest of the molecule, making comparisons possible between the different molecules in their respective solvent layers.

The oil-soluble fluorophore, 12-AS, was studied in heptane and at the heptane/water interface, and the water-soluble fluorophore, 4-ABA, was studied in a pH 7 phosphate buffer and at the heptane/phosphate buffer interface. As evidenced by Fig. 18, the emission maximum of each fluorophore shifts between the bulk solvent and the heptane/aqueous interface, showing that the polarity of the heptane/water interface is between the polarities of bulk heptane and water. Further, the fluorescence maxima for 4-ABA and 12-AS at the heptane/water interface differ from each other, with 4-ABA red-shifted relative to 12-AS. This indicates that the probe molecules are in interfacial environments with different polarities, and that the solvation environment of the fluorophores at the heptane/water interface is dependent on the structure of the probe molecule.

The time-dependent fluorescence spectra of the two fluorophores at their respective interfaces and in a binary mixture of 3 M ethanol in heptane were also measured.[26] At the heptane/water interface, the emission wavelength of 12-AS showed a time-dependent red-shift, while the center wavelength of 4-ABA at the heptane/buffer interface remained unchanged. When the two fluorophores were studied in the mixture of ethanol and heptane, they both showed a time-dependent spectral shift. The results of these studies indicate that 12-AS undergoes preferential solvation at the heptane/water interface, with the fluorophore solvated by more heptane than water in the ground state and by more water than heptane in the excited state. On the other hand, 4-ABA does not undergo preferential solvation. It is solvated by more water molecules than heptane molecules, and is located closer to the water phase than 12-AS.

Another application of TIRFS to the study of liquid/liquid interfaces has been in the area of molecular recognition. Molecular recognition is important in biological systems, where it proceeds at microscopic interfaces such as cell and protein surfaces in the aqueous phase. It occurs through noncovalent interactions such as hydrogen bonding or electrostatic or hydrophobic interactions. However, due to experimental difficulties, only a few spectroscopic studies have been done at liquid/liquid interfaces.[27–29]

Recently, time-resolved TIRFS was used to observe molecular recognition mediated by hydrogen bonding at a CCl_4/water interface.[30] Riboflavin was placed in the aqueous phase, and was studied at the CCl_4/water interface with and without N,N-dioctadecyl-[1,3,5]triazine-2,4,6-triamine (DTT) in the CCl_4 phase (Fig. 19). These molecules are capable of forming triple hydrogen bonds with each other, but since they are soluble in different phases, it was expected that such a process should only occur at the interface. The fluorescence decay profiles of riboflavin in the absence of DTT were fit with a single-exponential, while a double-exponential was necessary to fit the decay measured in the presence of DTT. Fluorescence anisotropy was also used to measure reorientation times, which showed that both riboflavin and DTT have faster reorientation times than the hydrogen-bonded complex, thereby indicating that molecular recognition mediated by the formation of a triple hydrogen bonded complex can take place at an interface.

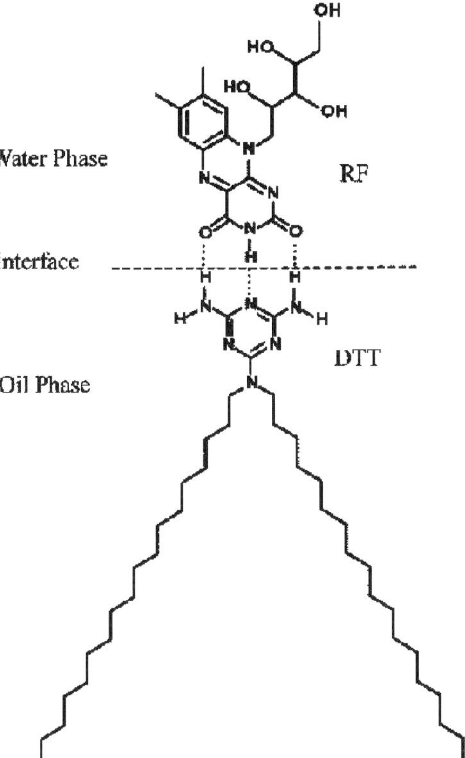

Fig. 19 Schematic of molecular recognition between riboflavin and DTT mediated by hydrogen-bonding interactions at an oil/water interface. Reprinted with permission from ref. 30. © American Chemical Society 2003.

X-Ray and neutron scattering

X-Ray and neutron scattering are increasingly being used to explore the liquid/liquid interface. Both scattering techniques are used to obtain information about the structure of thin films. This information is extracted by modeling the interface as a series of layers with different parameters that are adjusted until the model agrees with experimental results. Fig. 20 presents a vapor-tight stainless steel sample cell used for X-ray scattering measurements, which enables the study of the interface between two thin liquid layers in a vacuum environment.[31]

Li *et al.* used off-specular diffuse X-ray scattering to study monolayers of $CF_{12}OH$ at the hexane/water interface.[32] The system was studied previously with Brewster angle microscopy,[24] and has also been presented here. However, the use of X-ray scattering allowed access to domains that were smaller than the resolution of BAM. In this study, the size and distribution of the interfacial domains in $CF_{12}OH$ monolayers were probed as a function of temperature. It was found that the domain size and the mean separation distance remain constant over a range of temperatures, as can be seen in Fig. 21. In contrast, the interfacial coverage changes by a factor of ten over the same temperature range (Fig. 22). These results indicate that new domains are created or destroyed as the temperature is changed. Additionally, the domain sizes remain constant when heating and cooling through the solid–gas transition. This suggests that the domain sizes are at equilibrium, which can be established through the exchange of $CF_{12}OH$ molecules between the monolayer and the bulk hexane solution.

Another recent study of a liquid/liquid interface uses X-ray reflectivity to study the interface between water and *n*-alkanes with carbon numbers 6–10, 12, 16, and 22.[31] Fig. 23 presents the logarithm of the X-ray reflectivity, normalized to the Fresnel reflectivity, for each of the interfaces

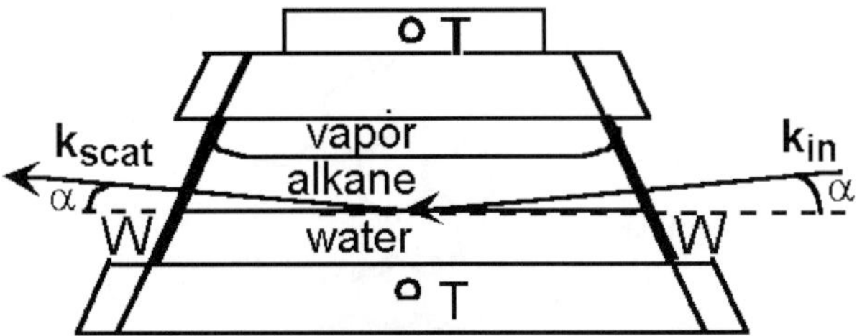

Fig. 20 Cross-sectional view of sample cell: W, Mylar windows; T, thermistors to measure temperature. The kinematics of surface X-ray reflectivity is also indicated: k_{in} is the incoming X-ray wave vector, k_{scat} is the scattered wave vector, and α is the angle of incidence and reflection. Reprinted with permission from ref. 31. © American Physical Society 2000.

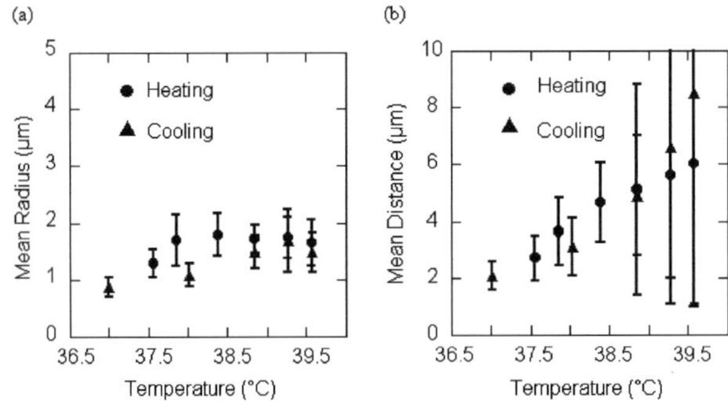

Fig. 21 (a) Mean radius of domains ($\bar{R}$) *vs.* temperature and (b) mean separation distance between domains ($\bar{D}$) *vs.* temperature. ●: heating curve; ▲: cooling curve. The error bars indicate $\pm\sigma_{R,D}$, a measure of polydispersity in $\bar{R}$ and $\bar{D}$. Reprinted with permission from ref. 32. © EDP Sciences 2002.

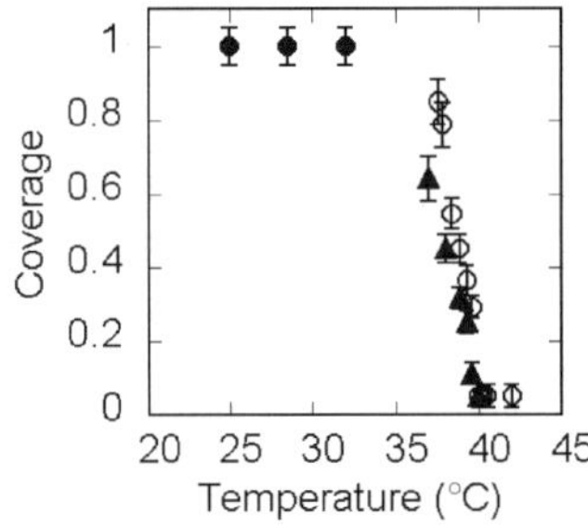

Fig. 22 Interfacial coverage (fraction of interface covered by the low-temperature solid phase) as a function of temperature. ○: heating curve; ▲: cooling curve; ●: from earlier reflectivity measurements. Reprinted with permission from ref. 32. © EDP Sciences 2002.

studied. The solid lines are one parameter fits to the reflectivity $R(Q_z)$

$$R(Q_z) = R_F(Q_z)e^{-Q_z Q_z^T \sigma^2} \cong \left| \frac{Q_z - Q_z^T}{Q_z + Q_z^T} \right|^2 e^{-Q_z Q_z^T \sigma^2} \tag{7}$$

where $R_F(Q_z)$ is the Fresnel reflectivity calculated for an ideal, zero width, step-like interface, Q_z is the wave vector transfer in the z direction, given by $Q_z = (4\pi/\lambda)\sin\alpha$, Q_z^T is the z component of the

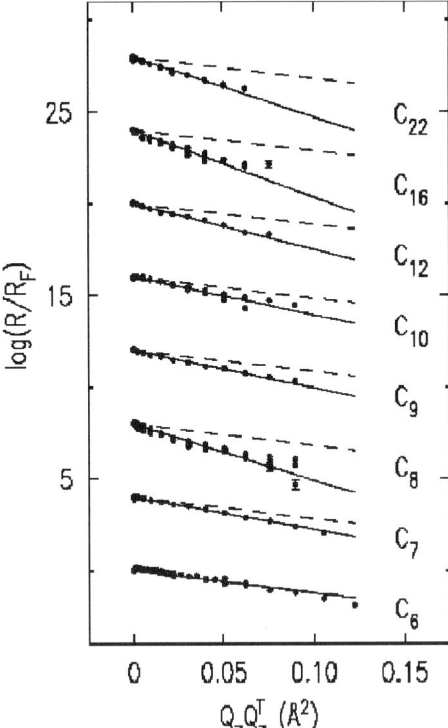

Fig. 23 Logarithm of X-ray reflectivity, normalized to Fresnel reflectivity $R_F(Q_z)$, as a function of $Q_zQ_z^T$, for water/alkane interfaces for eight different n-alkanes. Curves for the different interfaces are offset for clarity. The solid lines are one parameter fits of the measurements. The slopes of the fits are the interfacial width, σ. Also shown with dashed lines are predicted reflectivity curves for interfaces whose width is determined solely by capillary waves. Reprinted with permission from ref. 31. © American Physical Society 2000.

wave vector transfer with respect to the lower phase, and the fitting parameter σ represents the interfacial width. The dashed lines are the predicted reflectivity curves using only capillary-wave theory to predict the interfacial width. The measured interfacial width of the hexane/water interface agrees well with the prediction of capillary-wave theory. However, capillary-wave theory alone is not sufficient for predicting the interfacial width for the interfaces between water and the longer n-alkanes. Instead, these interfaces can be better described by a combination of capillary-wave theory and an intrinsic structural contribution. For short molecules, the gyration radius sets the length scale of the intrinsic interfacial structure, while the bulk correlation length sets the length scale for longer molecules.

A recent study by Bowers *et al.* has demonstrated how neutrons can be used to examine liquid/liquid interfaces.[33] This study involves interfaces of non-volatile oils and oil-soluble species. They have employed a spin–freeze–thaw technique to form an oil/water interface that is thin enough to transmit the neutron beam. The only assumption in this protocol is that the oil film is greater than 2000 Å, allowing the use of the thick film reflectivity formula. By using this new procedure, they were able to study the hexadecane/water interface, both neat and with the addition of polybutadiene (PB)–poly(ethylene oxide) (PEO) block copolymer. The neat hexadecane/water interface was found to be rougher than predicted by capillary-wave theory. Upon the addition of the PB–PEO copolymer, it was found that the copolymer segregates at the interface, as shown in Fig. 24. The blocks occupy a relatively thin, concentrated region immediately on either side of the interface that is about 20 Å thick and composed of about 20% polymer, followed, on each side, by a thicker, more dilute region which is about 50 Å thick and composed of less than 10% polymer.

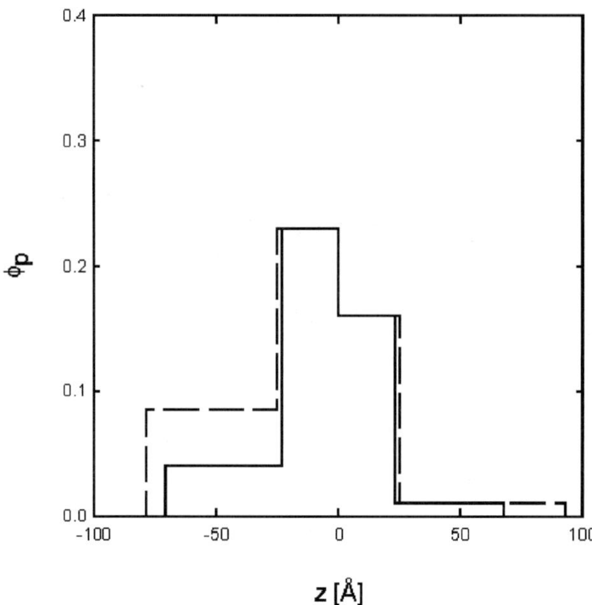

z [Å]

Fig. 24 Volume fraction profiles ϕ_p of the polymer distributions from modeled spectra. The dividing surface $z = 0$ is placed at the position of PB/PEO block segregation. Interfacial concentration $\Gamma = 4$ mg m^{-2} (–); $\Gamma = 10$ mg m^{-2} (--). Reprinted with permission from ref. 33. © American Chemical Society 2001.

Future directions

The molecular properties of liquid/liquid interfaces have long been a subject of interest. There is a clear need for an expanded understanding of the molecular structure and bonding at liquid/liquid interfaces because these properties determine the interfacial kinetic and thermodynamic processes. Only recently, however, has molecular level information about these systems been obtainable. A few recent studies have been presented here, highlighting several emerging experimental methods, as well as providing some examples of current systems of interest. The future of this research is promising, with many exciting areas to be explored.

Determination of the widths of these interfaces continues to be an area of interest, and is being studied with a number of techniques. Characterizing different interfaces as molecularly sharp or as a more diffuse, mixed interfacial region is important for a better understanding of those interfaces. Many of the studies that have been done on liquid/liquid interfaces have focused on the equilibrium properties. However, the dynamics that occur at interfaces are equally important to our understanding of interfacial processes. Recent advances in short pulse-width lasers have enabled the study of interfacial dynamics, such as reactions and adsorption processes.

Molecular adsorption to liquid/liquid interfaces is another area that has many exciting opportunities. Ongoing studies of surfactants and biologically relevant molecules at interfaces are focusing on the effect of the adsorbed molecules on the structure and orientation of the interfacial water molecules, and their interaction with the adsorbates. Also of interest is the interaction of the adsorbates with each other. Studies are focusing on understanding the environment of isolated adsorbates, as well as the overall interfacial structure in the presence of large numbers of adsorbed molecules and whether they form an even distribution or islands at the interface. Other studies are focusing on the effect of electrolytes on the structure and orientation of interfacial water molecules. The potentials across liquid/liquid interfaces provide yet another area for study, and modeling these interfacial potentials will further our understanding of these systems. Molecular transport across interfaces is also of great relevance, especially to biological applications, and it is important to continue to improve our understanding of the barriers to the transport across the liquid/liquid interface on a molecular level. Although this review has not addressed theoretical advances in modeling and understanding these interfaces, the coupling of theory and experiment to the study of liquid/liquid interfaces will be invaluable for further advances in the field.

Acknowledgements

The authors thank Cathryn McFearin and Adam Hopkins for their help with this review, and the National Science Foundation (CHE-0243856), the Department of Energy, Basic Energy Sciences (DE-FG-03-96ER45557), and the Office of Naval Research for funding. Prof. Richmond is very appreciative for being honored with the Spiers Lecture Award.

References

1 Y. R. Shen, *Annu. Rev. Phys. Chem.*, 1989, **40**, 327.
2 K. B. Eisenthal, *Chem. Rev.*, 1996, **96**, 1343.
3 J. C. Conboy, J. L. Daschbach and G. L. Richmond, *J. Phys. Chem.*, 1994, **98**, 9688.
4 R. M. Corn and D. A. Higgins, *Chem. Rev.*, 1994, **94**, 107.
5 S. G. Grubb, M. W. Kim, T. Rasing and Y. R. Shen, *Langmuir*, 1988, **4**, 452.
6 J. C. Conboy and G. L. Richmond, *Electrochim. Acta*, 1995, **40**, 2881.
7 T. Uchida, A. Yamaguchi, T. Ina and N. Teramae, *J. Phys. Chem. B*, 2000, **104**, 12091.
8 Y. R. Shen, *The Principles of Nonlinear Optics*, Wiley, New York, 1984.
9 H. Wang, E. Borguet and K. B. Eisenthal, *J. Phys. Chem. B*, 1998, **102**, 4927.
10 W. H. Steel and R. A. Walker, *Nature*, 2003, **424**, 296.
11 M. C. Messmer, J. C. Conboy and G. L. Richmond, *J. Am. Chem. Soc.*, 1995, **117**, 8039.
12 J. C. Conboy, M. C. Messmer and G. L. Richmond, *J. Phys. Chem.*, 1996, **100**, 7617.
13 M. R. Watry and G. L. Richmond, *J. Am. Chem. Soc.*, 2000, **122**, 875.
14 M. M. Knock, G. R. Bell, E. K. Hill, H. J. Turner and C. D. Bain, *J. Phys. Chem. B*, 2003, **107**, 10801.
15 L. F. Scatena, M. G. Brown and G. L. Richmond, *Science*, 2001, **292**, 908.
16 M. G. Brown, D. S. Walker, E. A. Raymond and G. L. Richmond, *J. Phys. Chem. B*, 2003, **107**, 237.
17 H. Jensen, D. J. Fermin and H. H. Girault, *Phys. Chem. Chem. Phys.*, 2001, **3**, 2503.
18 Z. Samec, *Chem. Rev.*, 1988, **88**, 617.
19 D. S. Walker, M. G. Brown, C. L. McFearin and G. L. Richmond, *J. Phys. Chem. B*, 2004, **108**, 2111.
20 L. F. Scatena and G. L. Richmond, *J. Phys. Chem. B*, 2004, **108**, 12518.
21 B. M. Ocko, X. Z. Wu, E. B. Sirota, S. K. Sinha, O. Gang and M. Deutsch, *Phys. Rev. E*, 1997, **55**, 3164.
22 Q. Lei and C. D. Bain, *Phys. Rev. Lett.*, 2004, **92**, 176103.
23 B. P. Binks, J. H. Clint, A. K. F. Dyab, P. D. I. Fletcher, M. Kirkland and C. P. Whitby, *Langmuir*, 2003, **19**, 8888.
24 S. Uredat and G. H. Findenegg, *Langmuir*, 1999, **15**, 1108.
25 S. Ishizaka, H.-B. Kim and N. Kitamura, *Anal. Chem.*, 2001, **73**, 2421.
26 T. Yamashita, T. Uchida, T. Fukushima and N. Teramae, *J. Phys. Chem. B*, 2003, **107**, 4786.
27 M. J. Crawford, J. G. Frey, T. J. VanderNoot and Y. Zhao, *J. Chem. Soc., Faraday Trans.*, 1996, **92**, 1369.
28 T. Kakiuchi, K. Ono, Y. Takasu, J. Bourson and B. Valeur, *Anal. Chem.*, 1998, **70**, 4152.
29 K. Nochi, A. Yamaguchi, T. Hayashita, T. Uchida and N. Teramae, *J. Phys. Chem. B*, 2002, **106**, 9906.
30 S. Ishizaka, S. Kinoshita, Y. Nishijima and N. Kitamura, *Anal. Chem.*, 2003, **75**, 6035.
31 D. M. Mitrinović, A. M. Tikhonov, M. Li, Z. Huang and M. L. Schlossman, *Phys. Rev. Lett.*, 2000, **85**, 582.
32 M. Li, A. M. Tikhonov and M. L. Schlossman, *Europhys. Lett.*, 2002, **58**, 80.
33 J. Bowers, A. Zarbakhsh, J. R. P. Webster, L. R. Hutchings and R. W. Richards, *Langmuir*, 2001, **17**, 140.

X-Ray studies of the interface between two polar liquids: neat and with electrolytes

Guangming Luo,[a] Sarka Malkova,[a] Sai Venkatesh Pingali,[a] David G. Schultz,[a] Binhua Lin,[b] Mati Meron,[b] Timothy J. Graber,[b] Jeffrey Gebhardt,[b] Petr Vanysek[c] and Mark L. Schlossman*[a]

[a] Department of Physics (M/C 273), University of Illinois at Chicago, 845 W. Taylor St. (Rm. 2236), Chicago, IL, 60607-7059 . E-mail: schloss@uic.edu
[b] The Center for Advanced Radiation Sources, University of Chicago, Chicago IL, 60637
[c] Department of Chemistry and Biochemistry, Northern Illinois University, DeKalb, IL, 60115

Received 14th April 2004, Accepted 5th May 2004
First published as an Advance Article on the web 22nd September 2004

We demonstrate the use of X-ray reflectivity to probe the electron density profile normal to the interface between two polar liquids. Measurements of the interfacial width at the neat nitrobenzene/water and the neat water/2-heptanone interfaces are presented. These widths are consistent with predictions from capillary wave theory that describe thermal interfacial fluctuations determined by the tension and bending rigidity of the interface. Variation of the temperature of the water/nitrobenzene interface from 25 °C to 55 °C indicates that the role of the bending rigidity decreases with increasing temperature. X-ray reflectivity measurements of the electrified interface between an aqueous solution of $BaCl_2$ and a nitrobenzene solution of TBATPB demonstrate the sensitivity of these measurements to the electrolyte distribution at the interface. A preliminary analysis of these data illustrates the inadequacy of the simplest, classical Gouy–Chapman theory of the electrolyte distribution.

Introduction

Understanding molecular ordering at the interface between two liquids is important in many chemical, biological, and condensed matter systems. During the past several years we have used the techniques of X-ray surface scattering to study the interface between non-polar oils and water as well as the interface between two aqueous solutions.[1–12] Studies of the neat water/alkane interface for a range of chain lengths have demonstrated that the interfacial width can be described as due to a capillary wave contribution plus an intrinsic structural contribution.[9,10,12] In this case, the intrinsic contribution to the width is determined by the two physically relevant length scales: the gyration radius of the alkane and the alkane bulk correlation length. Studies of the interface between water and a hexane solution of surfactants have demonstrated the ability of X-ray reflectivity to determine the surfactant ordering in solid and liquid monolayers on sub-nanometer length scales.[1,2,4,7,11] X-ray diffuse scattering has been used to probe the size and separation of micrometer sized domains in surfactant monolayers at the water/hexane interface.[4] These

DOI: 10.1039/b405555a

measurements of surfactant ordering and monolayer phase transitions have highlighted differences between the behavior at the water/vapor and water/oil interfaces.

For the past several decades electrochemical effects at the interface between two polar liquids have been studied.[13–17] The structure of this interface has been debated in the electrochemical community due to its importance in the understanding of electrochemical measurements. For example, the interfacial width (currently unmeasured) is necessary to determine association constants for ion-pairing from capacitance data.[18] Interfacial particle distributions (currently unmeasured) are necessary for understanding molecular reaction rates or the rate of assisted ion transfer from electrochemical measurements.[19]

Electrodes can be inserted into the bulk phases of two immiscible electrolyte solutions to impose and control a large electric field across the interface between the solutions. This field can be used to transfer electrolytes to or through the interface, control chemical reactions at the interface, and control electron transfer across the interface. In spite of their broad applicability and scientific importance, these interfaces have not been previously studied with X-ray scattering. Here, we present some of our initial experiments in this area. These include X-ray reflectivity studies of the following interfaces: neat nitrobenzene/water, neat 2-heptanone/water, and nitrobenzene/water with supporting electrolytes subject to an externally imposed electric potential.

Methods and materials

Materials, densities, and interfacial tension

Nitrobenzene (Fluka, puriss. p.a., ACS reagent) and 2-heptanone (Aldrich, 98%) were purified before use by filtering through basic activated alumina (Fisher). Tetrabutylammonium tetraphenylborate (TBATPB) was used as received from Fluka (puriss., electrochemical grade). Tetrabutylammonium chloride monohydrate for the reference electrode (Aldrich, 98%) was stored over a desiccant. Barium chloride dihydrate (Aldrich, 99% ACS reagent grade) was roasted at 160 °C for 12 h to remove volatile organics and hydration and hygroscopic water. Water was produced by a Barnstead Nanopure or Millipore Milli-Q source. Silver electrodes for reference electrodes were prepared from 99.9% silver wire, coated with chloride by anodic oxidation in 1 mol l^{-1} hydrochloric or sulfuric acid.

The two liquid phases in all the experiments were equilibrated by gentle shaking or swirling by hand, then the samples were allowed to sit for 24 h or more before measurement. Densities of the phases were measured by a pycnometer. Interfacial tensions were measured by the Wilhelmy plate method using a plate made from Teflon (for the nitrobenzene/water interface) or from filter paper (for the water/2-heptanone interface) and hung by a Pt wire to a Cahn microbalance.

X-ray sample cell

The liquids were contained in either a stainless steel sample cell[11] for studies of the neat liquid/liquid interface or in a glass sample cell for studies of the polarized interface. The former had mylar windows for the X-rays and mylar inserts arranged such that the liquid/liquid interface came in contact only with the mylar. The interfacial area was 76 mm × 100 mm (along the beam × transverse) with X-rays penetrating through the upper phase. The stainless steel sample cell was placed into a one-stage cylindrical aluminium thermostat and temperature controlled to ± 0.03 °C.

The glass cell contained a standard 4-electrode arrangement with two Luggin capillaries commonly used for studying electrochemistry at the liquid/liquid interface.[20] The inner diameter of the glass cell at the location of the interface was 70 mm. The interface position was fixed by the top edge of a thin Teflon sheet wrapped around the inner glass surface. X-rays penetrated through the glass to travel through the top aqueous phase and reflect off the liquid/liquid interface. A potentiostat and impedance analyzer from Solartron (1287 and 1260) were used to control the electric potential difference in the cell and measure the impedance of the interface.

X-ray methods

The kinematics of reflectivity is illustrated in Fig. 1. The incident and outgoing angles are equal for specular reflection ($\beta = \alpha$), therefore, the wave vector transfer, $\boldsymbol{Q} = \boldsymbol{k}_{\text{scat}} - \boldsymbol{k}_{\text{in}}$, is only in the

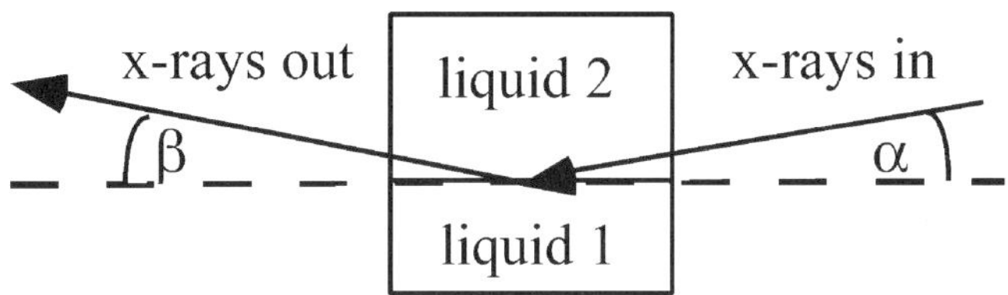

Fig. 1 Experimental geometry. Liquid 2 is either water or 2-heptanone for studies of the nitrobenzene/water or water/2-heptanone interface, respectively.

z-direction normal to the interface; $Q_x = Q_y = 0$ where x and y are in the plane of the interface, and $Q_z = (4\pi/\lambda)\sin\alpha$ (λ is the X-ray wavelength, typically 0.4077 Å). X-ray reflectivity probes the electron density in the z-direction, averaged over the interfacial plane.

X-ray reflectivity from the liquid/liquid interface was measured at the ChemMatCARS' sector 15-ID (Advanced Photon Source, Argonne National Laboratory, USA) with instrumentation and techniques previously described.[6,11,21] Briefly, a monochromatic X-ray beam was incident upon a pair of slits used to set the incident beam size. An ionization chamber after these slits and immediately prior to the sample measured the incident X-ray flux. The sample was followed by two pairs of slits to reduce the background scattering and set the angular resolution of the scattered X-rays measured by a scintillator detector. Scanning the detector angle β through the value for specular reflectivity α yields a peak that contains the reflected intensity and the background scattering. The reflectivity is given by subtracting the background from the reflected intensity, then normalizing by the incident flux.

X-ray data and analysis

The X-ray reflectivity measurements and analysis shown later in Fig. 3 are typical of those presented in this manuscript. The lines illustrate that below a critical wave vector transfer for total reflection, Q_c, the reflectivity is nearly one, above Q_c the reflectivity drops off rapidly with increasing Q_z. The critical wave vector transfer, Q_c, is calculated from the measured bulk densities, as $Q_c = 4[\pi r_e(\rho_b - \rho_t)]^{1/2}$ (where $\rho_{b,t}$ are the electron densities of the two bulk phases, bottom and top phases, respectively, and $r_e = e^2/mc^2$). Due to the very small values of Q_c for these interfaces, typically 0.007 Å^{-1}, it was not possible to reliably measure reflectivity for values of Q smaller than Q_c.

X-ray reflectivity probes the electron density profile normal to the interface. Measurements from neat liquid/liquid interfaces can be fit by an error function interfacial profile that describes the variation in electron density as a function of the normal coordinate z through the liquid/liquid interface, given by

$$\langle \rho(z) \rangle = \frac{1}{2}(\rho_b + \rho_t) + \frac{1}{2}(\rho_b - \rho_t)\mathrm{erf}\left[z\big/\sigma\sqrt{2}\right] \text{ with } \mathrm{erf}(z) = \frac{2}{\sqrt{\pi}}\int_0^z \varepsilon^{-t^2}\mathrm{d}t, \qquad (1)$$

where $\langle \rho(z) \rangle$ is the interfacial electron density averaged over the plane of the interface. For this interfacial profile the reflectivity can be written as, for $Q_z > Q_c$,[22,23]

$$R(Q_z) \approx \left|\frac{Q_z - Q_z^T}{Q_z + Q_z^T}\right|^2 \exp\left(-Q_z Q_z^T \sigma^2\right). \qquad (2)$$

where

$$Q_z^T \approx \sqrt{Q_z^2 - Q_c^2}$$

is the z-component of wave vector transfer with respect to the lower phase. Data fit to eqn. (2) has three fitting parameters, the interfacial width σ, the critical wave vector transfer Q_c, and a wave vector transfer offset Q_{offset} of the X-ray instrument that represents a small angular misalignment of the instrument. Our measurements of the densities of the bulk liquids allow us to calculate the critical wave vector transfer Q_c. This calculated value agrees with the fit values to within 10^{-4} Å^{-1}. The angular offset of the instrument also results in a very small adjustment of 10^{-3} Å^{-1} (or less) in values of the wave vector transfer. This leaves only one significant fitting parameter, the interfacial width σ. Values of error bars on the interfacial width include its dependence on the other two parameters.

Earlier measurements of water/alkane interfaces have successfully interpreted the measured interfacial width σ in terms of a hybrid model of the interface that describes an intrinsic structural profile roughened by capillary waves.[9,10,12] In this case, the interfacial width σ can be represented as a combination of an intrinsic structural width, σ_o, and a resolution dependent, capillary wave contribution,[24–29]

$$\sigma^2 = \sigma_o^2 + \frac{k_B T}{4\pi^2 \gamma} \iint \frac{d^2 q}{q^2 + \xi_\parallel^{-2}} \equiv \sigma_o^2 + \sigma_{cap}^2, \tag{3}$$

where $k_B T$ is Boltzmann's constant times the temperature, γ is the measured interfacial tension, the correlation length, $\xi_\parallel$, is given by $\xi_\parallel^2 = \gamma/\Delta\rho_m g$ and determines the exponential decay of the interfacial correlations given by the height–height correlation function of interfacial motion,[30] $\Delta\rho_m$ is the *mass* density difference of the two phases, and g is the gravitational acceleration. Integration is over in-plane capillary wave vectors q corresponding to the range of capillary waves that the measurement probes. After some simplifications, the capillary contribution evaluates to $\sigma_{cap}^2 = (k_B T/2\pi\gamma) \log(q_{max}/q_{min})$, where q_{max} (chosen to be $2\pi/5$ Å^{-1}) is determined by the cutoff for the smallest wavelength capillary waves that the interface can support and $q_{min} = (2\pi/\lambda)\,\Delta\beta \sin\alpha$ is determined by the incident angle α and the angular acceptance of the detector $\Delta\beta = 4.7 \times 10^{-4}$. Note that the effect of the intrinsic width is to always increase the value of the interfacial width from that of capillary waves alone.

If double layers of electrolytes or adsorbed surfactants are present at the interface, this additional interfacial structure is accounted for by calculating the reflectivity with the Parratt formalism.[31] In this formalism the electron density profile is divided into a series of layers each with a well defined thickness and electron density. The exact reflection and transmission coefficients at the interfaces of each layer are used to calculate the specular reflectivity from the entire interfacial profile.

Background on interfaces between two immiscible electrolyte solutions

An early prediction of the ionic distribution near interfaces under static conditions was the Gouy–Chapman theory for the double layer.[32,33] Based upon approximations for dilute solutions, point charge ions, and a dielectric continuum solvent, this theory predicts the electrolyte distribution near a metal/solution interface. However, the liquid/liquid interface is different from the metal/electrolyte interface because the metal has essentially infinite permittivity. Verwey and Niessen adapted the Gouy–Chapman model to describe two back to back diffuse double layers at the liquid/liquid interface.[34] There is currently good evidence that diffuse double layer models alone are inadequate to describe the charge distribution at the liquid/liquid interface. Deviations are believed to be due to the interfacial structure.

There are primarily two different views of the interfacial structure in the electrochemical literature. These are based upon either a molecularly sharp interface with capillary wave fluctuations or an interface that consists of a region where molecules from the two phases are mixed (see Fig. 2).

Kakiuchi and Senda used interfacial tension measurements to suggest the existence of a sharp boundary between the two solvents that contains an ion-free solvent layer.[35] In this modified Verwey–Niessen model a compact layer of one or two solvent molecules thick separates the back to back double layers, one containing an excess of positive charge and the other an equal amount of negative charge.[16,17,36] For the second view, also using interfacial tension measurements, Girault and Schiffrin disputed the existence of the compact inner layer and suggested that the boundary is a region of diffuse mixed solvent.[37–40]

In addition to studies of interfacial tension, numerous studies have shown that at low electrolyte concentrations the interfacial capacitance is higher than the value predicted from the Gouy–Chapman theory.[41–43] This discrepancy is interpreted as due to interfacial structure not included in

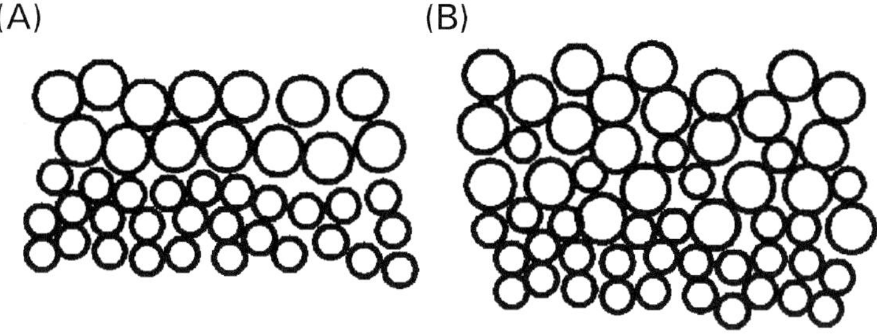

Fig. 2 Two models of liquid/liquid interfaces, solvents only (no ions). (A) Sharp interface, interfacial fluctuations due to capillary waves. (B) Mixing of solvents at interface.

the Gouy–Chapman theory. As one recent example, a study by Pereira *et al.* combined capacitance measurements with a lattice–gas model to provide support for the existence of a mixed boundary layer.[44]

The current debate on the interfacial structure is partially due to a lack of techniques that directly probe the structure. Interfacial tension and capacitance both probe macroscopic properties of the interface. A recent vibrational sum-frequency non-linear optical study of the neat dichloroethane/water interface demonstrated a very small optical response from polarization normal to the interfacial plane. This was interpreted as evidence for a mixed solvent interface.[45]

Different views of the interface also appear in the theoretical literature. Marcus calculated rate constants for an electron transfer for two redox species in opposite phases, using both the sharp and the diffuse boundary layer model.[46,47] The two results differ by two orders of magnitude. However, the scarce experimental data available and their uncertainty, as well as uncertainties in the theoretical assumptions, have failed to provide definitive answers to the question of the interfacial structure.[48]

Molecular dynamics simulations predict a locally sharp interface that is roughened by capillary waves.[49–52] This observation could be consistent with the experimentally observed electrical capacitance, which exceeds the value predicted by the Gouy–Chapman theory. The interfacial area could be simply larger than the assumed geometrical area. A molecular dynamics simulation on a thin layer liquid/liquid interface by Philpott *et al.* exhibited local surface fluctuations, consistent with a graded composition of the interface on the molecular scale and capillary waves on a larger length scale.[53] Pecina and Badiali used a modulated interface geometry to calculate interfacial capacitance and the roughness function of the interface.[54]

Numerical solutions to non-linear versions of the Poisson–Boltzmann equation have been used to incorporate ion–ion correlations, image forces, and finite ion size. These calculations model the capacitance of electrolyte solutions at the water/nitrobenzene and water/1,2-dichloroethane interfaces, providing better agreement with experiments than previous theories.[55,56] In disagreement with this modified Poisson–Boltzmann approach, a recent density functional approach leads to a mixed solvent layer at the interface with a thickness of several solvent diameters.[57] Lattice–gas calculations by Schmickler and coworkers incorporate a mixed solvent layer whose thickness can be related to the miscibility of the two solvents.[44,58] In this model, the capacitance is increased above the Gouy–Chapman value because more charge can be stored in the mixed solvent layer.

Here, we demonstrate the use of X-ray reflectivity to probe the microscopic interfacial structure. We first discuss measurements of the structure from the neat interface between two polar liquids that do not contain added electrolytes. Then, we discuss our recent measurements of the interface between two electrolyte solutions with an imposed external electric potential.

Nitrobenzene/water neat interfaces

X-ray reflectivity measurements from the neat nitrobenzene/water interface at 25 °C, 35 °C, 45 °C, and 55 °C are illustrated in Fig. 3. The solid lines are fits of these data to eqn. (2) to yield the interfacial widths shown in Table 1. These values are compared with calculations of σ_{cap} in eqn. (3) that use the measured values of interfacial tension (see Table 1). The dashed lines in Fig. 3 illustrate the reflectivity calculated using eqn. (2) and the calculated values of σ_{cap}. The interfacial width is

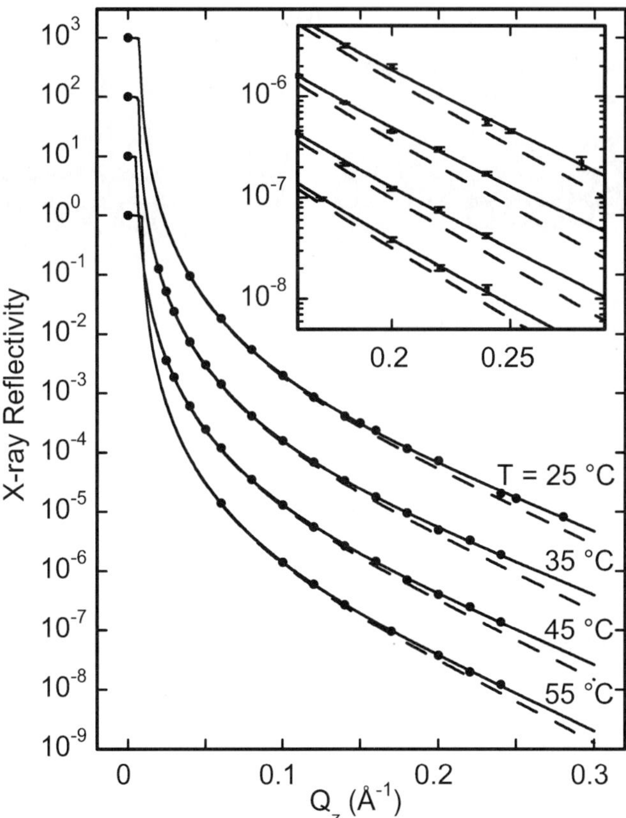

Fig. 3 X-ray reflectivity as a function of wave vector transfer Q_z from the neat nitrobenzene/water interface at four temperatures ($T = 25\,°C$, $35\,°C$, $45\,°C$, and $55\,°C$). Solid lines are fits to eqn. (2). Dashed lines are predicted by calculating the interfacial width as in eqn. (3) by using the measured tension, but not including the effect of bending rigidity as in eqn. (4). The inset illustrates error bars in the high Q_z region (same ordering of temperatures as in the main figure). Data are offset by factors of 10 in the main figure, factors of 3 in the inset.

slightly smaller than expected from capillary wave theory. This deviation from the capillary wave value cannot be explained by an intrinsic structural contribution to the interfacial width because the intrinsic contribution shown in eqn. (3) will only increase the width.

The capillary wave theory shown in eqn. (3) assumes that interfacial fluctuations are described by distortions that increase the interfacial area, and therefore raise the interfacial free energy in proportion to the interfacial tension. Alternatively, interfacial distortions can occur without increasing the interfacial area in a manner similar to flexing a piece of paper. The energy required for this distortion is proportional to the interface's bending rigidity.[59,60] In that case the capillary

Table 1 Nitrobenzene/water neat interfaces. $\sigma_{measured}$ is the interfacial width measured by X-ray reflectivity. σ_{cap} is the width calculated from the expression in eqn. (3). κ is the bending rigidity calculated from eqn. (4) that results in $\sigma_{gen\_cap} = \sigma_{measured}$. Note that error bars on κ are determined from the errors on $\sigma_{measured}$ and that κ does not vary linearly with σ

Temperature/°C	Interfacial tension/mN m^{-1}	$\sigma_{measured}/\text{Å}$	$\sigma_{cap}/\text{Å}$	$\kappa\ (k_BT)$
25	25.2	4.5 ($\pm$0.1)	5.2	6 ($+5/-3$)
35	24.5	4.5 ($\pm$0.1)	5.2	6 ($+6/-3$)
45	23.8	4.8 ($\pm$0.1)	5.4	3 ($+2/-1$)
55	22.9	5.1 ($\pm$0.1)	5.6	1 ($+1/-0.5$)

Fig. 4 (a) Lowest energy configuration of a line of athermal electric dipoles interacting only through dipole interactions. Dashed line is a mathematical dividing line representing the interface; arrows above and below represent the molecular dipoles closest to the interface for the two phases. (b) Interfacial distortion requires energy due to rearrangement of dipoles.

wave term in eqn. (3) is generalized to include both types of distortions. Neglecting the intrinsic structure, the interfacial width is then written as

$$\sigma^2_{\text{gen\_cap}} = \frac{k_B T}{4\pi^2 \gamma} \iint \frac{d^2 q}{(\kappa q^4/\gamma) + q^2 + \xi_\parallel^{-2}} \approx \frac{k_B T}{2\pi\gamma} \ln \frac{(\gamma/\kappa)^{1/2}}{q_{\min}}, \tag{4}$$

where $q_2 = (\gamma/\kappa)^{1/2}$ and κ is the bending rigidity.

Eqn. (4) indicates that the bending rigidity term is proportional to q^4, the capillary wave vector raised to the fourth power, whereas the interfacial tension term is proportional to q^2. This indicates that the bending rigidity is more important for larger values of q. Therefore, the effect of a bending rigidity is to reduce the smaller wavelength interfacial fluctuations and make the interface smoother. Values of the bending rigidity that yield values of $\sigma_{\text{gen\_cap}}$ in agreement with our measured values of σ are listed in Table 1. The largest value of the bending rigidity, $6k_B T$, occurs at the lowest temperatures. The value of the bending rigidity is reduced to $1k_B T$ at the highest temperature.

The existence of a bending rigidity at an interface between two polar fluids can be made plausible by considering a simple athermal arrangement of ideal electric dipoles with an interaction energy given by

$$U_{\text{dipole}} = \frac{1}{4\pi\varepsilon_0 r^3} [\vec{p}_1 \cdot \vec{p}_2 - 3(\vec{p}_1 \cdot \hat{r})(\vec{p}_2 \cdot \hat{r})], \tag{5}$$

where r is the distance between the dipoles, $\hat{r}$ is the unit vector from dipole 1 to dipole 2 and $\vec{p}_{1,2}$ are their dipole moments. The lowest energy configuration of a line of ideal dipoles has them arranged head to tail as illustrated in panel (a) of Fig. 4. This figure also illustrates a line of dipoles of the other phase in which the head to tail arrangement is in the opposite direction to minimize the energy of dipole interactions across the interface. If the interface is distorted, then the separation between dipoles on one side of the interface is increased, while on the other side the separation is decreased. This variation in dipole spacing requires energy and leads to a bending rigidity. In a real thermal system, the value of the bending rigidity will depend upon the distribution of dipole orientations as well as other molecular interactions. These other interactions may result in a dipole orientation normal to the interface, however, a bending rigidity will also be present for that situation.

These results are consistent with molecular dynamics simulations of the nitrobenzene/water interface that predict a rough interface that is locally sharp, *i.e.*, without solvent mixing.[52] Note that an interface with solvent mixing would also have thermal fluctuations. Therefore, solvent mixing could be accounted for by an intrinsic structural contribution σ_0 to the interfacial width (see eqn. (2)). This would increase the value of the measured interfacial width to be larger than the capillary wave value, contrary to our measurements of widths that are smaller than the capillary wave value given in eqn. (2).

Our suggestion that a bending rigidity is due to preferential alignment of interfacial dipoles within the plane of the interface is consistent with the results of a recent vibrational sum-frequency spectroscopy study of the water/dichloroethane interface.[45] In that study, the sum-frequency response was very small for dipole moments normal to the interface. Results for the response in the interface plane were not published.

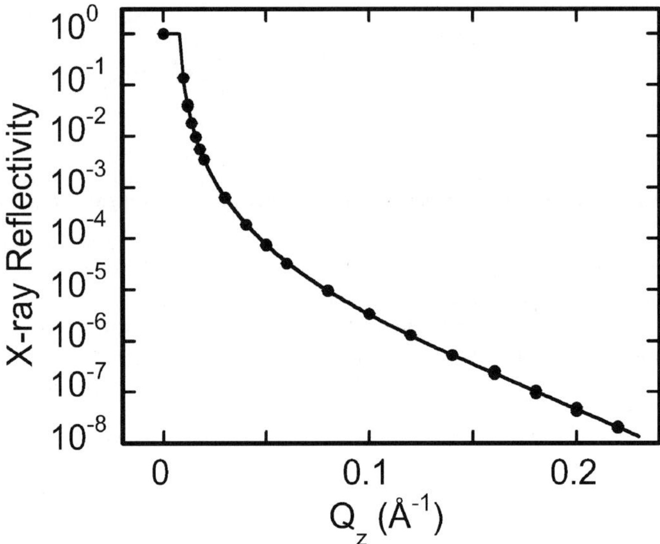

Fig. 5 X-ray reflectivity from the water/2-heptanone interface at 25 °C. Solid line is a fit to eqn. (2).

Water/2-heptanone neat interfaces

Fig. 5 illustrates data and a fit for X-ray reflectivity from the water/2-heptanone interface at 25 °C. The measured interfacial width is 7.1 ± 0.15 Å and can be compared to the value calculated from capillary wave theory of 7.3 Å using the measured interfacial tension of 12.6 mN m^{-1}. Again, the interfacial width is smaller than the capillary wave value, though the discrepancy is barely a two sigma effect. When interpreted in terms of a bending rigidity, this discrepancy yields a negligibly small value of κ. It is possible that the bending rigidity is negligible in this case since the polarization of 2-heptanone is much smaller than that of nitrobenzene. This result is consistent with a molecular dynamics simulation of the water/2-heptanone that predicted a sharp interface, roughened by capillary waves.[61]

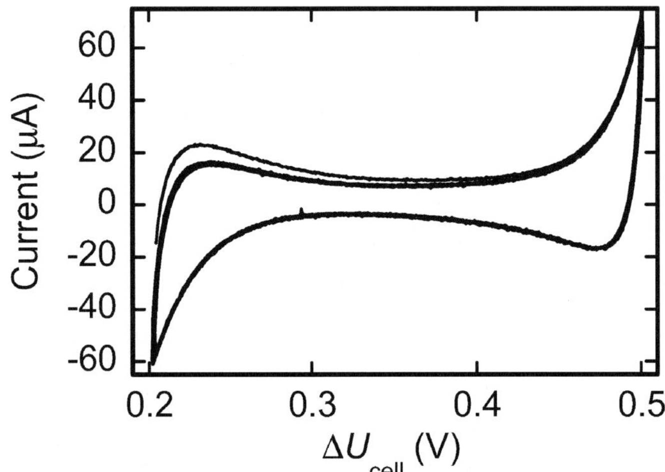

Fig. 6 Cyclic voltammogram from the interface between a 0.1 M solution of $BaCl_2$ in water and a 0.01 M solution of TBATPB in nitrobenzene.

Fig. 7 Electrochemical cell. Double line represents the liquid/liquid interface studied with X-rays.

Imposed external potential on the nitrobenzene/water interface with supporting electrolytes

Here, we present our recent, first X-ray measurements on the structure of the interface between a 0.1 M solution of $BaCl_2$ in water and a 0.01 M solution of TBATPB in nitrobenzene. With the caveat that further experimental work is required, we also present a preliminary analysis of these data. This analysis indicates that the simplest structural model of two classical Gouy–Chapman back-to-back double layers of point charge electrolytes does not adequately describe our X-ray reflectivity data. We recognize that this model is not realistic and plan to incorporate finite size effects into our analysis.

Fig. 6 illustrates the cyclic voltammogram for the electrochemical cell indicated in Fig. 7. The X-ray measurements were performed for electrochemical cell potentials (water minus oil) that varied from $\Delta U_{cell} = 0.25$ V to 0.51 V. The current at a fixed potential was typically a few microamperes except at 0.51 V where the current was slightly less than 10 µA. The potential of zero charge was determined by measuring both the impedance (to determine the potential that yields a minimum in capacitance) as well as the interfacial tension as a function of potential. The capacitance minimum was at 0.31 V and the maximum in interfacial tension was at 0.29 V, from which we chose the potential of zero charge to be $\phi_{pzc} = 0.31$ V. The potential difference across the interface is $\Delta\phi^{w-o} = \Delta U_{cell} - \phi_{pzc}$.

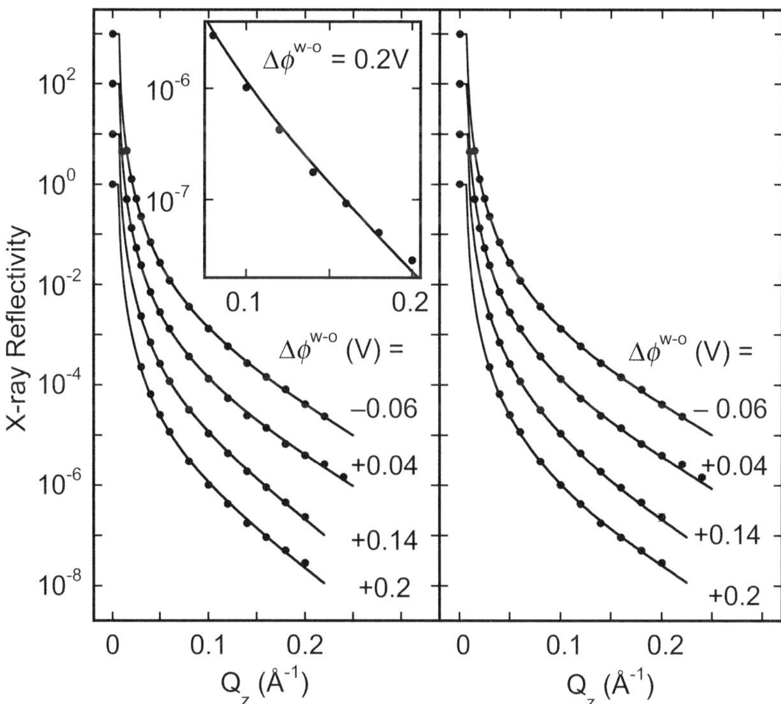

Fig. 8 X-ray reflectivity data for four different values of $\Delta\phi^{w-o}$ from the interface between a 0.1 M solution of $BaCl_2$ in water and a 0.01 M solution of TBATPB in nitrobenzene (data offset by factors of ten for viewing purposes). (A) The lines are fits from a simple interface model given by the electron density profile in eqn. (1). The inset illustrates the high Q_z data from $\Delta\phi^{w-o} = 0.2$ V to show the difference in the curvature of the data and the fit. (B) The lines are fits from the simplest Gouy–Chapman double layer model described in the text.

Table 2 Fitting parameters for the X-ray reflectivity data for four different values of $\Delta\phi^{\mathrm{w-o}}$ from the interface between a 0.1 M solution of $BaCl_2$ in water and a 0.01 M solution of TBATPB in nitrobenzene. Models are described in the text. The critical wave vector $Q_c = 0.0073$. $Q_{\mathrm{offset}} < 0.001$ Å^{-1} for all fits

			Simple interface model	Gouy–Chapman model
$\Delta\phi^{\mathrm{w-o}}$/V	Tension, γ/mN m^{-1}	σ_{cap}/Å	σ/Å	σ/Å
−0.06	25.5	5.1	4.9 ± 0.3	4.8
0.04	25.3	5.1	5.0 ± 0.4	5.0
0.14	23.4	5.3	6.4 ± 0.2	4.3
0.2	21.0	5.6	6.3 ± 0.3	6.8

Fig. 8A illustrates X-ray reflectivity data for four different values of $\Delta\phi^{\mathrm{w-o}}$ along with fits from a simple interface model given by the electron density profile in eqn. (2). This simple model does not include any double layers. It can be seen that the model fits well for the three smallest values of $\Delta\phi^{\mathrm{w-o}}$, but a systematic deviation at $\Delta\phi^{\mathrm{w-o}} = 0.2$ V produces a fit with a slightly different curvature than the data (Fig. 8A inset). The model also yields an interfacial width that is close to the capillary wave value for the two smallest values of $\Delta\phi^{\mathrm{w-o}}$ (see Table 2). The latter is calculated from the measured values of the interfacial tension (see eqn. (3) and Table 2). Therefore, these measurements are not very sensitive to the presence of the double layer for small values of $\Delta\phi^{\mathrm{w-o}}$. For $\Delta\phi^{\mathrm{w-o}} = 0.14$ V and $\Delta\phi^{\mathrm{w-o}} = 0.2$ V, the interfacial width is larger than expected for capillary waves. This indicates the presence of additional structure at the interface.

To include the effect of the double layer on our measurements we calculated the electrolyte distribution using the simplest classical Gouy–Chapman treatment that considers ions as point charges, and neglects ion polarization, variation of solvent dielectric constant, and image forces.[32,33] Solution of the Poisson–Boltzmann equation for the electric potential as a function of distance from the interface determines the distribution of electrolytes in the diffuse layers on either side of the interface. This distribution allows us to calculate the electron density profile through the interface. Given that we know the potential difference $\Delta\phi^{\mathrm{w-o}}$ across the interface, the electrolyte distribution is fully determined by this model. When $\Delta\phi^{\mathrm{w-o}} > 0$, the diffuse layers primarily contain Ba^{2+} ions on the water side of the interface and TPB^{-} ions on the nitrobenzene side of the interface. Convoluting the Gouy–Chapman electrolyte distribution with a Gaussian function allows us to model the thermal capillary wave fluctuations of this interface.

Fig. 8B illustrates X-ray reflectivity data for the four different values of $\Delta\phi^{\mathrm{w-o}}$ along with fits from this Gouy–Chapman model. In this model there are only two fitting parameters: the interfacial width and the angular offset of the spectrometer alignment. The latter is very small, less than 1×10^{-3} Å^{-1}. The interfacial width that results from the fitting can be compared to the values calculated from capillary wave theory (see Table 2). The interfacial widths for the two larger values of $\Delta\phi^{\mathrm{w-o}}$ differ greatly from the calculated values of σ_{cap} and also exhibit a non-monotonic behavior with increasing $\Delta\phi^{\mathrm{w-o}}$. This indicates that the Gouy–Chapman theory does not adequately describe the ion distributoin probed by these measurements.

These data indicate the necessity for more sophisticated structural models of the molecular distribution near a polarized interface between two electrolyte solutions. As previously mentioned, more traditional electrochemical measurements have also found discrepancies with the classical Gouy–Chapman model. Analysis of our X-ray data with more sophisticated structural models is ongoing.

Conclusions

We have demonstrated that X-ray reflectivity can be used to probe the liquid/liquid interface between polar liquids commonly used in electrochemical studies at the liquid/liquid interface. Our studies of the neat nitrobenzene/water and water/2-heptanone interfaces demonstrate that the widths of these interfaces are consistent with a generalized capillary wave theory. This theory accounts for thermal fluctuations of the interface controlled by the tension and bending rigidity of the interface. Solvent mixing, if present, would have increased the interfacial width to a value larger than our measurements.

We also used X-ray reflectivity to study the effect on the interfacial structure of imposing an external potential on the nitrobenzene/water interface with supporting electrolytes $BaCl_2$ and TBATPB. Our preliminary analysis of these data indicate that the simplest Gouy–Chapman model for the interfacial electrolyte distribution cannot adequately explain our structural measurements. Further analysis with more sophisticated structural models of the interface is planned. We anticipate that the sensitivity of the X-ray reflectivity technique will be even greater in studies of the adsorption of larger ions under electrochemical control than in these early studies of small ion adsorption.

Acknowledgements

Conversations with Ilan Benjamin are gratefully acknowledged. M. L. S. acknowledges the U.S. National Science Foundation (NSF)-DMR and NSF-CHE, and PV acknowledges NSF-CHE, for support of this work. ChemMatCARS is supported by NSF-CHE, NSF-DMR, and the U.S. Department of Energy (DOE). The Advanced Photon Source at Argonne National Laboratory is supported by the U.S. DOE, Office of Basic Energy Sciences.

References

1   A. M. Tikhonov, S. V. Pingali and M. L. Schlossman, *J. Chem. Phys.*, 2004, **120**, 11822.
2   A. M. Tikhonov and M. L. Schlossman, *J. Phys. Chem. B*, 2003, **107**, 3344.
3   M. Li, D. Chaiko and M. L. Schlossman, *J. Phys. Chem. B*, 2003, **107**, 9079.
4   M. Li, A. Tikhonov and M. L. Schlossman, *Europhys. Lett.*, 2002, **58**, 80.
5   M. L. Schlossman, *Curr. Opin. Coll. Int. Sci.*, 2002, **7**, 235.
6   D. M. Mitrinovic, S. M. Williams and M. L. Schlossman, *Phys. Rev. E*, 2001, **63**, 21 601.
7   A. M. Tikhonov, M. Li, D. M. Mitrinovic and M. L. Schlossman, *J. Phys. Chem. B*, 2001, **105**, 8065.
8   M. Li, A. M. Tikhonov, D. Chaiko and M. L. Schlossman, *Phys. Rev. Lett.*, 2001, **86**, 5934.
9   D. M. Mitrinovic, A. M. Tikhonov, M. Li, Z. Huang and M. L. Schlossman, *Phys. Rev. Lett.*, 2000, **85**, 582.
10 A. M. Tikhonov, D. M. Mitrinovic, M. Li, Z. Huang and M. L. Schlossman, *J. Phys. Chem. B*, 2000 **104**, 6336.
11 Z. Zhang, D. M. Mitrinovic, S. M. Williams, Z. Huang and M. L. Schlossman, *J. Chem. Phys.*, 1999, **110**, 7421.
12 D. M. Mitrinovic, Z. Zhang, S. M. Williams, Z. Huang and M. L. Schlossman, *J. Phys. Chem. B*, 1999, **103**, 1779.
13 C. Gavach, T. Mlodnicka and J. Guastalla, *C. R. Seances Acad. Sci., Ser. C*, 1968, **266**, 1196.
14 J. Koryta, P. Vanysek and M. Brezina, *J. Electroanal. Chem.*, 1976, **67**, 263.
15 J. Koryta, P. Vanysek and M. Brezina, *J. Electroanal. Chem.*, 1977, **75**, 211.
16 C. Gavach, P. Seta and B. d'Epenoux, *J. Electroanal. Chem.*, 1977, **83**, 225.
17 M. Gros, S. Gromb and C. Gavach, *J. Electroanal. Chem.*, 1978, **89**, 29.
18 C. M. Pereira, W. Schmickler, F. Silva and M. J. Sousa, *J. Electroanal. Chem.*, 1997, **436**, 9.
19 W. Schmickler, *J. Electroanal. Chem.*, 1997, **428**, 123.
20 P. Vanysek, *Electrochemistry on Liquid/Liquid Interfaces*, Springer, 1985.
21 B. Lin, M. Meron, J. Gebhardt, T. Graber, M. L. Schlossman and P. J. Viccaro, *Physica B (Amsterdam)*, 2003, **336**, 75.
22 L. Nevot and P. Croce, *Rev. Phys. Appl.*, 1980, **15**, 761.
23 S. K. Sinha, E. B. Sirota, S. Garoff and H. B. Stanley, *Phys. Rev. B*, 1988, **38**, 2297.
24 J. D. Weeks, *J. Chem. Phys.*, 1977, **67**, 3106.
25 F. F. Abraham, *Phys. Rep.*, 1979, **53**, 93.
26 A. Braslau, M. Deutsch, P. S. Pershan, A. H. Weiss, J. Als-Nielsen and J. Bohr, *Phys. Rev. Lett.*, 1985 **54**, 114.
27 A. Braslau, P. S. Pershan, G. Swislow, B. M. Ocko and J. Als-Nielsen, *Phys. Rev. A*, 1988, **38**, 2457.
28 D. K. Schwartz, M. L. Schlossman, E. H. Kawamoto, G. J. Kellogg, P. S. Pershan and B. M. Ocko, *Phys. Rev. A*, 1990, **41**, 5687.
29 M. L. Schlossman, in *Encyclopedia of Applied Physics*, ed. G. L. Trigg, VCH Publishers, New York, 1997, vol. **20**, p. 311.
30 M. P. Gelfand and M. E. Fisher, *Physica A (Amsterdam)*, 1990, **166**, 1.
31 L. G. Parratt, *Phys. Rev.*, 1954, **95**, 359.
32 G. Gouy, *C. R. Acad. Sci.*, 1910, **149**, 654.
33 D. L. Chapman, *Philos. Mag.*, 1913, **25**, 475.
34 E. J. W. Verwey and K. F. Niessen, *Philos. Mag.*, 1939, **28**, 435.
35 T. Kakiuchi and M. Senda, *Bull. Chem. Soc. Jpn.*, 1983, **56**, 1322.

36 Z. Samec, V. Marecek and D. Homolka, *J. Electroanal. Chem.*, 1985, **187**, 31.
37 H. H. Girault and D. H. Schiffrin, *J. Electroanal. Chem.*, 1983, **150**, 43.
38 H. H. Girault, *Electrochim. Acta*, 1987, **132**, 383.
39 H. H. Girault and D. J. Schiffrin, *J. Electroanal. Chem.*, 1984, **170**, 127.
40 H. H. Girault and D. J. Schiffrin, *J. Electroanal. Chem.*, 1988, **244**, 15.
41 Z. Samec, *Chem. Rev.*, 1988, **88**, 617.
42 Y. Chang, V. J. Cunnane, D. J. Schiffrin, L. Murtomaki and K. Kontturi, *J. Chem. Soc., Faraday Trans.*, 1991, **87**, 107.
43 C. M. Pereira, A. Martins, M. Rocha, C. J. Silva and F. Silva, *J. Chem. Soc., Faraday Trans.*, 1994, **90**, 143.
44 C. M. Pereira, W. Schmickler, A. F. Silva and M. J. Sousa, *Chem. Phys. Lett.*, 1997, **268**, 13.
45 D. S. Walker, M. G. Brown, C. L. McFearin and G. L. Richmond, *J. Phys. Chem. B*, 2004, **108**, 2111.
46 R. A. Marcus, *J. Phys. Chem.*, 1990, **94**, 4152.
47 R. A. Marcus, *J. Phys. Chem.*, 1990, **94**, 7742.
48 G. Geblewicz and D. J. Schiffrin, *J. Electroanal. Chem.*, 1988, **244**, 27.
49 I. Benjamin, *J. Chem. Phys.*, 1992, **97**, 1432.
50 I. Benjamin, *Science*, 1993, **261**, 1558.
51 I. Benjamin, *Chem. Rev.*, 1996, **96**, 1449.
52 D. Michael and I. Benjamin, *J. Electroanal. Chem.*, 1998, **450**, 335.
53 M. R. Philpott, T. T. Lin and J. N. Glosli, Molecular Dynamics Simulation on Liquid/Liquid Interfaces, Paper #1017, The Electrochemical Society 195th Meeting, 1999, Seattle.
54 O. Pecina and J. P. Badiali, *Phys Rev. E*, 1998, **58**, 6041.
55 Q. Cui, G. Zhu and E. Wang, *J. Electroanal. Chem.*, 1994, **372**, 15.
56 Q. Cui, G. Zhu and E. Wang, *J. Electroanal. Chem.*, 1995, **383**, 7.
57 D. J. Henderson and W. Schmickler, *J. Chem. Soc., Faraday Trans.*, 1996, **92**, 3839.
58 S. Frank and W. Schmickler, *J. Electroanal. Chem.*, 2000, **483**, 18.
59 W. Helfrich, *Z. Naturforsch., C: Biochem., Biophys., Biol., Virol.*, 1973, **28**, 693.
60 S. A. Safran, *Statistical Thermodynamics of Surfaces, Interfaces, and Membranes*, Addison–Wesley Publishing Co., Reading, MA, 1994.
61 P. A. Fernandes, M. N. D. S. Cordeiro and J. A. N. F. Gomes, *J. Phys. Chem. B*, 1999, **103**, 6290.

Orientational order of the water molecules at the vicinity of the water–benzene interface in a broad range of thermodynamic states, as seen from Monte Carlo simulations

Pál Jedlovszky,*[a] Ágnes Keresztúri[a] and George Horvai[b]

[a] *Department of Colloid Chemistry, Eötvös Loránd University, Pázmány Péter stny. 1/a, H-1117 Budapest, Hungary. E-mail: pali@para.chem.elte.hu*
[b] *Department of Chemical Information Technology, Budapest University of Technology and Economics, Gellért tér 4, H-1111 Budapest, Hungary*

Received 14th April 2004, Accepted 18th June 2004
First published as an Advance Article on the web 17th November 2004

Monte Carlo simulation of the water/benzene liquid–liquid interfacial system has been performed at six different thermodynamic state points, ranging from ambient conditions up to the vicinity of the critical point of water. The system has been found to consist of two immiscible liquid phases at every state point studied. The orientational preferences of the interfacial water molecules have been analysed in detail using the simulated configurations. The results obtained at ambient conditions are in agreement with previous results on various different water/apolar interfaces. Thus, interfacial water molecules have been found to have dual orientational preferences: the molecules located nearest to the organic phase prefer to stay perpendicular to the interface, pointing flatly toward the apolar phase by their dipole vectors, whereas the waters located somewhat farther from the organic phase prefer the parallel alignment with the interface. The observed orientational preferences are found to be rather stable with changing thermodynamic conditions: although the increase of the temperature has led, due to the increasing thermal motion of the molecules, to a gradual weakening of the orientational preferences, both preferences are found to exist up to at least 450 K, and found to be completely washed out at 575 K only. The pressure has not been found to influence the orientation of the water molecules noticeably.

1. Introduction

The orientational structure of water at the vicinity of planar interfaces with disordered apolar phases has been intensively studied in the past decade both by experimental methods[1–7] and by computer simulations.[8–20] However, the conclusions drawn from the numerous computer simulation analyses have often been conflicting with each other. Thus, among others, it has been concluded that interfacial waters do not have any particular orientational preference;[9] the dipole vector of the water molecules located at the apolar side of the interface points preferentially toward the apolar, while that of the waters being at the aqueous side of the interface toward the aqueous phase;[11] and the dipole vector of the interfacial water molecules lays preferentially parallel with the plane of the interface.[8,10,13] The latter conclusion has further been elaborated by stating that the

DOI: 10.1039/b405509h

deviation of the water dipoles from this preferential parallel alignment with the interface is not symmetric: the dipole vector of the water molecules located at the apolar side of the interface deviates from the preferred alignment by pointing toward the apolar phase with a slightly higher probability than toward the aqueous phase, whereas the deviation of the water dipoles from the preferred alignment at the aqueous side of the interface is just the opposite.[12,14] Several studies have gone beyond the analysis of the preferred alignment of just the molecular dipole vector; however, the conclusions drawn about the preferential orientation of the entire water molecule are also contradicting each other in some cases. Thus, in some studies the vector joining the two H atoms of the molecule (H–H vector),[10,13] whereas in some other studies one of the two O–H bonds[15] of the water molecules located closest to the apolar phase, has been found to be preferentially perpendicular to the plane of the interface. Moreover, about the orientation of the water molecules located beyond this layer, it has been found in some studies that their H–H vector does not have any particular orientational preference;[10] whereas in other studies this vector has been found to be preferentially parallel with the interface.[13]

Recently we have shown that the interfacial water molecules have two distinct orientational preferences: in the first of these preferred orientations, which is present in almost the entire interfacial region, the water molecule lays parallel with the plane of the interface, whereas in the second preferred orientation, present only among the water molecules penetrated deepest into the apolar phase, the plane of the molecule is perpendicular to the interfacial plane and one of the two O–H bonds points almost straight toward the apolar phase.[16] Moreover, we have demonstrated that this dual orientational preference can only be revealed by calculating the bivariate joint distribution of two independent parameters describing the orientation of the water molecule relative to the interface.[16,19] We have also shown that this orientational behaviour of water does not depend on the actual composition of the apolar phase[18] as long as its molecules cannot form hydrogen bonds with water. Thus, the orientational preferences of the water molecules have been found to be rather similar at the vicinity of the interface with liquid 1,2-dichloroethane,[16,18] CCl_4,[18,19] benzene,[18] n-octane,[17,18] vapour[17,18] and in the presence of various numbers of octane molecules adsorbed from the vapour phase at the interface.[17] On the other hand, when the apolar phase has consisted of an increasing number of 1-octanol molecules adsorbed at the interface, the second of the two preferred orientations (*i.e.*, in which the water molecule is perpendicular to the interface and points towards the apolar phase by one of its O–H bonds) is found to be gradually disappearing, due to the hydrogen bonding between the water and octanol molecules.[20]

In the present study we extend the analysis of the orientational order of interfacial water to a broad range of thermodynamic states from ambient conditions up to the vicinity of the critical point. For this purpose, we have performed computer simulations of the water–benzene system at six different thermodynamic state points, and analysed the temperature and pressure dependence of the orientational preferences of the interfacial water molecules on the basis of these simulations. The orientational order of the water molecules is analysed in terms of profiles of the average values of various orientational parameters, monovariate distributions of these single orientational parameters, and the bivariate distribution of two independent orientational variables. A full description of the orientational behaviour of the water molecules can only be expected at this latter level of description.

2. Monte Carlo simulations

Monte Carlo simulations of the water–benzene interfacial system have been performed on the isothermal–isobaric (N,p,T) ensemble at six different thermodynamic state points. The thermodynamic conditions under which the simulations have been performed are summarised in Table 1. The aqueous and organic phases of the systems consisted of 536 water and 108 benzene molecules, respectively. Water molecules have been modelled by the SPC/E potential,[21] whereas benzene has been described by a rigid six-site model, consisting of sites interacting through a Lennard–Jones potential with the parameters of $\sigma = 3.375$ Å and $(\varepsilon/k_B) = 78.3$ K.[22] The molecules have been placed in a rectangular basic box, the cross-section of which has been fixed at 25.216 Å $\times$ 25.216 Å, whereas the edge X perpendicular to the interface has been allowed to vary according to the pressure.

The simulations have been performed using the MMC program.[23] In a particle displacement step a randomly chosen molecule has been randomly translated and randomly rotated around a

Table 1 Temperature and pressure of the six systems simulated

System	I	II	III	IV	V	VI
T/K	300	300	375	450	450	575
p/MPa	0.1	10	10	10	100	100

randomly chosen space-fixed axis. The maximum distance of the translation and maximum angle of rotation have been set to 0.25 Å and 15°, respectively, for water, and to 0.4 Å and 25°, respectively, for benzene. Every 500 particle displacement steps have been followed by a volume change trial, in which the volume of the box has been attempted to change along the X-axis by no more than 400 Å^3. The rate of the accepted and tried water, benzene and volume change moves resulted in about 25%, 20% and 50%, respectively, at the 300 K state points, whereas higher acceptance rates have been obtained at higher temperatures.

Initial configurations have been prepared by performing two separate simulations of 536 water and 108 benzene molecules, respectively, at 300 K in two cubic simulation boxes of the edge lengths of 25.216 Å. Both boxes have been equilibrated by performing 10^8 Monte Carlo steps. Then the two boxes have been attached to each other, creating two interfaces between the two neat liquid phases. The interfacial systems have then further been equilibrated by performing another 5×10^8 Monte Carlo steps. Finally, 1000 configurations per system, separated by 10^6 Monte Carlo steps each, have been saved for the analyses. In order to define the location of the interfaces in terms of internal co-ordinates, the basic box has been set in each sample configuration in such a way that the centre-of-mass of the 536 water molecules has been placed at $X = 0$ Å, *i.e.*, at the middle of the box along the interface normal axis X.

3. Results and discussion

3.1. Density profiles

The profiles of the molecular number density ρ of water and benzene (represented by their O atoms and centres-of-mass, respectively) are shown in Fig. 1 as obtained at the six different thermo-dynamic state points investigated. In order to obtain better statistics, the resulted density profiles are averaged over the two interfaces present in the system. As is seen, the system simulated is separated to two liquid phases at each state point studied. As is expected, the density of both phases is found to be increased with increasing pressure and decreased with increasing temperature. The comparison of the density profiles obtained at state points lying along an isotherm or an isobar reveals that, similarly to the bulk phases, the properties of the interface also depend considerably on the thermodynamic conditions. Although the effect of the pressure is negligible at room temperature (at 300 K the increase of the pressure by a factor of 100 does not lead to noticeable changes in the density profiles), at higher temperatures its role in determining the properties of the interface becomes more important, as here the increase of the pressure makes the interfacial region noticeably thinner. Thus, at 450 K the width of the interfacial layer, defined as the X range within which the water density drops from 90% to 10% of its bulk value (90%–10% width) is about 15% larger at 10 MPa than at 100 MPa. The increase of the temperature has a more expressed effect on the width of the interface: along the 10 MPa isobar the 90%–10% width of the interface has found to be 2.6 Å, 3.7 Å and 5.0 Å at 300 K, 375 K and 450 K, respectively, whereas at 100 MPa this value resulted in 4.3 Å at 450 K and 6.7 Å at 575 K.

In order to analyze the orientation of the water molecules located at different parts of the interface separately, we have divided the region of the interfacial and near-interfacial water to four separate layers in each system. Thus, layer A, representing the waters penetrated deepest into the organic phase is defined as the X range within which the water density drops from 10% of its bulk value to zero. Layers B and C, defined as the X ranges where the water density drops from 50% to 10% and from 100% to 50% of the bulk phase value, respectively, are belonging to the interfacial region, whereas layer D, defined to be as wide as layer C, represents the subsurface water region. The definition of the four layers are shown in the inset of Fig. 1 in the case of the $\{T = 300$ K, $p = 0.1$ MPa$\}$ state point.

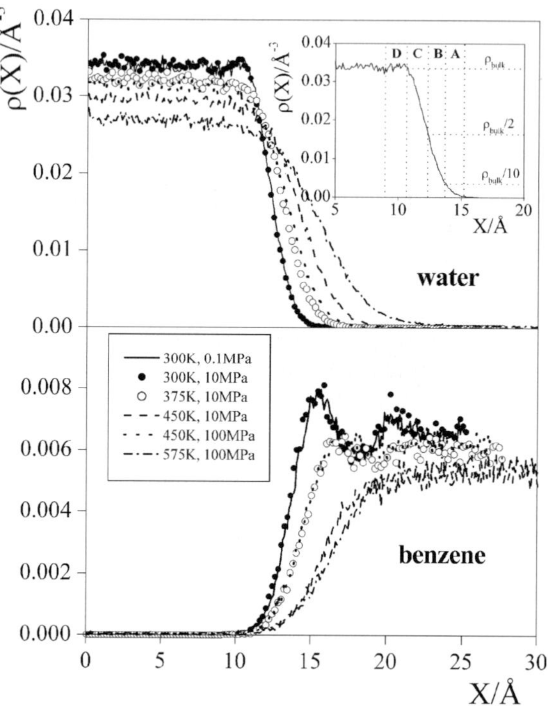

Fig. 1 Number density profile of the water O atoms (top) and benzene centers-of-mass (bottom) in the six systems simulated. All profiles are averaged over the two interfaces present in the system. Solid lines: system I, full circles: system II, open circles: system III, dashed lines: system IV, dotted lines: system V, dash-dotted lines: system VI. The inset illustrates the division of the interfacial and subsurface region to four separate water layers on the example of system I.

3.2. Orientational profiles

The change of the average orientation of the water molecules along the interface normal axis X can be monitored by calculating the profiles of various orientational parameters. It should be noted, however, that for molecules having lower symmetry than $C_{\infty v}$ single orientational parameters could only describe the orientation of various molecule-fixed vectors but not that of the entire molecule. In this study we have monitored the average orientation of the dipole vector d, normal vector n and the vector joining the two H atoms (H–H vector) r_{HH} of the water molecules by calculating the angles α, β and γ formed by these molecule-fixed vectors, respectively, with the interface normal vector pointing toward the benzene phase X. The definition of the three molecule-fixed vectors and the angles α, β and γ are illustrated in Fig. 2.

It should be noted that while the direction of the molecular dipole vector is determined by the charge distribution of the water molecule, the vectors n and r_{HH} can both point toward two opposite directions, and there is no physical way to distinguish between these two possibilities. Therefore these vectors can always be defined in such a way that the angles β and γ are not bigger than 90°. Thus, the angle α can vary in the angular range of $0° \leq \alpha \leq 180°$, and hence $-1 \leq \cos\alpha \leq 1$, whereas both β and γ are between 0° and 90°, and their cosines fall in the range between 0 and 1. Therefore, preferenceless orientation of the water molecules corresponds to the $\langle\cos\alpha\rangle$ value of 0 and to $\langle\cos\beta\rangle = \langle\cos\gamma\rangle = 0.5$, and hence the use of $\langle\cos\beta - 0.5\rangle$ and $\langle\cos\gamma - 0.5\rangle$ instead of $\langle\cos\beta\rangle$ and $\langle\cos\gamma\rangle$ as orientational parameters has the advantage of dropping to zero in the case when the orientation of the water molecules is not correlated with the interface. In order to take the statistical weight of the water orientations also into account the orientational parameters should also be multiplied by the number density of the water molecules ρ_w. Thus, the average orientation of the three molecule-fixed vectors along the interface normal axis X are described here by the profiles

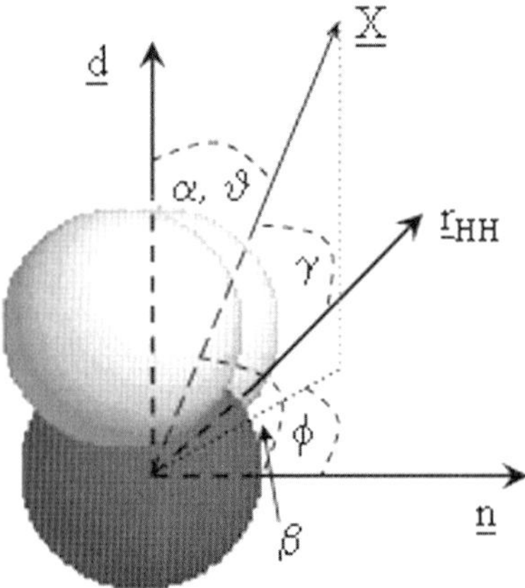

Fig. 2 Definition of the molecule-fixed vectors d, n and r_{HH}, their angles α, β and γ, formed with the interface normal vector pointing toward the organic phase X, respectively, and the angular polar coordinates ϑ and ϕ of the vector X in the local frame defined by the vectors d, n and r_{HH}.

$\Phi_\alpha(X)$, $\Phi_\beta(X)$ and $\Phi_\gamma(X)$, defined as

$$\Phi_\alpha(X) = \rho_w \langle \cos \alpha \rangle (X) , \tag{1}$$

$$\Phi_\beta(X) = \rho_w \langle \cos \beta - 0.5 \rangle (X), \tag{2}$$

and

$$\Phi_\gamma(X) = \rho_w \langle \cos \gamma - 0.5 \rangle (X), \tag{3}$$

respectively.

The orientational profiles obtained at the six different thermodynamic state points are shown in Fig. 3. As is seen, the $\Phi_\alpha(X)$ profiles have a deep negative peak above the X value of 10 Å, roughly at the boundary between layers C and D. This peak shifts towards larger values with decreasing density of the aqueous phase, reflecting simply the fact that the interface itself is shifted farther from the middle of the simulation box. The amplitude of the peak becomes smaller with increasing temperature, and becomes noticeably higher with increasing pressure at the high temperature state points. The negative values of Φ_α indicate that the dipole vector of the molecules points, on average, more likely towards the aqueous than towards the apolar phase in this X range. This minimum is followed by a positive peak of $\Phi_\alpha(X)$ around 14–15 Å, which is, however, washed out at the high temperature state points. In the region of this positive peak the water dipoles point toward the benzene phase with a higher probability than toward the aqueous phase. It should be noted, however, that positive and negative values of Φ_α do not necessarily mean that the water dipoles prefer to point straight toward the apolar and aqueous phase, respectively, they just indicate that outward and inward orientation (including every possible alignment from being parallel to perpendicular to the interface) of the water molecules, respectively, is more preferred than the opposite one.

The $\Phi_\beta(X)$ profile shows a large positive peak at the same position where the first (negative) peak of $\Phi_\alpha(X)$ is located, and damps to zero at the vicinity of the apolar phase, indicating that in regions B and C the average orientation of the molecules is closer to the parallel than to the perpendicular alignment relative to the plane of the interface, whereas this preference is lost in layer A. The $\Phi_\gamma(X)$ profile, similarly to $\Phi_\alpha(X)$ has two peaks at about the same positions than those of $\Phi_\alpha(X)$: a negative peak covering roughly the X range of layer C, and a positive peak in layers B and A. The peaks of $\Phi_\beta(X)$ and $\Phi_\gamma(X)$ show a rather similar dependence on the temperature and pressure than those of $\Phi_\alpha(X)$.

The obtained orientational profiles indicate that the average orientation of the water molecules changes dramatically upon approaching the apolar phase, as both $\Phi_\alpha(X)$ and $\Phi_\gamma(X)$ change sign,

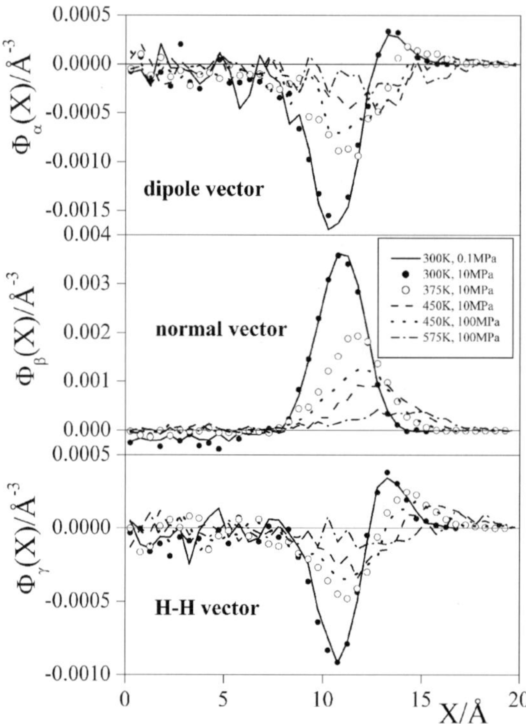

Fig. 3 Orientational profiles of the dipole vector (top), normal vector (middle) and H–H vector (bottom) of the water molecules across the six systems simulated. All profiles are averaged over the two interfaces present in the system. Solid lines: system I, full circles: system II, open circles: system III, dashed lines: system IV, dotted lines: system V, dash-dotted lines: system VI. For the definition of the $\Phi_\alpha(X)$, $\Phi_\beta(X)$ and $\Phi_\gamma(X)$ profiles, see the text.

indicating changes in the orientational preferences of the vectors d and r_{HH}, whereas $\Phi_\beta(X)$ shows the loss of the orientational preferences of the vector n roughly in the X range corresponding to layer B. Further analysis of these changes of the orientational preferences of the water molecules requires going beyond just calculating the average values of the orientational parameters, and their full distributions have to be calculated.

3.3. Monovariate distributions of single orientational parameters

The $P(\cos\alpha)$, $P(\cos\beta)$ and $P(\cos\gamma)$ distributions of the corresponding orientational parameters are shown in Figs. 4, 5 and 6, respectively, as obtained separately in layers A, B and C of the six systems simulated. As is seen, the $P(\cos\alpha)$ distributions have their peak at the vicinity of the cos α value of 0 in every case, however, the position of the peak is shifted from slightly negative to positive values upon going from layer C to A (*i.e.*, toward the apolar phase). This means that the water dipoles prefer a nearly parallel alignment with the plane of the interface, however, in their preferred orientation they point slightly toward the aqueous phase in layer C and toward the benzene phase in layer A. Thus, the negative and positive peaks of $\Phi_\alpha(X)$ do not correspond to the preference of the vector d for pointing straight toward the aqueous and apolar phase, respectively, instead they are simply due to this change of the slight deviation of the preferred orientation from the alignment parallel with the interface.

Contrary to $\Phi_\alpha(X)$, the peaks of the $\Phi_\beta(X)$ and $\Phi_\gamma(X)$ profiles indeed result from the preference of the corresponding molecule-fixed vectors for one of the two extreme alignments, as both the $P(\cos\beta)$ and $P(\cos\gamma)$ distributions are found to be monotonous in each case. Thus, the negative peak of $\Phi_\gamma(X)$ corresponds to the preferred perpendicular, and its positive peak to the preferred parallel alignment of the vector r_{HH} relative to the interface normal axis, whereas the positive peak of $\Phi_\beta(X)$

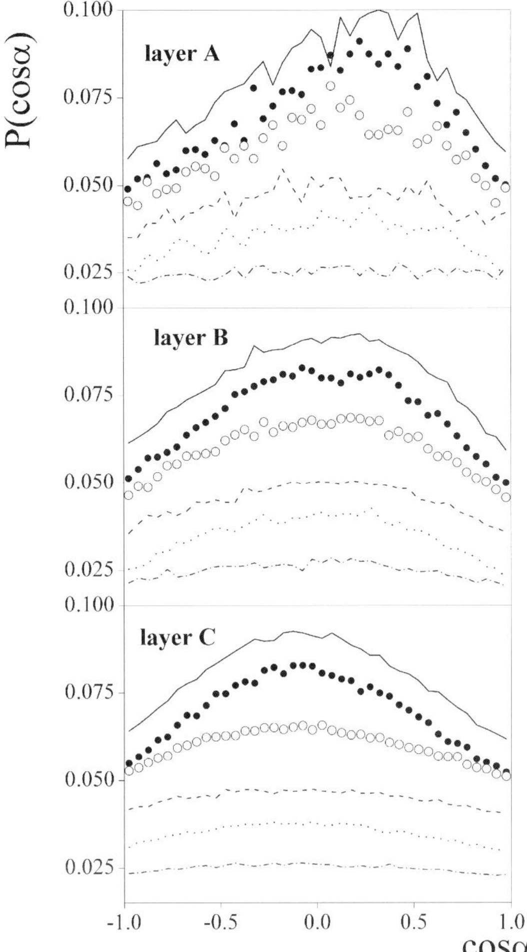

Fig. 4 Monovariate cosine distributions of the angle α formed by the dipole vector of the water molecules d with the interface normal vector pointing toward the benzene phase X in layers A (top), B (middle) and C (bottom) of the six systems simulated. Solid lines: system I, full circles: system II, open circles: system III, dashed lines: system IV, dotted lines: system V, dash-dotted lines: system VI. The results obtained in systems I, II, III, IV and V are shifted by 0.055, 0.045, 0.035, 0.02 and 0.01 units, respectively.

is due to the preference of the vector n for being parallel with X. Furthermore, the damping of $\Phi_\beta(X)$ to zero in the X range of layer A indeed corresponds to a random orientation of the normal vector of the water molecules relative to the interface, as the $P(\cos \beta)$ distributions in layer A are found to be uniform.

It should be noted, however, that although the monovariate distributions of single orientational parameters provide a full description of the orientation of the corresponding molecule-fixed vectors, they cannot describe unambiguously the orientational statistics of the entire water molecules. Such information cannot even be extracted by putting together the information contained by several of such monovariate distributions, since the preferred orientations of different molecule-fixed vectors do not necessarily correspond to the same alignment of the full molecule.[16,19] Such a situation occurs in layer B of the systems investigated here: as is seen from Figs. 5 and 6 the $P(\cos \beta)$ and $P(\cos \gamma)$ distributions both have their maximum at 1 in this layer, indicating that both the normal vector and the H–H vector of the water molecules—i.e., two vectors perpendicular to each other by definition—prefer to align parallel with the same external vector X. This finding clearly indicates

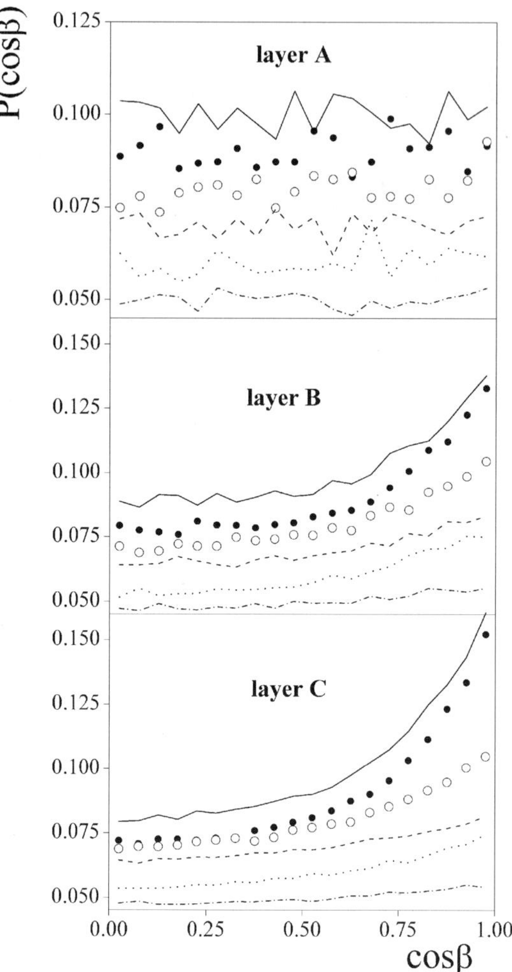

Fig. 5 Monovariate cosine distributions of the angle β formed by the normal vector of the water molecules n with the interface normal vector pointing toward the benzene phase X in layers A (top), B (middle) and C (bottom) of the six systems simulated. Solid lines: system I, full circles: system II, open circles: system III, dashed lines: system IV, dotted lines: system V, dash-dotted lines: system VI. The results obtained in systems I, II, III, IV and V are shifted by 0.05, 0.04, 0.03, 0.02 and 0.01 units, respectively.

that the water molecules contributing to the peak of $P(\cos \beta)$ at 1 must be different from the molecules that contribute to the peak of $P(\cos \gamma)$ at 1. The full description of the orientational statistics of the entire water molecules can only be given by calculating the bivariate distribution of two independent orientational parameters, as has been shown in our previous works.[16,19]

3.4. Full description of the water orientation—bivariate distribution of two independent orientational parameters

The full orientational statistics of rigid bodies relative to an external direction can, in general, be described by determining the bivariate joint distribution of two independent variables characterising the orientation of the rigid body. However, an appropriate choice of the parameters should also satisfy the condition that in the case when the alignment of the bodies is not correlated at all with the external vector the resulting distribution should be uniform. Such a choice of the parameters is a non-trivial task. For instance, any pair of the three orientational parameters used above (*i.e.*, cos α,

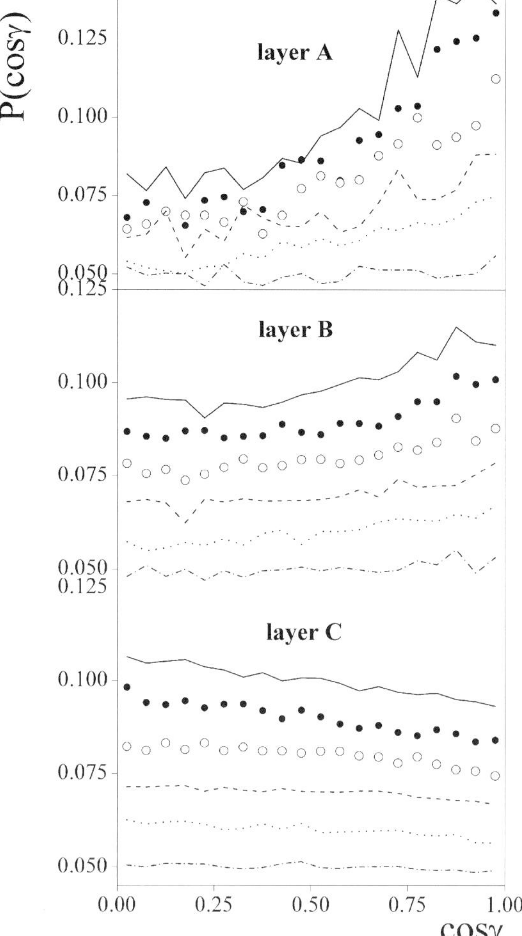

Fig. 6 Monovariate cosine distributions of the angle γ formed by the H–H vector of the water molecules r_{HH} with the interface normal vector pointing toward the benzene phase X in layers A (top), B (middle) and C (bottom) of the six systems simulated. Solid lines: system I, full circles: system II, open circles: system III, dashed lines: system IV, dotted lines: system V, dash-dotted lines: system VI. The results obtained in systems I, II, III, IV and V are shifted by 0.05, 0.04, 0.03, 0.02 and 0.01 units, respectively.

$\cos\beta$ and $\cos\gamma$) fails to satisfy this condition. This can be illustrated by considering the situation when the water molecule lies parallel with the plane of the interface, *i.e.*, the vector n is parallel with X. Then any vector lying in the plane of the water molecule, such as the vectors d and r_{HH}, must be parallel with the interface, and hence perpendicular to X. Thus, the choice of $\cos\beta = 1$ strictly determines the values of $\cos\alpha$ and $\cos\gamma$, being $\cos\alpha = \cos\gamma = 0$ (see Fig. 4 of ref. 19).

In previous works[16,19] we have demonstrated that an appropriate pair of independent orientational parameters are the angular polar coordinates of the interface normal vector X in a local coordinate frame defined by the geometry of the water molecule. Thus, choosing the x, y and z axes of the local frame as the vectors n, r_{HH} and d, the polar angle ϑ is the angle formed by the vectors d and X, whereas ϕ is the angle between the vector n and the projection of X to the plane determined by n and r_{HH}. Hence, the angle ϑ is equivalent with α, and in the case of $\vartheta = 90°$ (*i.e.*, when the water dipole lies parallel with the plane of the interface) ϕ becomes equivalent with β. The definition of the polar angles ϑ and ϕ is also illustrated in Fig. 2. It should be noted that while ϑ is an angle formed by two spatial vectors the directions of which are not restricted by any constraint, ϕ is the angle of two co-planar vectors. Therefore, uncorrelated orientation of the water molecules with X results in a

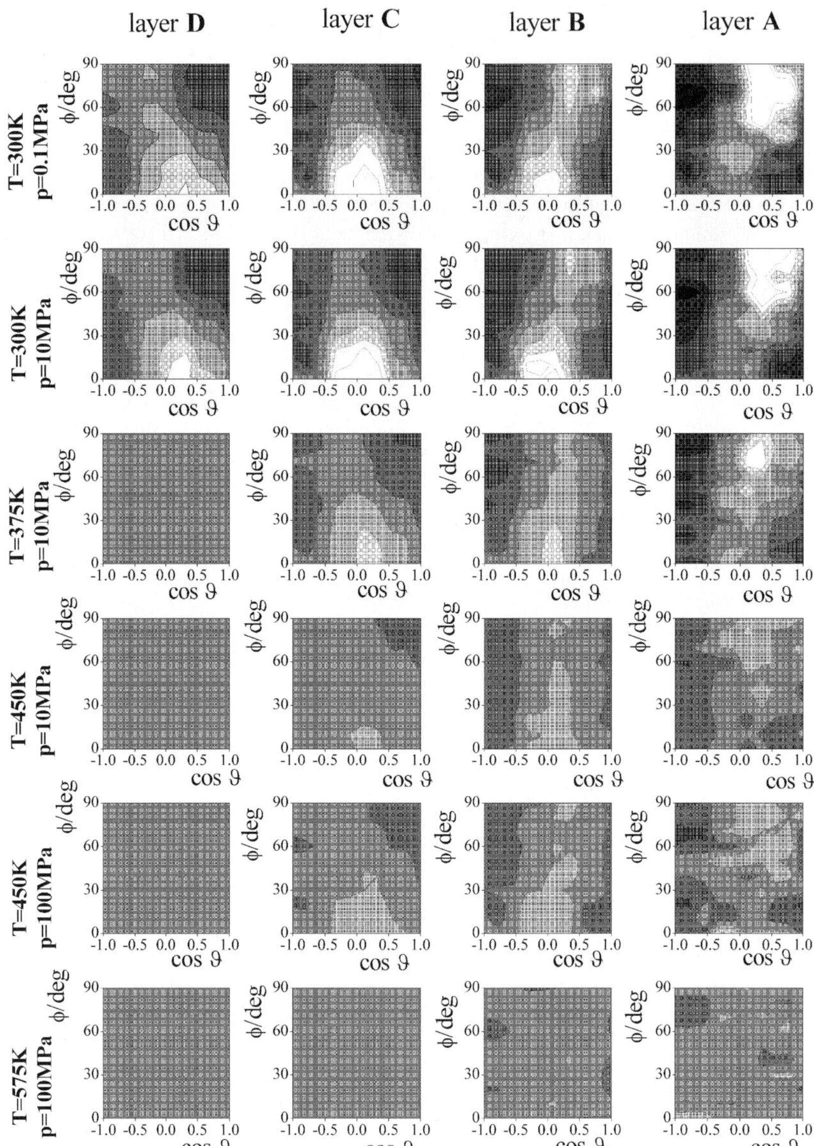

Fig. 7 Bivariate distributions of the angular variables $\cos \vartheta$ and ϕ, describing the orientation of the water molecule relative to the interface in the four separate layers of the six systems simulated. For the definition of the angles ϑ and ϕ, see the text. Lighter shades of gray indicate higher probabilities. Columns from left to right show results in layers D, C, B and A, respectively, whereas rows from top to bottom show results in systems I, II, III, IV, V and VI, respectively.

uniform distribution of $\cos \vartheta$ and ϕ. Hence, the orientation of the water molecules relative to the interface can be appropriately characterised by the $P(\cos \vartheta, \phi)$ bivariate distribution.

The $P(\cos \vartheta, \phi)$ distributions obtained in the four separate interfacial and subsurface water layers of the six systems simulated are shown in Fig. 7. The picture obtained under ambient conditions (*i.e.*, at the $\{T = 300$ K, $p = 0.1$ MPa$\}$ state point) is rather similar to the behaviour of water observed at interfaces with liquid 1,2-dichloroethane,[16] *n*-octane[17] and CCl_4[19] under the same thermodynamic conditions. Thus, in layers C and D the $P(\cos \vartheta, \phi)$ distribution has a peak at about $\cos \vartheta = 0$ and $\phi = 0°$, corresponding to the preferred parallel alignment of the water molecule with the plane of the interface. The preferred alignment of the water molecules changes dramatically upon approaching the apolar phase: in layer A another peak of the $P(\cos\vartheta,\phi)$ distribution emerges in a completely different position, *i.e.*, at about $\cos\vartheta = 0.5$ and $\phi = 90°$. The position of this peak corresponds to a perpendicular alignment of the molecular plane to the plane of the interface, in which one of the O–H bonds of the molecule points straight, whereas the dipole vector flatly, declining by about 30° from the plane of the interface, towards the apolar phase. As has been shown previously,[16] this latter orientation corresponds to the alignment of a H-donor hydrogen bonded neighbour of a water molecule that is located farther from the apolar phase than this neighbour (*i.e.*, in layer B or C), and is oriented parallel with the interface (see Fig. 7 of ref. 16). In layer B the preference of the water molecules for both orientations, as well as high probability of any kind of intermediate orientations between these two are observed.

Fig. 7 shows that these orientational preferences of the interfacial water molecules are remarkably stable when changing the thermodynamic conditions dramatically. Although the increase of the temperature gradually washes out the observed orientational preferences, the picture seen at ambient conditions remains basically unchanged up to temperatures as high as 450 K. The orientational preferences of the subsurface water molecules (*i.e.*, those of layer D) are already washed out at 375 K, however, in the three interfacial layers (*i.e.*, layers A–C) both of the peaks seen at ambient conditions are present, although their amplitude becomes smaller with increasing temperature, due to the increasing thermal motion of the molecules. The orientational preferences of the water molecules are found to be washed out completely at the 575 K state point only. Contrary to the temperature, the pressure of the system does not seem to have a considerable effect on the orientational preferences of the water molecules.

4. Conclusions

In this paper the orientational order of the water molecules at the vicinity of the water/benzene liquid–liquid interface has been studied in detail in a wide range of thermodynamic states from ambient conditions up to the vicinity of the critical temperature of water. The results obtained under ambient conditions are in full agreement with our previous results obtained for various other water/apolar liquid–liquid interfacial systems.[16–19] Thus, water molecules are found to have two distinct orientational preferences. The molecules located nearest to the organic phase prefer to align perpendicular to the plane of the interface, pointing toward the apolar phase flatly by the dipole vector and straight by one of the two O–H bonds. On the other hand, the preferred alignment of the water molecules located farther from the apolar phase in the interfacial region as well as in the subsurface water layer is parallel with the interface. The two preferred orientations correspond to the alignment of two water molecules forming a hydrogen bonded pair, in which the H-donor molecule, located closer to the apolar phase, has the first, whereas the molecule located farther from the apolar phase has the second orientation. It is demonstrated again that the dual orientational preferences of the water molecules can only be revealed by calculating the bivariate joint distribution of two independent orientational parameters, such as the angular polar coordinates of the interface normal vector in a coordinate frame fixed to the water molecule.

The extension of the study to high pressure and high temperature state points has revealed that the above dual orientational preferences of the water molecules are remarkably stable: although the increase of the temperature leads, due to the increasing thermal motion of the molecules, to the weakening of these preferences, they exist up to a temperature of 450 K, and are found to be completely washed out only at the 575 K state point studied. Considering the fact that several features of the orientational profiles of single orientational parameters and of the monovariate distributions of these parameters already disappear at lower temperatures, this finding stresses again the importance of using bivariate distributions in analysing the orientational preferences of molecules at interfaces.

Acknowledgements

P. J. is a Békésy György fellow of the Hungarian Ministry of Education, which is gratefully acknowledged.

References

1   D. Zhang, J. Gutow and K. B. Eisenthal, *J. Phys. Chem.*, 1994, **98**, 13729.
2   I. Tsuyumoto, N. Noguchi, T. Kitamori and T. Sawada, *J. Phys. Chem. B*, 1998, **102**, 2684.
3   D. E. Gragson and G. L. Richmond, *J. Phys. Chem. B*, 1998, **102**, 3847.
4   D. M. Mitrinovic, Z. Zhang, S. M. Williams, Z. Huang and M. L. Schlossman, *J. Phys. Chem. B*, 1999, **103**, 1779.
5   D. Zimdars, J. I. Dadap, K. B. Eisenthal and T. F. Heinz, *J. Phys. Chem. B*, 1999, **103**, 3425.
6   C. Fradin, A. Braslau, D. Luzet, D. Smilgies, M. Alba, N. Boudet, K. Mecke and J. Daillant, *Nature (London)*, 2000, **403**, 871.
7   A. J. Fordyce, W. J. Bullock, A. J. Timson, S. Hasalam, R. D. Spencer-Smith, A. Alexander and J. G. Frey, *Mol. Phys.*, 2001, **99**, 677.
8   P. Linse, *J. Chem. Phys.*, 1987, **86**, 4177.
9   I. L. Carpenter and W. J. Hehre, *J. Phys. Chem.*, 1990, **94**, 531.
10 I. Benjamin, *J. Chem. Phys.*, 1992, **97**, 1432.
11 A. R. van Buuren, S. J. Marrink and H. J. C. Berendsen, *J. Phys. Chem.*, 1993, **97**, 9206.
12 Y. Zhang, S. E. Feller, B. R. Brooks and R. W. Pastor, *J. Chem. Phys.*, 1995, **103**, 10252.
13 T. M. Chang and L. X. Dang, *J. Chem. Phys.*, 1996, **104**, 6772.
14 P. A. Fernandes, M. N. D. S. Cordeiro and J. A. N. F. Gomes, *J. Phys. Chem. B*, 1999, **103**, 6290.
15 I. Benjamin, *J. Chem. Phys.*, 1999, **110**, 8070.
16 P. Jedlovszky, Á. Vincze and G. Horvai, *J. Chem. Phys.*, 2002, **117**, 2271.
17 P. Jedlovszky, I. Varga and T. Gilányi, *J. Chem. Phys.*, 2003, **119**, 1731.
18 P. Jedlovszky, Á. Vincze and G. Horvai, *J. Mol. Liq.*, 2004, **109**, 99.
19 P. Jedlovszky, Á. Vincze and G. Horvai, *Phys. Chem. Chem. Phys.*, 2004, **6**, 1874.
20 P. Jedlovszky, I. Varga and T. Gilányi, *J. Chem. Phys.*, 2004, **120**, 11839.
21 H. J. C. Berendsen, J. R. Grigera and T. P. Straatsma, *J. Phys. Chem.*, 1987, **91**, 6269.
22 S. Gupta, W. B. Sediawan and E. McLaughlin, *Mol. Phys.*, 1988, **65**, 961.
23 M. Mezei, MMC program, http://www.fulcrum.physbio.mssm.edu/~mezei/mmc.

Photo-induced ion transfer across the liquid/liquid interface

Ilan Benjamin

Department of Chemistry, University of California, Santa Cruz, CA 95064, USA

Received 14th April 2004, Accepted 24th May 2004
First published as an Advance Article on the web 20th September 2004

Traditional studies of ion transfer across the interface between two immiscible liquids involve the establishment of a steady-state ion current across the interface. The data obtained from these studies are used to develop models of the interfacial kinetic. However, this approach, while straightforward to implement experimentally, is not sensitive to the microscopic structure of the interface region, which is only a few nanometers in size. We propose and examine the feasibility of a more direct approach to elucidating the ion dynamic by reporting the results of equilibrium and non-equilibrium molecular dynamics calculations in which an iodine ion is created at the interface between water and CCl_4 by photodissociation of an adsorbed I_2^-. In this paper, we describe the model, examine the associated free energy curves and discuss the recombination of the parent molecule and its vibrational relaxation as a competing process for the ion transfer process.

Introduction

Ion transfer across the interface between two immiscible liquids is a fundamental phenomenon of importance to many fields of science, such as phase transfer catalysis,[1] biophysics[2] and electrochemistry.[3] Over the last several decades, it has been the focus of many experimental and theoretical studies that have been extensively reviewed.[4-6] Most experimental studies have taken advantage of the ability to control ion transfer dynamics by an external electric field. The results of these experiments have been described using language and models that have been developed for electrochemical charge transfer at a metal electrode,[3] for example, in terms of kinetic rate constants. While this approach has been very successful in terms of its ability to summarize a large body of data,[5] the problems of understanding at a fundamental level the mechanism of ion transfer, and the ability to calculate the ion transfer rate constant, are still open.

To make progress, one must obtain a molecular level understanding of the structure of the neat interface, of the electric double layer and of the solvated ion as it crosses the interface. This is a formidable problem both experimentally and theoretically, and progress to date has been slow. While significant progress has been made on elucidating the structure of the interface using computer simulations,[7] non-linear spectroscopic techniques[8,9] and other experimental methods,[10-13] there is much less overlap between theory and experiments as far as the ion transfer process itself is concerned. Thus, while it is nowadays quite feasible to compute the structure and dynamics of ion transfer across a realistic (fully atomistic) liquid/liquid interface,[6,14-16] the corresponding experimental probe of the ion transfer at the molecular level is still in its infancy.

The ability of experimentalists to obtain molecular level information about the vibrational and electronic spectra of molecules adsorbed at the interface[8,9,17] has been extended in recent years to include time-resolved data on the pico-second time scale.[18] This suggests that additional progress on understanding the detailed molecular level mechanism of ion transfer across the interface can be

DOI: 10.1039/b405487c

made if these techniques can be utilized to investigate the ion transfer under non-equilibrium conditions.

In this paper, we present molecular dynamics calculations on a simple system, which can be used to study the non-equilibrium ion transfer following the photochemical creation of an ion at the interface. While the choice of the system in this work was dictated by its simplicity (and the fact that it was studied both theoretically and experimentally in bulk liquids), it is possible to modify the system in order to accommodate systems that are more amenable to experimental study.

A model for an ionization reaction at the liquid/liquid interface

As a result of a photoinduced ionization process in solution, a charge redistribution in the molecule produces one or more ions which may diffuse away or recombine. When this process takes place at the interface between two immiscible liquids, the ions may cross the interface. To study this process theoretically one needs to include both the free ion transport across the interface as well as the change in the electronic structure of the solute (as a result of the interactions with the solvents) during the recombination/dissociation. In this initial study we chose to consider a system that has been studied in the past in a number of bulk solvents (including water and organic solvents). This is the dihalide dissociation reaction $I_2^- \rightarrow I + I^-$, which has been studied extensively (in bulk liquids and in clusters) by experiments, theory and simulation.[19–24] Here we consider this reaction at the water/CCl_4 liquid/liquid interface. The neat interface has been studied in recent years by experiments[9] and by molecular dynamics computer simulations.[25]

A simple model, which captures most of the interesting features involved with the above reaction taking place in a condensed medium, is based on an empirical valence bond description of the I_2^- molecule and its interaction with the solvents.[26] This model has been used to study this reaction in bulk acetonitrile, and since the extension to the interfacial system is straightforward, we only briefly outline the model here and refer the reader to a previous publication[27] for additional details. Similar models have also been extensively used to study electron transfer[28] and proton transfer reactions,[29] as well as S_N1 ionization reactions.[30]

A minimal valence bond (VB) description of the ground electronic state of the I_2^- molecule (and its interaction with the water and CCl_4 solvents) is based on the mixing of two diabatic states: $I^- \cdots I$ and $I \cdots I^-$, coupled by the electronic resonance term $V_{el}(r)$, where r is the I–I distance. When r is small, the strong coupling gives rise to a stable state, corresponding to the I_2^- molecule with a delocalized charge distribution (a half unit charge on each atom). As r increases, the polar solvent will stabilize the localized charge of -1 a.u. on one of the iodine atoms. In bulk water, the two possible channels ($I + I^-$ and $I^- + I$) are identical. If the reaction takes place at the interface, the symmetry will be broken because the stabilization of the ion near the organic solvent will be greatly diminished.

We denote by H_1 and H_2 the classical Hamiltonians describing the system in the two diabatic states:

$$H_1 = E_{kin} + U_{liq} + U_{int}^{(1)} + U_{dia}(r)$$
$$H_2 = E_{kin} + U_{liq} + U_{int}^{(2)} + U_{dia}(r)$$

$$(1)$$

where E_{kin} is the total kinetic energy of all the atoms, U_{liq} is the total potential energy of the liquids (including intramolecular terms of the water and CCl_4, their individual intermolecular terms and the water–CCl_4 interactions), $U_{int}^{(1)}$ is the interactions of the water and the CCl_4 molecules with the two iodine atoms (including Lennard–Jones and electrostatic contributions) with the charge of -1 localized on one iodine atom, and $U_{int}^{(2)}$ is an identical term, except for the charge of -1 being localized on the second iodine atom. $U_{dia}(r)$ is the potential energy of the I_2 system with a localized charge distribution. In principle, both $U_{dia}(r)$ and the coupling $V_{el}(r)$ may be calculated using semi-empirical quantum calculations.[31] A simpler approach is to use the fact that in the gas phase the ground state potential energy is given by $U_{dia} - V_{el}$, and this can be set equal to the experimentally known Morse potential function ($De^{-2a(r-r_{eq})} - 2De^{-a(r-r_{eq})}$) of I_2^-. Taking both U_{dia} and V_{el} to be exponential functions fixes their value. This approach has been shown to give a reasonable description of the dissociation of I_2^- in bulk acetonitrile.[27] Additional details about the water flexible SPC potential energy functions, the CCl_4 flexible Lennard–Jones plus electrostatic terms and the interaction between the two liquids, as well as between the two liquids and the iodine atoms, can be found elsewhere.[27,32]

If the two diabatic states are taken to be orthonormal, the dynamic of the system is governed by the quantum Hamiltonian

$$H = \begin{pmatrix} H_1 & V_{el} \\ V_{el} & H_2 \end{pmatrix} \qquad (2)$$

In the Born–Oppenheimer approximation, one assumes that the system is restricted to be in the ground adiabatic state of this Hamiltonian, which corresponds to the lowest eigenvalue of the matrix H:

$$H_{ad} = \frac{1}{2}(H_1 + H_2) - \frac{1}{2}\left[(H_1 - H_2)^2 + 4V_{el}^2\right]^{1/2} = E_{kin} + U_{ad}. \qquad (3)$$

We thus approximate our system by a classical system moving on the adiabatic potential energy U_{ad}. Substituting eqn. (1) into eqn. (3) and utilizing the fact that the two diabatic states have identical form (except for the single charge on the iodine atom) show that U_{ad} has a particularly simple form:

$$U_{ad} = U_{I_2^-} + V_{CF}$$
$$V_{CF} = V_{el} - \left[V_{el}^2 + s^2/4\right]^{1/2} \qquad (4)$$
$$s = H_1 - H_2$$

where $U_{I_2^-}$ is the total potential energy of a system with fixed $-1/2$ charges on each of the iodine atoms, and s is the energy gap between the two diabatic states, taken to be the solvent coordinate (this is a standard choice in molecular dynamics studies of charge transfer reactions).[33]

Before continuing, it is worthwhile discussing the qualitative behavior expected from a system governed by the Hamiltonian H_{ad}. When the solvent is equilibrated to the state where a full -1 charge is localized on one iodine atom, $|s|$ is large (negative for one state and positive when the charge is on the other atom). We denote the equilibrium value of s by s_{eq}. In bulk water or other symmetric media, the two diabatic states have equal and opposite values of s_{eq}. However, if one of the iodine atoms is in a much more polar medium, the diabatic state corresponding to this case has a much larger value of $|S_{eq}|$. At short I–I bond distances, the strong electronic coupling delocalizes the charge, giving rise to a symmetric charge distribution and a stable I_2^- molecule. When the solvent is equilibrated to this charge distribution, $S_{eq} \approx 0$. At large I–I internuclear separation, the enhanced stabilization of the localized charge distribution and the weak electronic coupling give rise to a bi-stable state corresponding to the charge of -1 localized on one or the other iodine atoms. The probability of observing one or the other state is identical in symmetric media, but not at the liquid/liquid interface, where the more polar water phase will tend to localize the charge on the iodine atom close to the water molecule, and thus enhance the charge separation.

Simulation details

The system includes 2000 water molecules and 1256 CCl_4 molecules in a rectangular box of cross section 49.6688 Å × 49.6688 Å and a varying length which keep the normal pressure tensor fixed. After inserting the I_2^- solute at the interface and equilibrating the system at $T = 298$ K, two sets of calculations were performed. Equilibrium calculations at a fixed temperature at different fixed locations of the solute atoms and with different charge distributions on the two iodine atoms were used to compute the free energy surface of the system, as explained below. A second set of calculations was done by running multiple non-equilibrium trajectories at constant energy starting from an independent set of initial conditions. In each one of these trajectories, a single I_2^- molecule was given sufficient energy for dissociation and its fate (recombination or dissociation followed by ion transfer) was followed. The velocity version of the verlet algorithm with periodic velocity rescaling was used to maintain a fixed temperature in the equilibrium runs. The contribution of

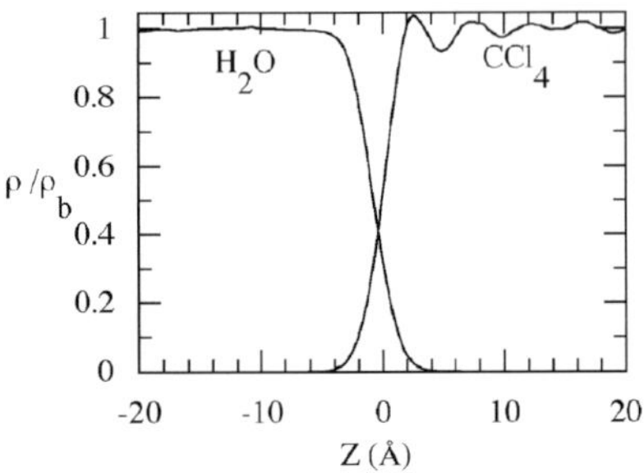

Fig. 1 The density profiles of water and CCl$_4$ at the liquid/liquid interface region (298 K). Each liquid's density is normalized by its bulk density ρ_b.

long-range forces was accounted for by the reaction field method, with a continuous force switching function applied at 23.5 Å over a distance of 1 Å.

The density profile of the system at the liquid/liquid interface region is shown in Fig. 1. We take the plan were the water density reached 50% of its bulk value to be at $Z = 0$ Å. This is approximately the location of the Gibbs dividing surface.

The free energy profile

An important characterization of the system is the equilibrium free energy map as a function of all relevant coordinates. For I$_2^-$ in bulk liquids, there are two important coordinates to consider: the obvious I–I bond distance, which we will denote by r, and the solvent coordinate, s, which indicates the degree to which the solvent orientational and translational polarizations deviate from equilibrium. As has been discussed extensively by Hynes[31,34,35] and mentioned qualitatively above, this coordinate is necessary because the time scale for the iodine atom's motion is comparable with (and even faster) than the solvent motion. Thus, one needs to explicitly consider situations where the solvent is out of equilibrium with the solute charge distribution. The choice of $s = H_1 - H_2$ has also been discussed.[33,36–38]

At the liquid/liquid interface, the locations z_1, z_2 of the two solute atoms along the interface normal are two important additional coordinates. Alternatively, one may use the center of the I–I bond, $z = (z_1 + z_2)/2$, and the orientation of the bond with respect to the interface normal, $\theta = \cos^{-1}[(z_2 - z_1)/r]$. The formal expression for the free energy surface in terms of the four coordinates r, s, z, θ is:

$$G(r,s,z,\theta) = -RT \ln P(r,s,z,\theta)$$

$$P(r,s,z,\theta) = \int e^{-\beta U_{\text{ad}}} \delta(r - r')\delta(s - H_1 + H_2)\delta(z - z')\delta(\theta - \theta')d\Gamma \tag{5}$$

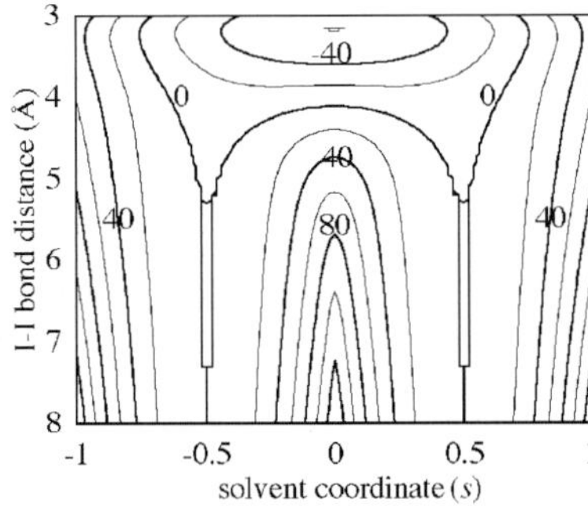

Fig. 2 Free energy map of I$_2^-$ in bulk water.

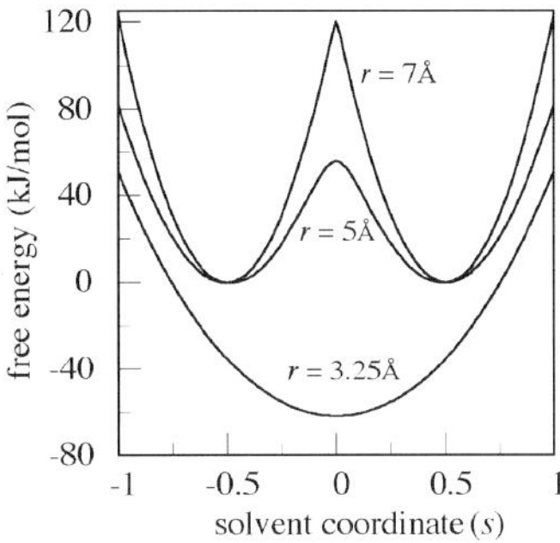

Fig. 3 Cuts in $G(r, s)$ at different I–I bond distances (bulk water).

where we have omitted a constant normalization factor in P (which means that G is determined up to a constant additive number), and the integral is over all nuclear coordinates. Clearly a full determination of G (which requires 5 dimensions) is out of the question, and a more tractable approach is to calculate several projections of G along r and s with fixed values of z and θ.

The methodology we use to compute G at some fixed values of z and θ is identical to the procedure used previously in bulk liquids, and we refer the reader to that paper for details.[27] As in that case, we normalize s by the equilibrium value of s when the system is in one of the diabatic states (such as I$^-\cdots$I) in bulk water, so that these two states correspond to $S_{eq} = \pm 1/2$.

Fig. 2 shows the free energy map in bulk water. Since the absolute value of G is arbitrary, we take the separate I + I$^-$ state to have $G = 0$. The single minimum at $r = 3.25$ Å corresponds to the I$_2^-$ state, and it is 61.5 kJ mol^{-1} lower. Fig. 3. shows cuts in G at different fixed I–I bond distances as a function of the solvent coordinate. The symmetric double well corresponds to the two diabatic states separated by a barrier to the (adiabatic) electron transfer process between them. Recombination of I and I$^-$ (and dissociation of I$_2^-$) is accompanied by significant solvent motion that favors the stabilization of the localized charge distribution.

Figs. 4 and 5 show that the situation at the interface is markedly different. These figures depict the free energy map for the case where the vector connecting the two iodine atoms is perpendicular to the interface, with the center of the bond located at the Gibbs surface. For I–I bond distances near the equilibrium bond length of I$_2^-$, the single minimum near $s = 0$ is now at -42.7 kJ mol^{-1}, reflecting the somewhat smaller free energy of solvation of I$_2^-$ at the interface compared with bulk water. The location of the minimum is at $s = -0.18$, indicating that one of the iodine atoms (with a charge of $-1/2$), is in a slightly more polar environment. A more dramatic change is observed at large I–I bond distances. The symmetric double-well picture is replaced by an asymmetric one. The deep minimum at $s = -0.5$ corresponds to one iodine ion in the water phase, and the other neutral atom in the organic phase. This minimum is only 7.1 kJ mol^{-1} higher than in bulk water. The other shallower minimum corresponds to the less stable reversed case where the ion is in the organic phase.

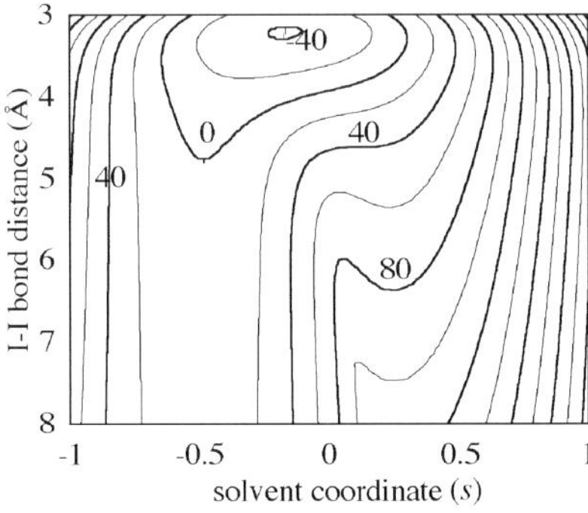

Fig. 4 Free energy map of I$_2^-$ at the water/CCl$_4$ interface.

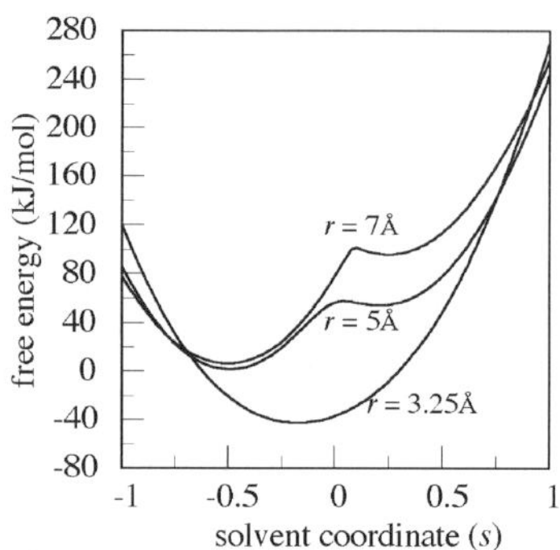

Fig. 5 Cuts in $G(r, s)$ at different I–I bond distances (water/CCl_4).

Other locations (along the interface normal) and orientations of the I–I bond give rise to free energy maps intermediate between those shown in Figs. 4 and 5 and those in bulk water. For example, when the I–I bond is parallel to the interface, G becomes symmetric with respect to the coordinate s (like in bulk water), except that the barrier separating the two diabatic states is lower, and it becomes a function of the location z along the interface normal, reaching the value in bulk water when $z = -9$ Å.

While the free energy map concisely describes the equilibrium behavior of the system and indicates the minimum energy pathways for ion transfer, electron transfer and recombination to produce I_2^-, the actual dynamic of the system may follow a different path, as the solvent motion can play an important role. The actual dynamics of the system are governed by the Hamiltonian given in eqn. (3), and these dynamics are discussed next.

Cage escape and ion transfer

Armed with a qualitative understanding of the free energy profile, we consider here the processes accompanied by dissociation of the I–I bond at the interface. Depending on the initial location (relative to the interface) of the adsorbed I_2^- molecule, the neutral iodine atom and the iodine ion may diffuse away from each other or recombine. The motion of the ions and the neutral atom may be followed by the position coordinates z_1, z_2 and r, while the solvent motion can be approximately followed by the coordinate s.

Typical trajectories are depicted in the different panels of Fig. 6. The panels on the left show the time-dependent solvent coordinate and the I–I bond distance, and the corresponding panels on the right show the position of the I and I^- fragments along the interface normal ($Z = 0$ is the average location of the Gibbs surface).

The top panels show a typical recombination trajectory: the two fragments initially obtain a large bond separation ($r \approx 5$ Å), which rapidly relaxes (line labeled "r" in the top left panel). The solvent coordinate remains near $s = 0$. The two fragments slowly diffuse toward the water side of the interface.

The middle panels show a typical dissociation followed by an ion transfer: the two fragments' bond separation rapidly increases to near $r \approx 8$Å, and then continues to increase as the I^- diffuses into the water phase and the neutral atom diffuses toward the CCl_4 phase. The solvent coordinate slowly reaches the equilibrium value of $s = -1/2$, as expected based on the free energy profile.

The bottom panels also correspond to a dissociation. However, in this case the ion remains at the interface, while the neutral atom is close by in the water phase. The two fragments' bond separation remains near $r \approx 7$ Å. The solvent coordinate slowly reaches the equilibrium value of $s = 0.4$, corresponding to the much shallower minimum shown in the free energy profile. Note that the free energy profile shown in Figs. 4 and 5 is not quite appropriate for discussing the dynamics since it corresponds to the case where the I–I bond is perpendicular to the interface with its center at $z = 0$.

It is interesting to consider that the hydration structure accompanies the ion transfer process. Fig. 7 depicts the hydration number (normalized by the value in bulk water) of the ion and the

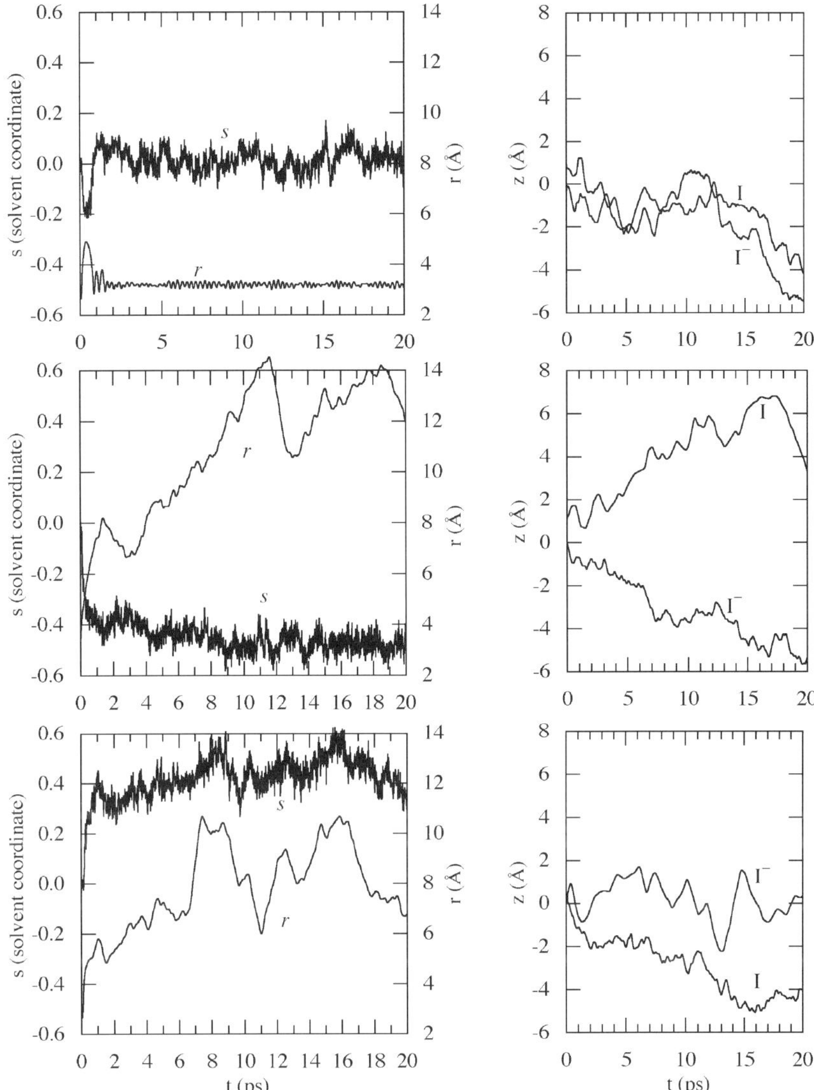

Fig. 6 Example of I_2^- dissociation trajectories at the water/CCl$_4$ interface. The three right panels give the positions of the two fragments along the interface normal. See text for additional details.

neutral atoms as a function of time, corresponding to the trajectory shown in the middle panels of Fig. 6. After a short transient period, as the ion crosses the interface, the hydration number very rapidly reaches the value in bulk water, while the neutral atom sheds off the water molecules as it crosses to the organic phase. The large fluctuations in the hydration number of the ion as it crosses the interface are similar to what is observed in other simulations.[14] These reflect the short lifetime of the water molecules in the first hydration shell of a relatively large ion. Work on understanding the rate and mechanism of the water exchange during the ion transport process is in progress.

Recombination and vibrational relaxation

Approximately 80% of the I_2^- molecules that were photo-excited (200 different trajectories) ended up recombining to form vibrationally hot parent molecules. The study of the vibrational relaxation

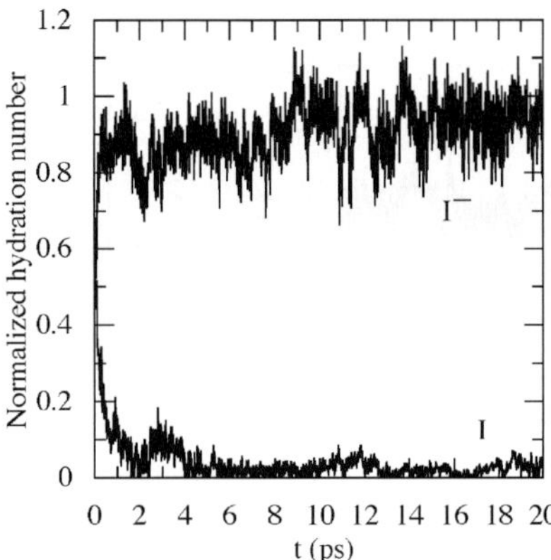

Fig. 7 Normalized hydration number as the ion and the neutral atoms cross the interface.

process is important for a complete understanding of the process considered in this work, as well as for gaining further insight into the nature of the interface. We have extensively studied the vibrational relaxation of I_2^- in bulk water and at the water liquid/vapor interface,[39] and a detailed study of the vibrational relaxation of this molecule at the water/CCl$_4$ interface as a function of location is in progress. Here we briefly comment on the relaxation rate and compare it with other systems.

If $E(0)$ is the average initial vibrational energy deposited in the I_2^- bond, and $E(\infty)$ is the final equilibrium value (classically, equal to RT), we can follow the relaxation dynamic by considering the following time-dependent non-equilibrium correlation:

$$C(t) = \frac{\bar{E}(t) - \bar{E}(\infty)}{\bar{E}(0) - \bar{E}(\infty)} \tag{6}$$

which varies from 1 to 0 as the system relaxes. Fig. 8 shows the result of $C(t)$ for the trajectories which recombine.

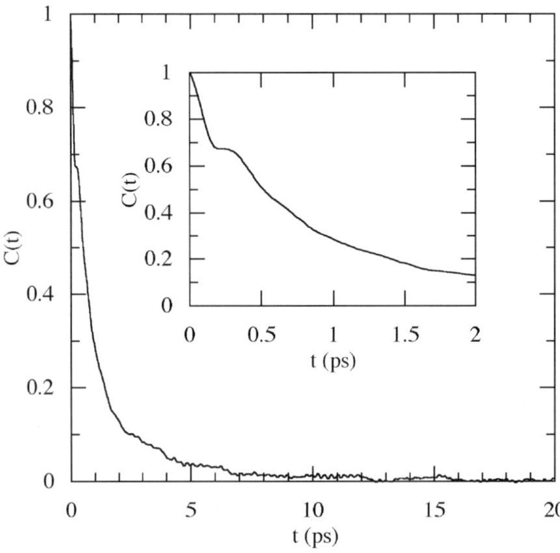

Fig. 8 Vibrational relaxation of the recombined I_2^- formed at the water/CCl$_4$ interface. The inset shows the short time behavior. See eqn. (6).

The initial shoulder in the data represents a transient period where the molecule explores the relatively flat region in the potential energy near dissociation. The vibrational relaxation after this period can be fit to a single exponential $e^{-t/\tau}$ with a time constant of $\tau = 0.9$ ps. Simulations of vibrational relaxation of I_2^- in bulk water[40] give a time constant of 0.6 ps (nearly independent of initial excitation), while the value at the water liquid/vapor interface[41] is 0.9 ps. This is in contrast with a significant slowing down of the vibrational relaxation rate of non-ionic solutes (such as I_2) when comparing interface to bulk. The reason for this was discussed elsewhere.[41] The ion at the interface tends to keep a hydration structure quite similar to that in bulk water, resulting in local friction on the bond similar to that in the bulk. The water/CCl$_4$ system is just another example of this effect. This is consistent with the observation that during an ion transfer from water to an organic phase, even large ions keep contact with a significant fraction of their hydration shell, as long as it is within 1–2 nm of the Gibbs surface.[14]

Conclusions

The model described above is capable of quantitative description of the several possible outcomes following the photodissociation of an ionic solute at the liquid/liquid interface. This includes the possible solute transfer of the ionic (and neutral) fragments and the recombination to form the vibrationally excited parent molecule. And although not discussed here, the electron transfer process is also included within an adiabatic approximation. Experimental observation of the effects discussed here will depend on the ability to monitor the electronic and vibrational spectra of the solute following the photoexcitation. For example, the sum frequency spectra of water molecules already enable detection of change in the interfacial hydration structure[9] and the electronic spectra of the solute already enable monitoring of the solute location.[17] These techniques need to be extended to the time-domain in order to understand the dynamics of solute and ion transfer processes across the liquid/liquid interface.

Acknowledgements

This work has been supported by the National Science Foundation (grant. No. CHE-0345361).

References

1   C. M. Starks, C. L. Liotta and M. Halpern, *Phase Transfer Catalysis*, Chapman & Hall, New York, 1994.
2   K. Arai, M. Ohsawa, F. Kusu and K. Takamura, *Bioelectrochem. Bioenerg.*, 1993, **31**, 65.
3   H. H. Girault and D. J. Schiffrin, *Electrochemistry of Liquid–Liquid Interfaces*, in *Electroanalytical Chemistry*, ed. A. J. Bard, Dekker, New York, 1989, p. 1.
4   P. Vanysek, *Electrochim. Acta*, 1995, **40**, 2841.
5   A. G. Volkov and D. W. Deamer, in *Liquid-Liquid Interfaces*, CRC Press, Boca Raton, 1996.
6   I. Benjamin, *Annu. Rev. Phys. Chem.*, 1997, **48**, 401.
7   A. Pohorille and M. A. Wilson, *J. Mol. Struct. (THEOCHEM)*, 1993, **103**, 271.
8   K. B. Eisenthal, *Chem. Rev.*, 1996, **96**, 1343.
9   G. L. Richmond, *Chem. Rev.*, 2002, **102**, 2693.
10 D. M. Mitrinovic, Z. Zhang, S. M. Williams, Z. Huang and M. L. Schlossman, *J. Phys. Chem. B*, 1999 **103**, 1779.
11 M. L. Schlossman, *Curr. Opin. Colloid Interface Sci.*, 2002, **7**, 235.
12 A. Grimm, K. Muhlfriedel and K. H. Baumann, *Chem.-Ing.-Tech.*, 2002, **74**, 1582.
13 P. Brodard and E. Vauthey, *Rev. Sci. Instrum.*, 2003, **74**, 725.
14 K. J. Schweighofer and I. Benjamin, *J. Phys. Chem. A*, 1999, **103**, 10274.
15 L. X. Dang, *J. Phys. Chem. B*, 1999, **103**, 8195.
16 B. Schnell, R. Schurhammer and G. Wipff, *J. Phys. Chem. B*, 2004, **108**, 2285.
17 W. H. Steel and R. A. Walker, *Nature*, 2003, **424**, 296.
18 A. V. Benderskii and K. B. Eisenthal, *J. Phys. Chem. A.*, 2002, **106**, 7482.
19 J. M. Papanikolas, V. Vorsa, M. E. Nadal, P. J. Campagnola, J. R. Gord and W. C. Lineberger, *J. Chem. Phys.*, 1992, **97**, 7002.
20 A. E. Johnson, N. E. Levinger and P. F. Barbara, *J. Phys. Chem.*, 1992, **96**, 7841.
21 D. A. V. Kliner, J. C. Alfano and P. F. Barbara, *J. Chem. Phys.*, 1993, **98**, 5375.
22 J. M. Papanikolas, P. E. Maslen and R. Parson, *J. Chem. Phys.*, 1995, **102**, 2452.

23 V. Vorsa, S. Nandi, P. J. Campagnola, M. Larsson and W. C. Lineberger, *J. Chem. Phys.*, 1997, **106**, 1402.
24 C. J. Margulis and D. F. Coker, *J. Chem. Phys.*, 1999, **110**, 5677.
25 T. M. Chang and L. X. Dang, *J. Chem. Phys.*, 1996, **104**, 6772.
26 R. Bianco and J. T. Hynes, *J. Chem. Phys.*, 1995, **102**, 7864.
27 I. Benjamin, P. F. Barbara, B. J. Gertner and J. T. Hynes, *J. Phys. Chem.*, 1995, **99**, 7557.
28 A. Warshel and R. M. Weiss, *J. Am. Chem. Soc.*, 1980, **102**, 6218.
29 D. Borgis and J. T. Hynes, *J. Chem. Phys.*, 1991, **94**, 3619.
30 R. E. Westacott, K. P. Johnston and P. J. Rossky, *J. Phys. Chem. B*, 2001, **105**, 6611.
31 B. J. Gertner, K. Ando, R. Bianco and J. T. Hynes, *Chem. Phys.*, 1994, **183**, 309.
32 D. Michael and I. Benjamin, *J. Chem. Phys.*, 2001, **114**, 2817.
33 G. King and A. Warshel, *J. Chem. Phys.*, 1990, **93**, 8682.
34 H. J. Kim and J. T. Hynes, *J. Chem. Phys.*, 1990, **93**, 5194.
35 H. J. Kim and J. T. Hynes, *J. Chem. Phys.*, 1990, **93**, 5211.
36 A. Warshel, *J. Phys. Chem.*, 1982, **86**, 2218.
37 E. A. Carter and J. T. Hynes, *J. Chem. Phys.*, 1991, **94**, 5961.
38 R. A. Marcus, *Rev. Mod. Phys.*, 1993, **65**, 599.
39 J. Vieceli and I. Benjamin, *J. Phys. Chem. B*, 2003, **107**, 4801.
40 I. Benjamin and R. M. Whitnell, *Chem. Phys. Lett.*, 1993, **204**, 45.
41 J. Vieceli, I. Chorny and I. Benjamin, *J. Chem. Phys.*, 2002, **117**, 4532.

Simulation of interfaces between room temperature ionic liquids and other liquids

R. M. Lynden-Bell,*[ab] J. Kohanoff[b] and M. G. Del Popolo[b]

[a] University Chemical Laboratory, Lensfield Road, Cambridge University, Cambridge, UK CB2 1EW. E-mail: rmlb@cam.ac.uk; Fax: +44 1223 336362; Tel: +44 1223 763874

[b] Atomistic Simulation Group, School of Mathematics and Physics, Queen's University, Belfast, UK BT7 1NN

Received 14th April 2004, Accepted 20th May 2004
First published as an Advance Article on the web 21st September 2004

The structure and properties of the interfaces between the room temperature ionic liquid dimethylimidazolium chloride ([dmim]Cl) and different Lennard–Jones fluids and between ionic liquid and water have been studied by molecular dynamics simulations, and compared to the ionic liquid–vapour interface. Two contrasting types of interface were investigated, thermodynamically stable interfaces between ionic liquid and vapour and between ionic liquid and Lennard–Jones fluids, and diffusing interfaces between miscible phases of different compositions involving water. The density profiles of different species through the interface are presented. The cations and water molecules near the former type of interface are aligned relative to the surface, but no orientational preference was found near or in the broad diffusing interface. The ionic liquid has a negative electrostatic potential relative to vapour or Lennard–Jones fluid, but is more positive than pure water. This contrast is explained in terms of the relative importance of orientation and concentration differences in the two types of interface.

Introduction

One of the key steps in chemical synthesis in ionic liquid media is the extraction of products and recovery of the solvent. Usually, this is achieved by means of biphasic extraction processes involving two immiscible liquids. The second liquid is chosen in such a way as to optimise the partitioning of the reactants from the ionic liquid into the second liquid and the further extraction of the reaction products from it. The passage of chemical species through the liquid–liquid interface is an important part of the process. The main factors that influence this phenomenon are the structure and dynamics of the interface. Apart from the obvious relevance for extraction, recovery and purification, liquid–liquid interfaces are not yet as well understood as are solid–liquid or liquid–vapour ones. Experiments on liquid–liquid interfaces are notably more difficult than on other interfaces and only very recently X-ray and neutron reflectivity studies[1,2] allowed for a microscopic insight into the structure of these systems. Therefore, besides the obvious relevance of the partitioning process, there is great interest on fundamental aspects of the liquid–liquid interface. Due to the complexity of the phenomenon involved, computer simulation plays a crucial role in enhancing this understanding.[3–5] One of the issues under debate is the role of the different

DOI: 10.1039/b405514d

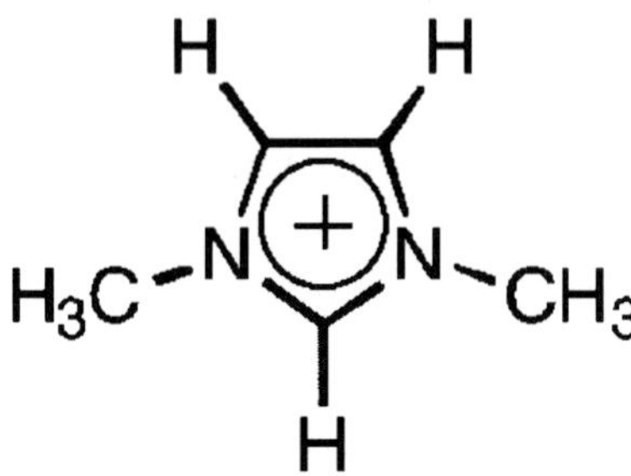

Fig. 1 Dimethylimidazolium cation

contributions to the width of the interface. On the one hand, there is the intermixing of the two species (intrinsic width). On the other hand, as predicted by the theory of capillary waves, there is a second component in the shape of the interface associated with thermal fluctuations.[3,6]

Room temperature ionic liquids have been advocated as potential green solvents for industrial chemical synthesis. The prospect of replacing traditional organic solvents has motivated a large experimental effort by several groups around the globe.[7] Room temperature ionic liquids are organic liquids formed solely of ions which, in contrast to their inorganic counterparts like NaCl, exhibit significantly lower melting temperatures. Besides being liquid at room temperature, they are non-volatile. Many industrially relevant processes have now been carried out in these ionic liquids. The aim of the present study is to analyse basic structural aspects of interfaces between these ionic liquids and other types of liquid using computer simulation.

In this paper we will study interfaces between a simple ionic liquid and two types of undissociated liquids. In each case the ionic liquid is the room temperature molten salt dimethylimidazolium chloride ([dmim]Cl), see Fig. 1. The first type of interface is between this ionic liquid and a Lennard–Jones atomic fluid. The Lennard–Jones fluid consists of spherical non-polar molecules and can be taken to represent a generic organic liquid. We have chosen two different Lennard–Jones liquids, 'strong' and 'weak', with different values of the self intermolecular interaction. These correspond to a normal dense liquid and to a supercritical fluid, respectively. The intermolecular potential between ionic liquid sites and the Lennard–Jones particles is taken to be the same in both cases. The interfaces appear to be thermodynamically stable with no sign of mixing of the ionic liquid with the non-polar fluids.

The second type of interface is formed between a slab of 50 : 50 water–[dmim]Cl mixture and pure water. It is not thermodynamically stable although it changes slowly on the time scale of the simulations, thus allowing for a study of the diffusing interface and the determination of the liquid junction potential.

In all cases we have investigated the density profiles and the preferred orientation of the cations in the interfacial region and compared the results with corresponding liquid–vacuum interfaces. We have also determined the change in electrostatic potential across the interface.

Computational details

Intermolecular potentials

The intermolecular interactions were modelled using site–site potentials with partial charges on the sites. The dimethylimidazolium ion consists of a five-membered aromatic ring with two nitrogen atoms carrying methyl groups. Each of the three carbon atoms carries a proton. The molecule is described by the rigid 10-site model used in our earlier work.[8–12] In this model the methyl group is described as a single site (united atom model). The total charge on the [dmim]$^+$ ion is +1e, distributed over all the atomic sites, while the chloride ion carries a charge of −1e. The short range and dispersion parts of the intermolecular potential are described by site–site Buckingham potentials.

Water was described by the widely used SPC/E model[13] in which there are partial charges on oxygen and hydrogen sites and a single Lennard–Jones interaction between oxygen sites. As in our

Table 1 Lennard–Jones parameters for site–site interactions

Interaction	$\varepsilon/\mathrm{kJ\ mol^{-1}}$	ε/k	$\sigma/\text{Å}$
Strong A–A	3.74	450 K	4.0
Weak A–A	0.374	45 K	4.0
A–N	0.7294		3.625
A–C(ring)	0.5188		3.670
A–Me	0.7505		3.930
A–H(ring)	0.2167		3.255
A–Cl	0.5594		4.0

earlier work the water–[dmim] and water–chloride interactions are described in terms of Lennard–Jones parameters.[11]

The Lennard–Jones particles interact with each other with a potential described by two parameters: ε which determines the energetics and σ which determines the interparticle spacing. The two Lennard–Jones fluids differ in their values of ε, but have the same σ (4 Å). Values are given in Table 1.

The ε/k values for strong and weak Lennard–Jones particles were chosen to be 450 K and 45 K, respectively. As the triple and critical temperatures of a Lennard–Jones fluid are 0.67 ε/k[14] and 1.3 ε/k[15] respectively, at 400 K the strong Lennard–Jones fluid corresponds to a normal dense liquid with $T = 0.89\ \varepsilon/k$, while the weak Lennard–Jones fluid is in the supercritical region of the phase diagram. In order to be sure that the strong Lennard–Jones fluid did not mix with the ionic liquid, the interactions of the Lennard–Jones particles with the [dmim] and chloride sites were chosen to be significantly smaller than the self-interactions. Values are given in Table 1. These values were used for both the Lennard–Jones fluids. We found no tendency for the Lennard–Jones particles to penetrate the ionic liquid even in the weak case.

Simulation parameters

Classical molecular dynamics simulations were carried out using a modified version of the DL_POLY program[16] in a NVT ensemble at 400 K (127 °C) using a Berendsen thermostat. A time step of 0.2 fs was used and the long-range interactions evaluated using the Ewald summation technique. In order to maximise the surface area, hexagonal slab boundary conditions were used. The lateral distance between nearest images was about 40 Å and the surface area about 1420 Å^2. The periodically repeated cells contained a slab of ionic liquid in contact with a slab of either Lennard–Jones fluid or water, so that each cell contained two liquid–liquid interfaces. Initial configurations were prepared from equilibrated slabs of ionic liquid with vacuum above and below the slabs taken from our earlier work.[12] After equilibration, data for the interfaces with Lennard–Jones were collected for about 1 ns. The interfaces with water are not stable and data was collected for 180 ps only. The characteristics of all the systems studied are summarised in Table 2, below.

Analysis

In order to investigate the properties of the interface, data were collected in histograms according to the position of the site along the z-axis perpendicular to the surface. Data collected included the density of cation centres, anions and Lennard–Jones or water molecules, the charge density, and

Table 2 Properties of the runs

Run	P_{zz}/kbar	$\gamma/\mathrm{mNm^{-2}}$	Number of ion pairs	Number of other mols	Cell height/Å
Strong LJ	0.820	107 ± 10	252	1000	90
Weak LJ	2.220	68 ± 10	252	1000	90
Vacuum	−0.029	84 ± 8	252	0	90
Water	−0.030	0 ± 10	231	1577	59
Water mixtures/vac	−0.024	58 ± 6	231	235	60

properties of molecular orientation. In our earlier work[12] we found that the latter was most easily described in terms of the Legendre polynomial function $P_2(\cos\theta)$, defined as

$$P_2(\cos\theta) = \frac{1}{2}\left(3\cos^2\theta - 1\right),$$

where θ is the angle between various molecule-fixed axes and the surface normal. The advantage of this function is that its average is zero in the absence of any preferred orientation, while the signs and magnitudes of the average values give information about the preferred orientation.

Error bars are shown in many of the graphs displayed in this paper. These were calculated by averaging over the two interfaces and over 5 portions of each run and represent the standard deviation of the mean. It should be remembered that this may not be a true representation of the error, as each complete run may not explore the whole of the relevant phase space.

The surface tension γ was measured from the components of the pressure tensor P_{ii} using the formula $\gamma = -b_z(P_{xx} + P_{yy} - 2P_{zz})/4$ where b_z is the length of the molecular dynamics simulation box in the direction parallel to the surface normal, and the factor of 4 in the denominator takes into account the presence of two equivalent interfaces.

Results

Ionic liquid–Lennard–Jones fluid interfaces

Fig. 2 shows the number densities of cation centres, anions and Lennard–Jones particles for slabs of ionic liquid in vacuum and in contact with the two Lennard–Jones liquids. There are two main effects to note. Firstly the density of the Lennard–Jones liquid near the interface changes as the strength of the interaction changes and secondly the region of increased density just below the interface of the ionic liquid becomes less prominent in the presence of the Lennard–Jones liquid, especially the weak one.

The deficit in density of the strong Lennard–Jones liquid near the interface can be interpreted as showing that the strongly interacting fluid with $\varepsilon/k = 450$ K probably does not wet the ionic liquid surface. On the other hand the weak fluid with $\varepsilon/k = 45$ K is strongly attracted to the interface giving an increase in density near it.

In our earlier work[12] we found that there was a high degree of alignment of the cations immediately below the vacuum interface. In their preferred orientation the imidazolium rings lie perpendicular to the surface with one methyl group pointing to the vacuum and one into the liquid. In this region of alignment the density is also enhanced. Fig. 3 shows the changes in the degree of alignment in the two Lennard–Jones-ionic liquid interfaces and the ionic liquid–vacuum interface. The bold curve shows the P_2 function (see the previous section for the definition) of the cosine of the angle between the normal to the interface and an axis through the two N atoms in the ring. The positive value of this function and the approximately equal negative values of the P_2 functions for the CH dipole axis (bold dotted) and the axis perpendicular to the ring (bold dashed) show that immediately below the interface the cations are aligned in the way described above and that the presence of the Lennard–Jones fluid makes little difference. The thin dotted line shows the cation density. The few cations on the outer edge of the interface near the vacuum (top graph) have a different preferred alignment, tending to lie flat on the surface of the ionic liquid. This preference decreases when the ionic liquid is in contact with a Lennard–Jones fluid as these outlying molecules interact with the Lennard–Jones particles. This effect is greater for the supercritical fluid, because it exhibits a higher density in the contact region.

The change in electrostatic potential on traversing the interface is shown in Fig. 4. This figure shows that there is a decrease in potential on entering the liquid. The potential change across the interface is similar for the ionic liquid–vacuum interface and the two ionic liquid–Lennard–Jones fluid interfaces. We note that a negative potential within a phase corresponds to outward pointing dipoles on the surface of that phase. This figure also shows that charge oscillations penetrate a long distance into the ionic liquid, with little attenuation.

Table 2 shows some properties of the runs including the pressure perpendicular to the surface and estimates of the surface tension. The total number of particles and the volume of the cell were kept constant for the runs with different Lennard–Jones liquids. As one would expect, reducing the

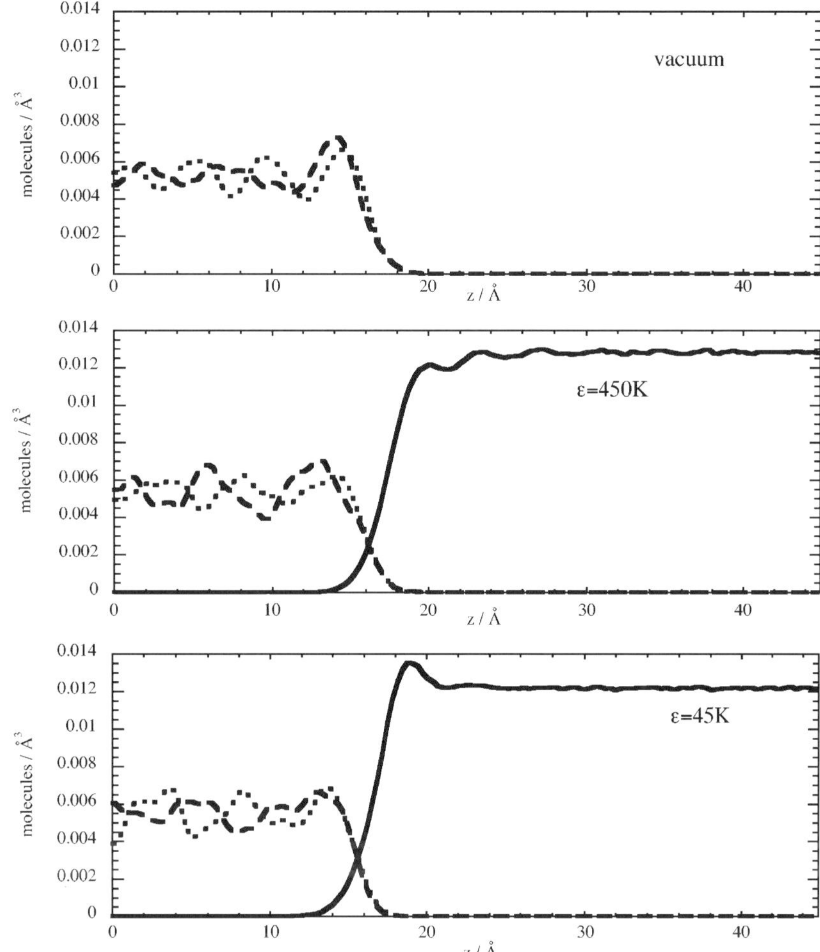

Fig. 2 Molecular densities for three interfaces: [dmim]Cl/vacuum (upper panel), [dmim]Cl/strong Lennard–Jones (middle panel), and [dmim]Cl/weak Lennard–Jones (lower panel). Dashed lines indicate the density of cation centres, dotted lines anion densities, and solid lines densities of Lennard–Jones particles. Note the increased density of the Lennard–Jones liquid near the interface in the supercritical Lennard–Jones liquid (bottom), and the reduction of the density maximum in the ionic liquid below the interface.

strength of the Lennard–Jones interaction leads to an increase in pressure. This is also reflected in a small increase of density in the ionic liquid, which can be seen in Fig. 2.

The values for the pressure and hence the surface tension are very sensitive to the treatment of the long-range parts of the forces. Although these results are not fully converged, the trends shown in Table 2 appear to be reliable. They show that the surface tension of the liquid–liquid interfaces is higher than the corresponding liquid–vapour interface and that the magnitude is largest for the interface with the strong Lennard–Jones liquid. We have already suggested the latter does not wet the ionic liquid.

Ionic liquid–water interfaces

In this section we compare properties of the interface between a slab of a 50 : 50 water–[dmim]Cl mixture and vapour with the interface formed when the previous slab is surrounded by water. Unlike the liquid–liquid interfaces studied in the previous section this is an unstable interface, as water and [dmim]Cl, as modelled in this work, are completely miscible.[11] The interface changes as

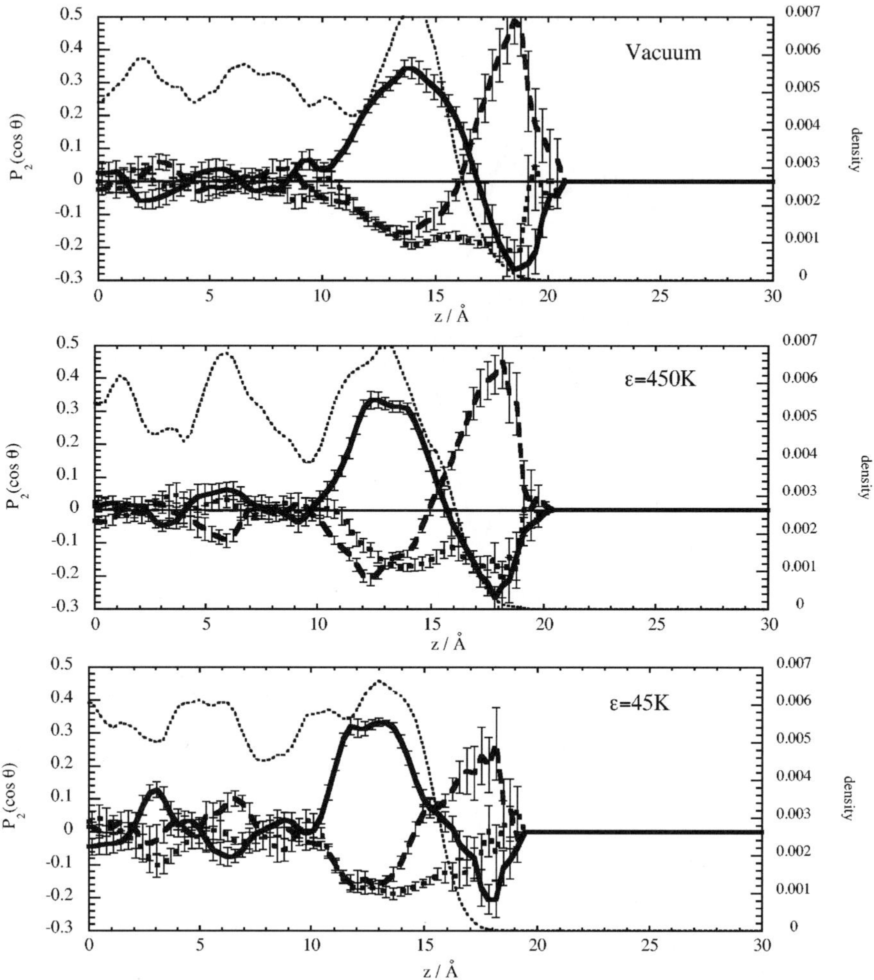

Fig. 3 Cation orientation in the absence and the presence of a Lennard–Jones liquid interface. The bold line is the Legendre polynomial P_2 of the N–N axis, the bold dashed line is the P_2 for the axis perpendicular to the imidazolium ring, and the bold dotted line is P_2 for the C_2 symmetry axis. The thin dotted line shows the cation density. The Legendre polynomials show that, on the ionic liquid side of the interface, the cations are aligned perpendicular to the surface with the NN axis parallel to the surface normal, while the few cations on the Lennard–Jones or vacuum side of the interface tend to lie flat.

the ions diffuse into the water and *vice versa*. Nevertheless, as the density profile changes slowly on the time scale of 200 ps, we can use our simulation to study this diffusing interface. The results shown are based on a 180 ps run. Fig. 5 shows density profiles of the two slabs. The upper graph relates to the slab in contact with vapour with a stable interface and the lower graph to the slab in contact with water with a diffusing interface. The slab in contact with vapour has a structured interface with water adsorbed on the outer edge.[12] When this is surrounded by water and equilibrated for about 60 ps one observes the density distribution shown in the lower graph. Water has penetrated the slab and the ions have begun to diffuse out into the water. The first process, which involves the diffusion of water into the ionic liquid, is very quick. After this, ions diffuse more slowly into the water. It is clear that our system is not in thermodynamic equilibrium and changes with time. Therefore, the results must be treated with caution. The density profiles, however, remain reasonably stable during times of the order of a few hundred ps, which allows for the analysis of the orientation of cations and water molecules, the electrostatic potential across the interface (the liquid junction potential), and the qualitative understanding of the latter.

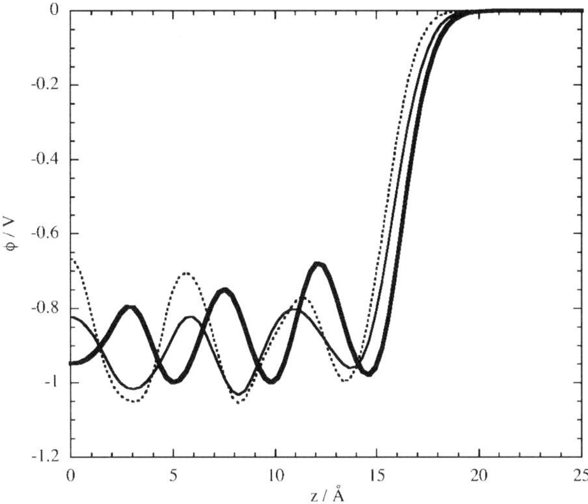

Fig. 4 Electrostatic potential variation across the interface. Ionic liquid slab on the left, vacuum or Lennard–Jones liquid on the right. The bold solid line corresponds to the vacuum interface, the thin solid and dotted lines to the strong and weak Lennard–Jones liquid interfaces.

In previous work[12] we found that, in the slab in contact with vapour, cations show the same preferred orientation as in the slab of pure ionic liquid. As shown in the upper panel of Fig. 6, [dmim]$^+$ ions tend to lie perpendicular to the surface while water molecules tend to have one proton pointing towards and one away from the interface. In the diffusing interface both the [dmim]$^+$ and

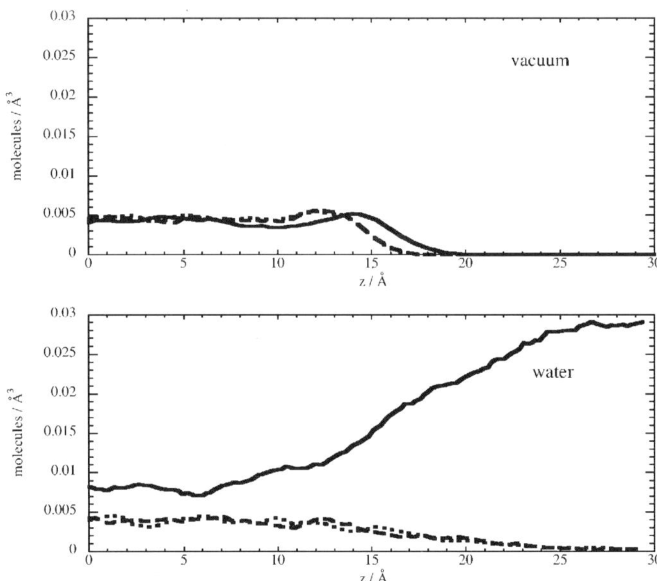

Fig. 5 Molecular densities for the two systems: 50 : 50 mixture of ionic liquid and water in vacuum (upper panel), and the same system embedded in water (lower panel). In both cases the ionic liquid/water slab is located on the left, and vacuum or neat water on the right. The dashed line represents the cation density, the dotted line is the anion density, and the solid line the water density.

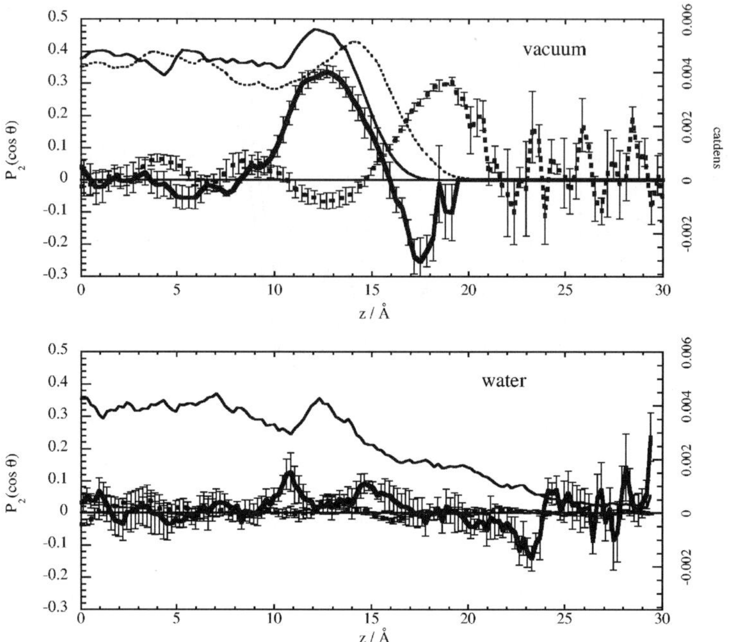

Fig. 6 Cation and water orientations in the (ionic liquid + water)/vacuum interface (upper panel) and diffusing liquid–liquid interface (lower panel). The solid lines with error bars represent the P_2 for the N–N axis in [dmim]. The dashed lines with error bars are the P_2 for the H–H axis in water. The thin solid and thin dashed lines show the cation and water densities, respectively.

water molecules have lost all orientational preference within the error bars of the observations. This is shown in the lower panel of Fig. 6.

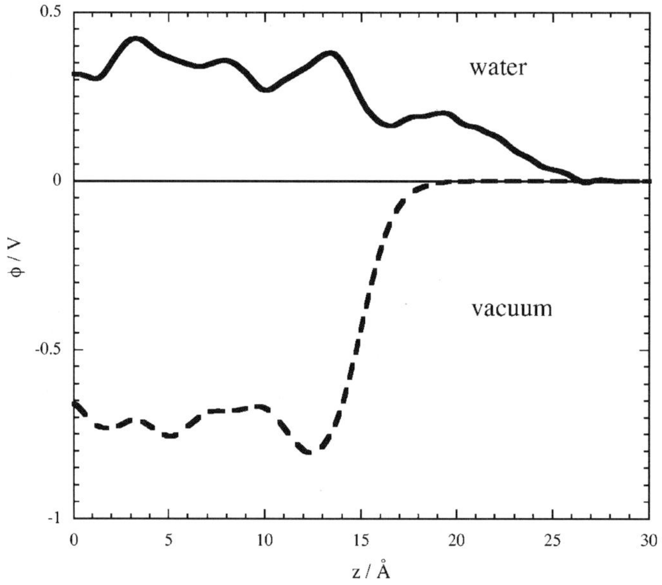

Fig. 7 Variation of electrostatic potential through the interface for both ionic liquid/water–vacuum interface (dashed line) and ionic liquid/water–water diffusing interfaces (solid line). Note the difference in signs of the change in electrostatic potential in the two cases.

Another striking difference between the stable and diffusing interfaces is shown in Fig. 7. There we show the potential profile for going from the vapour into the 50:50 liquid mixture (lower graph) and that for the liquid–liquid mobile interface (upper graph). Comparing this figure with Fig. 4 for the Lennard–Jones interfaces, one can see that the potential drop between the vapour and the 50:50 mixture is about 80% as that between the vacuum and pure ionic liquid. At the same time charge oscillations seem to be suppressed. Interestingly the potential change on going through the diffusing interface, from pure water into ionic liquid, has the opposite sign. This reflects a difference in the origin of the interface potential.

This difference is illustrated in Fig. 8. The bold line in the upper panel of this figure shows the charge density profiles for the 50:50 liquid–vacuum interface calculated taking into account the charge on every atom. This differs significantly from the thin line which is the difference between cation and anion densities. In contrast the lower panel shows that for the diffusing interface the charge profiles and the cation–anion density difference is very small. These results show that, for the stable liquid–vacuum interface, molecular orientations of both cations and water molecules provide the major contribution to the potential change across the interface, while in the second case it is the difference in cation and anion concentrations that cause the potential change.

In a diffusing interface a large part of the potential variation (the liquid junction potential) is due to the difference in the transport numbers and diffusion rates of the ions. In the present case the potential increase on going into the phase that is rich in ionic liquid can be explained if the chloride ions diffuse more rapidly into the water than the $[dmim]^+$ ions, that is if the transport number of the chloride ions is higher than for the cations. The observed potential profile is essentially a dynamic phenomenon related to the relative motion of the ions. In contrast, the potential decrease on going from vapour into a stable slab of ionic liquid–water is due to the arrangement and alignment of the molecules and cations near the interface.

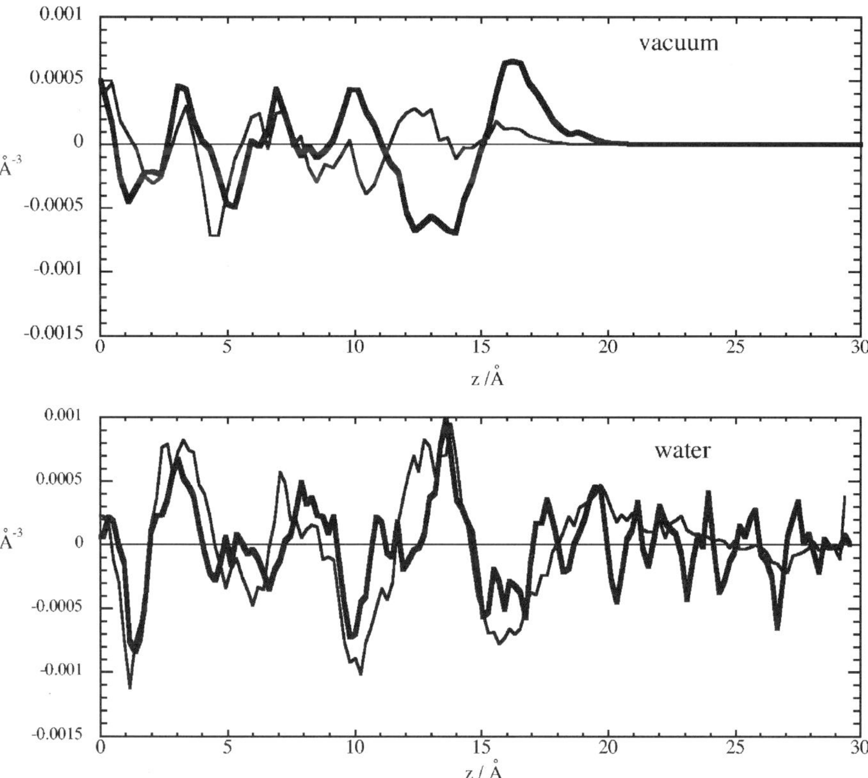

Fig. 8 Density of charge (thick line) and difference between cation and anion densities (thin line) for the 50:50 water–ionic liquid/vacuum interface (upper panel) and the diffusing ionic liquid–water interface (lower panel). Note the difference in the two curves in the upper panel and the similarity in the lower panel.

We note that the diffusing interface has no apparent surface tension as measured by the difference of the parallel and perpendicular components of the pressure.

Discussion

In this paper the structure of the interfaces between the room temperature ionic liquid dimethylimidazolium chloride ([dmim]Cl) and different Lennard–Jones fluids and water has been studied by molecular dynamics simulations. Lennard–Jones fluids are expected to behave in a similar way to non-hydrogen-bonding organic solvents. The two fluids used in this study did not mix with [dmim]Cl and formed thermodynamically stable interfaces with the ionic liquid. The two examples showed contrasting density changes near the interface. The 'strong' ionic liquid, which is similar to a normal liquid, has a deficit in density while the supercritical 'weak' Lennard–Jones fluid exhibits an enhanced density in the interface region.

On the ionic liquid side of the interface, the density increases to a constant value and the density maximum seen in the interface of ionic liquid with vacuum is no longer apparent. However, the presence of Lennard–Jones particles has little effect on the orientation of the cations, which tend to lie perpendicular to the surface. This preferred orientation of the cations contributes to the change of electrostatic potential across the interface, which in the present case corresponds to a dipole moment pointing outwards from the ionic liquid phase.

Water–ionic liquid mixtures were studied under two different conditions. The first system was a slab of a mixture of water–ionic liquid in contact with vacuum. This showed a stable interface with an increase in the water concentration on the outside of the interface. Cations and water molecules both showed preferred orientation in the interfacial region, which provides the major contribution to the change of electrostatic potential across the surface. The second system consisted of a diffusing interface formed when the ionic liquid–water slab was embedded in water. After a short time, during which water molecules diffused rapidly into the ionic liquid-rich phase, the concentration profile across the interface changed slowly. This interface was much broader than the thermodynamically stable interfaces and the molecules and ions were not oriented.

The potential change across this interface had the opposite sign to that between the slab and vacuum and, instead of arising from the orientation of molecules and cations, arose from the different mobility of the ions.

Acknowledgements

We thank members of QUILL for useful discussions and EPSRC (grants GR/M720401, GR/S41562 and GR/S18106) and the Leverhulme Trust for support.

References

1   D. M. Mitriniovic, Z. Zhang, S. Williams, Z. Huang and M. L. Scholman, *J. Phys. Chem. B*, 1999, **103**, 1779.
2   D. M. Mitriniovic, A. M. Tikhonov, M. Li, Z. Huang and M. L. Scholman, *Phys. Rev. Lett.*, 2000, **85**, 582.
3   S. Senapati and M. L. Berkowitz, *Phys. Rev. Lett.*, 2001, **87**, 176 101.
4   H. A. Patel, E. B. Nauman and S. Garde, *J. Chem. Phys.*, 2003, **119**, 9199.
5   J. P. Nicolas and N. R. Souza, *J. Chem. Phys.*, 2004, **120**, 2464.
6   J.-L. Barrat and J.-P. Hansen, *Basic Concepts for Simple and Complex Liquids*, Cambridge University Press, Cambridge, 2003.
7   *Green industrial applications of Ionic Liquids*, ed. R. D. Rogers, K. R. Seddon and S. Volkiv, NATO Science Series, Kluwer Academic Publishers, Dordrecht, 2002.
8   C. G. Hanke, S. L. Price and R. M. Lynden-Bell, *Mol. Phys.*, 2001, **99**, 801.
9   C. G. Hanke, N. Atamas and R. M. Lynden-Bell, *Green Chem.*, 2002, **4**, 107.
10 R. M. Lynden-Bell, N. A. Atamas, A. Vasilyuk and C. G. Hanke, *Mol. Phys.*, 2002, **100**, 3225.
11 C. G. Hanke and R. M. Lynden-Bell, *J. Phys. Chem. B*, 2003, **107**, 10 873.
12 R. M. Lynden-Bell, *Mol. Phys.*, 2003, **101**, 2625.
13 H. J. C. Berendsen, J. R. Grigera and T. P. Straatsma, *J. Phys. Chem.*, 1987, **91**, 6269.
14 J.-P. Hansen and L. Verlet, *Phys. Rev.*, 1969, **184**, 151.

15   J. J. Potoff and A. Z. Panagiotopoulos, *J. Chem. Phys*, 1998, **109**, 10 914; J. M. Caillol, *J. Chem. Phys.*, 1998, **109**, 4885.
16   W. Smith and T. R. Forester, *DL_POLY Package of Molecular Simulation Subroutines*, Copyright CCLRC, Daresbury Laboratory, 1996, http://www.cse.clrc.ac.uk/msi/software/DL_POLY.

Surface charge effects on solvation across liquid/liquid and model liquid/liquid interfaces

C. L. Beildeck, W. H. Steel and R. A. Walker*

Department of Chemistry and Biochemistry, University of Maryland, College Park, MD 20742-2021, USA

Received 14th April 2004, Accepted 1st July 2004
First published as an Advance Article on the web 13th October 2004

Resonance enhanced second harmonic generation (SHG) coupled with novel solvatochromic surfactants has been used to explore solvation across liquid/liquid interfaces in the presence of excess surface charge in the aqueous phase. The surfactants—dubbed "molecular rulers"—consist of hydrophobic, solvatochromic chromophores connected to charged headgroups *via* variable length alkyl spacers. Interfacial dipolar width is monitored as a function of chromophore/headgroup separation. Data show that cationic and anionic surfactants of equivalent lengths sample very different environments across a water/cyclohexane interface. The effective excitation wavelengths of cationic surfactants are shifted persistently to higher energies than in bulk cyclohexane while anionic surfactants solvation converges smoothly from the aqueous to the organic limit with increasing spacer length. To further evaluate the effect of surface charge on interfacial solvation, SHG was used to probe the environment surrounding molecular ruler chromophores adsorbed to the aqueous liquid/vapor interface in the presence of densely packed monolayers of 1-octanol. These monolayer systems are shown to reproduce qualitatively many of the features associated with bulk water/alkane interfaces and are logistically easier to assemble. Changing the ionic strength of the underlying aqueous sub-phase suggests that the headgroup of cationic molecular ruler surfactants alters the electronic structure of the chromophore rather than properties of the surrounding nonpolar environment.

I. Introduction

In the most general sense, solvation describes the interaction of a solute with its surroundings. These interactions can be strong (*e.g.* Coulomb) or weak (*e.g.* dispersion) and may extend over large distances (*e.g.* $1/r^2$) or short distances (*e.g.* $1/r^6$).[1] Solvation forces will influence a solute's eigenstate structure as well as its equilibrium conformation, dynamic behavior and, ultimately, its reactivity. While numerous studies have examined static and dynamic aspects of solvation in bulk solution,[2–5] solvation at surfaces has received less attention. In part, this situation has arisen because experiments probing interfacial properties are typically more difficult to execute than measurements of bulk solution phenomena. Furthermore, surface properties do not always scale in a way that can be predicted by bulk solution or mean field behavior.[6,7] Solvents having very similar bulk solvating properties can create markedly different environments at surfaces simply by virtue of the fact that they have different shapes.

DOI: 10.1039/b405552g

Of particular relevance to the experiments discussed below is how a solute in one phase is influenced by surface charges in an adjacent phase. Electrolytes having different surface activities will create an electrical double layer at an interface, and these double layers are particularly well defined if one charged species is a surfactant. Electrical double layers play prominent roles in controlling micelle formation and morphology, colloid stability and adsorption at membrane surfaces.[8] Important properties used to characterize double layers include the surface potential (ψ) and the spatial extent of the anisotropic charge distribution.[9] Both properties can be described in terms of classical electrostatics, and in the limit of low concentration, surface potential decays exponentially with distance in the electrolyte containing phase as described by Debye–Hückel theory.[8] Less clear is how a double layer in one phase will affect solute structure and solvation in an adjacent, immiscible phase. Understanding how electric fields propagate from double layers formed at a water surface into a low dielectric medium has important consequences in chemistry and biology. For example, double layers set up at the outer or inner walls of cell membranes will directly influence the local environment within the lipid bilayer and control concentration gradients and orientations of membrane components such as cholesterol.[10] Molecular dynamics simulations and continuum models have predicted how double layer electric fields in one phase propagate across liquid surfaces,[11–17] but simulations must overcome difficult technical issues associated with accurately modeling interaction potentials between charged species in anisotropic environments. In order to assist simulations and better refine our understanding of solvation at charged liquid surfaces, more extensive experimental data are needed. Of particular utility would be measurements that probe the effects of surface charge from a nominal interfacial boundary.

Studies described below examine the effect that surface charge has on interfacial solvation. Specifically, experiments begin to identify how charged species in one phase alter the local environment surrounding a solute in an adjacent, immiscible phase. Surface specific, second harmonic generation (SHG) is used to measure the solvatochromic response of hydrophobic chromophores anchored to an aqueous phase by formally charged anionic and cationic headgroups. Surface charge is varied by changing the ionic strength of the underlying, aqueous subphase. In addition, we explore the utility of neutral, self assembled monolayers as mimics for liquid/liquid interfaces. Given recent experiments that show the interface between bulk 1-octanol and water to be dominated by an exceedingly nonpolar region,[18] we wanted to determine whether or not mono-layers formed at liquid/vapor interfaces of aqueous solutions saturated with linear alcohols would approximate the environments observed at liquid/liquid boundaries. While these monomolecular films adsorbed to the liquid/vapor interface do not reproduce quantitatively the properties of boundaries between bulk liquids, they can serve as qualitative models of weakly associating liquid/liquid interfaces and help identify how surface charge effects extend into low dielectric media.

II.   Motivation and experimental considerations

Chromophores used in this work to probe solvation are sensitive to local solvent polarity. In the absence of a rigorous definition of polarity, we choose to describe this property in terms of the averaged, equilibrium electric field experienced by a solute due to polarization of its surrounding solvation shell. We employ this Onsager-based description of solvent polarity[19] because of its widespread use in solvation literature and its simple, physically-appealing interpretation.[20,21] The field induced within the solute cavity can affect a solute's electronic and nuclear structure. Consequently, solvent dependent shifts in solute excitation and emission energies can serve as sensitive indicators of local solvent polarity. When relevant, dipolar interactions between the solvent and solute are thought to be the dominant contribution to a solute's solvatochromic behavior. However, a recent, thorough examination of two common empirical solvation scales suggests that dispersion forces may play a more important role in solvation than previously believed.[22] In addition, high concentrations of electrolytes in solution also enhance the polarity experienced by a dipolar solute.

How these different forces influence solvation *across* an interface remains an area of active investigation. In the past several years, numerous studies have attempted to characterize interfacial polarity either through experimental measurements or through molecular dynamics simula-tions.[18,23–29] Eisenthal and coworkers reported that interfacial polarity at liquid/liquid interfaces as well as liquid/vapor interfaces could be modeled by considering equal contributions from the two adjacent phases.[27,28] This "average" polarity model arises from a series of SHG experiments that measure the excitation spectra of solvatochromic probes adsorbed to different liquid/liquid and liquid/vapor interfaces. Total internal fluorescence measurements carried out by Kitamura and coworkers support the original Eisenthal results although data show that the average polarity model begins to break down for polar organic solvents.[23] A molecular perspective of interfacial polarity comes from molecular dynamics simulations carried out by Michael and Benjamin.[24] These simulations demonstrate that a solute can report an average polarity when adsorbed to a sharp Gibbs dividing surface that is thermally roughened. However, simulations suggested that results should be exceedingly sensitive to the equilibrium distribution of solutes across the interface. Steel and Walker tested this premise using two solutes sharing similar electronic structure—p-nitrophenol (PNP) and 2,6-dimethyl-p-nitrophenol (dmPNP)—but having very different partition ratios.[25] PNP partitions $\sim 100:1$ between water and cyclohexane, while dmPNP partitions $\sim 1:2$ between the same two solvents. Resonance enhanced SHG spectra show that PNP samples a dielectric environment at the water/cyclohexane interface that is very close to that of bulk water, while the SH response of dmPNP indicates that this solute samples an alkane-like environment *at the same water/cyclohexane interface.*

A limitation faced by all of the experiments described above is their inability to control the equilibrium distribution of solutes across the interfacial region. Solutes adsorbed to interfaces will not only align themselves to accommodate polar and nonpolar functional groups, they will also distribute themselves across the interfacial region in order to minimize interfacial free energy.[30] The need to alter systematically solute location across different liquid/liquid interfaces motivated us to create families of surfactants dubbed molecular rulers.[31,32] Molecular ruler surfactants consist of hydrophobic, solvatochromic probes attached to charged headgroups by alkyl spacers of varying length. (Fig. 1) The probe itself is closely related to p-nitroanisole (PNAS), an aromatic molecule having a solvatochromic window of 22 nm as solvent polarity varies between that of alkanes ($\lambda_{\mathrm{exc}} = 294$ nm) and that of water ($\lambda_{\mathrm{exc}} = 316$ nm). PNAS is ~ 20 times more soluble in alkane solvents than in water thus satisfying the requirement that the probe of interfacial polarity be hydrophobic. Surfactant headgroups can be either anionic or cationic, although the cationic surfactants appear to be much more sensitive to local electric fields, as discussed below. For this reason, experiments described in this work examine the behavior of cationic surfactants in the presence of excess surface charge rather than anionic surfactants. Throughout the discussion below, surfactants are identified

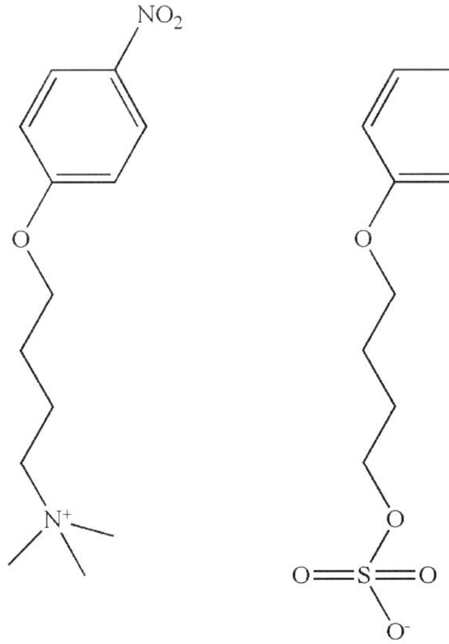

Fig. 1 Cationic and anionic ruler structures. The cationic rulers (left) contain a tetramethylammonium headgroup while the anionic rulers (right) utilize a sulfate group. Details regarding the syntheses of these surfactants can be found in refs. 25 and 31, respectively.

by the length of their alkyl spacer and the charge of the headgroup. For example, C_4^+ denotes a cationic surfactant having four methylene groups between the chromophore and the headgroup.

Resonance-enhanced SHG spectroscopy was used to monitor the solvatochromic response of surfactants adsorbed to liquid surfaces. By measuring how SHG spectra vary with surfactant length and aqueous phase ionic strength, we were able to assess the effects of surface charge on the local environment experienced by the probe solvated in the adjacent phase. Because of its origins, the resonance-enhanced response is both surface and molecularly specific, meaning that spectra result only from solutes influenced by interfacial anisotropy.[33–37] In a typical experiment, a single coherent optical field of frequency ω is incident upon an interface having a sub-monolayer coverage of a given ruler surfactant. A nonlinear polarization of frequency 2ω and intensity $I(2\omega)$ is detected, where the intensity of this second harmonic is proportional to the square of the second-order susceptibility, $\chi^{(2)}$,[33,34,36]

$$I(2\omega) \propto |\chi^{(2)}|^2 I^2(\omega)$$

(1)

and $\chi^{(2)}$ is a third rank tensor that under the electric dipole approximation is zero in isotropic environments. The $\chi^{(2)}$ tensor, then, imparts to the technique its inherent surface specificity. The tensor itself contains both resonant and nonresonant contributions:

$$\chi^{(2)} = \chi_R^{(2)} + \chi_{NR}^{(2)}$$

(2)

For dielectric systems, such as the interfaces considered here, the resonant term is typically much larger than the nonresonant contribution and can be related to the microscopic hyperpolarizability:

$$\chi_R^{(2)} = N \sum_{k,e} \frac{\langle \mu_{gk} \mu_{ke} \mu_{eg} \rangle}{(\omega_{gk} - \omega - i\Gamma)(\omega_{eg} - 2\omega + i\Gamma)}$$

(3)

where N denotes the total number of molecules contributing to the response, μ_{ij} is the transition matrix element between state i and state j (where g stands for the ground state, k for an intermediate, virtual state, and e for the first excited state), and the brackets represent an orientationally averaged distribution in the laboratory frame of reference. The ω_{ij} refers to the transition energies between the ground state and states k and e, and Γ is the transition's line width. When 2ω is resonant with ω_{eg}, $\chi^{(2)}$ becomes large, leading to a strong resonance enhancement in the observed intensity at 2ω. Thus, measuring the scaled intensity $(I(2\omega)/I^2(\omega))$ as a function of 2ω records an *effective* excitation spectrum of solutes adsorbed to an interface.[28] We note that while $\chi_{NR}^{(2)}$ is generally small, this term can lead to either constructive or destructive interference so that the actual excitation wavelength may not coincide with the observed intensity maximum in a SH spectrum. With the exception of data recorded to determine solute orientations, spectra in this work were acquired under $P_\omega P_{2\omega}$ polarization conditions, where P polarized light describes light that is polarized in the plane defined by the direction of propagation and the surface normal. Varying the incident and detected polarizations enables us to determine the average chromophore orientation using methods described previously.[38] Different polarizations did not lead to qualitatively different SHG spectra.

The SHG apparatus is built around a Ti : sapphire regeneratively amplified, fs laser (Clark-MXR CPA 2001) that produces 130 fs pulses with energies of ~ 700 μJ at a wavelength of 775 nm and a repetition rate of 1 kHz. The output of the Ti : sapphire laser pumps a commercial optical parametric amplifier (OPA, Clark-MXR). The visible output of the OPA is tunable from 550 to 700 nm, with a bandwidth of 2.5 ± 0.5 nm. The polarization of the incident beam is controlled using a Glan–Taylor polarizer and a half-wave plate. A series of filters blocks the fundamental 775 nm and any second harmonic light generated from the preceding optical components. Second harmonic photons are detected in the reflected direction using photon-counting electronics. Typical signal levels average 0.01–0.1 photon per shot. A second polarizer selects the polarization of the SH signal, and a short pass filter and monochromator serve to separate the second harmonic signal from background radiation.

Because the visible OPA cannot be synchronously tuned, acquisition of a complete SHG spectrum requires multiple hours. A typical procedure entails letting the sample equilibrate followed by manual tuning of ω_{vis} to each desired wavelength. System alignment is reoptimized at every wavelength to account for the wavelength-dependent changes in the angle of the reflected SH signal. At each wavelength, SH data are collected for four 10-s intervals and normalized for incident power. Although tedious, this procedure ensures that spectra are reproducible. A single wavelength might be sampled three separate times several hours apart (beginning, middle, and end of an acquisition sequence). If the normalized SH signal from each of these three samplings does not fall within experimental uncertainty (typically $\pm$ 15%), data acquisition is halted and the spectrum discarded. In addition, data at the same wavelength were often acquired using several different incident powers and then normalized to confirm quadratic dependence of the SH signal intensity on the incident field intensity predicted by eqns. (1)–(3). Predicted quadratic behavior was always observed.

Interfacial systems probed by cationic surfactants typically had surface coverages of less than 20% of a full monolayer. These fractional coverages lead to surface excess concentrations of 1×10^{13} per cm^2. Spectra of anionic surfactants were acquired from monolayers having similar fractional coverage, although absolute surface concentrations were $\sim 5 \times 10^{13}$ per cm^2 due to the reduced surface activity of the anionic species. Liquid/liquid interfaces were generated by first placing aqueous solutions into a cylindrical kel-F cell having a reservoir 4 cm in diameter. Then an application of a thin layer (~ 1–3 mm) of organic solvent atop the aqueous solution creates the aqueous–organic interface. A trapezoidal fused silica prism ($50 \times 50 \times 30$ mm, JDSU Casix) is secured atop the reservoir, preventing evaporation of the solvent. Prior to use the prism is cleaned in a $50:50$ mixture (by volume) of concentrated sulfuric and fuming nitric acids. Prisms cleaned in this way have been shown to be hydrophilic, as demonstrated by complete wetting of the surface. All SHG spectra were acquired at room temperature, 22 ± 1.5 °C. Experiments carried out at the liquid/vapor interfaces of octanol-containing aqueous solutions took advantage of the 1-octanol monolayer that formed spontaneously at the liquid/vapor interface. At room temperature, 1-octanol is soluble in water up to concentrations of ~ 2.8 mM ($x_{1\text{-oct}} \sim 5 \times 10^{-5}$).[39] 1-Octanol in saturated aqueous solutions spontaneously self assembles to form monolayers that are reasonably compressed (4.8×10^{14} per cm^2 or 21 Å^2 molecule^{-1}), and surface vibrational spectra show that these monolayers possess a high degree of conformational order.[40]

III. Results and discussion

1. Cationic *vs.* anionic rulers adsorbed to water/cyclohexane interfaces

The initial report describing cationic molecular ruler surfactants noted behavior that contrasted with that shown by related anionic surfactants.[31] At the same interface (water/cyclohexane), anionic and cationic surfactants having similar fractional monolayer surface coverages appear to sample very different dielectric environments. The excitation wavelength of the C_2^- surfactant exhibits a distinct maximum at 301 nm in its resonant SHG spectrum. This value lies between the bulk aqueous and organic limits of 316 and 294 nm, respectively. In contrast, the spectrum of the C_2^+ surfactant shows an excitation maximum at 294 nm, a wavelength that is slightly lower than the bulk alkane limit of 296 nm.[41] This effect appears general as evidenced by the spectra acquired from C_4^+ and C_4^- surfactants adsorbed to the water/cyclohexane interface (Fig. 2). Again, the cationic excitation wavelength is blue-shifted slightly from the bulk cyclohexane limit while the anionic excitation wavelength is still longer than the bulk solution limit. Fig. 2 summarizes the water/ cyclohexane interfacial behavior of cationic and anionic surfactants having 2, 4 and 6 carbon spacers. From the data, one readily sees that the behavior of cationic and anionic species converges for longer length surfactants. For both species of surfactants (cationic and anionic) the polarization dependent SHG signal was measured in order to calculate the average chromophore orientation. Data showed that chromophores in cationic and anionic surfactants shared equivalent orientations of $44 \pm 3°$ relative to the interfacial normal. Chromophore orientation exhibited no systematic dependence on alkyl spacer length. Despite uncertainties regarding the width of the distribution about this average orientation, similarities between the cationic and anionic results are reassuring because they imply that chromophores from both surfactants are solvated in the same manner, regardless of headgroup identity. There appears to be no charge dependent effect forcing the chromophore of the cationic surfactant to adopt an orientation that is more perpendicular or more parallel to the interface, relative to the anionic species.

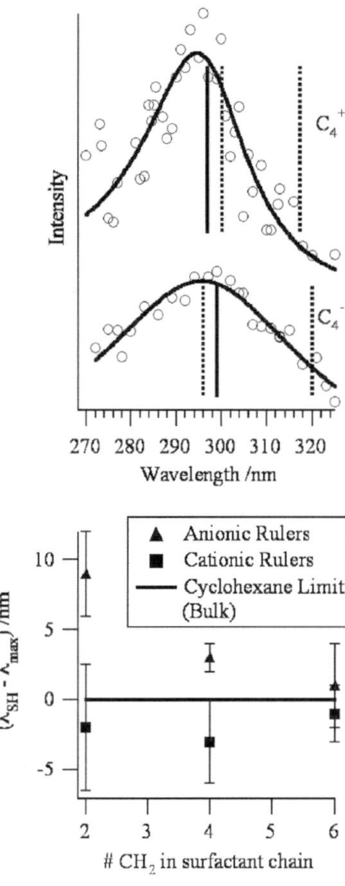

Fig. 2 SH behavior of anionic and cationic rulers at the water/cyclohexane interface. The top panel shows representative SHG data for the C_4^+ and the C_4^- rulers at the water/cyclohexane interface. The spectra have been fit according to eqns. (1)–(3), and the solid vertical lines correspond to the chromophore's interfacial excitation wavelength maximum. Dashed vertical lines correspond to excitation wavelength maxima in bulk aqueous (long wavelength) and organic (short wavelength) solvents. The bottom panel shows the solvatochromic behavior of the PNAS chromophores attached to different length cationic and anionic molecular ruler surfactants at the water/cyclohexane interface.

The disparity between cationic and anionic solvatochromic behavior for the shorter surfactants is especially significant. These results imply that the same probe (PNAS) integrated into the same surfactant architecture (C_2 or C_4) adsorbed to the same interface (water/cyclohexane) with the same relative surface coverage (~ 0.20 monolayer) samples markedly different polarities. One possible source of the disparity might be the difference in the probe distributions across the interface, an effect we refer to as "float depth." While neither headgroup is soluble in the organic phase, the cationic, trimethylammonium headgroup is considerably more hydrophobic than the anionic sulfate headgroup. *If* the interface were molecularly sharp and *if* the headgroups were solvated differently (with the ammonium headgroup being located closer to the interfacial Gibbs dividing surface), then we would expect the PNAS probe attached to the cationic surfactant to sample a less polar environment than the probe attached to the anionic headgroup.

This explanation can not be ruled out as a contributing cause of the headgroup-dependent behavior, although it fails to account for the small but persistent blue shift in cationic SHG spectra relative to the bulk alkane limit. A more likely source of the observed differences is the charge of the headgroup. For cationic surfactants, bulk solution Nuclear Overhauser Enhancement (NOE) NMR experiments clearly demonstrate that the cationic headgroup interacts strongly with the electronic structure of the aromatic chromophore as evidenced by through-space coupling between methyl protons on the headgroup and the aromatic protons *meta* to the nitro group.[31] This interaction is strongest for the shortest surfactants. Surfactants having alkyl chain lengths of four methylene groups and longer show no evidence of chromophore/headgroup association. The charge–dipole interaction between the headgroup and the aromatic chromophore reduces the change in electronic state dipole experienced by the solute in solution, leading to less pronounced (or blue-shifted) solvatochromic behavior relative to the behavior of neutral species. Anionic surfactants show no such intra-molecular interaction, and anionic surfactant solvatochromic behavior mimics that of neutral surfactants.

A final cause for the unexpected behavior of cationic surfactants adsorbed to the water/ cyclohexane interface could be aggregation. If adsorbed surfactants were interacting with each other, then the observed solvatochromic responses would reflect probe–probe interactions or, perhaps, interactions between the probe of one surfactant and the cationic headgroup of an adjacent surfactant. However, even at very low surfactant concentrations adsorption isotherms show no characteristic signature of surfactant aggregation. In addition, surfactant concentrations are several orders of magnitude below the critical micelle concentration of quarternized ammonium surfactants having longer hydrophobic tails.[8] Finally, five-fold changes in surface coverage (full monolayer to <0.20 monolayer coverage) lead to no observed changes in the SHG response of adsorbed chromophores. Thus we believe that differences in interfacial solvation sampled by cationic and anionic surfactants having similar lengths arises from having different charges located at well defined distances relative to the solvatochromic probe.

2. Polarity *vs.* surface charge at model liquid/liquid interfaces

The headgroup dependent water/cyclohexane results lend themselves to two possible interpretations. Either the cationic headgroup influences the environment surrounding the chromophore leading to a lower apparent interfacial polarity *or* the headgroup alters the electronic structure of the chromophore making it less sensitive to the local dielectric environment. In order to investigate further the effect of surface charge on solvation in an adjacent phase, we sought to mimic the weakly associating water/alkane interface with a system that proved logistically easier to assemble during day-to-day operation. Aqueous solutions saturated with 1-octanol appear to capture most of the important features of the water/alkane systems studied previously. SHG studies of the boundary created between bulk water and bulk 1-octanol show that the interface is dominated by a very nonpolar region having a local dielectric environment similar to that of alkanes.[18,42] Furthermore, this region appears to extend for the length of a single, extended octanol molecule suggesting that the surface octanol species hydrogen bond to the underlying sub-phase and create a dense Langmuir film at the water/1-octanol interface. Surface tension data coupled with surface specific vibrational spectra have been used to characterize the liquid/vapor interface created by aqueous solutions saturated with 1-octanol.[40] Not surprisingly, the solvated organic species spontaneously self-assemble at the liquid/vapor interface forming a tightly packed monolayer (21 Å^2 molecule^{-1}) in

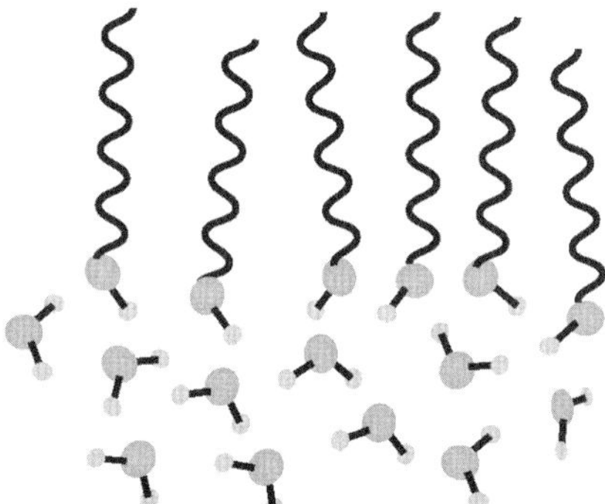

Fig. 3 Schematic representation of the formation of a self-assembled monolayer of 1-octanol at the water/ vapor surface. Details about surface coverage and orientation are discussed in the text and in ref. 40.

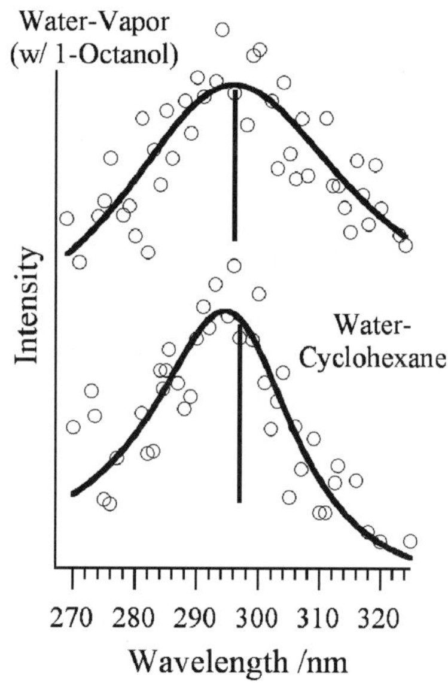

Fig. 4 SHG spectra of C_4^+ adsorbed to the water/cyclohexane interface and the water/vapor interface where the water has been saturated with 1-octanol. The spectra have been fit according to eqns. (1)–(3), and the solid vertical lines correspond to chromophore interfacial excitation maxima.

which the 1-octanol species are very well ordered (Fig. 3). Fig. 4 shows the resonance enhanced SHG spectra of the C_4^+ adsorbed to both the water/cyclohexane interface and the liquid/vapor interface formed by an aqueous solution saturated by 1-octanol. The spectrum from the liquid/vapor interface is broader (45 nm $vs.$ 27 nm FWHM) suggesting a more heterogeneous environment, but the actual excitation wavelengths of the same surfactant in these two different systems are equivalent. In addition, the C_4^+ chromophore is deflected $44 \pm 3°$ off of surface normal in both systems.

Water/1-octanol interfacial mimics were used to investigate the effects of surface charge on interfacial solvation. Solutions of C_4^+ surfactants (60 μM) were prepared in aqueous solvents saturated with 1-octanol. The surface charge was controlled by dissolving NaI into the aqueous solution in order to achieve the desired ionic strength. The SHG spectrum of the adsorbed C_4^+ surfactant was then acquired and fit according to eqns. (1)–(3). Representative spectra and a summary of SH wavelength maxima for systems having varied NaI concentrations appear in Fig. 5. Though subtle, changes in the interfacial environment (as sampled by the surfactant chromophores) clearly depend on the electrolyte concentration of the aqueous subphase. As the NaI concentration increases, chromophore excitation shifts to longer wavelengths and approaches the C_4^- limit. Given the solvatochromic properties of the probe, this behavior could signify increasing polarity within the 1-octanol film. Alternatively, a red-shift in excitation wavelength could represent a weakening of probe-headgroup interactions that allows the electronic structure of PNAS to become more sensitive to local solvent polarity.

Deconvoluting these two effects—polarity $vs.$ intramolecular alterations of chromophore electronic structure—is not trivial, and both are likely to contribute to the observed variations in chromophore behavior that accompany changes in the aqueous ionic strength. Nevertheless, we believe that the intramolecular interactions bear more responsibility for the observed behavior than do changes in the local dielectric properties of the film. Based on thermodynamic,[43] spectroscopic[6,18,40,42] and neutron and X-ray scattering data,[44–46] we assume that the interior of the monolayer films are alkane-like in nature, meaning that external fields (from the electric double layer in the aqueous phase) are unable to induce a large polarization within the organic interior.[47] Consequently, the intrinsic dielectric character of the film is unlikely to change significantly. However, chromophore-headgroup interactions can be attenuated if the cationic headgroup charge is screened effectively.

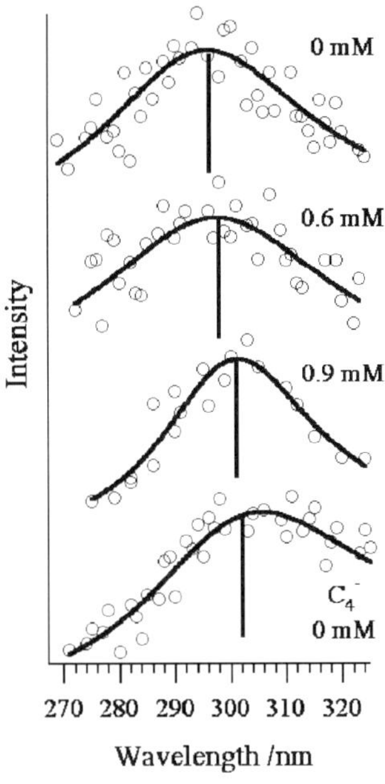

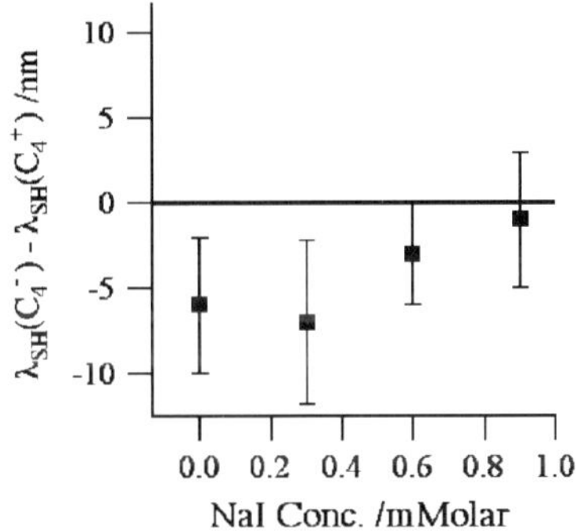

Fig. 5 Interfacial solvation of the C_4^+ ruler adsorbed to the 1-octanol saturated water/vapor interface as a function of NaI concentration. The top panel displays representative spectra for the C_4^+ with 0 mM, 0.6 mM, and 0.9 mM NaI concentrations in the aqueous subphase as well as a comparison to the SHG spectrum of the C_4^- ruler adsorbed to the same interface with no additional salt present. The bulk aqueous concentration of the cationic surfactant is 60 μM while that of the anionic surfactant is 10 mM. The bottom panel summarizes the results of fitting the spectra to eqns. (1)–(3), and the solid vertical lines correspond to chromophore interfacial excitation maxima.

Several considerations suggest that increasing aqueous NaI concentration should inhibit the headgroup's ability to interact with the attached chromophore. First, large, monovalent anions such as Br^- and I^- bind more effectively than smaller anions (*e.g.* Cl^-) to quarternized ammonium surfactants in solution. This effect shows up most dramatically in a lowering of the surfactant critical micelle concentration as ionic strength increases.[48–50] Stronger cation–anion association at the interface would necessarily reduce the strength of cation–chromophore interaction (Fig. 6). Adding excess salt (NaI) to a solution of cationic surfactants reduces the intramolecular interactions, and the surfactant's solvatochromic behavior would begin to resemble that of analogous anionic species. For example, aqueous solutions of C_2^- and a neutral C_2^+ analog containing a dimethylamine headgroup rather than the quarternized ammonium both exhibit bulk solution excitation wavelengths of 316 nm in water. The excitation wavelength of C_2^+ in water is 310 nm, but this value shifts to 316 nm in aqueous solutions saturated with NaI. Again, bulk solution absorption

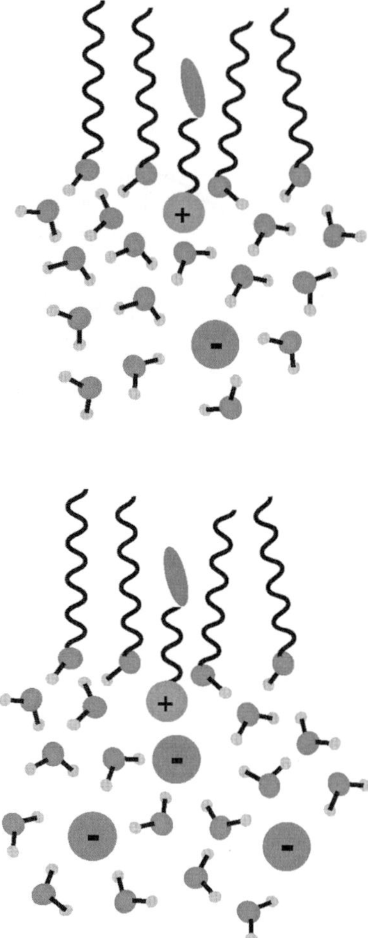

Fig. 6 Schematic representation of the interfacial environment in the absence (top panel) and presence (bottom panel) of excess surface charge in the form of additional I^-. As the iodide concentration in aqueous solution increases (by addition of NaI), the cationic headgroups can be more effectively screened from the chromophore solvated in the adjacent phase. The dimensions of the interfacial region are not to scale. Debye–Hückel theory predicts a double layer screening distance (κ^{-1}) of ~40 nm in the absence of excess salt and ~10 nm at salt concentrations of ~1 mM. (See ref. 8, ch. 11 for more details.)

data support this model as do preliminary SHG results using NaCl rather than NaI as the aqueous electrolyte. The effect of increasing chloride concentration on the chromophore's solvatochromic behavior is much less pronounced than for equivalent concentrations of NaI.

Another source of headgroup screening may arise from the surface activity of the iodide anion itself. Recent molecular dynamics simulations predict that smaller anions such as F^- and Cl^- are depleted at the liquid/vapor surface, but larger ions such as Br^- and I^- preferentially partition to the interface.[51] Such surface activity of the I^- would again help "neutralize" the ammonium headgroup from the perspective of the aromatic chromophore in the adjacent, nonpolar medium. Comparisons between the simulations and results presented above can not be quantitative given the large disparity in electrolyte concentrations—simulations modeled 1 M salt solutions while salt solutions used in experiments were all sub-millimolar. We also note that there exists a degree of ambiguity associated with this picture in light of recent experiments[52–54] examining the vibrational structure of different aqueous/electrolyte interfaces, but any difference in anion surface activities will necessarily influence the structure of the electrical double layer and solvation in an adjacent phase.

IV. Summary

Results presented in this work have begun to identify how surface charges can "reach" across interfaces to influence solvation in an adjacent phase. Two effects must be considered: the effect of surface charge on the properties of the adjacent medium and the effect of surface charge on the electronic structure of the solvated chromophore. Results show that having oppositely charged headgroups in close proximity to a hydrophobic probe leads to qualitatively different pictures of interfacial solvation based on the probe's bulk solution solvatochromic behavior. The effect of headgroup charge on interfacial solvation falls off with increasing headgroup/chromophore separation. Neutral, self-assembled monolayers can serve as useful models of weakly associating, liquid/liquid interfaces. These systems readily lend themselves to studying the effects of surface charge on interfacial solvation. By changing the ionic strength of the aqueous phase, we believe that the headgroup dependent differences in interfacial solvation result from direct headgroup–chromophore interactions between positively charged alkylammonium headgroups and the aromatic chromophore. Screening the headgroup with an excess of I^- ions allows cationic surfactant behavior to converge to anionic surfactant limits. Clearly, considerable work remains to be done, but the data presented above provide an enticing picture of how charges influence solvation across liquid/liquid interfaces.

Acknowledgements

The authors gratefully acknowledge support from the Research Corporation (RI0362) and the National Science Foundation (CHE0094246). R. W. thanks the organizers of FD 129 for all of their efforts that culminated in such a scientifically rewarding meeting.

References

1 *Quantitative Treatments of Solute/Solvent Interactions*, ed. P. Politzer and J. S. Murray, Elsevier, Amsterdam, 1994, p. 368.
2 C. J. Bardeen, S. J. Rosenthal and C. V. Shank, *J. Phys. Chem. A*, 1999, **103**, 10506.
3 G. R. Fleming and M. H. Cho, *Annu. Rev. Phys. Chem.*, 1996, **47**, 109.
4 M. L. Horng, J. A. Gardecki, A. Papazyan and M. Maroncelli, *J. Phys. Chem.*, 1995, **99**, 17311.
5 C. Laurence, P. Nicolet, M. T. Delati, J. M. Abboud and R. Notario, *J. Phys. Chem.*, 1994, **98**, 5807.
6 X. Zhang and R. A. Walker, *Langmuir*, 2001, **17**, 4486.
7 X. Zhang, M. M. Cunningham and R. A. Walker, *J. Phys. Chem. B*, 2003, **107**, 3183.
8 P. C. Hiemenz and R. Rajagopalan, *Principles of Colloid and Surface Chemistry*, 3rd edn., Marcel Dekker, New York, 1997.
9 G. Makov and A. Nitzan, *J. Phys. Chem.*, 1994, **98**, 3459.
10 S. A. Safran, *Statistical Thermodynamics of Surfaces, Interfaces and Membranes*, Addison-Wesley Publishing Co., New York, 1994.
11 L. X. Dang and T. M. Chang, *J. Phys. Chem. B*, 2002, **106**, 235.
12 H. Dominguez, *J. Phys. Chem. B*, 2002.
13 D. Michael and I. Benjamin, *J. Phys. Chem.*, 1995, **99**, 16810.
14 L. R. Pratt, *J. Phys. Chem.*, 1992, **96**, 25.
15 M. A. Wilson, A. Pohorille and L. R. Pratt, *J. Phys. Chem.*, 1987, **91**, 4873.
16 I. Benjamin, *J. Phys. Chem. A*, 1998, **102**, 9500.
17 L. R. Pratt, G. J. Tawa, G. Hummer, A. E. Garcia and S. A. Corcelli, *Int. J. Quantum Chem.*, 1997, **64**, 121.
18 W. H. Steel and R. A. Walker, *Nature*, 2003, **424**, 296.
19 L. Onsager, *J. Am. Chem. Soc.*, 1936, **58**, 1486.
20 P. Suppan and N. Ghoneim, *Solvatochromism,* Royal Society of Chemistry, Cambridge, UK, 1997.
21 K. Takahashi, K. Abe, S. Sawamura and C. D. Jonah, *Chem. Phys. Lett.*, 1998, **282**, 361.
22 D. V. Matyushov, R. Schmid and B. M. Ladanyi, *J. Phys. Chem. B*, 1997, **101**, 1035.
23 S. Ishizaka, H. B. Kim and N. Kitamura, *Anal. Chem.*, 2001, **73**, 2421.
24 D. Michael and I. Benjamin, *J. Chem. Phys.*, 1997, **107**, 5684.
25 W. H. Steel and R. A. Walker, *J. Am. Chem. Soc.*, 2003, **125**, 1132.
26 A. A. Tamburello-Luca, P. Hebert, P. F. Brevet and H. H. Girault, *J. Chem. Soc., Faraday Trans.*, 1996 **92**, 3079.
27 H. Wang, E. Borguet and K. B. Eisenthal, *J. Phys. Chem. A*, 1997, **101**, 713.
28 H. Wang, E. Borguet and K. B. Eisenthal, *J. Phys. Chem. B*, 1998, **102**, 4927.

29   Z. Xu and Y. Dong, *Surf. Sci.*, 2000, **445**, L65.
30   T. J. Su, J. R. Lu, R. K. Thomas and J. Penfold, *J. Phys. Chem. B*, 1997, **101**, 937.
31   C. L. Beildeck, W. H. Steel and R. A. Walker, *Langmuir*, 2003, **19**, 4333.
32   W. H. Steel, F. Damkaci, R. Nolan and R. A. Walker, *J. Am. Chem. Soc.*, 2002, **124**, 4824.
33   R. M. Corn and D. A. Higgins, *Chem. Rev.*, 1994, **94**, 107.
34   K. B. Eisenthal, *J. Phys. Chem.*, 1996, **100**, 12997.
35   G. L. Richmond, *Annu. Rev. Phys. Chem.*, 2001, **52**, 357.
36   Y. R. Shen, *Nature*, 1989, **337**, 519.
37   X. Zhuang, P. B. Miranda, D. Kim and Y. R. Shen, *Phys. Rev. B*, 1999, **59**, 12632.
38   X. Zhang, W. H. Steel and R. A. Walker, *J. Phys. Chem. B*, 2003, **107**, 3829.
39   L. Scheflan and M. B. Jacobs, *The Handbook of Solvents*, Van Nostrand Co., Inc., New York, 1953.
40   O. Esenturk, D. Mago and R. A. Walker, submitted.
41   Due to the insolubility of charged surfactants in nonpolar media, bulk solution limits were acquired using neutral analogs. The neutral analog of the anionic surfactant terminated in an alcohol rather than a sulfate, while the cationic analog terminated in a dimethyl amine. In principle, the actual excitation wavelength of the charged species in nonpolar solvents may be slightly different than the reported values if the charged headgroup alters the electronic structure of the solvatochromic chromophore.
42   D. T. Cramb and S. C. Wallace, *J. Phys. Chem. B*, 1997, **101**, 2742.
43   G. L. Gaines, *Insoluble Monolayers at Liquid–Gas Interfaces*, John Wiley and Sons Inc., New York, 1966.
44   J. Alsnielsen, D. Jacquemain, K. Kjaer, F. Leveiller, M. Lahav and L. Leiserowitz, *Phys. Rep.*, 1994, **246**, 252.
45   J. Penfold, R. M. Richardson, A. Zarbakhsh, J. R. P. Webster, D. G. Bucknall and A. Rennie, *J. Chem. Soc., Faraday Trans.*, 1997, **93**, 3899.
46   M. Schalke and M. Losche, *Adv. Colloid Interface Sci.*, 2000, **88**, 243.
47   C. J. Böttcher, *Theory of Electric Polarization*, Elsevier, New York, 1973, Vol. 1.
48   L. J. Magid, Z. Han, G. G. Warr, M. A. Cassidy, P. D. Butler and W. A. Hamilton, *J. Phys. Chem. B*, 1997, **101**, 7919.
49   S. B. Velegol, B. D. Fleming, S. Biggs, E. J. Wanless and R. D. Tilton, *Langmuir*, 2000, **16**, 2548.
50   V. E. Haverd and G. G. Warr, *Langmuir*, 2000, **16**, 157.
51   P. Jungwirth and D. J. Tobias, *J. Phys. Chem. B*, 2002, **106**, 6361.
52   R. Weber, B. Winter, P. M. Schmidt, W. Widdra, I. V. Hertel, M. Dittmar and M. Faubel, *J. Phys. Chem. B*, 2004, **108**, 4729.
53   D. Liu, G. Ma, L. M. Levering and H. C. Allen, *J. Phys. Chem. B*, 2004, **108**, 2252.
54   E. A. Raymond and G. L. Richmond, *J. Phys. Chem. B*, 2004, **108**, 5051.

NMR study of lithium ion in nitrobenzene/water system

Greg Moakes, Leslie T. Gelbaum, Johannes Leisen and Jiri Janata*

School of Chemistry and Biochemistry, Georgia Institute of Technology, Atlanta, GA 30332-0400 USA. E-mail: jiri.janata@chemistry.gatech.edu; Fax: 404 894 8146; Tel: 404 894 4828

Received 14th April 2004, Accepted 6th May 2004
First published as an Advance Article on the web 16th September 2004

The goal of this project has been to elucidate the solvation states and kinetics involved in the transfer of Li ions between nitrobenzene and water. A large interface area has been realized by confining one phase in polymer beads and suspending them in the other phase. A 2D ^{7}Li-NMR spectroscopic result of such a system obtained on a 400 MHz instrument has shown that the rate of the Li ion transfer exchange between nitrobenzene and water is too fast to resolve individual solvatomers. A pair of well-resolved ^{7}Li-NMR peaks corresponding to different solvation environments of Li$^+$ in "wet" nitrobenzene has been observed. The most surprising result is that the ratio of those two peaks and their separation changes over a period of days, but it can be *reversibly* changed by mechanical disturbance, such as ultrasonication. This effect is reproducible and suggests spontaneous but slow formation of metastable solvatomers of Li$^+$. Another surprising observation has been slow and irreversible incorporation of lithium ion from wet nitrobenzene into the walls of the silica-containing vessels.

Introduction

Nitrobenzene/water is the most extensively studied system in the area of interfaces of two immiscible electrolyte solutions (ITIES). Both electrochemical[1–7] and spectroscopic[8–11] techniques have been used. The questions of interest involve both thermodynamics and kinetics of ion transfer between the two phases. The role of water in the ion transfer has been the focus of work of several groups. Among these studies the work of Osakai's group occupies the most prominent role.[12–16]

When an ion exchanges between aqueous and organic phase it must change its solvation shell or retain it in some metastable solvatomeric arrangement with the new environment.[11,12,14] It stands to reason that the solvation exchange processes are not infinitely fast, in fact some of them may be unexpectedly slow, leading to the existence of metastable solvation states of the exchanging ion. The details of the solvation mechanism depend on the location at which the solvation/resolvation process takes place, on the concentration of the ions and on the local activity of water. That, in turn, depends on the structure of the L/L interface.[9,11,17] The rate determining step governing the ion exchange at ITIES and the existence and kinetics of the metastable states are of prime interest. Proton NMR spectroscopy has been a standard technique of study of solvation in homogeneous phases[12] and at interfaces.[18] Electrochemical *in situ* experiments at solid electrodes have been investigated by solid phase NMR as a new powerful auxiliary technique.[19] Here we report the results of ^{7}Li NMR studies that may contribute some new information about the Li$^+$/nitrobenzene/ water system.

DOI: 10.1039/b405409a

Experimental

Materials

Lithium bromide (99.95%, Strem Chemicals) was used as received. Nitrobenzene (99%, ACROS Organics) was dried before use by molecular sieves (4A, Sigma–Aldrich). Glass and Quartz 10 mm NMR sample tubes were obtained from Wilmad-Labglass. All water used was de-ionized to 18 MΩ using a US-Filter Modulab water system (US Filter, Warrendale, PA). The Sephadex beads used in this work were purchased from Sigma–Aldrich (Milwaukee, WI) and the polystyrene beads were obtained from Duke Scientific, Corp. (Palo Alto, CA). The polystyrene–DVB beads (PS) were polydispersed 1–50 µm diameter, CAT # 445. The 25–100 µm, "moderately hydrophobic" lipophilic Sephadex beads (LS), CAT# LH 20 100 and the regular 100–300 µm hydrophilic Sephadex beads (HS) CAT# G 25 300 were used.

Procedures

Saturated solutions of LiBr in dry NB were prepared by shaking LiBr(s) with dry NB for several hours. Saturated solutions of LiBr in water were prepared by adding LiBr to de-ionized H_2O until solid salt remained. For preparation of "wet" NB solution of LiBr, 5 mL of saturated aqueous solution was added to 20 mL of nitrobenzene. The two phase mixture was magnetically stirred for 24 h. When the stirring stopped the "experimental clock" was arbitrarily set to $t = 0$. The phases were allowed to separate for 3 h and after that 2–3 mL of the upper clear nitrobenzene phase were transferred into a clean vial and then to the NMR sample tube. The double-transfer step was performed to prevent accidental contamination of the sample with free water.

The two phases were also used in the polymer bead experiments. The hydrophilic beads were first saturated with an aqueous solution of LiBr and then mixed with the LiBr-saturated nitrobenzene phase to form a thick paste. Similarly, nitrobenzene phase saturated with LiBr was confined to the lipophilic Sephadex or polystyrene beads, respectively. The aqueous phase saturated with LiBr was used to form a thick paste.

The Li^+ content of the sample was determined by atomic absorption spectroscopy as 1.43×10^{-5} M. The content of water in "wet" nitrobenzene solution of LiBr was determined by Karl Fischer titration using a Metrohm 831 KF coulometer (Metrohm Inc.) as 60 mM. The published concentration of water in water-saturated "wet" nitrobenzene is between 170 and 190 mM.[12] This apparent discrepancy requires further experiments.

NMR experiments

The NMR spectra were recorded using a Bruker AMX-400 spectrometer (Bruker BioSpin) with a 10 mm broadband probe. The ^7Li signal was observed at 155.5 MHz. with a 5 s pulse (30°) and a 2.3 s pulse delay. A ^7Li-NMR spectrum of LiBr in H_2O was recorded before each experiment and the corresponding signal calibrated to 0 ppm. Benzene-d_6 (99.96 atom.%d, Sigma–Aldrich), was sealed in a glass capillary which was inserted into the 10 mm sample tube as a field frequency lock. Since the acquisition of multiple spectra took several hours the "experimental time" for each set indicates the beginning of the NMR session. For sonication experiments the ultrasonic bath Aquasonic Model 150 HT, delivering an average sonic power of 135 W, was used.

Results

The ^7Li nucleus has a spin of 3/2 and a sensitivity which is 1540 times higher than ^{13}C. NMR spectra of $^7Li^+$ in dry NB, water and "wet" NB are shown in Fig. 1. The chemical resonance of $^7Li^+$ in dry NB is shifted 1.280 ppm downfield, relative to the resonance in water. The $^7Li^+$ resonance peak in "wet" NB is significantly broader and lies between these two values.

NMR at the water–nitrobenzene interface

In order to obtain a sufficiently strong ^7Li-NMR signal a high interphase area has been realized by confining LiBr aqueous or NB phases in polymer beads (Fig. 2).

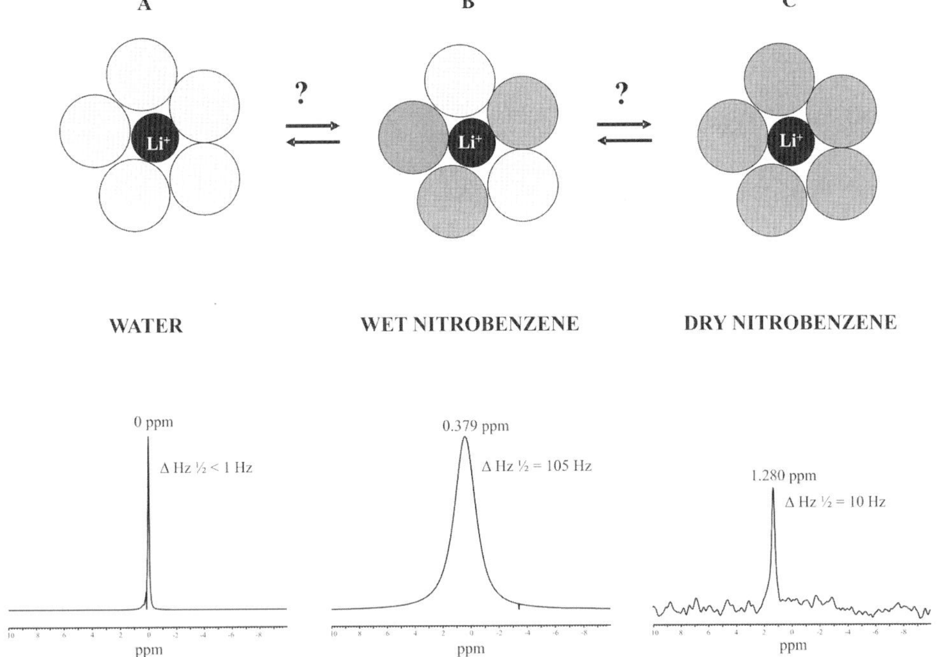

Fig. 1 NMR resonance spectra of lithium ion solvated in three different environments and representation of the solvatomeric equilibria. (A) Water; (B) wet nitrobenzene; and (C) dry nitrobenzene.

The spectra of $^7Li^+$ in dry NB and in water show that it is possible to distinguish between the two distinctly different solvation environments. Hence, for the systems containing dispersed polymer beads one would expect spectra which are largely a superposition of spectra observed for Li dissolved in the aqueous phase (*cf.* Fig. 1A) and Li dissolved in the wet nitrobenzene phase (Fig. 1B).

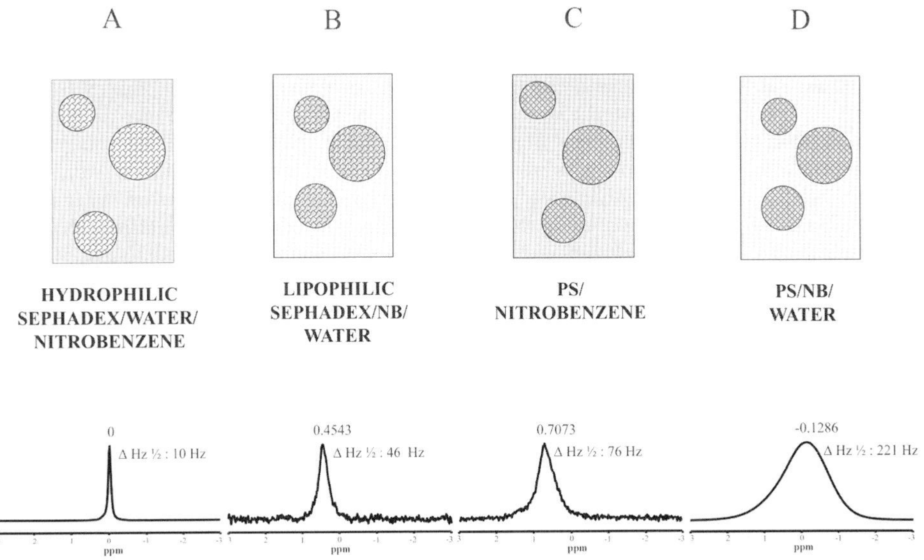

Fig. 2 Realization of high interface area between nitrobenzene and water. One phase is confined in polydispersed polymeric beads which are then suspended in the other phase. Corresponding ^7Li NMR spectra showing different solvation states of the lithium ion.

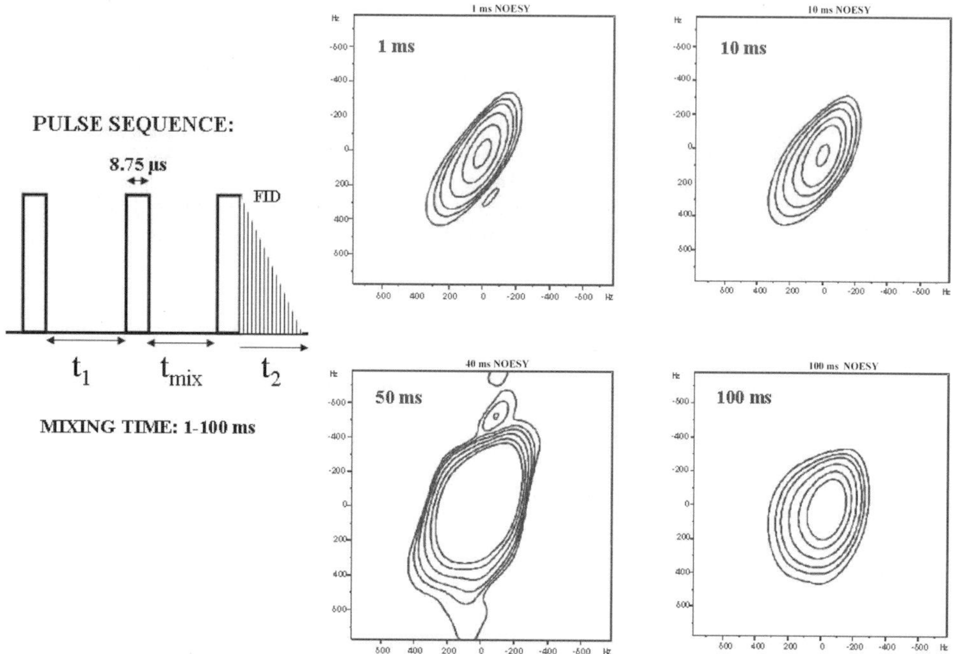

Fig. 3 2D exchange NMR spectroscopy of the nitrobenzene/water interface realized in the bead experiment and performed according to the scheme polystyrene/NB/water, shown in Fig. 2.

Experimentally, for the system containing hydrophilic sephadex a narrow line was observed at 0 ppm, which is largely identical to the spectrum observed for Li^+ dissolved in pure water. Hence, in this system the majority of Li^+ ions will be dissolved in the water. For systems containing hydrophobic beads loaded with nitrobenzene, rather broad peaks are observed (Fig. 2D). For the most hydrophobic polymer beads (Fig. 2B, C, D) a peak is observed, which is broader than the one observed for Li^+ ions dissolved in wet nitrobenzene (Fig. 1C). The rather broad peak may be viewed as Li^+ species experiencing different states of solvation. Each state of solvation will lead to a distinct chemical shift, however individual peaks, corresponding to different chemical solvation states are not resolved leading to a rather broad peak. In order to quantify the exchange rate between these species we have attempted to use 2D exchange spectroscopy (NOESY/EXSY). In this experiment, after the initial frequency labeling, the nuclei are given longitudinal Z magnetization with a second $90\times$ pulse. These vectors are now allowed to exchange through their dynamic process during the mixing time, leading to cross peaks in the 2D spectrum. The relative intensity of the off diagonal signal observed as a function of the mixing time allows the evaluation of the exchange rate. We chose to carry out these experiments at four different mixing times (1 ms, 10 ms, 50 ms and 100 ms) since we had no prior idea of the exchange rate. The results of these experiments are shown in Fig. 3. It is clear that the exchange rate must be slower that 10 ms since we do not see any evidence of an off diagonal intensity (broadened shape) until we reach a 50 ms mixing time. Unfortunately we cannot distinguish a unique cross peak from these spectra on the 400 MHz instrument.

Solvatomers of the lithium ion in "wet" nitrobenzene phase

The NMR spectra were taken at $t = 0$ (see procedure for description of the time scale). The absence of laser light scattering (632.8 nm) indicated that the NB was homogeneous and that there were no microemulsions present, at least on the µm scale. Relatively narrow resonant peaks were obtained for $^7Li^+$ in pure water ($\Delta Hz_{1/2} < 1$ Hz) and in dry nitrobenzene ($\Delta Hz_{1/2} = 10$ Hz) in contrast to the $^7Li^+$ peak in "wet" nitrobenzene ($\Delta Hz_{1/2} = 105$ Hz), indicating the presence of multiple solvatomers (Fig. 1).

It has not been recognized in the initial experiments (Fig. 1B) that the dynamics of this mixed solvation is exceptionally slow, spanning hundreds of hours. It has only been noticed when the NMR spectrum of a one week old sample was compared to the spectrum of the sample taken within 3 h of its preparation. The spectrum taken at $t = 3$ h shows two well-resolved peaks at 0 ppm and 1.38 ppm (Fig. 4A). After 90 h, the peak at 1.38 ppm disappears (Fig. 4B). A short, 5 min immersion of the NMR sample in an ultrasonic bath causes the reappearance of the same peak (Fig. 4C). This experiment has been repeated five times, always yielding the reproducible pattern of behavior although the peak separation was slightly different in different sets of experiments. This variability can be attributed to the different amount of sonic energy absorbed by the sample. The observed

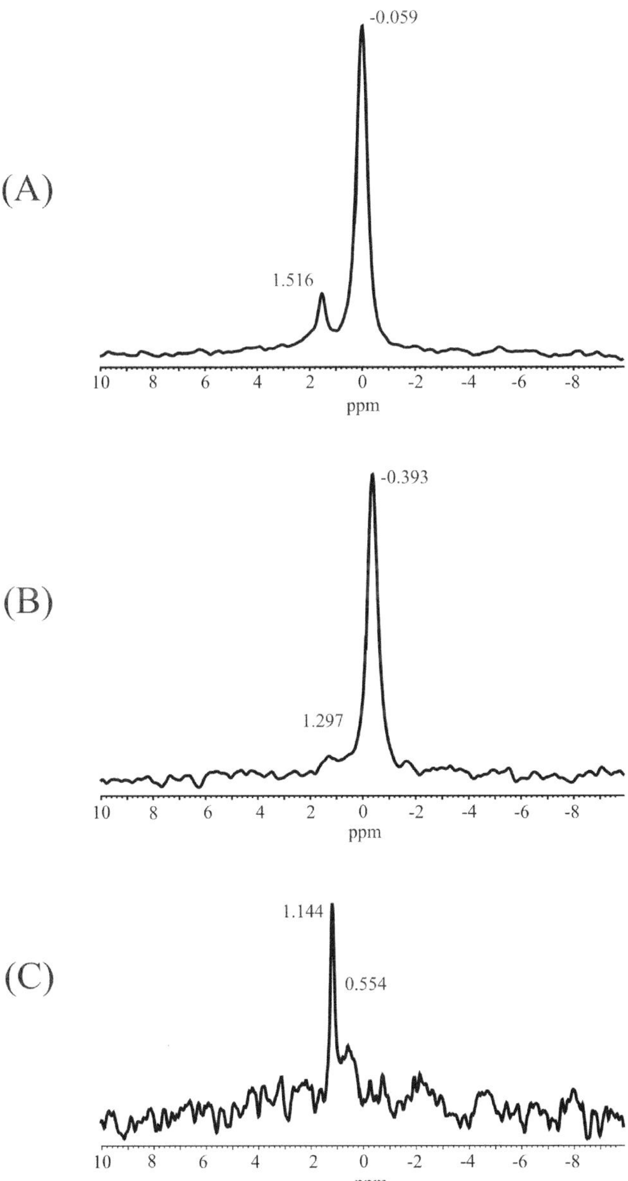

Fig. 4 Dynamics of the self-organization in Li$^+$/nitrobenzene (water) system. ^{7}Li NMR spectrum in wet nitrobenzene recorded at (A) $t = 0$ h (20 000 scans), (B) $t = 90$ h (5500 scans); (C) as in (B), but after sonication for 5 min (12 100 scans).

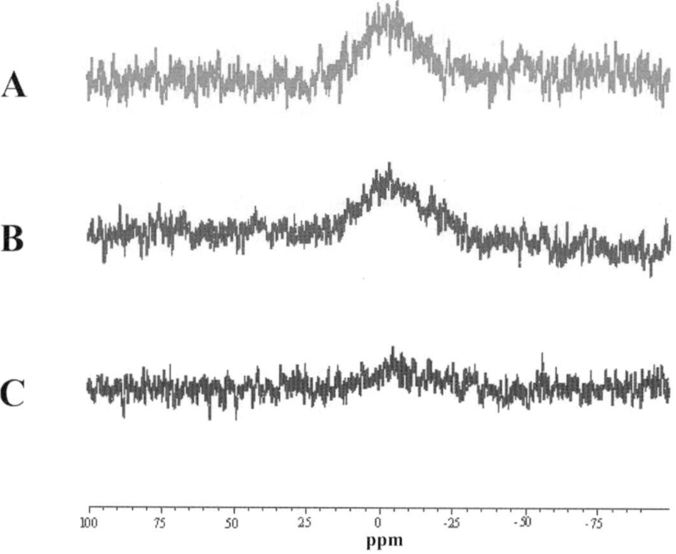

Fig. 5 NMR spectrum of the Li$^+$ present in the walls of the NMR tube and in the glass enclosure of the NMR probe. 20000 scans have been acquired. (A) ^{7}Li-NMR spectrum of glass NMR tube filled with pure nitrobenzene; (B) ^{7}Li NMR spectrum of the "empty" 10 mm glass NMR tube; (C) ^{7}Li-NMR spectrum of the "empty" NMR probe. The weak ^{7}Li signal is coming from the glass of the probe. Each spectrum represents 10000 scans.

changes indicate that a reorganization of the solvated lithium ion is taking place on a very slow time scale. We find it quite remarkable that solvent reorganization can be reversed by *mechanical disruption*. No noticeable changes in the ^{1}1H-NMR spectra taken simultaneously have been observed.

Interaction of lithium ion with glass

Until recently, all the experiments were performed in standard 10 mm diameter glass NMR tubes. The continuous recording of the spectrum of the ^{7}Li$^+$ in "wet" NB revealed an interesting experimental artifact. After four weeks of storage in the glass NMR tube, the ^{7}Li$^+$ peak broadened and could no longer be restored into the original two peak pattern by sonication or by reflux. The Li$^+$ assumed some "immobile" form and virtually disappeared from the solution. Moreover, the concentration of Li$^+$ in that NB solution dropped below the detection limit (*i.e.* < 10^{-6} M) of the atomic absorption spectroscopy analysis. The cumulative spectra acquired over 20000 scans of the empty glass NMR tube, or of the probe *without the* NMR tube also produced a broad ^{7}Li$^+$ signal that became indistinguishable from the background (Fig. 5). The glass NMR tube was then replaced by a quartz NMR tube, and the spectrum of a newly prepared LiBr solution in "wet" NB was acquired. The two peaks of the differently solvated lithium ion were again obtained indicating the presence of Li$^+$ in two different solvation environments. The spectra of the solutions stored in Teflon ware have maintained their "free" solution appearance. On the basis of these experiments we conclude that the wall of the glass NMR tube sorbs the lithium ion. These observations are also corroborated by the fact that the concentration of the lithium ion in solution left in a glass NMR tube dropped below the detection limit of the AAS instrument (<10^{-6} M) while the freshly prepared saturated solution of LiBr in "wet" NB could be comfortably determined as [Li$^+$] = 1.43 × 10^{-5}M.

Discussion

The kinetics of the equilibrium exchange of most ions between water and nitrobenzene, determined electrochemically is known to be very fast[5] ($k^0 \sim 10^{-1}$ cm s^{-1}). Our preliminary *in situ* 2D NMR experiment described in this paper indicates that this is indeed so. In that experiment the required high interface area was realized by stabilizing one phase inside the polymeric beads. Further bead experiments are in progress. It appears that both very slow (on the scale of days) as well as very fast (milliseconds) processes are involved in the exchange of lithium ion between nitrobenzene and the aqueous phase. Lithium ion profoundly affects the overall behavior of the NB/water system. Due to its high affinity for water Li^+ promotes transfer of water into NB. The coextraction of water into NB was determined by the Karl Fisher titration[14] and the hydration numbers of a series of anions have been determined[12] from spin–lattice relaxation times measured by ^1H NMR. It has been found that six water molecules are associated with a lithium ion and two with a bromide ion. Since the two ions do not appear to form an ion pair[14] there are at least eight moles of water bound by one mole of LiBr in "wet" nitrobenzene. With Li^+ concentration being determined as 1.43×10^{-5} M and water concentration 60 mM there appears to be an excess of water to fully hydrate the LiBr salt. The NMR results presented here indicate that hydrated Li^+ in wet NB exists as a metastable species which slowly exchanges its solvation sphere through a series of equilibria involving multiple solvatomers over a period of days. The striking observation is the gradual merging of the two, initially well-resolved NMR resonances into one broad peak that *can be reversed* by mechanical disruption (sonication). To our knowledge the slow changes attributable to solvation shell reorganization have not been observed or reported in ^1H-NMR studies of the Li^+/NB/water system. It is not surprising because there is a much higher molar concentration of water in NB than of the lithium ion and the slow changes on the proton NMR signal originating from the hydration water of Li^+ are obscured by the excess, "free", uncoordinated water.

Another complicating factor is the slow but significant incorporation of lithium ion from "wet" nitrobenzene into hydrated silica-containing walls of the storage vessels. Since the molar concentration of water in NB is relatively low (1.8 mol% relative to NB), the activity of "free" water and thus of the hydrated Li^+ is expected to be high. This situation is analogous to high acidity (or basicity) functions found in mixed organic solvents where the activity of poorly solvated proton or hydroxyl ions can reach abnormally high values at low concentrations of free water in the organic solvent.[20] This high activity of Li^+ apparently drives lithium ion into the hydrated glass (and also possibly quartz) walls of the vessels in which this solution is stored. This effect can result in depletion of Li^+ from the solution of the separate "wet" nitrobenzene phase when it is in prolonged contact with silica containing materials.

References

1 F. Reymond, D. Fermin, Hye Jin and H. H. Girault, *Electrochim. Acta*, 2000, **45**, 2647.
2 K. Kontturi, J. A. Manzanares, L. Murtomaeki and D. J. Schiffrin, *J. Phys. Chem. B*, 1997, **101**, 10 801.
3 F. Scholz, R. Gulaboski and K. Caban, *Electrochem. Commun.*, 2003, **5**, 1388.
4 T. Kakiuchi and Y. Teranishi, *Electrochem. Commun.*, 2001, **3**, 168.
5 Z. Samec, *Electrochim. Acta*, 1998, **44**, 85.
6 Z. Samec, Y. Kharkats and Y. Y. Gurevich, *J. Electroanal. Chem.*, 1986, **204**, 257.
7 P. Vanysek, *Lecture Notes in Chemistry*, Springer–Verlag, New York, 1985, **vol. 39**.
8 L. F. Scatena, M. G. Brown and G. L. Richmond, *Science (Washington, D. C.)*, 2001, **292**, 908.
9 G. L. Richmond, *Annu. Rev. Phys. Chem.*, 2001, **52**, 357.
10 M. R. Watry, M. G. Brown and G. L. Richmond, *Appl. Spectrosc.*, 2001, **55**, 321A.
11 H. H. Girault and D. J. Schiffrin, *Electrochim. Acta*, 1986, **31**, 1341.
12 T. Osakai, M. Hoshino, M. Izumi, M. Kawakami and K. Akasaka, *J. Phys. Chem. B*, 2000, **104**, 12 021.
13 T. Osakai, H. Ogawa, T. Ozeki and H. H. Girault, *J. Phys. Chem. B*, 2003, **107**, 9829.
14 T. Osakai, A. Ogata and K. Ebina, *J. Phys. Chem. B*, 1997, **101**, 8341.
15 T. Osakai, A. Tokura, H. Ogawa, H. Hotta, M. Kawakami and K. Akasaka, *Anal. Sci*, 2003, **19**, 1375.
16 T. Osakai and K. Ebina, *J. Phys. Chem. B*, 1998, **102**, 5691.
17 D. Michael and I. Benjamin, *J. Electroanal. Chem.*, 1998, **450**, 335.
18 J. Grandjean, *Annu. Rep. NMR Spectrosc.*, 1998, **35**, 217.
19 YuYe Tong, E. Oldfield and A. Wieckowski, *Anal. Chem.*, 1998, **70**, 518A.
20 C. Reichardt, *Solvents and Solvent Effects in Organic Chemistry*, VCH Publishers, New York, 1990.

General discussion

Dr Bain welcomed everyone on behalf of the Chairman and Organising Committee to Faraday Discussion 129 on Dynamics and Structure at Liquid–Liquid Interfaces.

In order to provide some structure to the discussions, Dr Bain had put together a summary of some of the generic issues that underlie the papers presented. Dr Bain said: while there will be ample opportunity to address specific issues on each paper, I hope that you will also look at the broader picture and ask how the work described helps us to understand the structure and properties of liquid/liquid interfaces. The following list is not exclusive, nor are the issues precisely defined or in any particular order, but it does at least provide a guide to where the key issues lie.

1. Are L/L interfaces sharp or diffuse? How does the profile of an interface depend on the property being measured? How are the profiles of different properties related?

2. How does one extrapolate from bulk to interfacial properties (experiment & simulation)? How good is the liquid/air interface as a model for the L/L interface? Which properties of the interfaces are transferable?

3. What influence do electrolytes have on interfacial structure?

4. Can we develop better models for the potential across the L/L interface?

5. What are the microscopic origins of kinetic barriers to transport across the L/L interface (electrostatic, local dielectric environment, solvent reorganisation, steric hindrance...)?

6. To what extent does the equilibrium structure of an interface allow us to predict dynamical processes at the interface? Does the presence of an ion/neutral molecule passing through an interface have a major effect on the structure of the interface?

7. What are the microscopic mechanisms of emulsion coalescence or bilayer fusion? What are the roles of long-range forces, surface elasticity, Marangoni effects, *etc.*?

8. How do we connect macroscopic mechanical and kinetic properties of interfaces with the structure and dynamics of an interface on a molecular scale?

Prof. Deutsch commented: Much of the literature on liquid/air and the liquid/liquid interfaces is concerned with the "width" of the interface. Focusing on only one attribute (not always the most important) of the interface is like addressing only, say, the blue colour in a Rembrandt painting. Moreover, this "width" may be different for different quantities. For example, the width of the concentration profile of a particular solute species at an interface may be different from that of other species and/or the solvent. The *shape* of the distribution of a certain quantity across the interface is not less, or even more, important than the width. For example, a monotonic low-to-high variation of the concentration of a particular ion across an interface may well impart electrochemical properties to the interface which could differ significantly from those arising from an oscillatory concentration profile, *i.e.* a layered interface, even though the two interfaces may have the same "width". Thus, I suggest that rather than considering the "width" one should focus on the distribution profiles across the interface of the various physical and chemical properties and study their interrelations to extract the fundamental physics and chemistry which determine them all.

I would also like to suggest adding one more fundamental question to the list. Almost all of the work performed to date on interfaces in general, either uses a molecular-level approach (*e.g.* atomistic simulations) or a continuum description of the interface (*e.g.* surface tension). To some extent these approaches can be characterized as "bottom up" and "top down" approaches. The interface between these two regimes has only been very little explored to date, certainly experimentally. One example that comes to mind is the work of Jacob Klein and coworkers,[1,2] who used a surface force balance to study the shear behaviour in a thin water layer confined within two solid walls as the thickness of the layer is reduced from a macroscopic to a few-molecules thickness. They

DOI: 10.1039/b416300c

observed the transition from a low-shear, liquid-like continuum behaviour at macroscopic thicknesses to a solid-like, granular slip-and-stick motion for a thickness of 6–7 molecular diameters, where the continuum description of the layer breaks down and the discrete molecular nature of the layer starts dominating. Thus, I suggest to add to the list of important issues the question of exploring and understanding the transition regime between the continuum and the molecular-level descriptions of the interface.

1   J. Klein and E. Kurnacheva, *Science*, 1995, **269**, 816.
2   U. Raviv, S. Perkin, P. Laurat and J. Klein, *Langmuir*, 2004, **20**, 5322.

Dr Bain replied: Prof. Deutsch's suggestions are both well-made. The profile, rather than merely the width, and the how the profile differs for different molecular properties are clearly of importance. The transition from a continuum to a molecular description of an interface can certainly be added to the list of 'Generic Issues' in the published proceedings.

Prof. Beattie addressed Dr Walker: We have recently established that surfactant-free O/W emulsions are stabilised by the preferential adsorption of hydroxide ions.[1] The idea has been around for over 60 years, but we appear to be the first to demonstrate it quantitatively. A coarse emulsion of 2–5 vol% of an insoluble oil such as hexadecane is homogenised at pH 9. The pH would drop unless base were added to maintain a constant pH in a pH-stat experiment. The quantity of base added corresponds to the total charge on the newly created surface as the emulsion droplets become smaller. The surface area of the emulsion is obtained from the droplet size and size distribution. We use electroacoustics to measure these quantities directly on the concentrated emulsion during the homogenisation, but a similar experiment would have been possible previously by diluting a sample of the emulsion and measuring the size by light scattering.

The results for three different oils are shown in Fig. 1. The surface charge density is 6 ± 1 μC cm^{-2}, almost independent of the identity of the oil, which suggests that it reflects the structure of water at the interface.

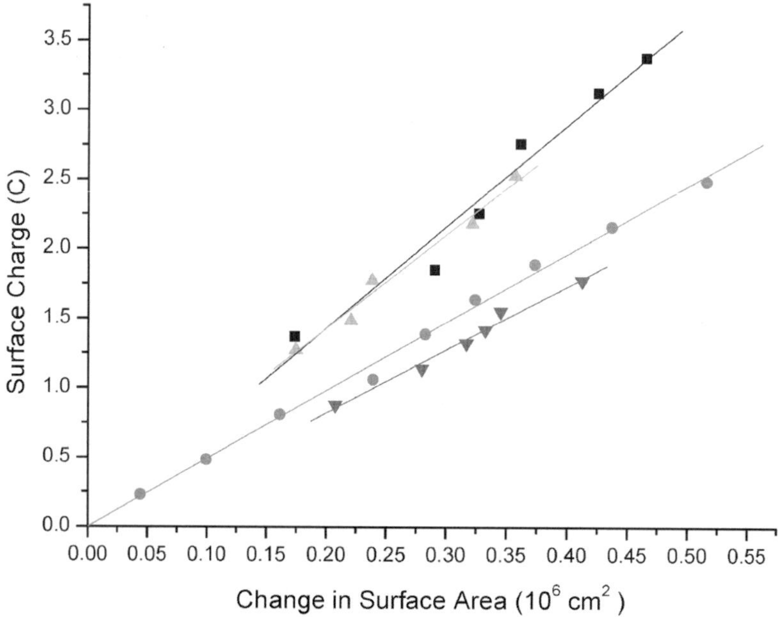

Fig. 1 The surface charge density calculated from measured values of surface charge and increase in surface area for 2 vol% oils in 0.2 mM NaCl: (■) perfluoromethyldecalin, pH 9.0, -7.3 μC cm^{-2}; (▲) squalene, pH 9.0, -6.7 μC cm^{-2}; (●) hexadecane; pH 9.0, -4.9 μC cm^{-2}; (▼) hexadecane; pH 7.0, -4.6 μC cm^{-2}

When a hexadecane emulsion prepared at pH 9 is titrated with acid, the zeta potential drops below pH 7, approaching an 'isoelectric point' at pH 3–4 (Fig. 2). The titration is reversible and the original zeta potential and droplet size are regained when the pH is readjusted to pH 9. The pH dependence of the zeta potential is independent of the identity of the anion, implying that OH^- is the charge determining anion and that dispersion effects are not important.

If a surfactant-free hexadecane emulsion is prepared at pH 9 and SDS is then added the pH increases. The SDS displaces some but not all of the surface hydroxide ions. This implies that hydroxide and SDS anions compete for adsorption at this pH. This is confirmed if emulsification in the presence of SDS is conducted in a pH-stat experiment. In this case the pH would drop unless base were added to maintain a constant pH, but the amount required is less than in the absence of SDS (Fig. 3). The results show that there is an intrinsic charge at these O/W interfaces, unless the pH is controlled at the 'isoelectric point'. This charge must be considered in interpreting other experiments.

For example, it could explain the observed difference between the cationic and anionic surfactants described by Walker, and leads to a proposed experimental test. The hypothesis is that an intrinsic negative charge on the interface repels the anionic surfactant and attracts the cationic one. The experimental test is to examine the pH dependence of the difference. The point of zero charge is around pH 3–4, where the difference between the two surfactants should disappear if the hypothesis is correct.

1   J. K. Beattie and A. M. Djerdjev, *Angew. Chem., Int. Ed. Engl.*, 2004, **43**(27), 3568–3571.

Dr Walker replied: Prof. Beattie raises several interesting issues regarding the population and identity of charged species adsorbed to liquid/liquid interfaces. However, the hypothesis that an intrinsic negative charge on the interface repels the anionic surfactant and attracts the cationic one is inconsistent with the observed behaviour of the molecular ruler surfactants used in experiments probing polarity across liquid/liquid interfaces. The terminal monolayer concentration of anionic surfactants adsorbed to water/alkane interfaces is ~ 1–2×10^{14} molecules cm^{-2} (increasing slightly with surfactant length) whereas the terminal monolayer concentration of cationic surfactants adsorbed to the same interfaces is lower by a factor of ~ 2–3. This behaviour does not support a model where cationic surfactants are attracted to a water/alkane interface and anionic surfactants are repelled.

In addition, any charged species present at the interface not associated with the adsorbed surfactants would not be correlated directly with the aromatic probe and therefore would have little effect on the probe's solvatochromic behaviour. Ongoing studies in our laboratory have examined

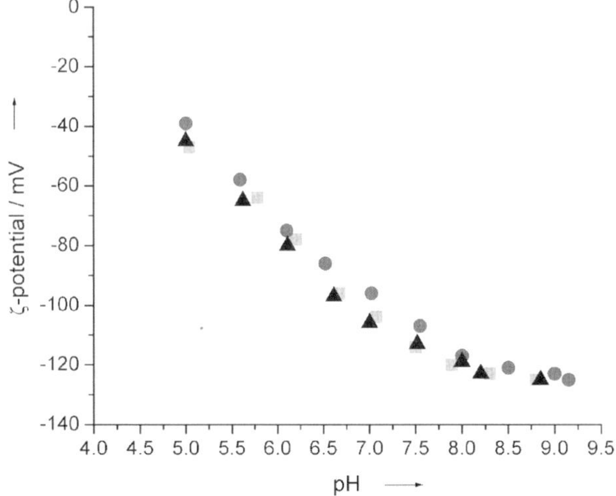

Fig. 2   pH dependence of zeta potential of 2 vol% hexadecane emulsions prepared at pH 9 in 0.4 mM $NaClO_4$ (■), NaI (●) or NaCl (▲) solutions.

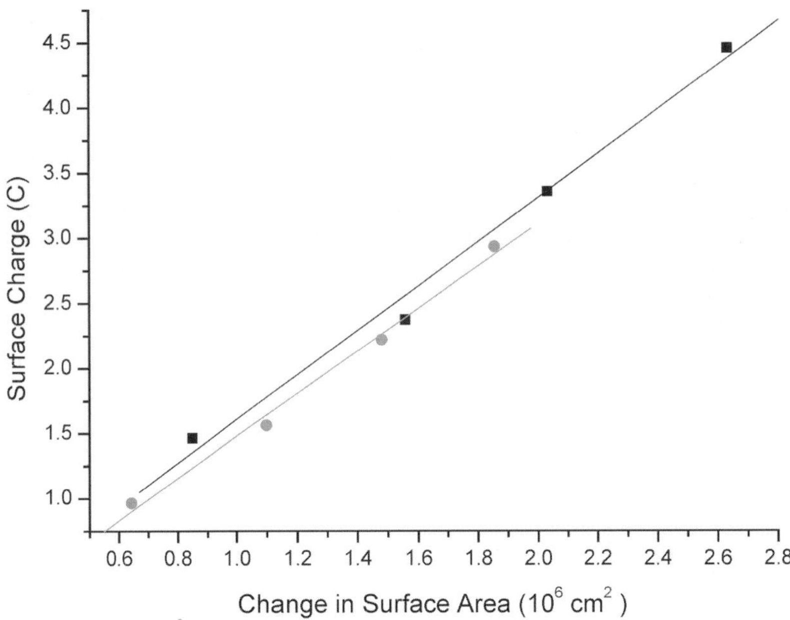

Fig. 3 Surface charge density of 2 vol% hexadecane (■) ($-1.7\ \mu C\ cm^{-2}$) and squalene (●) ($-1.6\ \mu C\ cm^{-2}$) emulsified in 2 mM SDS.

the solvatochromic behaviour of the probe attached to both anionic and cationic headgroups in aqueous solutions having very high ionic strengths (>1 M) adjusted with a variety of salts, acids and bases. In no instance have we been able to reproduce the differences observed between *short* anionic and cationic surfactants adsorbed to water/alkane interfaces. These results suggest strongly that charged species *other than those correlated directly with the solvent sensitive probe* have only a minor effect on the probe itself.

We note that the systems described by Prof. Beattie—coarse emulsions of alkanes in water—presumably have much higher surface areas (and corresponding absolute surface charge) than the systems used in our studies of interfacial polarity using molecular ruler surfactants. The polarity studies used planar interfaces having a surface area of $\sim 20\ cm^{2}$. Surface areas in the emulsions studied by Prof. Beattie and co-workers are likely to be much larger. Depending on the size of the emulsion droplets and the amount of adsorbed charge (reported to be 6 $\mu C\ cm^{-2}$), the effects of charged species on solutes adsorbed in the organic phase might be much more pronounced than in the planar systems studied in the presented work.

Dr Bain addressed Prof. Richmond: Prof. Beattie has shown evidence for adsorption of OH⁻ at the pure alkane–water interface. The resulting interfacial electric field would orient water molecules close to the surface. Do you see any evidence for this in your experiments on pure alkane–water surfaces and in the presence of very low adsorbed amounts of surfactant?

Prof. Richmond replied: We do not see any evidence of this in our experiments but I would not discount the possibility that OH-adsorption is occurring.

Prof. Eisenthal said: I have been asked to comment on the assertion that there is a very large hydroxyl ion, OH⁻ concentration, 10^{-3} M, at the air/water and alkane/water interfaces. The presence of OH⁻ at such a high concentration is contrary to our experimental findings. We have measured the population of the organic cation, malachite green (MG) at the air/water, the octane/water, and the pentadecane/water interfaces. The second harmonic generated (SHG) signal is very strong because there is a two photon resonance at twice the frequency of the incident light, 820 nm.[1] If the interfacial form of MG was the neutral carbinol form, MG–OH, the SHG signal would be very weak because the carbinol form is not even close to being two photon resonant at 420 nm. The SHG signal would be slightly above that of water, which yields a very weak signal. The issue then is whether MG is in its charged or neutral form at an air/water and alkane/water interfaces that has an OH^- concentration of 10^{-3} M. From our measurements of interfacial acid–base equilibria of a long chain phenol, a long chain aniline, and a long chain amine, we found that the pK_a changes from its bulk water value in the direction that favors the neutral form.[2,3] This is consistent with the fact that charges at the air/water interface experience repulsive image forces, which raise its free energy, and which thereby favor the neutral form compared with bulk water. The changes in pK_a values that we obtained ranged from 0.5 to 1.6 pK_a units. Even if the pK_a of the MG–OH$\cdots$MG acid base equilibrium constant was to increase by $2pK_a$ units above its bulk water value of 6, it would remain far below the interface pH value ≥ 11, which corresponds to an OH^- concentration of 10^{-3} M. We therefore conclude that the population of charged MG at the water interface is negligible. If this is the case then we should have obtained a very weak SHG signal, somewhat above that of water, rather than the very strong SHG signals that we obtained.[1]

1   X. Shi, E. Borguet, A. N. Tarnovsky and K. B. Eisenthal, *Chem. Phys.*, 1996, **205**, 167.
2   H. Wang, X. Zhao and K. B. Eisenthal, *J. Phys. Chem. B*, 2000, **104**, 8855.
3   X. Zhao, S. Subrahmanyan and K. B. Eisenthal, *Chem. Phys. Lett.*, 1990, **171**, 558.

Prof. Karyakin opened the discussion of Prof. Richmond's paper: Are you considering an amount of water in the organic phase?

What is critical concentration for micelles formation for the surfactant systems you've shown? With 5 mM surfactant you seem to exceed this and absorb water to the organic phase.

Prof. Richmond replied: Yes we do take that into account as it is an important factor in our experiments.

At the highest concentrations of surfactants shown we are near the c.m.c. However, our focus has primarily been on understanding surfactant behavior at much lower concentrations where we are several orders of magnitude lower than the c.m.c.

Prof. Eisenthal asked: Among the factors that determine the strength of the H_2O SFG signal is the orientation of the H_2O molecules, the "flatter" their orientation the smaller their contribution. Beyond this the SFG signal is proportional to the square of the difference in population of H_2O molecules with dipole moments pointing "up" minus the population with dipole moments pointing "down". This difference in population is determined by the energy difference in up *vs.* down. This relates to your observation that the SFG signal is so much weaker for the H_2O/DCE interface than the H_2O/CCl$_4$ interface. DCE is more polar than CCl$_4$. Does this suggest that the "up" *vs.* "down" energy is smaller for DCE than CCl$_4$, or is it that the molecules are "flatter", or neither of those factors. Temperature dependent studies could shed some light on this.

Prof. Richmond replied: There are many factors that go into the relative strength of the SFG signal from water at the DCE and CCl$_4$ interface. At this point we have not measured quantitatively that difference. We believe that the water molecules are displaying more freedom of motion for the DCE interface which would result in more random orientation of water and consequently a more isotropic distribution of orientations that SFG cannot detect. Temperature dependent studies would indeed be very interesting and we are in the midst of pursuing such studies.

Prof. Samec asked: One of conclusions of your recent study[1] that was also mentioned in your talk has been that a mixed phase interfacial region (diffuse layer) consisting of randomly oriented water molecules is formed at a water–1,2-dichloroethane interface. Can you estimate what the thickness of such a region could be?

1   D. S. Walker, M. G. Brown, C. L. McFearin and G. L. Richmond, *J. Phys. Chem. B*, 2004, **108**, 2111.

Prof. Richmond replied: Unfortunately we cannot measure the thickness of the region but are working to derive an estimate of the thickness through our experiments and MD simulations.

Prof. Girault asked: What do we know, *e.g.* from SFG measurements, about water dissociation $(2H_2O \leftrightarrow H_3O^+ + OH^-)$ at the interface with respect to the bulk?

Prof. Richmond responded: No significant understanding to date has been derived from SFG measurements.

Prof. Beattie added: Various pieces of evidence, including our zeta potential pH titrations described above, suggest that the point of zero charge of the oil/water (and the air/water) interface is in the region pH 3–4. If this is correct then the pK_w of the surface water is <8, reduced by six orders of magnitude from 14 in bulk water. This corresponds to 8 kcal mol^{-1}, or 33 kJ, or 0.35 V. This is comparable to the binding energy of 0.4 eV calculated by density functional theory for the adsorption of OH^- on a methyl terminated *n*-alkane chain,[1] or the potential of 0.5 V arising from the orientation of the water dipoles at the interface calculated from a molecular dynamics simulation.[2] It would be very useful to have an experimental measure of this potential.

1   H. J. Kreuzer, R. L. C. Wang and M. Grunze, *J. Am. Chem. Soc.*, 2003, **125**(27), 8384–8389.
2   S. I. Mamatkulov, P.K. Khabibullaev and R. R. Netz, *Langmuir*, 2004, **20**(11), 4756–4763.

Dr Walker asked: Nonlinear vibrational studies have created a very compelling picture of solvent structure and organization in the out-of-plane dimension. Scattering experiments and MD simulations imply that this picture should be just as detailed and rich in the in-plane dimension.

Do experiments sampling the in-plane vibrational structure ($C_{sum}P_{vis}S_{ir}$ polarization conditions) give rise to reproducible, well defined spectra? If so, what challenges must be overcome before these spectra can be interpreted with the same level of detail as has been performed with the out-of-plane spectra (ssp)?

Prof. Richmond replied: Yes we now have very reproducible well defined spectra of the "in-plane" response that is referred to here as the sps polarization conditions. The analysis of that polarization and others that shed light on the water molecules with in-plane components will be forthcoming in a publication in the next few months. We are now obtaining an understanding of these spectra at the same level that we have with the out-of-plane spectra (referred to here as ssp).

Prof. Rusling opened the discussion of Prof. Schlossman's and Dr Jedlovszky's papers: Can Prof. Schlossman and Dr Jedlovszky compare and contrast their results to those presented by Prof. Richmond?

Prof. Schlossman answered: The spectroscopic results presented by Prof. Richmond provide complementary information to the structural results from X-ray reflectivity. The former probes transitions between electronic energy levels, whereas the latter probes the electron density as a function of interfacial depth. The results of both types of experiments can then be interpreted to yield a picture of the molecular organization at the interface. Application of both of these techniques to the same system would be ideal, however, technical difficulties have prevented this to some extent. In the study of interfaces between two polar liquids, the reflectivity results indicate that the electron density profiles at the nitrobenzene/water and water/2-heptanone interfaces are sharp on the molecular length scale. A related study by Prof. Richmond on the 1,2-dichloroethane/water interface observed a very small sum frequency response for polarization normal to the interface. Unfortunately, technical barriers have prevented us from observing X-ray reflectivity from the neat 1,2-dichloroethane/water interface, though we have been able to observe a molecularly sharp interface between electrolyte solutions of 1,2-dichloroethane (0.005 M BTPPATPBCl) and water (0.01 M Li_2SO_4). The spectroscopic observations of the neat 1,2-dichloroethane/water interface could be interpreted in terms of solvent mixing at the interface though this would imply a thicker interface than observed by X-ray reflectivity on these other systems. An alternative interpretation of the spectroscopy, consistent with the presence of a sharp interface, would be that the interfacial molecular dipoles lie preferentially within the interfacial plane. More work remains to resolve these issues.

Dr Jedlovszky responded: The comparison of our results with those presented by Prof. Richmond is a nice example of the interplay between experiments and simulations. I was particularly happy to hear about Prof. Richmond's results on the existence of free (non-hydrogen bonded) O–H groups of waters located at the interface, since this finding clearly confirms our simulation results. On the other hand, as Prof. Richmond has said, from the SFG spectra they could not unambiguously determine how sharply these free O–H groups prefer to point toward the apolar phase. This is a point where simulations can add something to the experimental results, being able to see the system investigated at atomic resolution. Thus, we have seen that the free O–H groups point almost straight, whereas the water dipole vectors flatly toward the apolar phase. The relevance of this extra information can be confirmed by the good agreement with the experimental data.

Prof. Hampe asked: From the papers of Jedlovszky *et al.* and Lynden-Bell *et al.* we learned that the orientation of molecules in the immediate vicinity of the interface differs from the random orientation within the bulk phase. Jedlovszky *et al.* state that two parameters are needed to describe the orientation of molecules at an interface. Lynden-Bell *et al.* characterise orientation by one parameter only. My question is a very general one: When describing processes we do this within the frame of field theories and use field quantities: concentration, pressure tensor, temperature, *etc.* None of these takes orientation into account. What would be an appropriate field quantity in order to describe orientation at an interface?

Prof. Lynden-Bell replied: I believe that a selection of spherical harmonics would be appropriate field quantities. Dr Jedlovszky's figures show probability distributions and provide more detail than would be appropriate for a field theory. The instantaneous orientation of an imidazolium ring at height z in the interface is described by two parameters, which could be chosen to be the spherical polar angles of the surface normal in the molecular coordinates, or a selection of two spherical harmonics. Field quantities are averages of some local property (*e.g.* the concentration is the average probability density) and here the averages of, for example, $Y_{2,0}$, and $Y_{1,1}$ would provide a sensible set of field parameters taking into account the ring symmetry. This choice is not unique.

Prof. Penfold asked: You have presented some remarkably accurate specular reflectivity data. Of course, we know that the capillary waves transfer intensity from the specular expand into the diffuse or "off-specular" direction (surface diffuse scattering). Could you measure the surface diffuse scattering with sufficient accuracy to correlate with the changes in the specular reflectivity that you observe?

Prof. Schlossman replied: The surface diffuse scattering would provide an important check on the reflectivity results because this scattering directly probes the height-height correlation function of the interfacial fluctuations. Work of this nature on a liquid surface was first reported for the water-vapor interface by Schwartz *et al.*[1] Unfortunately, up till now, we have been unable to measure surface diffuse scattering data from the nitrobenzene/water interface with high enough statistics to make any meaningful conclusions. The difference between measurements on this system and the earlier studies of liquid-vapor interfaces is the additional diffuse scattering that arises from the upper bulk liquid phase (in this case, water). This is not to say that X-ray surface diffuse scattering is not useful for the study of liquid-liquid interfaces. In fact, we have used this technique to study surfactant domains at the water/hexane interface.[2] In that case, the domains provide enhanced diffuse scattering that allows the diffuse scattering from the surface to be separated from that due to the bulk.

1   D. K. Schwartz, M. L. Schlossman, E. H. Kawamoto, G. J. Kellog, P. S. Pershan and B. M. Ocko, *Phys. Rev. A*, 1990, **41**, 5687–5690.
2   M. Li, A. M. Tikhonov and M. Schlossman, *Europhys. Lett.*, 2002, **58**, 80–86.

Prof. Deutsch opened the discussion of Prof. Schlossman's paper: The very carefully measured X-ray reflectivity curves in Fig. 3 (of your paper) are systematically higher than the curves calculated using the "standard" capillary wave model (eqn. (3), of your paper), which neglects bending rigidity. The observations are interpreted as arising from a non-zero bending rigidity, which damps the capillary waves, yielding a lower effective interfacial roughness and hence a higher reflectivity. This interpretation yields an interface-normal (electron) density profile which changes *monotonically* (over a distance given by the effective roughness of the interface) between the density values deep within the two bulk phases.

While this interpretation of the observations is certainly plausible, consistent with the observations, and has a strong physical motivation (the electric dipole interactions), an alternative interpretation, that of a layered interface, is also possible. A layered interface, demonstrated to exist at the liquid/vapour interface of liquid metals[1,2] has a *non-monotonic* interfacial density profile, which exhibits decaying oscillations over a distance of a few atomic diameters from the interface (see solid line in the inset of Fig. 4). The decaying oscillations yield a quasi-Bragg peak in the X-ray reflectivity curve at some $Q_z^{\text{peak}} \approx 2\pi/d$, where d is the distance between adjacent layers. However, if the surface tension is low, as is the case for the present liquid/liquid interface, the unfavourable signal/noise ratio of the reflected intensity often does not allow to reach Q_z^{peak}, and only the rising part of the quasi-Bragg peak is observed. A qualitative example demonstrating this effect for potassium[3,4] is shown in Fig. 4.

Except for the much-larger measurable Q_z-range, due to the higher surface tension and greatly reduced background at the liquid/vapour interface as compared to the liquid/liquid one, Fig. 4 is very similar to Fig. 3 (Prof. Schlossman's paper), and shows the same effect of a reflectivity higher than that predicted from the "standard" capillary wave model. Diffuse scattering measurements presented in the above-cited papers make it clear that in this case the increase is due to interfacial layering and no deviation from the "standard" capillary wave model is detected. Unfortunately, since Q_z^{peak} cannot be reached, a definitive fit by this model of the potassium reflectivity, as well as that of the liquid/liquid interface in Fig. 3 (Prof. Schlossman's paper), is not possible without some additional external data on the expected layering. In the case of potassium, this additional information and support is provided by the observation that the intrinsic structure factor of the potassium interface, over its measurable range, is almost identical with that of gallium and indium, which were measured well beyond their respective Q_z^{peak}.

X-ray diffuse scattering measurements from the interface, which probe directly the capillary waves, could, in principle, distinguish between the two alternatives. However, the high parasitic scattering by the upper liquid phase (water in this case) preclude such measurements at the liquid/liquid interface. Perhaps a measurement of the variation of the reflectivity with temperature, or with some of the parameters of the theory presented by Prof. Schlossman to account for the non-zero rigidity, could be used to favour one of these two alternatives over the other.

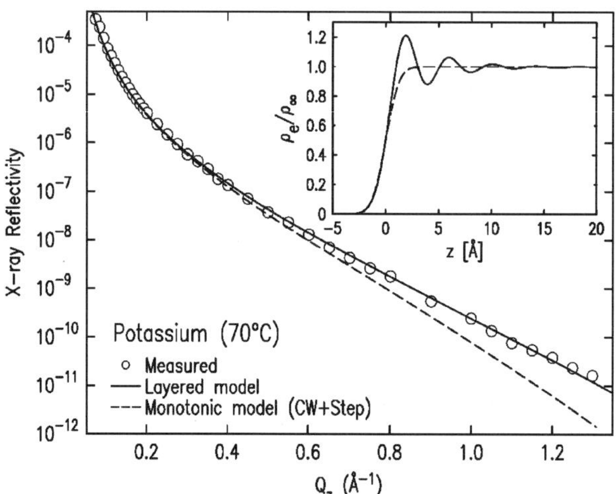

Fig. 4 Measured (○) X-ray reflectivity of the liquid/vapour interface of liquid potassium under UHV conditions. The -- line is the reflectivity calculated assuming a simple monotonic electron density profile, broadened by capillary waves as calculated from the "standard" capillary wave model, eqn. (3). The — line is for a layered interface model. The inset shows the corresponding electron density profiles ρ_e/ρ_∞, normalized to the electron density of the bulk ρ_∞, across the vapour ($z < 0$)/liquid ($z > 0$) interface.

Contributions by Dr O. Shpyrko (Harvard) to this comment are gratefully acknowledged.

1   O. Magnussen and B. Ocko, *Phys. Rev. Lett.*, 1995, **74**, 4444.
2   M. J. Regan, E. H. Kawamoto, S. Lee and P. S. Pershan, *Phys. Rev. Lett.*, 1995, **75**, 2498.
3   O. Shpyrko, P. Huber, A. Grigoriev, P. Pershan, B. Ocko, T. Tostmann and M. Deutsch, *Phys. Rev. B*, 2003, **67**, 115405.
4   O. Shpyrko, M. Fukuto, P. Pershan, B. Ocko, I. Kuzmenko, T. Gog and M. Deutsch, *Phys. Rev. B*, 2004, **69**, 245423.

Prof. Schlossman replied: I agree with Prof. Deutsch that layering can be an alternative explanation of these data. The disk-like structure of nitrobenzene, as well as dipole interactions, may favor layering on the nitrobenzene side of the interface. Reflectivity measured at four temperatures (25 °C, 35 °C, 45 °C and 55 °C, see Fig. 3 of our paper) demonstrated that the discrepancies with the predictions of eqn. (3) (our paper) are progressively reduced with temperature. This can indicate either a decreased bending rigidity or a decreased amplitude of layering. Recent, unpublished computer simulations by Prof. Ilan Benjamin and co-workers also provide some justification for weak layering at this interface, though the possibility of finite-size effects in these simulations must be considered. Further analysis of these data is ongoing, including comparison with the computer simulations.

Prof. Samec said: Your results and conclusions concerning the structure of the polarized liquid–liquid interface are consistent with our most recent capacitance measurements[1] pointing to existence of an inner or compact layer at the interface between two immiscible electrolyte solutions (ITIES). Fig. 5 of our paper shows the plot of the inverse capacity C^{-1} of the polarized water–1,2-dichloroethane interface *vs.* the inverse surface charge density $(q^0)^{-1}$ on the organic side for two organic supporting electrolytes. Since the capacitance of the diffuse double layer is likely to be controlled by the surface charge density, the difference in the inverse capacitance for the two electrolytes can be ascribed to the inner (compact) layer, *i.e.* to a difference in the ion size

$$\Delta\left(\frac{1}{C}\right) = \Delta\left(\frac{1}{C_i}\right) = \frac{\Delta r_i^o}{\bar{\varepsilon}^o \varepsilon_0} \tag{1}$$

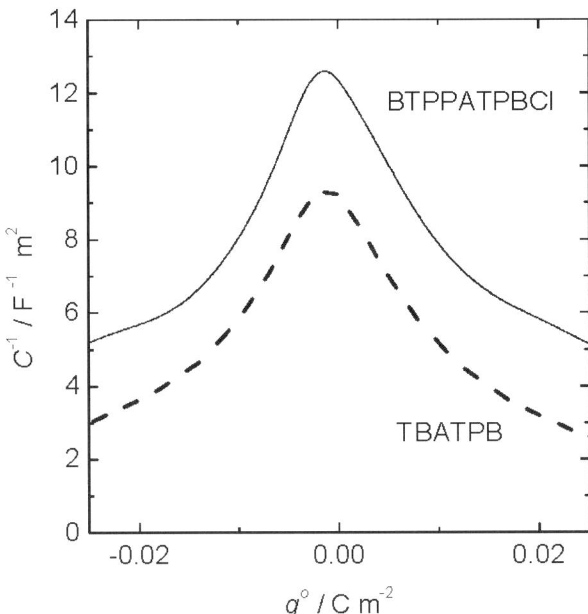

Fig. 5   Inverse differential capacity *vs.* the surface charge density on the organic solvent side for the interface between 0.1 M LiCl in water and 5 mM BTPPATPBCl (A) or TBATPB (B) in 1,2-dichloroethane (adapted from ref. 1).

In eqn. (1), C_i represents the capacity of the inner layer, Δr_i^0 is the difference in ionic radii and $\bar{\varepsilon}^0$ is the relative dielectric permittivity. Fig. 5 shows that the data for BTPPATPBCl are shifted by an almost constant value of *ca.* 3 F^{-1} m^2 with respect to data for TBATPB composed of ions with smaller radii. Assuming that the permittivity of the inner organic layer takes the bulk value of 10.23, the difference in the inverse capacitance corresponds to a decrease in the ion radius by $\Delta r_i^0 = 0.27$ nm, which appears to be a reasonable value. Our conclusion has been that the description of ITIES as an ion-free solvent layer separating two space charge regions comprising finite-size ions is plausible.[1,2]

1   A. Trojánek, A. Lhotský, V. Mareček and Z. Samec, *J. Electroanal. Chem.*, 2004, **565**, 243.
2   T. Wandlowski, K. Holub, V. Mareček and Z. Samec, *Electrochim. Acta*, 1995, **40**, 2887.

Prof. Schlossman responded: Prof. Samec cites results of his recent study of the 1,2-dichloroethane/water (with electrolytes) interface and then refers to the results of a preliminary analysis of X-ray reflectivity measurements from the interface of an aqueous solution of BaCl$_2$ (0.1 M) and a nitrobenzene solution of TBATPB (tetrabutylammonium tetraphenylborate, 0.01 M) that used a modified Poisson–Boltzmann theory, MPB5.[1] Since this analysis was not discussed in my manuscript, but I presented it at the Faraday Discussions meeting, I will present it here for the benefit of the readers though I emphasize that it is preliminary. Our tactic for this preliminary analysis was to use the MPB5 theory to predict the ion distribution for the experimental interfacial potential, map this distribution onto a mathematically sharp interface between bulk nitrobenzene and water, convert this distribution to electron density, then convolute the electron density profile with a Gaussian to model the influence of capillary waves on the interface. Since the width of this Gaussian was determined by the capillary wave theory using our measured values of the interfacial tension, this procedure contains no fitting parameters. This calculated electron density profile is then used to compute an X-ray reflectivity which is compared to experiment.

The ion distribution, final electron density, and experimental comparison are shown in Fig. 6 for an interfacial potential $\Delta\phi^{w-o} = 0.2$ V. Here it is seen that the TPB$^-$ ion density is unrealistically high (high enough to exclude all nitrobenzene in this region) and that the rigid wall aspect of the MPB models excludes approach of the ion center to regions within an ionic radius of the interface. Clearly, the fit to the data is poor.

To improve agreement between the prediction and the data, we limited the maximum value of the TPB$^-$ concentration near the interface. The value shown in Fig. 7 allows nitrobenzene to mix with the TPB$^-$ immediately adjacent to the interface. It is seen that the match between data and prediction is greatly improved.

We have also modified the ion density profiles by adding additional ion density in the region immediately adjacent to the interface (modelled by a Gaussian). This allows ions to penetrate into the region within a ion radius of the interface. We have found that this modification does not make much difference and yields predictions for the reflectivity indistinguishable from that shown in Fig. 7. We suspect this is primarily because the final electron density is smeared by the presence of capillary waves which have a similar effect on the profile probed by the X-rays.

We have performed many variations of this type of analysis and consistently find that variations that closely approximate the data all have electron density profiles similar to that in Fig. 7. This profile has a low plateau on the nitrobenzene side of the interface that corresponds to increased concentration of TPB$^-$ ions mixed with nitrobenzene. This conclusion is different from that mentioned by Prof. Samec, in his comment, of an ion-free solvent layer separating two space charge regions. For example, an interfacial region consisting of one or two full layers of nitrobenzene (at the density of bulk nitrobenzene) without any ions would produce a non-monotonic electron density profile exhibiting a local maximum just before the crossover to the aqueous phase (in contrast with the monotonic electron density profile shown in Fig. 7). However, it is possible that a sub-monolayer interfacial region of primarily nitrobenzene could be consistent with the electron density profile shown in Fig. 7. At this stage, it is necessary to re-emphasize the preliminary nature of this analysis.

Prof. Samec's observations on the dependence of capacity on ion size suggests the importance of performing X-ray measurements using ions of different size. This might lead to electron density

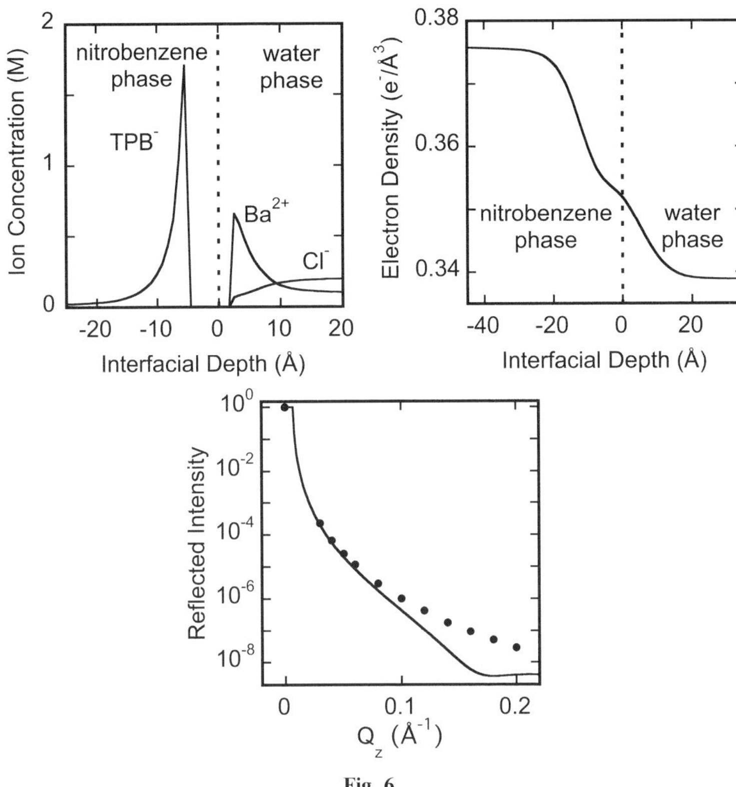

Fig. 6

profiles, similar to that shown in Fig. 7, that contain a plateau region whose thickness scales with the ion size.

1   C. W. Outhwaite and L. B. Bhuiyan, *J. Chem. Soc., Faraday Trans. 2*, 1983, **79**, 707–718.

Prof. Benjamin said: Molecular dynamics simulations of the water/nitrobenzene interface indeed suggest that water and nitrobenzene molecular dipoles lie parallel to the interface. However, it is not clear how one may use the information about the orientational distribution to compute the bending force constant.[1]

1   D. Michael and I. Benjamin, *J. Electroanal. Chem.*, 1998, **450**, 335.

Prof. Schlossman replied: Computation of the bending modulus would require consideration of all the interactions at the interface, not just the dipole interactions. The cartoon in Fig. 4 of my manuscript, and accompaning discussion, was intended as a plausibility argument to motivate the possibility that the dipole interactions at the interface between two polar liquids could generate a bending modulus beyond that expected at the interface between polar and non-polar liquids. An example of the latter is the water–alkane interface for which we have not had to consider the possibility of a bending modulus to explain X-ray reflectivity data.[1]

1   D. M. Mitrinovic, A. M. Tikhonov, M. Li, Z. Huang and M. Schlossman, *Phys. Rev. Lett.*, 2000, **85**, 582–585.

Prof. Kornyshev stated: There are elements in interface "smearing". One is atomic smearing due to interpenetration of molecules of one liquid into the other; for immiscible liquids it extends only to molecular dimensions. The other element is the corrugation due to capillary waves. (In the limit of high wave-vectors the latter merges with the former.) Building the theory of capacitance one should

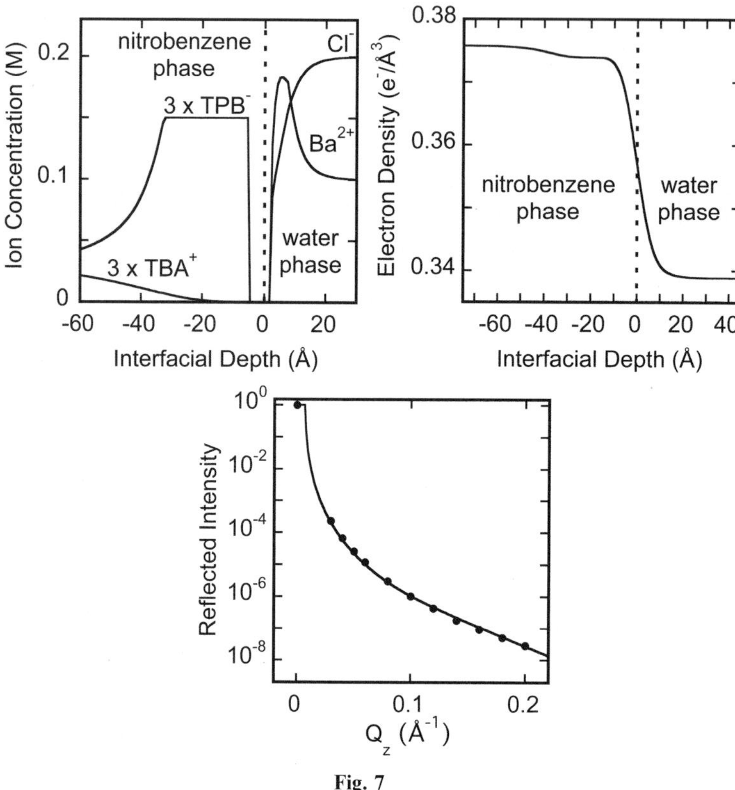

Fig. 7

take into account both effects. On the other hand, it has been shown[1] that surface corrugation (and the whole spectrum) of capillary waves depends on the potential drop and interfacial capacitance. Thus the corrugation and capacitance should be considered self-consistently, as it has been performed in the quoted paper. It predicts, in particular, that for a small potential drop, V, across the interface the corrugation amplitude scales as $d^2 = d_o^2 + \sim V^2$, *i.e.* at small V it is a parabolic law. A nonlinear theory shows faster development of $|d(v)|$ at large V. It might be interesting to treat your data using these concepts. The trends you have shown are in this direction, but whether the absolute value of the effect will be as large as predicted, this we need to check.

Note that molecular smearing and the ion penetration into the unfriendly medium has been taken into account in the theory of capacitance at ITIES[2] which seems to reproduce well the existing experimental data, specifically due to avoiding the assumption of atomically immiscible interface.

1   L. I. Daikhin, A. A. Kornyshev and M. I. Urbakh, *J. Electroanal. Chem.*, 2000, **483**, 68.
2   L. I. Daikhin, A. A. Kornyshev and M. I. Urbakh, *J. Electroanal. Chem.*, 2001, **500**, 461.

Prof. Schlossman replied: I agree with Prof. Kornyshev that a full theory of interfacial ion distributions must account for the capillary waves, ion penetration, and the interfacial electric field self-consistently. Clearly, the MPB5, and earlier versions of the modified Poisson–Boltzmann theories, were written to describe systems with hard walls and do not include capillary waves or ion penetration. Perhaps that is why our preliminary analysis with the MPB5 theory has not led to a satisfactory conclusion, in spite of our *ad hoc* attempts to include these features. It is my understanding that our method of including capillary waves, which calculates their effect using the measured interfacial tension under conditions of imposed electric potential, will account for the linear variation of capillary waves under a small imposed potential as described by Prof. Kornyshev. Further analysis will consider seriously the two theories mentioned by Prof. Kornyshev, one that considers the interplay of capillary waves and applied potential and the other that considers the effects of ion penetration. Since these reflectivity data are sensitive to both capillary waves and the ion distribution down to sub-nanometer length scales, a theory that combines the capillary wave effects with a realistic ion distribution would be most welcome in their analysis.

Prof. Kornyshev added: But your smearing parameter, extracted from the data, does seem to be rather small and its variation in the full range of potential does not exceed 20%. I am intrigued, whether our theory will be able to explain this. To be sure that we are doing the right thing we must know where the potential of zero charge lies.

Prof. Schlossman responded: We determine the potential of zero charge in two ways, by measuring the maximum in the interfacial tension as a function of imposed potential and by measuring the minimum in the capacitance as a function of potential. These independent measurements agree to with 0.02 V. These measurements of the potential of zero charge allow us to determine the interfacial potential $\Delta\phi^{w-0}$ and should allow us to make a direct comparison with Daikhin, Kornyshev and Urbakh theories.

Prof. Schlossman opened the discussion of Dr Jedlovszky's paper: Have you applied your simulation method to the interface between water and a polar oil such as dichloroethane or nitrobenzene? If so, how do the orientations of interfacial water compare with the results of recent experiments?

Dr Jedlovszky replied: We have applied it for the water/1,2-dichloroethane interface[1] and we have seen the same picture. Different results have been, however, obtained at the water/1-octanol interface,[2] *i.e.*, when the molecules of the apolar phase are able to form hydrogen bonds with water. Here the peak corresponding to the second preferred orientations (*i.e.*, when one of the OH bonds points almost straight to the apolar phase) is missing. We have interpreted this result as the octanol molecules are able to replace these water molecules in the hydrogen bonding positions connected to the waters laying parallel with the interface.

1 P. Jedlovszky, A. Vincze and G. Horvai, *J. Chem. Phys.*, 2002, **117**, 2271–2280.
2 P. Jedlovszky, I. Varga and T. Gilányi, *J. Chem. Phys.*, 2004, **120**, 11839–11851.

Prof. Benjamin said: In some cases one may be able to infer the existence of different orientational populations of water molecules by examining the OH vector orientational distribution.

Dr Jedlovszky responded: In some lucky cases this might be true, but it is not for sure. For instance, we have analysed the orientation of the O–H vector at the water/1,2-dichloroethane interface. In the first of the two preferred orientation (*i.e.*, when the water lays almost parallel with the interface) the two O–H vectors should give rise to a peak at about the cosine value of −0.3, whereas in the other preferred orientation one of the O–H vectors at 1 and the other one at about −0.4. The observed distribution (see Fig. 4 of the paper by Jedlovszky *et al.*[1]) has indeed a peak at 1, and another broad one at about −0.4, to which both preferred orientations contribute. Thus, without using the information obtained from the bivariate distribution, the orientational distribution of the O–H vectors can also be interpreted as a sign of solely the second of the two preferred orientations.

1 P. Jedlovszky, A. Vincze and G. Horvai, *J. Chem. Phys.*, 2002, **117**, 2271–2280.

Prof. Urbakh opened the discussion of Prof. Benjamin's paper: Recently we have done Brownian dynamics calculations of ionic transfer across a liquid/liquid interface.[1,2] These calculations demonstrated that a coupling between the ionic motion and capillary wave fluctuations can significantly slow down the transfer across the interface. The Brownian dynamics is a semiphenomenological approach, and my question is: Do you observe a similar effect in microscopic molecular dynamics simulations?

1 A. A. Kornyshev, A. M. Kuznetsov and M. Urbakh, *J. Chem. Phys.*, 2002, **117**, 6766.
2 C. G. Verdes, M. Urbakh and A. A. Kornyshev, *Electrochem. Commun.*, 2004, **6**, 693.

Prof. Benjamin replied: Some time ago we demonstrated[1] that molecular scale "capillaries" may couple to ion motion. Whether these density fluctuations can slow down or speed up ion motion depends on the ion solvation free energy difference between the two liquids. For example, if a small ion is moving from the aqueous phase to the organic phase, water "fingers" will slow down the ion motion.[2]

1 I. Benjamin, *Science*, 1993, **261**, 1558–1560.
2 K. Schweighofer and I. Benjamin, *J. Phys. Chem. A*, 1999, **103**, 10274–10279.

Prof. Kornyshev asked: In your simulations, have you balanced the difference in chemical potentials of ions in the bulk of the two solvents by the electric field?

Prof. Benjamin replied: In the simulations reported here, the ion is produced as a result of photodissociation reaction, so there is no external potential to balance the driving force.

Prof. Samec asked: Some time ago[1] we had been attracted by the possibility of studying the transfer of an ion in its excited state (*i.e.* a hot ion) across the polarised liquid–liquid interface. It has turned out however that the photocurrent measured upon the UV excitation of a tetraaryl ion (tetraphenylarsonium cation or tetraphenylborate anion) is associated rather with a product of photodecomposition of this ion. Can you estimate whether the photoexcitation of an I_2^- anion could lead to a significant change in the solvation energy, and thereby the standard Gibbs energy of ion transfer, which would drive the transfer of the hot ion to the organic phase?

1 Z. Samec, A. R. Brown, L. J. Yellowlees, H. H. Girault, *J. Electroanal. Chem.*, 1990, **288**, 245.

Prof. Benjamin replied: The ion current observed in our simulation is also the result of a product of photodecomposition. Photoexcitation of the parent ion could also lead to a more polarizable, and thus a more surface active ion.

Prof. Deutsch asked: I was wondering how the whole picture presented for the di-iodine dissociation reaction at the H_2O/CCl_4 interface would change if one of the phases, say the CCl_4, would be replaced by a vapour phase, so that the dissociation reaction takes place at a free surface, rather than the interface between liquids.

Obviously, the symmetry breaking due to the surface is still there, but both of the dissociation products stay at the water phase, since none can move into the vapour (or can they?) In this case one would still expect asymmetric free energy maps and cuts similar to those in Figs. 4 and 5 (in Prof. Benjamin's paper). Is this correct?

If this is indeed the case, how would the maps and cuts for the H_2O/vapour interface differ from those of the H_2O/CCl_4 interface?

Prof. Benjamin answered: At the water liquid/vapor interface, all the species will remain at the interface. The free energy maps presented in this work correspond to a situation where one ion is in the organic phase and one in the water phase. We expect that at the water liquid/vapor interface, the free energy corresponding to the ion on the vapor side will be higher than in the case of CCl_4, and higher if the neutral atom is in the vapor phase. Thus, the cut will still look asymmetric.

Prof. Kornyshev asked: Ion slowing down, when moving across the interface, determined by coupled ion-protrusion Brownian dynamics[1] has been shown[2] to be much less dramatic with lowering surface tension. Obviously, when surface tension is low the coupling between protrusion fluctuations and ion motions is negligible and the corresponding mechanism of slowing down vanishes. Testing this prediction systematically will be a good test of this "surface polaron mechanism" of slowing down.[3] But is there any possibility to vary surface tension in real experiments, without changing other parameters of the system, such as viscosity, capacitance, *etc.*? An easy thing will be to play with surfactants, but a surfactant layer itself may act as a "fence" for the moving ion? Any ideas what we can do in this respect?

1 A. A. Kornyshev, A. M. Kuznetsov and M. Urbakh, *J. Chem. Phys.*, 2002, **117**, 6766.
2 C. G. Verdes, M. Urbakh, A. A. Kornyshev, *Electrochem. Commun.*, 2004, **6**, 693.
3 A. A. Kornyshev, A. M. Kuzuetsov and M. Urbakh, *Russ. J. Electrochem.*, 2003, **39**, 119.

Prof. Benjamin replied: I am not aware of a way to experimentally vary surface tension between two liquids without changing other properties. However, one may be able to test the theory you refer to by computer simulations. We have recently[1] demonstrated that one can build a series of liquid/liquid interfaces with the same bulk viscosity but a variable surface tension.

1 A. W. Hill and I. Benjamin, *J. Phys. Chem. B*, 2004, **108**, 15443–15445.

Dr Walker addressed Prof. Kornyshev: This question is addressed to another speaker, but I will suggest an experimentally feasible solution to the question: use solvent isomers or different solvents that are closely related in size and composition. Using different solvent isomers (*e.g.* 1-octanol *vs.* 2-octanol) in contact with water leads to changes in surface tension that are small in magnitude but large on a relative scale. (For example, the water/1-octanol interfacial tension is 8.6 mN m^{-1} and the 2-octanol/water interfacial tension is 7.5 mN m^{-1}—a change of 1.1 mN m^{-1} or $\sim 12\%$. Larger changes could be realized by using shorter or longer alcohols. Alternatively, using haloalkanes (*e.g.* 1-chlorodecane) *vs.* saturated alkanes (*e.g.* decane) will lead to changes in interfacial tension of ~ 3–5 mN m^{-1} (with absolute tensions of ~ 45–50 mN m^{-1}). Perhaps Prof. Kornyshev could comment whether or not these changes in surface tension are large enough to change the rate of ion transport across an interface by an observable amount.

Prof. Kornyshev responded: The effect could be detectable. Our estimates in Fig. 3,[1] show this.

1 C. G. Vordes, M. Urbakh and A. Kornyshev, *Electrochem. Commun.*, 2004, **6**, 693.

Prof. Janata addressed Prof. Kornyshev: It is possible to change the surface energy at the L/L interface without changing the composition of the two liquids by applying external electric field across it from a tip of the tunneling microscope. The surface energy then varies according to the Gibbs–Lippmann equation. Experiments like this have been carried out and published.[1,2] It should be noted that increase of surface energy induces the surface waving as the absolute value of voltage across the interface is increased from zero.

1 J. F. T. Conroy, K. Caldwell, C. J. Bruckner-Lee and J. Janata, *J. Phys. Chem.*, 1996, **100**, 18222–18228.
2 J. F. T. Conroy, V. Hlady, C. J. Bruckner-Lee and J. Janata, *J. Phys. Chem.*, 1996, **100**, 18229–18233.

Prof. Kornyshev responded: This is all true. But what I was talking about, is that the problem should be solved self-consistently. Indeed not only the field affects surface tension, but surface undulations (corrugations) induced by a reduction of surface tension affect the capacitance which in turn affects the resulting electric field. This is what ref. 1 is about.

1 L. I. Daikhin, A. A. Kornyshev and M. Urbakh, *J. Electroanal. Chem.*, 2000, **483** 68.

Dr Webster opened the discussion of Prof. Lynden-Bell's paper: Using neutron reflection we have probed the structure of the air–ionic liquid interface and have found evidence for an oscillatory density profile close to the interface. This work is published in Langmuir.[1]

1 J. Bowers, M. C. Vergara-Gutierrez and J. R. P. Webster, *Langmuir*, 2004, **20**, 309.

Prof. Lynden-Bell replied: The work[1] describes neutron reflection results on butyl methyl and octyl methyl imidazolium ionic liquids. The interfacial structure normal to the surface is inhomogeneous and the results are interpreted in terms of distinct layers of alkyl chains and ionic headgroups. This is consistent with our results on cations with shorter side chains as it implies an orientational ordering of the imidazolium rings.

1 J. Bowers, M. C. Vergara-Gutierrez and J. R. P. Webster, *Langmuir*, 2004, **20**, 309.

Prof. Beattie asked: In view of the remark just made that the O/W interface thickness increases with miscibility, I would like to ask Prof. Schlossman to comment on his previously published X-ray reflectivity results[1] on the alkane/water interface that indicated the opposite trend, *i.e.* thickness increasing with alkane chain length from C_6 to C_{16}.

1 D. M. Mitrinovic, A. M. Tikhonov, M. Li, Z. Huang and M. Schlossman, *Phys. Rev. Lett.*, 2000, **85**(3), 582–585.

Prof. Schlossman replied: If the variation in interfacial width is governed solely by capillary wave fluctuations, then one expects that the width will vary inversely with the interfacial tension. To the extent that the tension varies inversely with the miscibility it is appropriate to expect an interfacial width that increases with miscibility. In the work cited by Prof. Beattie, X-ray reflectivity measurements from the interface between bulk liquid alkane and bulk water demonstrated a significant increase in the interfacial widths with increasing chain length of the alkane. Since the miscibility falls rapidly with increasing chain length, this common interface provides a counter-example to the idea that interfacial width increases with miscibility. In this case, the rapid fall in miscibility is accompanied by only a slight change in the interfacial tension ($\Delta\gamma = 3$ mN m^{-1}) as the chain length varies from C_6 to C_{16}. The small change in tension leads to a capillary wave prediction for the interfacial widths that are nearly constant with chain length (3.45 Å for hexane to 3.35 Å for hexadecane). These predictions are very different from the measured interfacial widths that vary progressively from 3.5 ± 0.2 Å for hexane to 6.0 ± 0.2 Å for hexadecane. In addition, we measured the width for docosane (C_{22}) to be very similar to that of hexadecane. These results can be explained phenomenologically by considering that the interface has an intrinsic width (due to interfacial molecular ordering) that is further broadened by capillary wave roughening. Our proposal that the intrinsic widths are determined by the alkane gyration radii for the shorter alkanes (16 carbons or less), and by the bulk correlation length for the longer alkanes (16 carbons or more), is in quantitative agreement with the experimental dependence of interfacial width on chain length.

Prof. Kornyshev asked: Can you combine the room temperature molten salt (ionic liquid) with aqueous electrolyte making the interface ideally polarisable?

Prof. Lynden-Bell answered: This is an interesting suggestion. One would have to choose an ionic liquid which is not completely miscible with water (for example an imidazolium salt with a longer sidechain and with PF_6^- as the counterion). There is still the possibility that the ions from the electrolyte would dissolve in the ionic liquid.

Prof. Fermín asked: The MD simulations of the interfaces between dimethlyimidazolium chloride and Lennard–Jones fluids show a well defined orientation of the cation at the onset of the ionic-liquid density profile. Are these results consistent with the description of these interfaces as molecularly sharp and roughened by capillary wave fluctuations?

Prof. Lynden-Bell replied: The use of periodic boundaries in our simulations suppresses capillary waves with wavelengths greater than the box length (around 40 Å in this case). There is certainly some surface roughness on the molecular scale which we have not attempted to quantify. However the continuum capillary wave description is inappropriate at wavelengths which are comparable to the molecular size. The interface density does change over a distance comparable to the molecular size, but other properties, notably the fluctuations in charge density (see Fig. 4 of our paper), propagate more deeply into the liquid.

Prof. Bopp asked: The question concerns the orientation and density profiles for the cations in the ionic liquid in contact with (a) vacuum; (b) a Lennard–Jones fluid; and (c) a supercritical Lennard–Jones fluid.

It was stressed in the lecture that the orientational profiles are quite similar in all three cases. In contrast, it would seem to me, especially from Fig. 3 in your paper, that the ionic liquid shows a stronger layering in case (b). Could this be due to the increased density of the Lennard–Jones fluid?

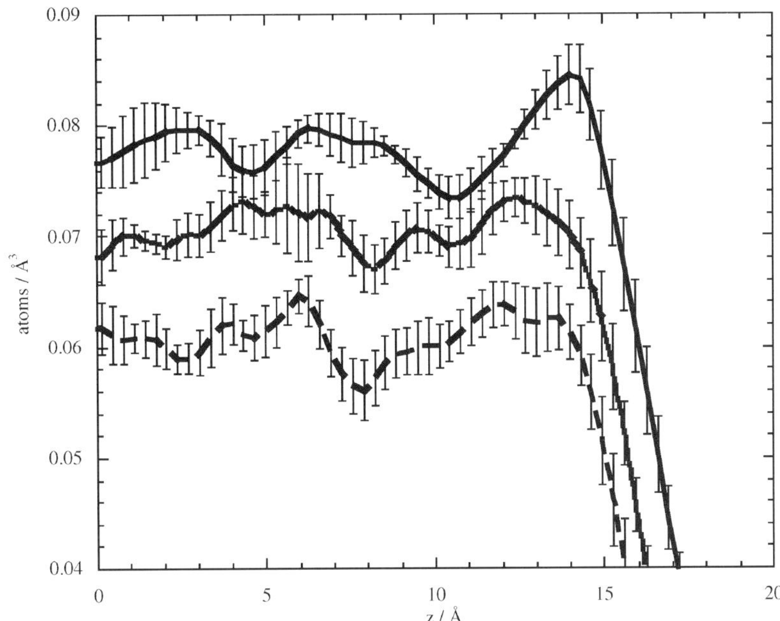

Fig. 8 Comparison of the total atomic densities near the ionic liquid interfaces. From top to bottom: Interface with vacuum—offset by 0.02; Interface with Lennard–Jones liquid—offset by 0.01; Interface with supercritical Lennard–Jones fluid. Error bars have been plotted at 2σ.

Prof. Lynden-Bell responded: The density layering near the ionic liquid interfaces is shown in Fig. 8. It differs from the figures in our paper in giving the average *total* density in atoms $\mathring{A}^{-3}$. The density fluctuations below the surface of the ionic liquid seem to be weakest in the interface with the normal Lennard–Jones liquid, but the error bars are large. The amplitude of the maximum immediately below the surface is certainly largest for the vacuum interface. It should be borne in mind that the pressure perpendicular to the interface differs in the three systems (see Table I of the paper).

Dr Leermakers opened the discussion of Dr Walker's paper: (1) The positively charged and negatively charged rules will not anchor in the same way in l/l interface because the effective hydrophobic nature of the two charged groups is not the same. The trimethylammonium is more hydrophobic than the anion. As a result one should not expect the same effect felt by the probe for the two cases. When the spaces in the ruler are very large the effect of difference in the anchoring may be lost as the probe can find the hydrophobic surrounding independent of the exact anchoring of the charged headgroups. The local electric field generated by the charged head group will also depend on the depth of anchoring at the interface also because the local dielectric constant depends on the position in the interface (with corresponding effects on the electrostatic potential). Why are you surprised that the two oppositely charged rulers are so different?

Additionally, the counterion also contributes to the effective anchoring of the ruler to the interface. A more polar counterion cannot come as close to the ruler as a less polar one. Again for both rulers the counterions differ. So there should also be a counterion effect. Is there evidence for this?

Dr Walker responded: The different affinities that the surfactant headgroups have for the organic phase are likely to impact what we call surfactant "float depth". Quarternized ammonium headgroups are clearly more hydrophobic than anionic sulfate headgroups and are likely to have a spatial distribution centered closer to the water/alkane interface. Consequently, the hydrophobic probes attached to the end of cationic surfactants will be able to "float" further into the organic phase in order to minimize its solvation energy. (The aromatic probes closely resemble *p*-nitroanisole, a small aromatic species that is $\sim 20\times$ more soluble in alkanes than in water.)[1] Assuming a sharp boundary at a water/alkane interface, one would expect the influence of "float-depth" to be most pronounced for short surfactants and less important for longer surfactants as Dr Leermakers correctly surmises.

The local polarity felt by the aromatic probe will depend upon its immediate solvation environment and, consequently, one might expect results from the cationic surfactants to report a sharper or more abrupt interface because the cationic headgroup will be localized close to the organic phase. However, the local polarity will also depend sensitively on the presence of nearby charged species. The fact that oppositely charged ruler surfactants (having the same alkyl spacer length) adsorbed to a liquid/liquid interface behave differently is not, by itself, a surprising observation. The way in which oppositely charged, adsorbed surfactants behave *is* surprising. The fact that the probes of short cationic surfactants (C_2 and C_4) consistently sample a local polarity that is *less than* the bulk cyclohexane limit implies that the close proximity of the cationic headgroup plays an important role in probe solvation. Furthermore, convergence of anionic and cationic surfactant behavior for longer rulers (C_6) provides for the first time quantitative information about how charges affect solvation across liquid/liquid interfaces.

In answer to the question: The counterion also contribute to the effective anchoring of the ruler to the interface. A more polar counterion cannot come as close to the ruler as a less polar one. Again for both rulers the counterions differ. So there should also be a counterion effect. Is there evidence for this?

One might expect the identity of a counterion to influence the ability of a charged headgroup (in the aqueous phase) to interact with the chromophore (in the organic phase). Currently, we have no data to test this hypothesis, but studies of interfacial solvation as a function of ionic strength and ion identity are planned for the near future.

1 W. H. Steel, Y. Y. Lau, C. L. Beildeck and R. A. Walker, *J. Phys. Chem. B*, 2004, **108**, 13370–13378.

Prof. Rathman asked: The paper and also one other questioner stated that the cationic ammonium headgroup is "more hydrophobic" than the anionic sulfate headgroup. For soluble surfactants, however, the cationic species behaves as if it is less hydrophobic. For example, comparing $C_{12}SO_4^-Na^+$ and $C_{12}N^+(CH_3)Cl^-$, the cationic has higher c.m.c. and lower Krafft ptc temperature—both of these suggest the cationic is more water soluble than the corresponding anionic. Are these observations consistent with the results presented in this paper for these species adsorbed at an interface?

Dr Walker replied: While the bulk solution properties (c.m.c., Krafft point) of cationic surfactants imply that these surfactants are less hydrophobic than anionic surfactants of similar chain length, the cationic surfactants remain much more surface active than their anionic analogs, forming full monolayers at bulk concentrations that are considerably lower than the concentrations required for full monolayer coverage of anionic surfactants. In the case of the molecular ruler surfactants used in these studies, the full monolayer coverage for cationic surfactants required bulk cationic surfactant concentrations of $\sim 100\ \mu M$. (See ref. 31 of our paper.) Full monolayer coverage for anionic surfactants required bulk concentrations of ~ 10 mM. (See ref. 32 of our paper.) This large disparity in surface activity is the basis for the statement that the cationic surfactants are "more hydrophobic" than their anionic counterparts.

Dr Wakisaka commented: In the cationic surfactant system, it is reasonable that the polar headgroup interacts with the aromatic chromophore. As for the effect of solvent on the SHG peak (shown in Fig. 2 of your paper), I would like to suggest the effect of solvation for the counter anion. In cyclohexane, the cationic head group form a contact ion pair with the counter anion, therefore, the cationic headgroup–aromatic chromophore interaction is not so strong. On the other hand, at the water–cyclohexane interface, the counter anion should be hydrated preferentially, which makes the cationic headgroup free from the counter anion. Accordingly the cationic headgroup can interact with the chromophore relatively strongly (see below).

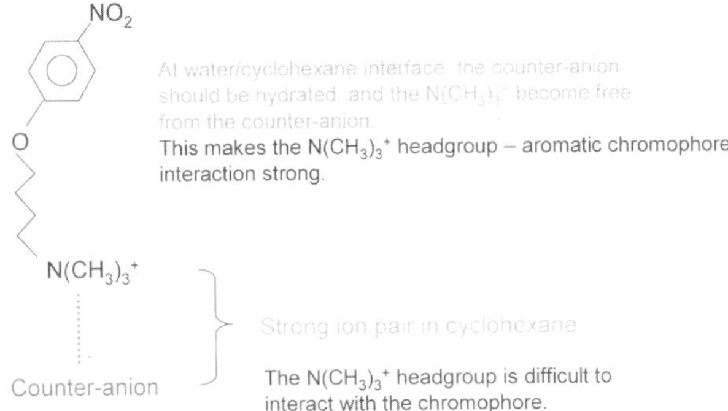

Dr Walker responded: The suggestion that the identity of the counterion (in the aqueous phase) may influence solvation around the probe (in the organic phase) is a good one, but one of the premises motivating this question needs to corrected. The short (C_2, C_4, and C_6) surfactants used in this work are *insoluble* in alkanes, regardless of their headgroup identity. Here, "insoluble" means not soluble enough to detect *via* standard UV-Vis spectrometry (with a sensitivity of ~ 1 mOD). Thus the idea of the headgroup forming a contact ion pair with the counterion *in cyclohexane* cannot be tested in our laboratory. In earlier work (ref. 32 of our paper), we demonstrated by NOE-NMR spectroscopy that the headgroup of the cationic surfactant interacted strongly with the aromatic probe *in aqueous solution* when the alkyl spacer was short (C_2, C_3). This interaction was no longer measurable for cationic surfactants having longer alkyl spacers ($\geq C_4$). No such probe/headgroup interaction was observed for anionic surfactants of any length. As noted in the response to Dr Leermakers, experiments examining the effects of counterion identity on solvation across water/alkane interfaces have not yet begun but are planned for the near future.

Prof. Girault asked: In your paper, you discuss the possibility of probe–probe interactions. Did you try spectroscopic techniques such as fluorescence measurements to investigate these interactions?

Dr Walker replied: Over the years, fluorescence spectroscopy has proven to be a very powerful technique for examining the properties of liquid surfaces and interfacial solvation.[1–3] Unfortunately, fluorescence is not surface specific. Given the relatively large concentration of surfactants dissolved in the aqueous phase, any fluorescence from surface species (individual or aggregated) is likely to be overwhelmed by the response arising from surfactants solvated in the bulk water. Even in a total internal reflection geometry, the evanescent wave of the excitation field would penetrate many solvent layers into the low-index aqueous phase exciting bulk-solvated chromophores. For this and other reasons, we have not attempted to acquire fluorescence spectra from molecular ruler surfactants adsorbed to liquid/liquid interfaces. The one experimental observation that suggests probe–probe interactions is an unexpected drop (of $\sim 20\%$) in the resonance enhanced second harmonic (SH) signal at surface coverages >0.50 full monolayer. Anionic ruler surfactants show this effect clearly. The data are less clear for cationic surfactants. Anionic surfactants have terminal surface coverages of $\sim 2 \times 10^{14}$ molecules cm^{-2}. Full monolayers of the cationic surfactants have surface coverages of $\sim 6 \times 10^{13}$ molecules cm^{-2}. This drop in signal intensity is not accompanied by a measurable change in probe orientation, meaning that molecular reorganization (*e.g.* a 2-dimensional phase transition) within the monolayer is not a likely source of the effect. A separate cause for the observed reduction in intensity could be interaction (aggregation? dimerization?) of the probes so that the effective excitation wavelength shifts outside the window of the experiment or the nonlinear susceptibility diminishes leading a lower overall SH response. Efforts to identify aggregation effects in bulk solution (at surfactant concentrations above the c.m.c.) have been unsuccessful thus far.

1   J. M. Kovaleski and M. J. Wirth, *J. Phys. Chem.*, 1996, **100**, 10304–10309.
2   S. Ishizaka, H. B. Kim, and N. Kitamura, *Anal. Chem.*, 2001, **73**, 2421–2430.
3   K. Nakatani, H. Nagatani, D. J. Fermín and H. H. Girault, *J. Electroanal. Chem.*, 2002, **518**, 1–5.

Prof. Schlossman commented: The interpretation of the SHG spectra relies upon the assumption that the chromophores exist in similar environments. This assumption is likely to be valid if the interface is homogeneous. Many interfaces are spatially inhomogeneous due to, *e.g.*, the presence of domains. Have you considered adapting your technique to probe inhomogeneous interfacial environments?

Dr Walker responded: Using NLO spectroscopy in combination with surfactants specifically designed to probe interfacial properties stands out as an attractive method for examining aggregation and domain formation at liquid/liquid interfaces. However, unambiguous data could come only from surfactants that incorporate a probe whose response changes noticeably between that from an adsorbed monomer and that of adsorbed aggregates. Designing and refining the synthetic procedures for new families of molecular ruler surfactants (involving different chromophore probes) sensitive to domain formation is not a trivial undertaking.

As detailed in the response to Prof. Girault, the second harmonic (SH) response from anionic molecular ruler surfactants at high surface coverage show behaviour that is inconsistent with predictions based on a simple analysis of SH intensities (*e.g.* $I(\omega_{SH}) \propto N^2$). Specifically, the intensity of the SH response actually diminishes slightly from a maximum at surface coverages of ~ 0.5 ML and the resonance response becomes much less pronounced. One possible explanation is that the probes are aggregating, and this aggregation either reduces the second order susceptibility (thereby reducing observed SH intensity) or shifts the excitation outside of the wavelength window spanned in these experiments. This explanation is only speculative, however, and efforts to determine the response of aggregated nitrophenol probes (using absorbance spectroscopy at high solution phase concentrations and in the solid state) have proven inconclusive.

Dr Wakisaka opened the discussion of Prof. Janata's paper: In the wet nitrobenzene (saturated by water), the water molecules will be existing as monomeric molecules. By the addition of LiBr to this wet nitrobenzene, the water molecules should be localised around the Li^+ and Br^-. Since there will be activation energy for this hydration, it takes a long time to achieve the equilibrium. After the sonication, the resulting hydrated ions are broken back to the dispersed monomeric water molecules (see below).

Such a change of microscopic structure will be monitored by NMR.

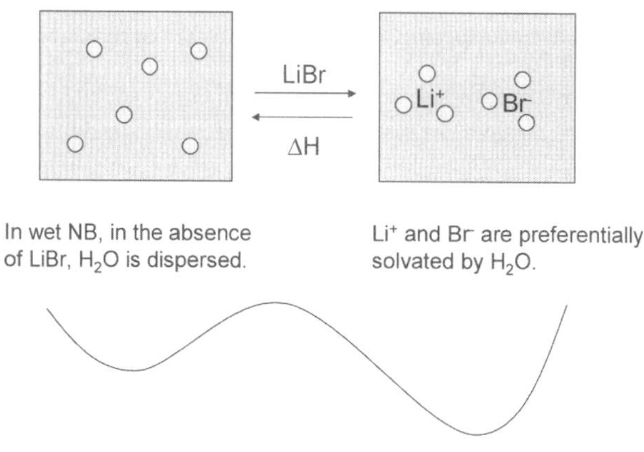

In wet NB, in the absence of LiBr, H_2O is dispersed.

Li^+ and Br^- are preferentially solvated by H_2O.

With sonication, H_2O goes back to the dispersed state.

Prof. Janata answered: I doubt that the water molecules exist as "monomers" in wet nitrobenzene. My guess is that they are aggregated in clusters, however, we have no evidence for it since we are following only the ^{7}Li-NMR signal and not the proton NMR spectrum. I agree with the idea that sonication temporarily breaks up the water clusters around Li^+ and the that reformation of hydrated Li^+ is a slow process, requiring tens of hours.

Prof. Beattie said: It is known that sonication can produce very small emulsion droplets, which then disappear by Ostwald ripening.[1] This could explain the slow change observed after sonication. Such an explanation requires the presence of similar droplets in the initial sample, however, created somehow in the sample preparation, perhaps through the 24 h stirring procedure.

1 K. Kamogawa, M. Matsumoto, T. Kobayashi, T. Sakai, H. Sakai and M. Abe, *Langmuir*, 1999, **15**(6), 1913–1917.

Prof. Janata responded: It is unlikely that water droplets exist in the NB phase as a result of stirring. The "ultrasonication effect" is reversible, albeit very slow (days). The Li-NMR spectrum goes back and forth on that timescale after repeated sonication, indicating that the water clusters (or droplets) are formed around Li^+ by reorganization of free water in NB.

Dr Dagastine asked: The basis of sonochemistry is the presence of nanobubbles in bulk solution from any number of sources ranging from dust in solution to dissolved gas. The main point is that they are always present and upon sonication these bubbles will grow and shrink in size at specific frequencies. In fact at the proper frequencies and energies these bubbles will collapse creating local regions of very high temperature and pressure.

The author questioned how simple sonication of the wet nitrobenzene sample resulted in a remixing of the hydrated lithium ions, since energy transfer from sonication requires an interface. If the nitrobenzene system has any amount of dissolved gas or other possible bubble nucleation sites, would this provide the necessary interface to induce mixing in the solution?

Prof. Janata replied: As I understand it, ultrasonic energy deposits at any interface between two materials of different density. Large water clusters around Li^+ could provide such an interface. We have not outgassed our samples and we have not observed any outgassing during the sonication step. However, I rather doubt that interfaces between gas and liquid (*i.e.* microbubbles) are involved. We have no proof of it one way or another.

Prof. Tominaga asked: The Li-NMR signal is broader in wet nitrobenzene than in water or in dry nitrobenzene. Can't you estimate the rotational correlation time in wet nitrobenzene?

Prof. Janata replied: The rotational correlation time can be estimated from the linewidth (*i.e.* $\Delta Hz_{1/2} \sim T_2^{-1}$) provided that the non-homogeneity effects can be neglected. However, the presence of solvatomers also complicates that kind of estimate. We are planning to do proper T_2 experiments, but have not done it yet.

Interaction forces between oil–water particle interfaces—Non-DLVO forces

Raymond R. Dagastine,[b] Tam T. Chau,[b] D.Y.C. Chan,[c] Geoffrey W. Stevens[b] and Franz Grieser*[a]

[a] School of Chemistry, Particulate Fluids Processing Center, University of Melbourne, Parkville, Victoria 3010 Australia. E-mail: franz@unimelb.edu.au; Fax: 61-3-93475180; Tel: 61-3-83446476
[b] Department of Chemical Engineering and Biomolecular Engineering, Particulate Fluids Processing Center, University of Melbourne, Parkville Victoria 3010 Australia
[c] Department of Mathematics and Statistics, Particulate Fluids Processing Center, University of Melbourne, Parkville, Victoria 3010 Australia

Received 20th April 2004, Accepted 28th May 2004
First published as an Advance Article on the web 5th October 2004

The interaction force between a rigid silica sphere and a butyl or octyl acetate droplet was measured in an aqueous environment using atomic force microscopy (AFM). The force measurements were performed without added stabilizers and the observed force behavior was found to be dependent on the type of inorganic electrolyte present, where the interfacial tension was constant over the electrolyte concentration range used. Force measurements in the presence of sodium nitrate showed repulsion at all concentrations. Force measurements in the presence of calcium nitrate or sodium perchlorate exhibited an initial repulsion followed by an attraction resulting in a mechanical instability in the AFM cantilever, termed jump-in. The force behavior observed was independent of the water solubility of the organic liquid, in that the same force–distance characteristics were obtained for slightly water soluble butyl acetate and the water sparingly soluble octyl acetate droplets. Modeling of the drop profile during particle–droplet interactions for this type of AFM measurement showed that the force–distance data for the sodium nitrate system obeys typical DLVO interactions. The disagreement between the DLVO predictions for the sodium perchlorate and calcium nitrate systems is attributed to a specific ion effect at the liquid–liquid interface, which gives rise to an attraction force that is greater than the electrostatic double layer repulsion over the length scale of 5 to 10 nm.

Introduction

The study of the interaction forces between liquid–liquid or liquid–vapor interfaces has been of interest for quite some time. Experimental approaches have employed a variety of methods. They have ranged from the classical measurements of Derjaguin *et al.*[1] to study the disjoining pressure of thin films, to more recent methods developed for studying rigid systems, and now extended to deformable systems. These later techniques include total internal reflection microscopy, a modified

Faraday Discuss., 2005, **129**, 111–124

surface forces apparatus,[2,3] alignment of droplets in magnetic fields[4,5] and atomic force microscopy (AFM).[6–20] All of these measurements are motivated by the study of how the structure of the molecules at liquid interfaces mediates the interaction forces between liquid–liquid or liquid–vapor interfaces. Understanding the interaction forces between liquid interfaces leads to insight into the properties of complex fluids commonly encountered in industrial applications. They include formulation, stability, and rheological properties of emulsions in areas such as food processing or the control of drop coalescence times in solvent extraction processes for applications as diverse as hydrometallurgy and pharmaceutical processing.

The present work uses AFM to study the interaction forces between silica spheres and the liquid–liquid interface of polar organic liquids in aqueous solution (see Fig. 1). This study is motivated by an earlier drop coalescence study between organic liquid drops in water . The previous work of Stevens et al.[21] investigated the effect of inorganic electrolytes on the coalescence times of organic liquid drops in water; the coalescence times were shown to increase for polar liquids, such as butyl acetate, whereas the coalescence times were independent of electrolyte concentration for non-polar liquids. The rate-limiting step in drop coalescence is commonly the film drainage between the interfaces where viscosity and surface tension have been shown to be crucial parameters for flow and interfacial deformation, respectively, as revealed by both theoretical and experimental studies (to name a few[22–25]). Interaction forces are significant to the thinned-film stability prior to rupture. Stevens et al.[21] concluded that a repulsive force helped to stabilize the polar organic liquids during coalescence. Furthermore, whereas electrophoresis measurements showed the droplets were charged, the origin of the force was not expected to be from an electrostatic double layer (EDL). An EDL force would be screened by the addition of electrolytes. This effect would result in a decrease in coalescence times, yet increases in the coalescence times were observed. Stevens et al.[21] suggested possible origins of the repulsive force as being due to increases in interfacial viscosity or an effect due to solvent structure or ordering at the liquid–liquid interface due to the polar nature of the organic liquids.

The extension of colloidal probe AFM (first developed by Ducker et al.[26] and Butt et al.[27] for rigid surfaces) to deformable interfaces has an added difficulty in the force analysis, due to interface deformation at close approach of the surfaces. The data analysis methods for rigid systems are not applicable to deformable surfaces.[10,11,14] In the analysis for rigid systems, measurement of force *versus* motion of the piezo distance actuator is converted to force *versus* intersurface separation. For this to be accomplished in deformable systems, one must separate the effects of interface deformation from changes in the interaction forces. A number of theoretical studies have examined this type of measurement and proposed several methods to analyze this type of data.[10–12,14,16,19,20,28] All the approaches involve modeling the deformation of the interface *via* the Young–Laplace equation where the drop shape is perturbed by the presence of the rigid probe. The drop size employed in the AFM experiments is typically below the capillary length of the system, thus surface tension and surface forces govern the interface shape where the effects of gravity are

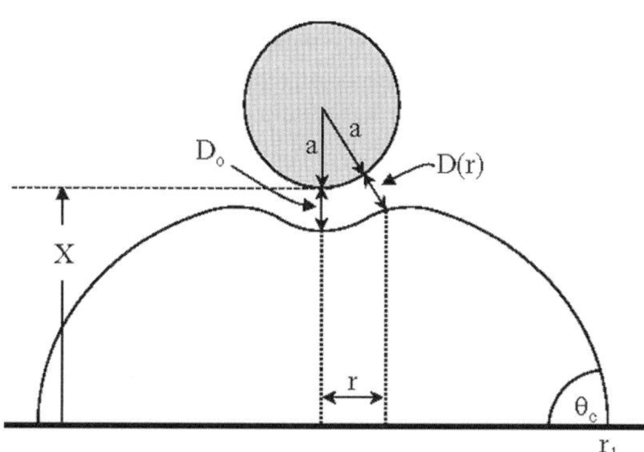

Fig. 1 A schematic (not to scale) of the AFM experiment between a rigid silica sphere and an oil droplet immobilized on a surface. Note: the deformation is exaggerated and only on the order of 100 nm to 200 nm.

negligible. The capillary length, λ, is defined as

$$\lambda = \left(\frac{\gamma}{\Delta\rho g}\right)^{1/2} \qquad (1)$$

where g is gravity, γ is interfacial tension of the fluid interface and $\Delta\rho$ is the difference in density between the fluid phases.[10]

Previous work has measured the forces between silica probes and non-polar liquids (such as decane[8,18] and hexadecane[11]) in the presence of aqueous inorganic electrolyte solutions. The force behavior showed an EDL repulsion followed by a short range van der Waals attraction leading to either engulfment or formation of a three phase contact line with the silica bead and droplet. Both the silica surface and the oil drop were charged and the overall force behavior could be explained using DLVO theory,[29,30] where the total potential energy is defined as the sum of the repulsive and attractive components from EDL and van der Waals forces. In the present study colloidal probe AFM is used to directly probe the interaction between a rigid silica sphere and a droplet (a spherical cap of oil in water immobilized on a surface, see Fig. 1) of a polar organic liquid immersed in aqueous salt solutions. The polar liquids, butyl and octyl acetate, have been examined in a series of inorganic electrolytes. The observed interaction forces follow the expected DLVO type force model for sodium nitrate, but a non-DLVO attraction is observed for sodium perchlorate and calcium nitrate regardless of the polar organic liquid used.

Methods

A Digital Instruments Multimode AFM and Nanoscope IIIa AFM controller were used for all experiments. Silica particles with a diameter of 5 μm were attached to Thermomicroscopes Microlevers (Digital Instruments) AFM cantilevers.[26] Cantilever spring constants were measured according to the method of Hutter and Bechhoefer[31] and ranged between 0.025 to 0.081 N m^{-1}. The inorganic salts used in these experiments were sodium nitrate (BDH), calcium nitrate (Ajax Chemicals), and sodium perchlorate (Ajax Chemicals). All glassware was bathed in surfactant cleaning solution for 1 h followed by 1 h in 10% nitric acid. Butyl acetate (Aldrich Chem. Co.) and octyl acetate (Aldrich Chem. Co.) were used as the oil droplet immobilized on a Melinex polymer film (semicrystalline PET film with no additives) according to the procedure outlined in Dagastine *et al.*[12] All measurements were performed with either butyl or octyl acetate saturated aqueous solutions. The colloidal probe was centered above the oil droplet (spherical caps with radii of the order of 100 μm to 300 μm) *via* transitional stages and then positioned above the droplet, first using the course stepping motor, and then using the piezo tube for fine control. Force curves (an approach and retract force–distance cycle) were taken at a series of approach speeds (100 nm s^{-1} to 2 μm s^{-1}), but no dependence of the force curve on the speed was observed. At the end of the experiments, the cantilever was pressed onto the underlying substrate to determine the detector sensitivity of the AFM photodiode. The radius of the spherical cap, r_1, was measured from above with a CCD camera to an accuracy of 5 μm. The undistorted radius of the drop, R_0, was calculated according to $\sin\theta_c = r_1/R_0$, where the contact angle, θ_c, and r_1 are defined in Fig. 1.

The AFM photodiode voltage was converted to cantilever deflection, Δd, using the detector sensitivity determined at the end of the experiment and then converted to force *via* $F = k_c\Delta d$, where k_c is the spring constant of the cantilever. A simple distance balance of the AFM measurement shows that the piezo motion, Δl, is equal to the changes in deflection, Δd, separation, ΔD, and deformation, Δz, of the system.[10,11,14] Using this distance balance, the deflection was subtracted from the piezo motion leading to

$$\Delta X = \Delta l - \Delta d \qquad (2)$$

where ΔX is the change in separation distance and deformation. Defined in Fig. 1, X is the distance from the bottom of the sphere to the top of the substrate. The definition of X is somewhat arbitrary because only changes in ΔX are measured and an absolute measure of the distance X is not known. It is standard practice to set an origin for X and display $F(\Delta X)$ relative to this origin.[10,14] In this respect, the curves have been shifted for convenience to clearly show specific features.

Interfacial tension was measured using the pendant drop method with a DataPhysics OCA 20 Tensiometer and axisymmetric drop shape analysis software. Sessile drop contact angle measurements were made using the same system. Solubility measurements of butyl acetate in different electrolyte solutions were measured using a Varian UV-Spectrophotometer. The measured molar absorptivity of butyl acetate was 53.4 M^{-1} cm^{-1} at a wavelength of 202 nm.

Results

Independent physico-chemical characterization of the drop properties is required for the quantitative analysis of the force curves. The interfacial tension for butyl acetate and octyl acetate were determined as 13.5 mN m^{-1} and 19.6 mN m^{-1} $\pm$ 0.5 mN m^{-1}, respectively, and were found to be independent of both the type of inorganic electrolyte and the electrolyte concentration over the range used in this study. The droplet contact angle on Melinex film was measured independently to be $140° \pm 2°$ using the sessile drop method. Butyl acetate is slightly soluble in water (measured to be 58.5 mM in MilliQ water at saturation), while octyl acetate is sparingly soluble. The solubility of butyl acetate decreases as expected to 7.18, 4.7 and 0.58 mM at 1 mM sodium nitrate, calcium nitrate, and sodium perchlorate, respectively.

The interaction between a silica sphere and a butyl acetate droplet in the presence of sodium nitrate solutions is shown in Fig. 2. The same cantilever and probe were used for each curve, but three different drops of similar radii ($\sim$ 300 µm) were employed. Butyl acetate is slightly soluble in water and whereas all experiments were done in saturated solutions, the solubility of a curved droplet is enhanced by the increased Laplace pressure. This results in droplets that are stable over the course of 30 to 60 min, but not stable over long enough times to perform measurements for a range of electrolyte concentrations. As discussed above, ΔX, is not the intersurface separation distance, thus the relative curve positions are not informative and have been spaced to show changes in force structure.

The range of the repulsion at low forces decreases with increasing ionic strength, as expected for an EDL force. EDL repulsive forces follow an approximate exponential form where the characteristic decay length, the Debye length, κ^{-1}, can be calculated based on solution ionic strength or regressed from a plot of the natural logarithm *versus* surface separation.[32] The natural logarithm for the force *versus* ΔX is displayed in the inset; where the limiting slope was compared to the expected

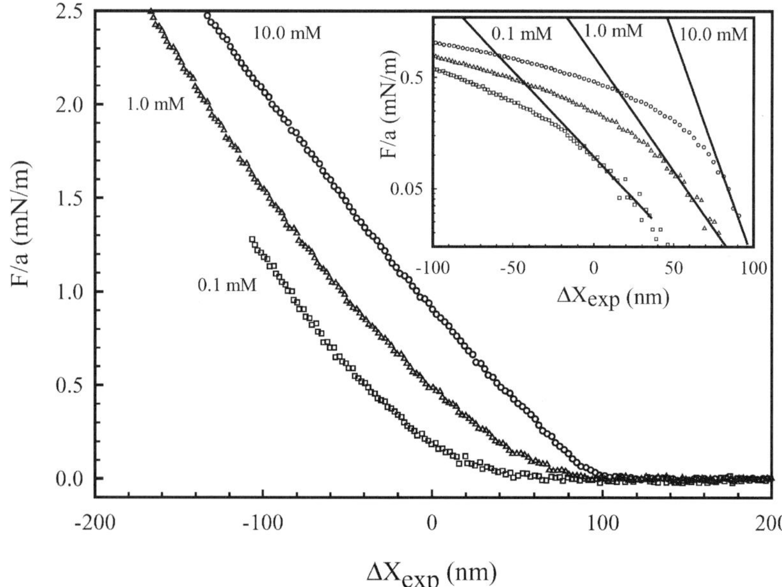

Fig. 2 The measured interaction force, F, *versus* relative piezo motion, ΔX_{exp}, between a silica sphere (with radius, a = 2.5 µm) and an immobilized butyl acetate droplet on a surface in the presence of sodium nitrate (NaNO$_3$) measured using AFM. The inset shows the logarithm of the force curves where the limiting slopes, marked by the solid line, are given in Table 1.

Table 1 The regressed slopes from the insets of Figs. 2–4 compared to the expected Debye length, κ^{-1}, based on solution ionic strength

Concentration/ mM	κ^{-1}/nm (NaNO$_3$)	κ^{-1}/nm (NaClO$_4$)	κ^{-1}/nm (Theory, 1:1)	κ^{-1}/nm (Ca(NO$_3$)$_2$)	κ^{-1}/nm (Theory, 2:1)
0.1	26.8	26.7	30.4	15.5	17.6
1.0	17.0	15.4	9.6	16.4	5.6
10.0	9.6	7.7	3.04	—	—

ionic strengths given in Table 1. The large disagreement in the regressed κ^{-1} values is due to interfacial deformation because the distance ΔX is a nonlinear function of intersurface separation.[10,12] Only repulsion was observed in the force measurements in sodium nitrate solution over a large range of the piezo drive distance. This behavior was *only* observed for sodium nitrate solutions.

The interaction force between butyl acetate droplets in the presence of sodium perchlorate and calcium nitrate are shown in Figs. 3 and 4, respectively. The range of the repulsion follows the same behavior as above for the decay lengths of force *versus* ΔX. The notable difference from the sodium nitrate case is the presence of an attractive force resulting in the termination of the force curve denoted with arrows in Figs. 3 and 4. The termination of the force curve is a consequence of a process referred to as "jump-in". Here, the gradient of the force curve exceeds the spring constant of the cantilever causing a mechanical instability, where the dynamics of the jump-in process are faster than the sampling rate of the AFM. Therefore, this point is the termination of the measurement. Attraction was observed intermittently for the sodium perchlorate results in Fig. 3 for the lowest concentration of 0.1 mM. This is consistent with previous results on non-polar liquids where jump-in is not always observed at low ionic strengths.[8,18] At all higher concentrations of sodium perchlorate and all cases for calcium nitrate the force curves all terminate in jump-in.

The results for octyl acetate for the same three inorganic electrolytes (sodium nitrate, sodium perchlorate and calcium nitrate) are given in Figs. 5–7. While octyl acetate is still polar it is sparingly

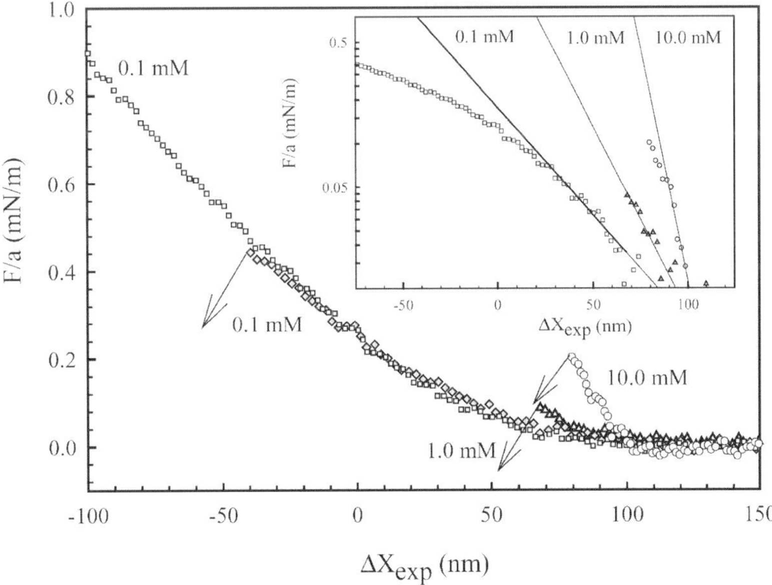

Fig. 3 The measured interaction force, F, *versus* relative piezo motion, ΔX_{exp}, between a silica sphere (with radius, $a = 2.5$ μm) and an immobilized butyl acetate droplet on a surface in the presence of sodium perchlorate (NaClO$_4$) measured using AFM. The inset shows the logarithm of the force curves where the limiting slopes, marked by the solid line, are given in Table 1.

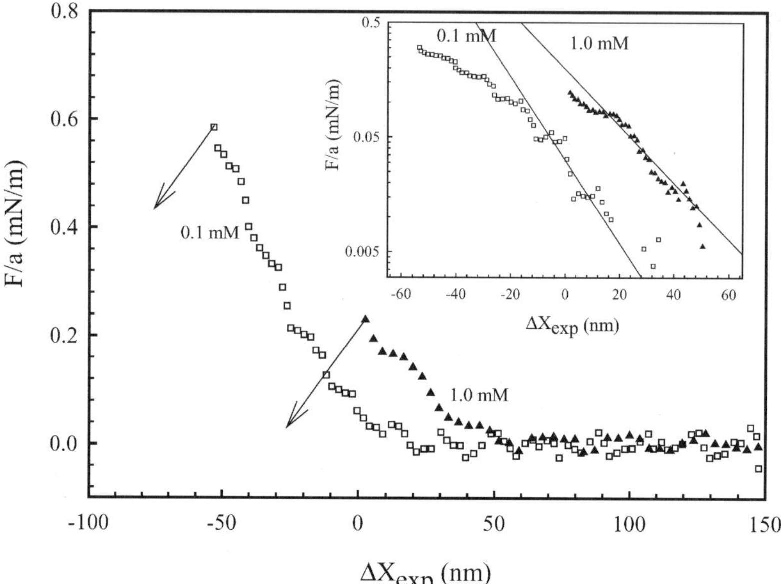

Fig. 4 The measured interaction force, F, *versus* relative piezo motion, ΔX_{exp}, between a silica sphere (with radius, $a = 2.5$ μm) and an immobilized butyl acetate droplet on a surface in the presence of calcium nitrate $(Ca(NO_3)_2)$ measured using AFM. The inset shows the logarithm of the force curves where the limiting slopes, marked by the solid line, are given in Table 1.

soluble in water, thus the immobilized drops are stable over the course of hours and all the data for each electrolyte was taken on the same drop with sizes of 200 μm to 300 μm. Only a repulsive

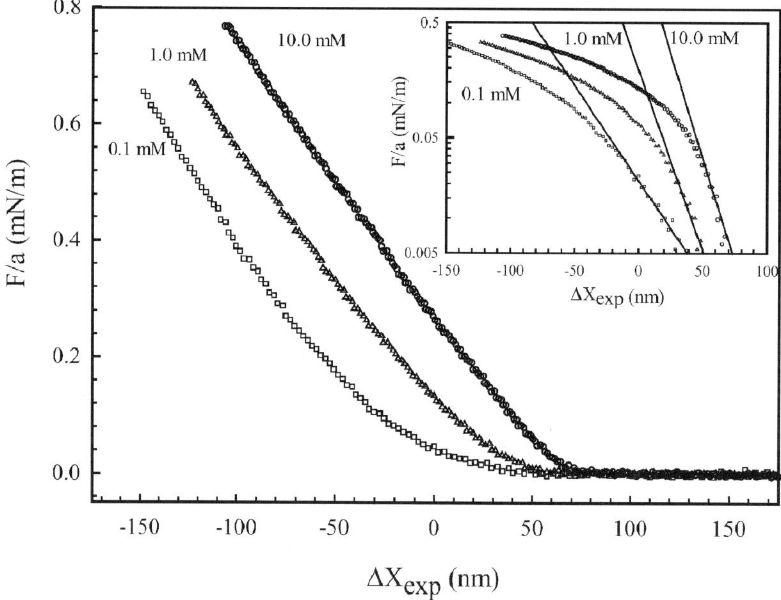

Fig. 5 The measured interaction force, F, *versus* relative piezo motion, ΔX_{exp}, between a silica sphere (with radius, $a = 2.5$ μm) and an immobilized octyl acetate droplet on a surface in the presence of sodium nitrate $(NaNO_3)$ measured using AFM. The inset shows the logarithm of the force curves where the limiting slopes, marked by the solid line, are given in Table 2.

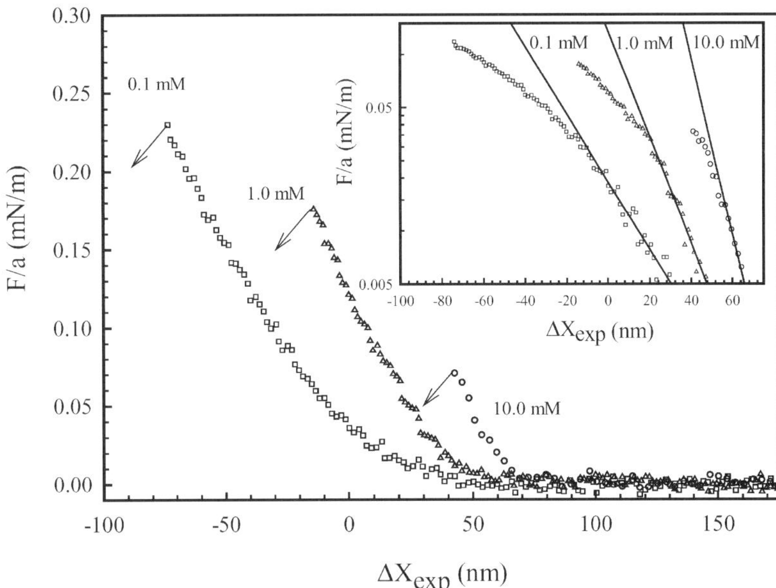

Fig. 6 The measured interaction force, F, *versus* relative piezo motion, ΔX_{exp}, between a silica sphere (with radius, $a = 2.5\ \mu m$) and an immobilized octyl acetate droplet on a surface in the presence of sodium perchlorate ($NaClO_4$) measured using AFM. The inset shows the logarithm of the force curves where the limiting slopes, marked by the solid line, are given in Table 2.

interaction was observed in the presence of sodium nitrate shown in Fig. 5 and the range of the repulsion at low forces again scales with ionic strength. Initial repulsion followed by jump-in was observed for all concentrations with the remaining two electrolytes in Figs. 6 and 7. Again, the range

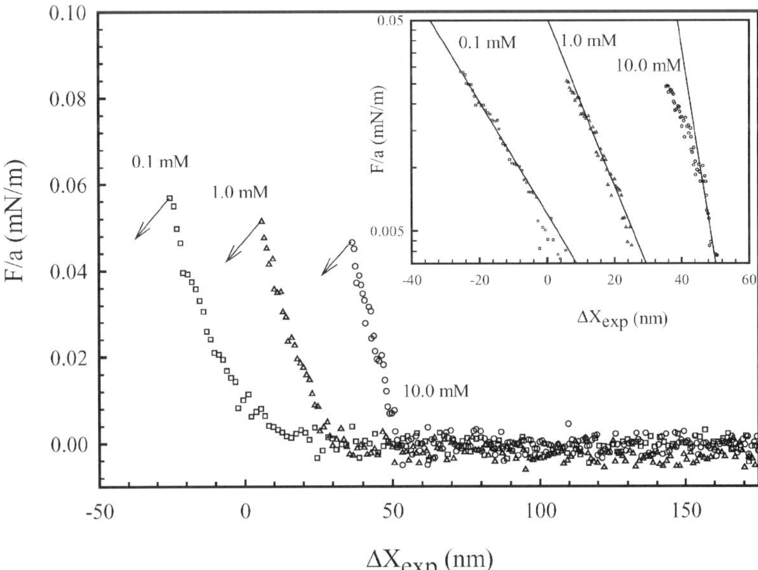

Fig. 7 The measured interaction force, F, *versus* relative piezo motion, ΔX_{exp}, between a silica sphere (with radius, $a = 2.5\ \mu m$) and an immobilized octyl acetate droplet on a surface in the presence of calcium nitrate ($Ca(NO_3)_2$) measured using AFM. The inset shows the logarithm of the force curves where the limiting slopes, marked by the solid line, are given in Table 2.

Table 2 The regressed slopes from the insets of Figs. 5–7 compared to the expected Debye length, κ^{-1}, based on solution ionic strength

Concentration/ mM	κ^{-1}/nm (NaNO$_3$)	κ^{-1}/nm (NaClO$_4$)	κ^{-1}/nm (Theory, 1:1)	κ^{-1}/nm (Ca(NO$_3$)$_2$)	κ^{-1}/nm (Theory, 2:1)
0.1	23.4	21.5	30.4	15.5	17.6
1.0	11.7	14.6	9.6	11.0	5.6
10.0	11.0	7.7	8.0	4.1	1.8

of the repulsion at low forces scaled with ionic strength and the regressed limiting slopes of the logarithm of the force *versus* ΔX, do not correlate with the expected Debye length as shown in Table 2.

Discussion

The above force data show a clear dependence on the type of electrolyte present in solution where all other variables appear to be constant. The interfacial tension and contact angle are independent of the type and concentration of the electrolyte. The undistorted drop radii are similar and theoretical analysis has shown that the force is not very sensitive to the variation of the drop radii used in these experiments.[12] The oil solubility does not appear to result in any difference in the force behavior.

The only variable unaccounted for is the surface force between the interfaces, which appears to be electrolyte dependent. To propose that one of the above force–separation behaviors (attraction or continued repulsion) is due to a non-DLVO force requires a convincing description of what is the expected DLVO behavior. In order to address this issue a semi-analytical model developed for analysis of this type of AFM measurement is used to predict the force curve, *i.e.* $F(\Delta X)$, where *a priori* knowledge of the disjoining pressure is required for this calculation.[10,17,19] This is a semi-quantitative approach to demonstrate the expected DLVO force behavior for the system in question. This treatment is separated into three parts. First a description of the physical situation required for either jump-in or repulsion only are explained in terms of the drop deformation and DLVO interaction forces. Second, using literature estimates for the surface potential for the silica and butyl acetate surfaces the force curves in this study are predicted based on a DLVO force model. Finally, the model results are compared to the experimental measurements to determine whether or not the force behavior follows a DLVO force model.

Jump-in or wrapping

An in-depth theoretical analysis of the occurrence of jump-in or alternatively, "wrapping" can be found in Dagastine and White,[19] only the relevant points for discussion are summarized here. As mentioned above, jump-in occurs from an attractive force, which leads to a mechanical instability in the cantilever. If jump-in does not occur irrespective of the amount of deformation in the droplet, a different physical situation occurs referred to as wrapping. When wrapping occurs, a film of effectively constant thickness, is maintained between the two interacting interfaces. This can be shown analytically by describing the gap between the interfaces, where the disjoining pressure is significant, according to:[10]

$$\left.\begin{array}{c} D'' + \dfrac{1}{t}D' - \left(2\left(1+\dfrac{a}{R_0}\right) - \dfrac{a\Pi(D)}{\gamma} \right)D_0 = 0 \\[2mm] D(0) = D_0 \quad D'(0) = 0 \end{array}\right\} \tag{3}$$

where a is the particle radius, $\Pi(D)$ is the disjoining pressure, and $D(t)$ is intersurface separation, and t is the radial distance in dimensionless form ($t = r/(aD_0)^{1/2}$). When predicting the force curve, the above equation is solved numerically, but it reduces to a constant ($D(t) = D_w$) if the disjoining pressure satisfies the condition

$$\Pi(D_w) = 2\gamma \left(\frac{1}{a} + \frac{1}{R_0} \right) \tag{4}$$

This condition is easily satisfied if the disjoining pressure has a large short-range force and the interface is sufficiently pliable. If this is true, instead of the two interfaces becoming closer, the observed increase in force is primarily from the deformation of the drop, and an increase in the interaction area.

The next step is to understand how changing parameters in a disjoining pressure using a DLVO force model results in wrapping or jump-in. The jump-in is from the relatively short range attractive van der Waals force component of the force curve, where the jump-in distance, D_j, is on the order of a few nanometers. Therefore, to suppress jump-in and cause wrapping of the deformable liquid–liquid interface around the colloidal probe only requires a film a few nanometers thick between the interfaces. The van der Waals force between an oil drop and a silica particle in an electrolyte solution is effectively constant for a specific oil,[33] thus the variable portion of the surface force is the EDL force which can alter through changing the surface potential and Debye length. The surface potential represents the overall magnitude of the repulsion while the Debye length represents the range of the force. If the Debye length is held constant, a decrease in surface potential results in a decrease in the force at which jump-in occurs. For the same drop parameters with a constant surface potential of at least 30 mV, a decrease in the Debye length results in an increase in the jump-in force and eventually leads to wrapping.

A comparison of these effects independently is not always applicable since in many systems the surface potential, Ψ_0, is coupled to the Debye length, κ^{-1}, where a decrease in κ^{-1} corresponds to a decrease in Ψ_0 (as is the observed case for silica[34] and butyl acetate[21]). This leads to competing effects on the force behavior, where holding Ψ_0 constant and decreasing κ^{-1} leads to wrapping, whereas holding κ^{-1} constant and decreasing Ψ_0 leads to jump-in.[19] While the above two competing effects can determine the jump-in or wrapping behavior it is important to note from eqn. (4) that the interfacial tension can overshadow the EDL effects. In the presence of a short range attraction and a large interfacial tension, jump-in is expected because the interface does not easily deform (this occurs when the EDL disjoining pressure is too small to satisfy eqn. (4)). Yet, for the same disjoining pressure, wrapping may be observed if the interfacial tension is sufficiently low. When all three effects are comparable, the expected DLVO behavior may be system dependent.

Predicting the force curve

The predictive model employed here uses the semi-analytical theoretical analysis of Chan *et al.*[10] and Dagastine and White[19] to calculate the force curve, $F(\Delta X)$, parametric in the minimum surface separation, D_0 (*i.e.* $F(D_0)$ and $\Delta X(D_0)$ are calculated, then $F(\Delta X)$ can be plotted. The modeling approach requires the wetting properties of the drop (interfacial tension, γ, and contact angle, θ_c) and the particle and undistorted drop radii. To predict the force curve requires the construction of a force model in the form of a disjoining pressure (or the pressure between two flat interfaces) based on some type of force model with adjustable parameters. This approach is limited if the observed force does not have a well characterized force model, *e.g.* a specific ion effect.

The construction of a DLVO based disjoining pressure for the butyl acetate–silica interaction was accomplished by using literature values for the surface potential for each interface, given in Table 3. Electrophoresis measurements for butyl acetate drops as a function of sodium chloride concentration are available from Stevens *et al.*[21] The analysis of electrophoretic mobility measurements of liquid droplets without any added stabilizer or surfactant is complicated by the possibility of internal flow in the drop. This effect results in a larger electrophoretic mobility for a droplet compared to a rigid particle of the same size and zeta potential.[35] Whereas the presence of stabilizers has been shown to arrest this phenomenon,[36] Stevens *et al.*[21] did not address this possibility in their

Table 3 The infinite separation surface potentials, Ψ_0^∞, for silica[34] and butyl acetate[21] as a function of electrolyte concentration used for the EDL calculations

Concentration/mM	Ψ_0^∞/mV, butyl acetate	Ψ_0^∞/mV, silica
0.1	30.0	65
1.0	29.6	58
10.0	26.5	43

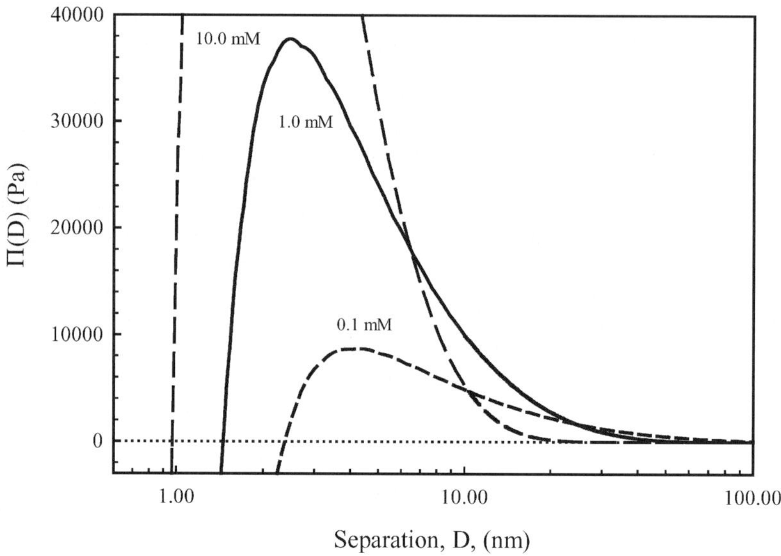

Fig. 8 The DLVO disjoining pressure between flat interfaces constructed from the sum of the retarded van der Waals interaction calculated using Lifshitz theory between an alkane and silica across a water with electrolyte and an EDL force from a numerical solution to the Poisson–Boltzmann equation with a constant surface charge boundary condition. The short dashed, solid, and long dashed lines correspond to 0.1 mM, 1.0 mM and 10.0 mM ionic strengths.

study, thus their mobility measurements may have overestimated the zeta potential on the droplets. Surface potential data for the silica probe were used from AFM colloidal probe measurements[34] in inorganic electrolyte solutions. It is possible that the saturated solutions of butyl or octyl acetate

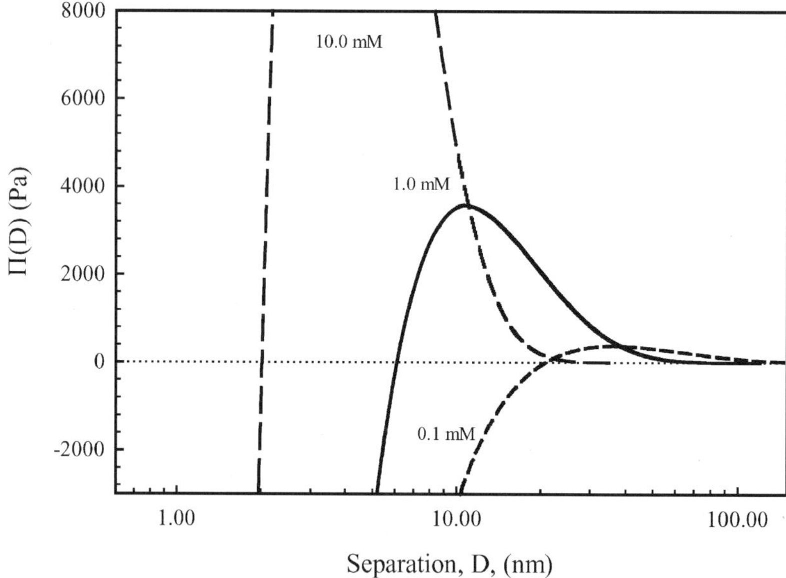

Fig. 9 The DLVO disjoining pressure between flat interfaces constructed from the sum of the retarded van der Waals interaction calculated using Lifshitz theory between an alkane and silica across a water with electrolyte and an EDL force from a numerical solution to the Poisson–Boltzmann equation with a constant surface potential boundary condition. The short dashed, solid, and long dashed lines correspond to 0.1, 1.0 and 10.0 mM ionic strengths.

used in this study may have lowered the silica zeta potential, but this should be a small effect. The EDL force was calculated by solving the Poisson–Boltzmann equation numerically using the algorithm of Chan *et al.*[37] for both constant charge and constant potential boundary conditions. These two cases bracket most types of EDL behavior. The van der Waals force was predicted by calculating the retarded Hamaker constant for silica interacting with an alkane across water in an electrolyte solution, using Lifshitz theory.[33] An alkane was used for the butyl acetate in the van der Waals calculation because the complete dielectric spectrum of butyl acetate was not available. The dielectric spectra used in the van der Waals calculation for water, silica and the alkane were from Dagastine *et al.*,[38] Hough and White,[39] and Parsegian and Weiss.[40] The resulting disjoining pressures are given in Figs. 8 and 9 for the constant surface charge (CP) and constant surface potential (CP) boundary conditions, respectively.

Predicted force–separation curves for three ionic strengths are given in Figs. 10 and 11 for the CC and CP boundary conditions. The ΔX_{th} ($= X - X_\infty$) in Figs. 10 and 11 is in terms of an absolute distance where X_∞ is the height of the undistorted drop interface at the center. This differs form ΔX for the experimental curves by an unknown constant. For the CC case, jump-in is only seen at 0.1 mM ionic strength and wrapping occurs at the other ionic strengths. In the constant potential case, jump-in is observed for the 0.1 mM and 1 mM ionic strengths, but wrapping occurs for the 10 mM ionic strength. The constant potential boundary conditions result in larger attractive disjoining pressures from the dissimilar surface potentials, where charge reversal occurs, thereby increasing the total attractive force. The wrapping thickness and jump-in distances are not known for the experimental data, but the model calculations allow estimations of their values because the predicted force curves were calculated parametric in separation distance, D_0, (*i.e.* $F(D_0)$ and $\Delta X(D_0)$). The calculated jump-in distances range from 2 nm to 4 nm and the wrapping distances range from 5 nm to 8 nm for the model force curves.

The wrapping behavior for the 1 mM and 10 mM ionic strengths in Fig. 10 is a product of the change in Debye length coupled with the low interfacial tension. The surface potential on the butyl acetate drop does not vary significantly with ionic strength (See Table 3). Therefore, this is similar to the discussion in the previous section where the surface potential was held constant and wrapping occurred as the Debye length decreased. It was also noted earlier that the surface potentials for the

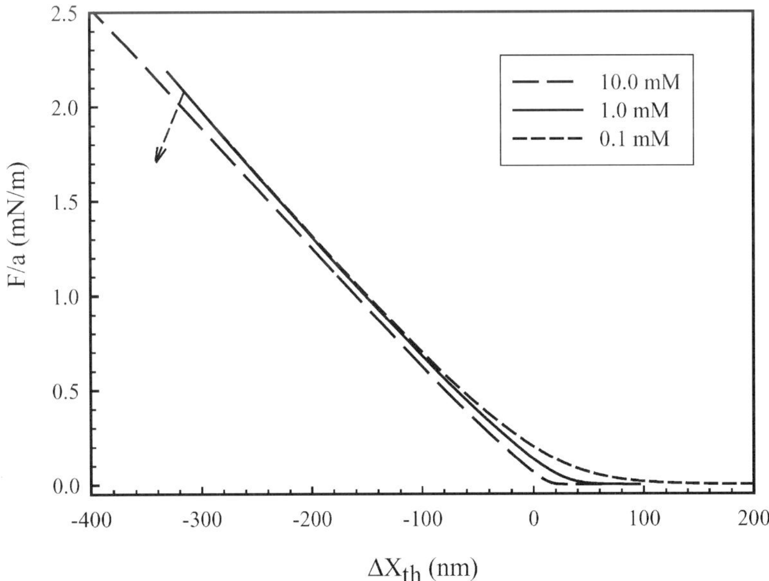

Fig. 10 The predicted force curves between a butyl acetate droplet and a silica sphere as a function of ionic strength were calculated using the semi-analytical model with independently measured drop parameters ($\gamma = 13.5$ mN m^{-1}, $\theta_c = 140°$) and the DLVO disjoining pressure given in Fig. 8 for constant surface charge boundary conditions for the EDL force. The arrow denotes jump-in for the 0.1 mM ionic strength.

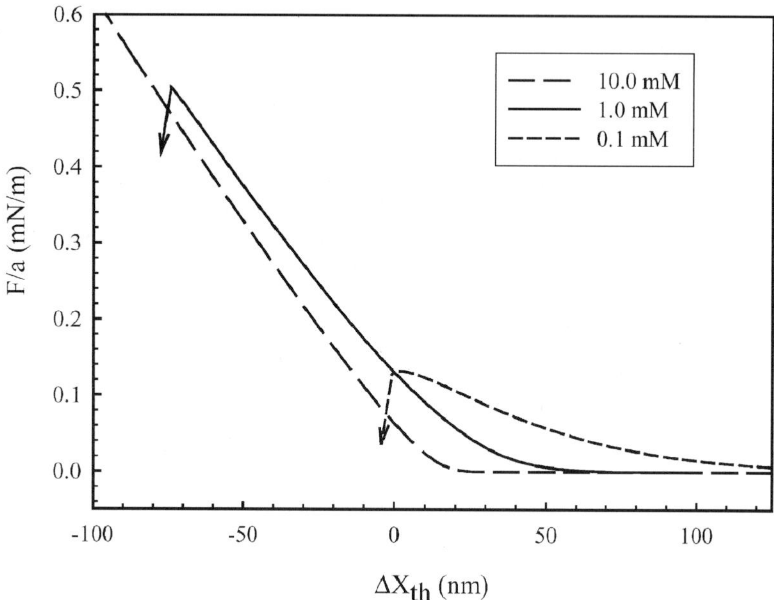

Fig. 11 The predicted force curves between a butyl acetate droplet and a silica sphere as a function of ionic strength were calculated using the semi-analytical model with independently measured drop parameters ($\gamma = 13.5$ mN m^{-1}, $\theta_c = 140°$) and the DLVO disjoining pressure given in Fig. 9 for constant surface potential boundary conditions for the EDL force. The arrows denote jump-in for the 0.1 mM and 1.0 mM ionic strengths.

butyl acetate may be an over estimate, but decreasing the surface potential by approximately 10 mV did not vary the outcome for any of the force curves where wrapping was observed. A significant increase in interfacial tension (on the order of 20 mN m^{-1}) is required to see jump-in behavior.

These predictions are in contrast to the DLVO behavior observed for force measurements with alkane droplets, where jump-in was observed experimentally and the force at jump-in decreased with ionic strength.[8,11,18] A van der Waals attraction is the expected source of the jump-in and this attraction alone leads to jump-in distances of only several nanometers. The force curve termination behavior in the alkane systems was governed by the change in surface potential as discussed in the previous section. The interfacial tension of alkanes is on the order of 50 mN m^{-1}, which is four times the interfacial tension for butyl acetate in water. Thus, wrapping would be observed if the interfacial tension had been significantly lowered. Wrapping behavior has been observed with alkanes by several investigators upon the addition of anionic surfactant to the alkane–water interface.[8,11,12,17,18] The present work is the first to study the wrapping and jump-in behaviors with organic liquids without stabilizers at considerably lower interfacial tensions than previously investigated.

Model-experiment comparison

The force measurements in sodium nitrate solution compare well to the predicted force curves with a constant charge boundary condition. The experimental force curve at 0.1 mM electrolyte does not show jump-in, but the predicted jump-in force for the CC boundary condition is larger than the experimental range measured. Wrapping was observed for 1 mM and 10 mM concentrations at all measured forces. This semi-quantitative agreement does not guarantee that the CC boundary condition is met and charge regulation may be possible (resulting in a surface boundary condition intermediate between the CC and CP cases), but the presence of wrapping implies that the CP boundary condition behavior is not observed. Therefore, the comparison of the sodium nitrate data to the model for butyl acetate does indicate that the force follows a DLVO type behavior. This force behavior for octyl acetate is also consistent with the DLVO force model, where the interfacial tension is only 5 mN m^{-1} greater and the polar organic liquid is insoluble in water.

Whereas the sodium nitrate solution force data follows the expected DLVO behavior, simply changing the electrolyte leads to jump-in at all concentrations for both the calcium nitrate and sodium perchlorate systems with both butyl and octyl acetate. As discussed above, all the other experimental parameters were constant other than the type of electrolyte used. Although the predicted force curves for the CP boundary condition predict jump-in for the lower two ionic strengths, the force at jump-in increases in the model calculations, whereas the force at jump-in decreases in the experimental data. The observed force behavior for these two electrolytes is similar to the behavior observed in alkanes, but to explain this behavior in terms of the DLVO force model would require an increase, by a factor of 3 to 4, in the interfacial tension. This was not observed in pendant drop measurements that were made on these solutions. Therefore, with the available electrokinetic data, it is clear that the behavior of the sodium perchlorate and calcium nitrate force measurements do not follow the expected DLVO behavior for either the butyl or octyl acetate force measurements.

The presence of the strong attraction forces with the calcium nitrate and sodium perchlorate systems cannot be explained by just DLVO forces alone. The multivalent calcium ion will screen the EDL double layer forces to a larger extent at the same concentrations as the other electrolytes, but the ionic strengths are still of the same order of magnitude. The sodium perchlorate data would be expected to match the sodium nitrate data without significant variation. The calcium nitrate and sodium perchlorate behavior gives rise to an unexplained attraction that must be large enough to overcome the EDL force. Since the predicted wrapping distances are on the order of 5 nm to 10 nm, this force is not necessarily long range. It is difficult to speculate on the origins of this force behavior, other than it is a specific ion effect that results in a non-DLVO attraction. The origin of this effect may be from molecular ordering at either interface or charge regulation driven by a specific ion effect on either interface, but this cannot be determined for the force measurements. It should be noted that non-DLVO forces as found in this study have not been observed in silica–silica interactions; therefore the source of the attraction probably resides with the liquid–liquid interface.

Conclusions

The attractive forces observed in aqueous solutions between a silica sphere and butyl or octyl acetate droplets in the presence of sodium perchlorate and calcium nitrate cannot be explained with the current DLVO theory for colloidal forces. Force measurements for the same systems with sodium nitrate were consistent with DLVO predictions. This indicates that the origin of this non-DLVO force on the length scale of 5 nm to 10 nm is manifestation of a specific ion effect. The origins of this effect may be a specific ion charge regulation mechanism or some type of molecular structure at the liquid–liquid interface. However, this speculation cannot be elucidated from the force data and is an area for further study.

Acknowledgements

This work was supported by the Australian Research Council and by the National Science Foundation under Grant No. INT-0202675.

References

1   B. V. Derjaguin, N. V. Churaev and V. M. Muller, *Surface Forces*, Consultant Bureau, New York, 1987.
2   R. Aveyard, B. P. Binks, W. G. Cho, L. R. Fisher, P. D. I. Fletcher and F. Klinkhammer, *Langmuir*, 1996, **12**, 6561.
3   B. P. Binks, W. G. Cho and P. D. I. Fletcher, *Langmuir*, 1997, **13**, 7180.
4   P. Omarjee, A. Espert, O. Mondain-Monval and J. Klein, *Langmuir*, 2001, **17**, 5693.
5   O. Mondain-Monval, A. Espert, P. Omarjee, J. Bibette, F. Leal-Calderon, J. Philip and J. F. Joanny, *Phys. Rev. Lett.*, 1998, **80**, 1778.
6   W. A. Ducker, Z. G. Xu and J. N. Israelachvili, *Langmuir*, 1994, **10**, 3279.
7   M. L. Fielden, R. A. Hayes and J. Ralston, *Langmuir*, 1996, **12**, 3721.
8   P. Mulvaney, J. M. Perera, S. Biggs, F. Grieser and G. W. Stevens, *J. Colloid Interface Sci.*, 1996, **183**, 614.
9   M. Preuss and H.-J. Butt, *J. Colloid Interface Sci.*, 1998, **208**, 468.

10 D. Y. C. Chan, R. R. Dagastine and L. R. White, *J. Colloid Interface Sci.*, 2001, **236**, 141.
11 D. E. Aston and J. C. Berg, *J. Colloid Interface Sci.*, 2001, **235**, 162.
12 R. R. Dagastine, D. C. Prieve and L. R. White, *J. Colloid Interface Sci.*, 2004, **269**, 84.
13 R. R. Dagastine, G. W. Stevens, D. Y. C. Chan and F. Grieser, *J. Colloid Interface Sci.*, 2004, **273**, 339.
14 D. Bhatt, J. Newman and C. J. Radke, *Langmuir*, 2001, **17**, 116.
15 H.-J. Butt, *J. Colloid Interface Sci.*, 1994, **166**, 109.
16 G. Gillies, C. A. Prestidge and P. Attard, *Langmuir*, 2001, **17**, 7955.
17 S. A. Nespolo, D. Y. C. Chan, F. Grieser, P. G. Hartley and G. W. Stevens, *Langmuir*, 2003, **19**, 2124.
18 P. G. Hartley, F. Grieser, P. Mulvaney and G. W. Stevens, *Langmuir*, 1999, **15**, 7282.
19 R. R. Dagastine and L. R. White, *J. Colloid Interface Sci.*, 2002, **247**, 310.
20 P. Attard and S. J. Miklavcic, *J. Colloid Interface Sci.*, 2002, **247**, 255.
21 G. W. Stevens, H. R. C. Pratt and D. R. Tai, *J. Colloid Interface Sci.*, 1990, **136**, 470.
22 S. Abid and A. K. Chesters, *Int. J. Multiphase Flow*, 1994, **20**, 613.
23 E. Klaseboer, J. P. Chevaillier, C. Gourdon and O. Masbernat, *J. Colloid Interface Sci.*, 2000, **229**, 274.
24 J. N. Connor and R. G. Horn, *Langmuir*, 2001, **17**, 7194.
25 D. G. Goodall, M. L. Gee, G. Stevens, J. Perera and D. Beaglehole, *Colloids Surf., A*, 1998, **143**, 41.
26 W. A. Ducker, T. J. Senden and R. M. Pashley, *Nature*, 1991, **353**, 239.
27 H. J. Butt, *Biophys. J.*, 1991, **60**, 1438.
28 P. Attard and S. J. Miklavcic, *Langmuir*, 2001, **17**, 8217.
29 E. J. W. Verwey and J. T. G. Overbeek, *Theory of Stability of Lyophobic Colloids*, Elsevier, Amsterdam, 1948.
30 B. V. Derjaguin and L. Landua, *Acta Physicochim. USSR*, 1941, **14**, 633.
31 J. L. Hutter and J. Bechhoefer, *Rev. Sci. Instrum.*, 1993, **64**, 1868.
32 R. J. Hunter, *Foundations of Colloid Science*, Clarendon Press, Oxford, 1995.
33 J. Mahanty and B. W. Ninham, *Dispersion Forces*, Academic Press, London, 1976.
34 W. A. Ducker, T. J. Senden and R. M. Pashley, *Langmuir*, 1992, **8**, 1831.
35 H. Ohshima, T. W. Healy and L. R. White, *J. Chem. Soc., Faraday Trans. 2*, 1984, **80**, 1643.
36 S. A. Nespolo, M. A. Bevan, D. Y. C. Chan, F. Grieser and G. W. Stevens, *Langmuir*, 2001, **17**, 7210.
37 D. Y. C. Chan, R. M. Pashley and L. R. White, *J. Colloid Interface Sci.*, 1980, **77**, 283.
38 R. R. Dagastine, D. C. Prieve and L. R. White, *J. Colloid Interface Sci.*, 2000, **231**, 351.
39 D. B. Hough and L. R. White, *Adv. Colloid Interface Sci.*, 1980, **14**, 3.
40 V. A. Parsegian and G. H. Weiss, *J. Colloid Interface Sci.*, 1981, **81**, 285.

Surface rheology as a tool for the investigation of processes internal to surfactant adsorption layers

Libero Liggieri,* Michele Ferrari, Daniela Mondelli and Francesca Ravera

*CNR-Istituto per l'Energetica e le Interfasi-Sezione di Genova, via De Marini 6,
16149 Genova, Italy. E-mail: l.liggieri@ge.ieni.cnr.it*

Received 14th April 2004, Accepted 16th June 2004
First published as an Advance Article on the web 16th September 2004

The description of adsorption at liquid interfaces has been largely improved after incorporation into models of dynamic processes internal to the adsorption layers, such as surfactant re-orientation, aggregation and chemical reactions. Evidence for most of these processes has been given by qualitative studies utilising direct imaging techniques or by tensiometric investigations. These processes strongly influence the dilational rheological features of the adsorption layer, *i.e.* the response of interfacial tension to perturbations of the interfacial area. The investigation of the dilational rheology is then very effective to assess the existence of these processes and to characterise their kinetic parameters and equilibrium properties. To this aim specific models and experimental techniques have been recently developed. Here the surface visco-elasticity of adsorbed surfactant layers, ε, at liquid–air and liquid–liquid interfaces is measured by means of different oscillating drop/bubble methods, which are based on the interfacial tension response to harmonic perturbations of the interfacial area and have been implemented in capillary pressure and in drop shape tensiometers. The comparison of this data with theoretical predictions allows at the same time the validation of the models and the quantification of the kinetic features of the internal processes.

1.   Introduction

Surfactants are utilised to control the properties of liquid–fluid interfaces in many technologically relevant phenomena, such as the stability of liquid films and foams, because they play a fundamental role in the physics of the liquid film and consequently in droplet or bubble coalescence.

In these systems, when the interface is perturbed, different processes occur which contribute to the re-equilibration of the system, *i.e.* to the adsorption dynamics. Among the mechanisms involved in the adsorption dynamics, there are the diffusion in the bulk phases and kinetic processes inside the adsorbed layer such as, re-orientation,[1,2] aggregation[3] and other re-arrangements of the layer or of the molecular structure. Moreover the compressibility of the interfacial layer has been recently hypothesised[4] to play an important role in determining the mechanical behaviour of the adsorbed layer at large surface coverage. Finally, for liquid–liquid, the surfactant transfer across the interface, with consequent depletion of one solution can be fundamental for the dynamic behaviour of the system.[5]

The characterisation of the adsorption at liquid interfaces is based on the measurement of macroscopic quantities such as the equilibrium and dynamic interfacial tension, γ, and the surface dilational visco-elasticity, ε.

DOI: 10.1039/b405538a

From one side, measurements of γ *versus* time during the ageing of a fresh surface allow the adsorption kinetics to be accessed, while experimental studies of the dilational visco-elasticity can be used to investigate the specific processes in the adsorbed layer such as surfactant aggregation and re-orientation.[6–8]

The dilational visco-elasticity, ε, expresses the surface tension response to dilational stresses and, for small amplitude perturbation, is given by

$$\varepsilon = \frac{\Delta\gamma}{\Delta \ln A} \tag{1}$$

where A is the surface area. The presence of relaxation processes, such as the diffusional exchange in soluble monolayers and the transformations inside the adsorbed layer, introduces a delay in such a response. This is the reason for the visco-elastic behaviour of the interface, which makes ε a complex function of the perturbation frequency ν,

$$\varepsilon(\nu) = \varepsilon_r(\nu) + i\varepsilon_j(\nu) \tag{2}$$

In this work an experimental study is presented about some non-ionic surfactants (polyoxyethylene glycol ethers C_iE_j) at water/air and water/hexane interfaces. In addition to assessing and comparing the characteristics of these surfactants in water–air and water–oil systems, the study is aimed at showing the potentialities of the investigation of surface rheology.

C_iE_j's are linear surfactants showing different solubility in water/oil systems, depending on the relative length of hydrophobic and hydrophilic chains. For these reasons they are utilised in different technological surfactant blends and are interesting for application as liquid dispersion stabilizers.[9] The possibility of tuning their surface activity also offers interesting opportunities for their utilisation as models for fundamental investigations.

These surfactants also present interesting peculiarities. In fact direct investigations of the structure of adsorbed C_iE_j layers by neutron reflection techniques[10,11] have evidenced a change of the average orientation distribution of the surfactant, from parallel to perpendicular to the interface, as the surface coverage increases. This feature has been confirmed by previous works[1,2,6,12] showing that, the equilibrium properties and the dynamic surface tension during the ageing of fresh interfaces of these surfactants are very well described in the framework of a two-state model,[13] which simulates the effects of the molecular re-orientation.

The possibility of different orientations for adsorbed C_iE_j molecules can be explained accounting for the partial hydrophobic/lipophilic character of the ethylene groups in the hydrophilic chain,[14] which at small surface coverage allows the molecules to lie along the surface direction. The circumstance is confirmed by previous studies,[2,12] showing that the surface areas predicted by the two-state model agree very well with those that can be calculated by the surfactant structure, considering the molecule adsorbing parallel or perpendicular to the interface.

In refs. 1, 2 and 12 it has been shown that the dynamic surface tension of freshly formed interfaces of C_iE_j solutions can be accurately predicted by a diffusion controlled adsorption model considering the two-state surface equation of state. This shows that the re-orientation process exists but it is much faster than bulk diffusion. Adsorption kinetics is then inadequate to gain information about the kinetic features of the re-orientation process.

In the present work the attention is thus focused on the dilational properties of these surfactant monolayers. In fact, besides its potential in applications, the investigation of dilational rheology provides a unique tool for accessing the kinetics features of the re-orientation process.

In fact, the modern experimental techniques developed for surface rheology allows the response of the interface as a function of the frequency of applied perturbations, accessing kinetic properties in specific time scales to be checked.

2. Experimental methods

Two drop tensiometers were used working at two different time/frequency scales. The first one is a drop shape tensiometer (DST) which was used for interfacial tension measurements, either of equilibrium or for slow kinetics, and for dilational elasticity measurements in the frequency range from 0.005 to 0.16 Hz. Dilational elasticity measurements for frequencies spanning from 0.1 to 100 Hz were instead carried out by the capillary pressure tensiometer (CPT).

All measurements were performed at the controlled temperature of 20 °C.

2.1. Drop shape tensiometry

By the drop shape tensiometer (DST) the surface/interfacial tension is calculated from the acquired drop profile, fitting the Bashforth–Adams equation for axis-symmetric drops to the acquired drop profile. Measurements were performed by the ASTRA[15] (Automatic Surface Tension Real time Acquisition) set-up developed in our laboratory, equipped with the PAT-1 software by Sintech-Germany.[16] The surface tension is calculated according to the Maze–Burnet procedure,[17] using as fitting parameters the shape factor, $\beta = \Delta\rho g b^2/\gamma$ (where g is the acceleration gravity and $\Delta\rho$ is the density difference), the curvature radius at the drop apex, b, and the error on the apex vertical co-ordinate. γ is measured with an accuracy of the order of 0.1 mN m^{-1} and a minimum sampling time of the order of 0.3 s. The droplet area and volume are also calculated from the drop profile, to be used as control parameters to impose automatic area variation cycles by means of a micro-syringe pump (Hamilton, Switzerland).

The dilational visco-elasticity, γ is measured while applying harmonic small area perturbations of amplitude $\tilde{A}$. The module of the dilational visco-elasticity is then calculated as

$$|\varepsilon| = A^0 \frac{\tilde{\gamma}}{\tilde{A}} \tag{3}$$

where A^0 is the average surface area and $\tilde{\gamma}$ is the amplitude of the dynamic surface tension response. The phase of the elasticity is the phase shift between the surface tension response and the imposed area oscillation.

Amplitudes and phases are calculated from the experimental γ and A signals by a FFT algorithm.

2.2. Capillary pressure tensiometry

The capillary pressure tensiometer (CPT) allows the interfacial tension response to different imposed changes of the interfacial area to be obtained through the direct measurement of the capillary pressure. CPT is a versatile technique which can be used with different experimental methodologies,[18] to investigate different physicochemical aspects of pure and surfactant systems. For example, by the pressure derivative method[19] it is possible to measure the interfacial tension of pure liquids, while the growing drop/bubble[20,21] has been used to investigate the dynamic aspects of the adsorption. The rheological properties of the interface can be studied by the Oscillating Bubble[22–24] and the low amplitude stress/relaxation. The Fast Formed Drop[25] and the Expanded Drop[26,27] methods were developed to study the adsorption kinetics during the ageing of a freshly formed interface, allowing adsorption processes with characteristic times from a few seconds to several minutes to be investigated.

Recently, a CPT was used[28] to measure the dilational modulus with the oscillating drop/bubble technique. Such a method, based on the acquisition of the pressure response to harmonic perturbations of the interfacial area of a drop/bubble, was introduced several years ago[22] and used by many authors with different experimental set-ups, progressively upgraded and with different approaches for the interpretation of acquired experimental data.[23,28–31]

In a typical CPT a droplet (or a bubble) is formed inside a liquid, at the tip of a capillary. The pressure difference across the interface is monitored by a pressure transducer while the drop radius, R, is measured by direct imaging or it is calculated from the injected liquid volume, allowing the interfacial tension to be derived according to the Laplace equation.

$$\gamma(t) = \frac{P_c(t)R(t)}{2} \tag{4}$$

where P_c is the capillary pressure, which is derived by the measured pressure.

The scheme of the CPT cell used in this work is shown in Fig. 1. It consists in two stacked chambers connected through a glass capillary. The lower chamber is closed and contains one liquid, while the upper chamber is open to the atmospheric pressure, and contains the other liquid in case liquid/liquid systems are studied. The bubble/droplet is formed at the internal edge of the glass capillary tip, whose diameter is about 0.5 mm, and which protrudes in the closed chamber. In this

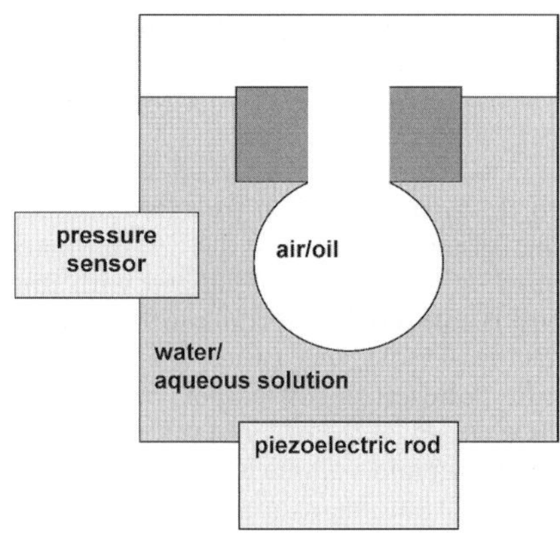

Fig. 1 Sketch of the capillary pressure tensiometer experimental cell.

chamber, immersed in the liquid, there are also the pressure sensor and a piezoelectric rod, utilised to vary and control the drop volume. A coarser volume control is also provided by a syringe pump.

Such a tensiometer, has already been utilised in previous experimental investigations of adsorbed layers based on the measurements of the dynamic interfacial tension[2,12] and of the dilational visco-elasticity[28] both at water/air and water/oil interfaces. Utilising the CPT according to the oscillating drop methodology, a sinusoidal variation of the volume of the cell, is applied by means of the variation of the volume of the piezoelectric rod

$$V_{pz} = V_{pz}{}^0 + \tilde{V}_{pz}\sin(2\pi\nu t). \tag{5}$$

This provides, in response, a corresponding harmonic oscillation of the interfacial area and of the surface tension. In fact, under small oscillation amplitude conditions the system behaves as a linear system and in a stationary regime all the quantity involved oscillates with the same frequency.[32] Thus the measured pressure is an oscillating quantity as well, with a certain phase shift, φ, with respect to the forcing V_{pz}, *i.e.*

$$P = P^0 + \tilde{P}\sin(2\pi\nu t + \varphi) \tag{6}$$

$\tilde{P}$ and φ are obtained as amplitude and phase of the component of frequency ν, extracted by the experimental signal *via* FFT algorithms.

In order to calculate the elasticity modulus from the experimental data, it is necessary to know the compressibility of the cell, and the reference surface tension and drop radius, just before the oscillation. For this reason, complementary measurements are executed *in situ* before starting an oscillating drop experiment. The reference equilibrium surface/interfacial tension is measured according to the drop radius steps method described in ref. 28, while the reference radius R_0 is measured by using a drop profile acquisition technique implemented in the CPT control software.

The compressibility of the system (cell plus liquid) can not be neglected because of the large ratio between the droplet and the cell volume. In this approach, because of the small pressure variations involved, such compressibility is simulated as caused by an effective amount of ideal gas, V_g, contained in the bottom chamber and measured as described in ref. 28. In other words, the phase in the closed part of the cell is modelled as incompressible liquid containing a gaseous phase whose volume is related to the pressure P through the perfect gas law.

Concerning the data treatment to obtain the visco-elasticity, the new approach reported in Appendix 1 is applied, based on a simplified analysis of the cell/bubble system, valid for not too large frequencies (≤ 100 Hz, in the present experimental set-up). The methodology is also suitable for the measurement of ε at liquid–liquid interfaces. In this case, however, the viscous flow of the liquid through the capillary sets a limitation on the frequency span to about 10 Hz.

This approach presents some advantages with respect to that previously utilised,[28] where reference calibration measurements on the pure system were used to eliminate the terms not depending on the surfactant presence. In fact, that calibration procedure gives satisfactory results provided that the same mechanical configuration is kept during all measurements. This latter is however a severe constraint, often difficult to achieve and can introduce error amplifications.

The treatment introduced here does not need a calibration but provides the ε values directly from measured P and V_{pz}.

3. Materials

The water used for the study was produced by a Millipore (Elix plus MilliQ) purification chain. Hexane was obtained from Merck-Uvasol and utilised without further purification. The solvent purity grade was checked by long time measurement of the dynamic surface/interfacial tension. At 20 °C a stable interfacial tension of 72.5 mN m^{-1} and 51.0 mN m^{-1} was found for water/air and water/hexane interfaces, respectively.

To avoid any possible pollution factor, all glassware and parts of the cells in glass and Teflon coming into contact with the measurement liquids were cleaned with sulfuric acid and rinsed several times with distilled water and, finally, with purified MilliQ water. The stainless steel parts, such as the pressure sensor and the piezoelectric case, were washed by isopropyl alcohol, then dried and rinsed with MilliQ water.

The effectiveness of the cleaning procedure was checked, before each measurement set, by on-site measurement of interfacial tension of the surfactant-free solvents.

The surfactants were water-soluble polyoxyethylene glycol ethers, C_iE_j, with different ratios between the length of the hydrophobic and hydrophilic chains which allows different partitioning properties and surface activity to be considered.

$C_{10}E_8$ and $C_{10}E_5$ were high purity grade samples with monodisperse polyoxyethylene chain length produced by Nikko (Japan). Brij35 ($C_{12}E_{23}$) and Brij52 ($C_{16}E_2$) were purchased by Sigma-Aldrich. All surfactants were used without further purification.

In a previous work,[5] the values of the partition coefficients $K_p = c_h/c_w$, for $C_{10}E_5$ and $C_{10}E_8$ were found to be 13.9 and 0.83, respectively, by a method[33] utilising water–air isotherms as calibration curves for the measurement of small surfactant concentrations. By the same method K_p of Brij35 was determined here to be smaller than 0.1. Indeed, owing to its molecular structure, this surfactant is expected to be nearly insoluble in hexane.

This methodology can not be used with Brij52 which is nearly insoluble in water, nevertheless a bounding value for K_p can be estimated on the basis of indirect considerations. In fact, it is known[34] that the dynamic interfacial tension, under surfactant transfer conditions can show a minimum when the ratio between the volumes of the supplying and the recipient phases is small enough. The appearance of the minimum depends then on the surfactant partition coefficient. Since for Brij52 a monotonic dynamic interfacial tension was observed during adsorption kinetics at water/hexane, on the basis of the approach described in ref. 34, it was possible to estimate $K_p > 10^2$.

4. Results and discussion

4.1. Interfacial equilibrium properties

The adsorption equilibrium properties were investigated by the drop shape tensiometer. The obtained γ–c isotherms at water/air are plotted in Fig. 2, while those for water/hexane are plotted in Figs. 3–4. They agree with the data reported in previous work.[2,12,28]

The isotherm of Brij52 refers to concentrations in hexane.

From Fig. 2 it is apparent that the surface activity of Brij35 is rather large compared to $C_{10}E_5$ and $C_{10}E_8$, but owing to its low critical micellar concentration the minimum interfacial tension achievable is quite high with respect to the other surfactants. This is coherent with the presence of a long oxyethylene chain.

The γ–c isotherms have been interpreted according to the two-state adsorption model.[13] This is an extension of the Langmuir model in which a second state for the molecules inside the adsorbed layer is considered, characterised by a different surface molar area. The two states coexist with an occupation degree depending on the surface pressure, the state at larger surface area being predominant at low surface pressure.

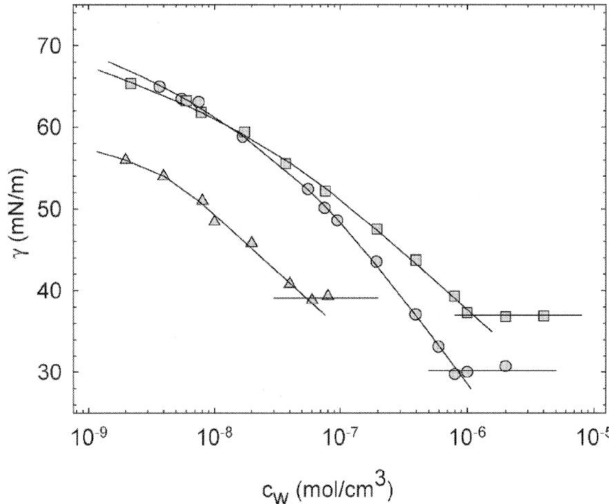

Fig. 2 Equilibrium interfacial tension *versus* surfactant concentration for $C_{10}E_5$ (●), $C_{10}E_8$ (■) and Brij35 (▲), at water/air, and best fit adsorption isotherm (solid line) by the two-state model. The best fit parameters are reported in Table 1.

In the framework of this model, a specific relationship is also assumed between the surface activities characterising the two states, b_1 and b_2, and the corresponding surface areas Ω_1 and Ω_2.

$$b_2 = b_1 \left(\frac{\Omega_2}{\Omega_1}\right)^{\alpha} \tag{7}$$

Then, the γ–c isotherm is

$$c = \frac{1 - \exp\left(-\frac{\Pi\Omega}{RT}\right)}{b_2\left[\left(\frac{\Omega_1}{\Omega_2}\right)^{\alpha}\exp\left(-\frac{\Pi\Omega_1}{RT}\right) + \exp\left(-\frac{\Pi\Omega_2}{RT}\right)\right]} \tag{8}$$

where Π is the surface pressure, $\gamma_0 - \gamma$.

Fitting of eqn. (8) to the γ–c data provides the values of the parameters reported in Table 1. As shown in Figs. 2–4 the corresponding isotherm curves provide a good description of the equilibrium data. The best fit values agree with those reported previously.[2] As discussed elsewhere,[1,12] the values of Ω_1 and Ω_2 confirm a picture where the exchange of molecules between the two states reflects a re-orientation process of the adsorbed molecules from a nearly parallel (state 1) to nearly perpendicular (state 2) orientation with respect to the interface.

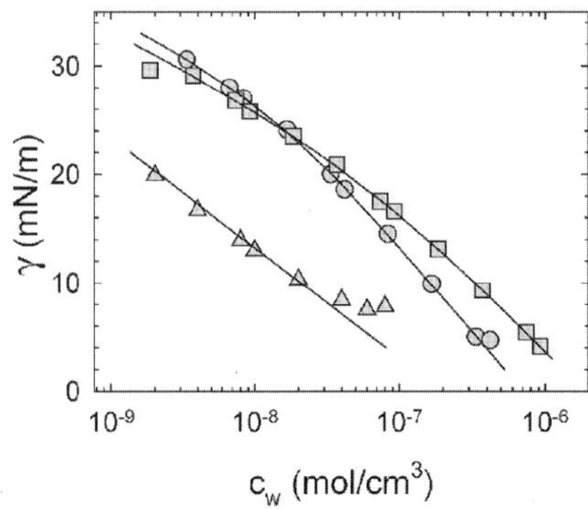

Fig. 3 Equilibrium interfacial tension *versus* surfactant concentration for $C_{10}E_5$ (●), $C_{10}E_8$ (■) and Brij35 ($C_{12}E_{23}$) (▲), at water/hexane, and best fit adsorption isotherm (solid line) by the two-state model. The best fit parameters are reported in Table 1.

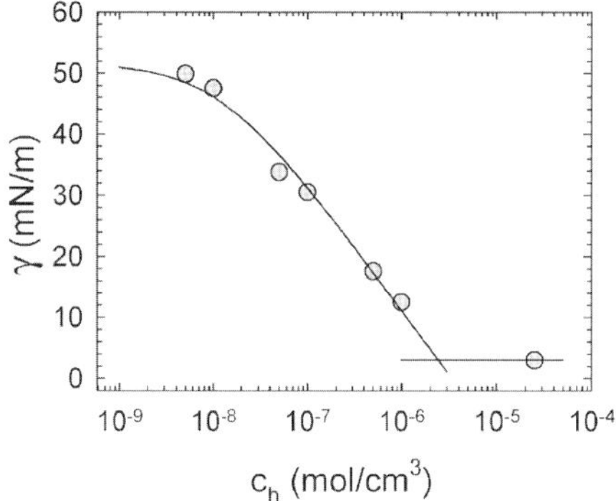

Fig. 4 Equilibrium interfacial tension *versus* surfactant concentration in hexane for Brij52 at hexane/water, and best fit adsorption isotherm (solid line) by the two-state model. The best fit parameters are reported in Table 1.

This picture is also sound for water–oil interfaces, at least for the C_iE_j's surfactants with the length of the carbon and oxyethylene chains not too different. In this case, the weak lipophilic character of the oxyethylene groups can become evident at low surface pressure, promoting the adsorption of molecules parallel to the water/hexane interface.

Indeed, by using the best fit values reported in Table 1 for water/hexane, eqn. (7) provides for $C_{10}E_8$ and $C_{10}E_5$ values of the ratio (b_1/b_2) of the order of a few hundreds, showing a significant adsorption in the state 1, even at rather large surface pressures. These values are about 2 orders of magnitude larger than those of Brij52 ($C_{16}E_2$) and Brij35 ($C_{12}E_{23}$).

Thus, for these latter, adsorption in the state 2 is predominant already at low surface pressure. This is however due to different causes. For Brij52 the oxyethylene chain is simply too short to compete with the lipophilic character of the carbon chain, while for Brij35 the adsorption in the state perpendicular to the interface is strongly surface active, as shown by the large value of b_2, owing to the long oxyethylene chain.

The comparison of the best fit isotherm parameters at water/air and water/hexane also shows that molar areas ω_1 and ω_2, are practically unchanged for water–air and water–hexane, as one would expect for the two-state model, and correspond with the prediction of molecular simulation.

4.2. Surface rheology at water/air interface

The measured dilational visco-elasticity at water/air for $C_{10}E_5$, $C_{10}E_8$ and Brij35, are shown in Figs. 5–7, where the real and imaginary parts of ε are plotted *versus* the imposed harmonic perturbation.

Table 1 Equilibrium parameters obtained by the analysis of the experimental equilibrium data for water–air and water–hexane: two-state adsorption isotherm parameters, Ω_1, Ω_2, α and b_2, critical micellar concentration, c.m.c. with the corresponding interfacial tension $\gamma_{c.m.c.}$, and partition coefficient K_p

	Ω_1/cm² mol⁻¹	Ω_2/cm² mol⁻¹	α	b_2/cm³ mol⁻¹	c.m.c./mol cm⁻³	$\gamma_{c.m.c.}$/mN m⁻¹	K_p
$C_{10}E_5$ at w/a	7.0×10^9	2.6×10^9	2.6	1.1×10^8	8.4×10^{-7}	30.2	
$C_{10}E_8$ at w/a	9.9×10^9	4.0×10^9	3.2	2.9×10^8	1.1×10^{-6}	37.0	
$C_{12}E_{23}$ (Brij35) at w/a	2.1×10^{10}	4.0×10^9	6.8	4.3×10^9	7.6×10^{-8}	39.1	
$C_{10}E_5$ at w/h	9.4×10^9	3.6×10^9	5.6	2.7×10^9	—	4.7	13.9
$C_{10}E_8$ at w/h	8.7×10^9	4.1×10^9	7.1	3.0×10^9	—	3.0	0.83
$C_{12}E_{23}$ (Brij35) at w/h	2.5×10^{10}	5.5×10^9	0.26	4.2×10^{11}	—	4.0	<0.1
$C_{16}E_2$ (Brij52) at h/w	3.5×10^{10}	2.6×10^9	0.6	7.0×10^7	2.3×10^{-6} [a]	3.0[a]	>100

[a] Inverse micelles c.m.c. in hexane.

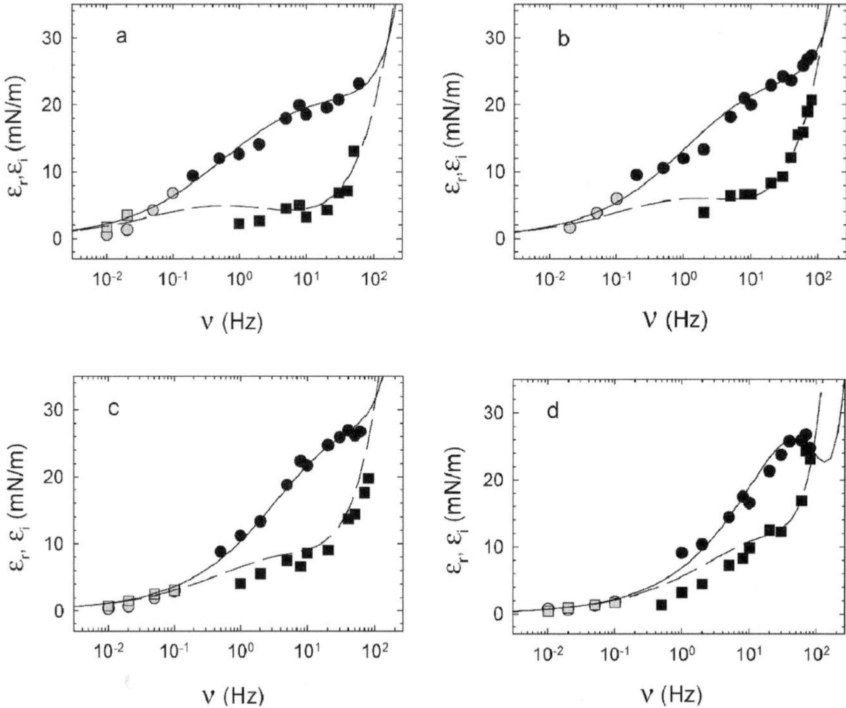

Fig. 5 Real ($\bullet$) and imaginary ($\blacksquare$) parts of ε for $C_{10}E_5$ at water/air for different bulk concentrations: (a) 1×10^{-8} mol cm^{-3}, (b) 3×10^{-8} mol cm^{-3}, (c) 1×10^{-7} mol cm^{-3} and (d) 3×10^{-7} mol cm^{-3}. Experimental data obtained with the CPT (black points) and the DST (grey points) and theoretical curves calculated by eqn. (10), with the values of the parameters reported in Tables 1 and 2.

The two experimental techniques described above, allow the frequency range from 0.01 to 100 Hz to be studied. The two sets of data obtained with these techniques show a good match for all the systems studied, both for the real and imaginary parts.

The real part of the elasticity *versus* frequency shows a weak dependence on the surfactant concentration. For $C_{10}E_5$ and $C_{10}E_8$, the frequency dependence of the imaginary part presents a larger slope when the surfactant concentration increases. For Brij35 the concentration dependence is even weaker. For this reason only one concentration has been reported. Both the real and the

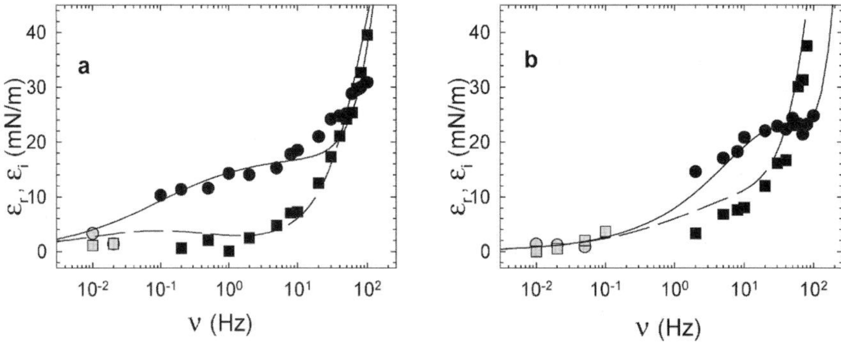

Fig. 6 Real ($\bullet$) and imaginary ($\blacksquare$) parts of ε for $C_{10}E_8$ at water/air for different bulk concentrations: (a) 2×10^{-8} mol cm^{-3} and (b) 1×10^{-7} mol cm^{-3}. Experimental data obtained with the CPT (black points) and the DST (grey points) and theoretical curves calculated by eqn. (10), with the values of the parameters reported in Tables 1 and 2.

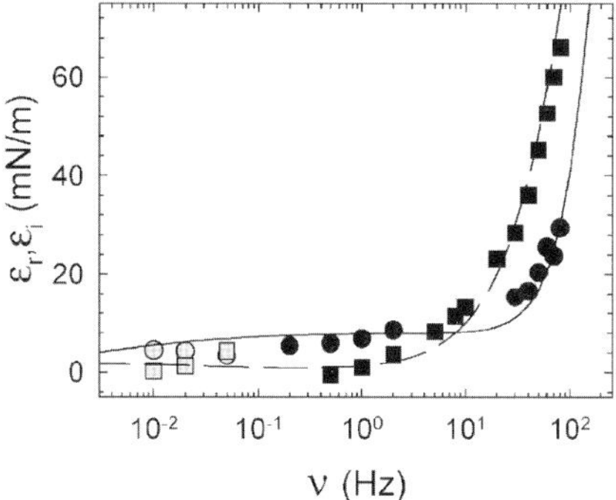

Fig. 7 Real ($\bullet$) and imaginary ($\blacksquare$) parts of ε for Brij 35 at water/air for $c = 10^{-8}$ mol cm^{-3}. Experimental data obtained with the CPT (black points) and the DST (grey points) and theoretical curves calculated by eqn. (10), with the values of the parameters reported in Tables 1 and 2.

imaginary part of the visco-elasticity tend to increase more rapidly with frequency with respect to the other surfactants. This could be caused by the larger molecular size of Brij35, which makes less effective the relaxation processes.

The expression of the visco-elasticity as a function of the frequency of the applied harmonic oscillation can be found[6,7,32] by analysing the perturbation response of the reference equilibrium state of a soluble monolayer.

The re-orientation process is described by a linear kinetic equation,[1,6]

$$\frac{\mathrm{d}\Gamma_2}{\mathrm{d}t} = \nu_{or}\left(\Gamma_1 - \frac{\Gamma_1^0}{\Gamma_2^0}\Gamma_2\right) \tag{9}$$

where Γ_1 and Γ_2 are the adsorption in states 1 and 2, respectively.

According to this approach an expression for the visco-elasticity in the framework of the two-state model has been obtained,[6] which accounts for both the diffusive exchange with the bulk and for the kinetics of the re-orientation process.

Here the experimental data are interpreted according to the expression of ε *versus* ν, found in ref. 6. Moreover an additional term accounting for a possible intrinsic surface viscosity, κ, is introduced, as discussed by other authors.[35,36] Therefore

$$\varepsilon = \varepsilon_{01}\frac{1 + (1-i)\xi - \frac{1+g_0}{g_0}\left[q_1(1-i)\xi + i\lambda\left(g_0 + g'\Gamma_2^0\right)\right]}{(1 + (1-i)\xi)(1 - i\lambda(1+g_0))} + \\ - \varepsilon_{02}g_0\frac{1 + (1-i)\xi - \frac{1+g_0}{g_0}\left[1 + q_1(1-i)\xi - i\lambda\left(1 - g'\Gamma_2^0\right)\right]}{(1 + (1-i)\xi)(1 - i\lambda(1+g_0))} + 2\pi\nu i\kappa \tag{10}$$

where, $\lambda = \nu_{or}/\nu$ and $\xi = \sqrt{\nu_D/2\nu}$, ν_{or} and ν_D are the characteristic frequency of the orientation and of the diffusion processes, respectively. ν_{or} is identified with the re-orientation rate,[1] while $\nu_D = D/2\pi(\mathrm{d}c/\mathrm{d}\Gamma|_0)^2$, where D is the diffusion coefficient in the bulk, c is the bulk surfactant concentration, Γ the adsorption.

Γ^0, ε_{0j}, g_0 and g' are equilibrium quantities, referring to the reference state and linked to the surface isotherm, q_1 is the probability that a molecule arriving from the bulk initially adsorbs in the state 1. Further details about the model and the meaning of the different parameters can be found elsewhere[1,6,12]

Eqn. (10), can be considered valid as far as the ratio Γ_1/Γ_2 is not too small (>0.01), which is the case of the reported investigation. An upgrade of the model is under way to extend its applicability.

Fitting eqn. (10), to the experimental data provides the parameters characterising the kinetics of the system, *i.e.* the surface re-orientation rate, the diffusion coefficient and the intrinsic surface viscosity. The corresponding best fit parameters are reported in Table 2.

All the constants in eqn. (10) have been calculated from the values of the adsorption isotherm parameters provided by the equilibrium measurements described in the previous section.

Table 2 Best fit kinetic parameters obtained by the fitting of eqn. (10) to the surface visco-elasticity data for water–air. Diffusion coefficient in water, D; re-orientation rate (or characteristic re-orientation frequency), intrinsic surface viscosity, κ. The values of the characteristic frequency of orientation, ν_D, are calculated from D and from the isotherm parameters. The diffusion coefficients evaluated from adsorption kinetics experiments, D_{adk}, are also reported for comparison

	C/mol cm^{-3}	D /cm^2 s^{-1}	ν_{or}/Hz	$\nu_D{}^a$/Hz	κ/s mN m^{-1}	D_{adk}/cm^2 s^{-1}
C$_{10}$E$_5$ at w/a	1×10^{-8}	5.8×10^{-6}	400	0.5	0.03	4.6×10^{-6}
	2×10^{-8}			1.2	0.07	
	1×10^{-7}			5.8	0.08	
	3×10^{-7}			110.6	0.09	
C$_{10}$E$_8$ at w/a	2×10^{-8}	2.0×10^{-6}	110	0.1	0.10	3.7×10^{-6}
	1×10^{-7}		400	29.1	0.11	
C$_{12}$E$_{23}$ (Brij35) at w/a	1×10^{-8}	9.0×10^{-6}	250	27.6	0.05	

^{a} Calculated from D and from the isotherm parameters: $\nu_D = (D/2\pi)*(\mathrm{d}c/\mathrm{d}\Gamma)^2$.

In agreement with its definition, the best fit re-orientation rate, ν_{or}, is nearly independent on the surface coverage, for given surfactant. The corresponding characteristic times, $1/\nu_{or}$, are orders of magnitude larger than the characteristic times of single molecular re-orientation events. In fact, these values refer to the rate of a collective process in the adsorption layer, described by eqn. (9).

The best fit values of the diffusion coefficients are in good agreement with those obtained in previous investigations[2] of the diffusion controlled adsorption kinetics and with those reported elsewhere for similar surfactants.[1,12,37,38] As an example, Fig. 8 shows the dynamic interfacial tension measured during the ageing of freshly formed C$_{10}$E$_5$ aqueous solutions at different concentrations. There is a good agreement with the prediction of the diffusion controlled adsorption with the two-state isotherm, for the parameters given in Table 1 and a diffusion coefficient of 4.6×10^{-6} cm^2 s^{-1}, which is very close to the best fit value of D from eqn. (10), reported in Table 2.

The order of magnitude of κ is found to be the same for all the samples, of the order of a few 0.01 s mN m^{-1}. These values are not far from what has been reported previously[35,36] for other non-ionic and ionic surfactants. Moreover, a slight increase of κ is observed with the surfactant concentration. This seems sound according to its physical meaning which should represent the proper viscosity of the interfacial layer. In fact, besides depending principally on the mechanical

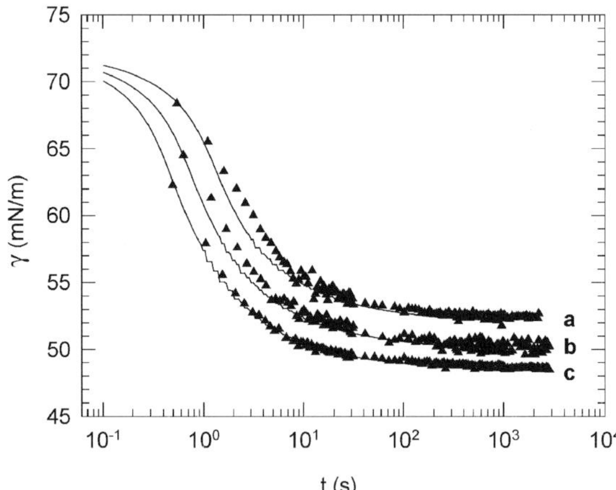

Fig. 8 Dynamic surface tension (▲) of C$_{10}$EO$_5$ during the ageing of a freshly formed interface at concentrations (a) 6.0×10^{-8}; (b) 8.0×10^{-8}; and (c) 1×10^{-7} mol cm^{-3}. The curves are the prediction of the diffusion controlled adsorption kinetics with the two-states isotherm. The isotherm parameters are those reported in Table 1 and the diffusion coefficient is 4.6×10^{-6} cm^2 s^{-1}.

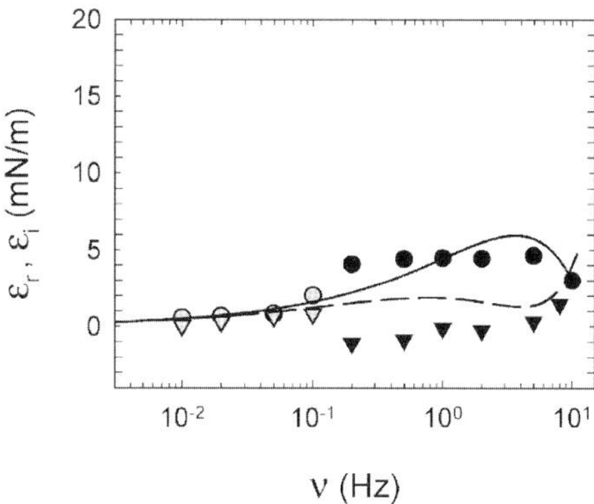

Fig. 9 Real (●) and imaginary (▼) parts of ε for $C_{10}E_5$ at water/hexane for bulk concentration in water $c = 10^{-7}$ mol cm^{-3}. Experimental data obtained with the CPT (black points) and the DST (grey points). ε_r (solid line) and ε_i (dashed line) theoretical curves calculated by eqn. (10), with the values of the parameters reported in Tables 1 and 3.

properties of the two bulk fluids (water and air in this case) and on their coupling, κ should also be affected by the nature and by the quantity of the adsorbed surfactant.

4.3. Surface rheology at water/hexane interface

The dilational visco-elasticity data obtained for water soluble and oil soluble surfactants at water/ hexane interfaces, are shown in Figs. 9–12, together with the best fit curves obtained from eqn. (10). Due to the limitations of the oscillating drop method for liquid–liquid, the maximum investigated frequency was 10 Hz.

At a given frequency, the measured visco-elasticity was weakly dependent on the surfactant concentration. For this reason, for all the investigated systems, only one surfactant concentration is reported in the plots.

For $C_{10}E_5$, $C_{10}E_8$ and Brij52, eqn. (10) describes the experimental data with a physically acceptable value of the best fit parameters (see Table 3). A satisfactory agreement was not found for Brij35. This could be due to the difficulty of determining accurately the adsorption isotherm, either because of the very low concentrations dealt with, or the probable presence of impurities adsorbing at long times. This surfactant was indeed not a high purity grade like other C_iE_j.

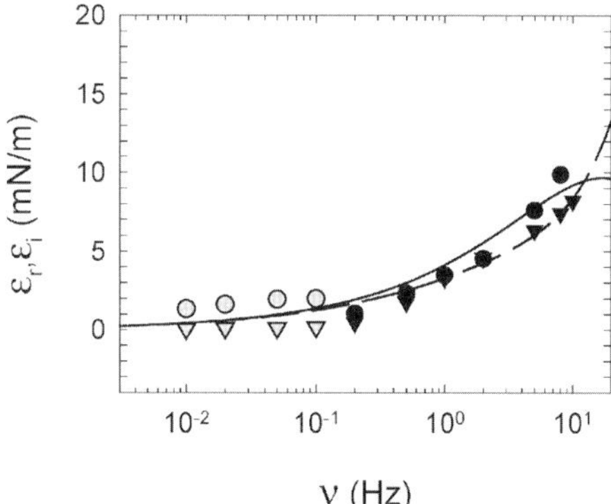

Fig. 10 Real (●) and imaginary (▼) parts of ε for $C_{10}E_8$ at water/hexane for bulk concentration in water $c = 10^{-7}$ mol cm^{-3}. Experimental data obtained with the CPT (black points) and the DST (grey points). ε_r (solid line) and ε_i (dashed line) theoretical curves calculated by eqn. (10), with the values of the parameters reported in Tables 1 and 3.

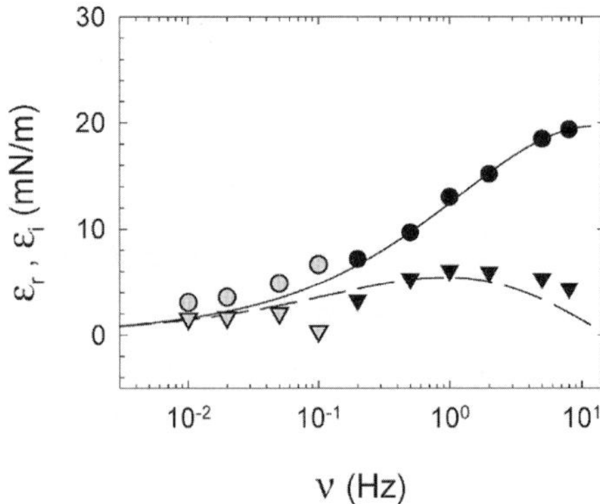

Fig. 11 Real (●) and imaginary (▼) parts of ε for Brij52 at hexane/water and bulk concentration in hexane $c = 10^{-7}$ mol cm^{-3} obtained with the CPT (black points) and the DST (grey points). ε_r (solid line) and ε_i (dashed line) theoretical curves calculated by eqn. (10), with the values of the parameters reported in Tables 1 and 3.

In all the systems, the intrinsic surface viscosity was found negligible at water/hexane. This could be indicative of a relationship between κ and the discontinuity of the mechanical properties (viscosity, density) between the two bulk phases.

While for water/air it is difficult to see a trend in the orientation rates, ν_{or}, for water/hexane there is definitely an increase with the length of the oxyethylene chain. This seems also coherent with the partial lipophilic character of this chain at water/hexane. In fact such a feature should on one side increase the energetic gap between the two states, while on the other side dump the energetic barrier between these two energy minima, resulting in a faster exchange between them. The longer the chain, the larger the dumping, the faster the exchange.

The theoretical approach leading to eqn. (10) considers the surfactant adsorbing from one phase only. Since this approach is based on the assumption of small perturbations of the equilibrium state, eqn. (10) can be extended to consider surfactant adsorbing from the two phases by assuming an effective diffusion coefficient which accounts for the surfactant partitioning[5,39]

$$D_{\text{eff}} = D_w \left(1 + K_p \sqrt{\frac{D_h}{D_w}} \right)^2 \tag{11}$$

where D_w and D_h are the diffusion coefficients in water and hexane, respectively.

Thus, the best fit diffusion coefficients for $C_{10}E_5$ and $C_{10}E_8$ in Table 3 correspond to D_{eff}, and, in fact, they are larger than those obtained for water/air (D_w in Table 2).

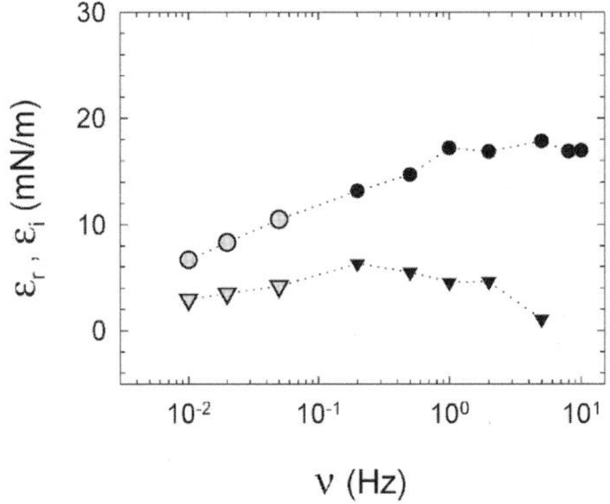

Fig. 12 Real (●) and imaginary (▼) parts of ε for Brij35 at w/h and bulk concentration in water $c = 10^{-8}$ mol cm^{-3}. Experimental data obtained with the CPT (black points) and the DST (grey points).

Table 3 Kinetic parameters obtained by fitting of eqn. (10) to the surface visco-elasticity experimental data for water–hexane. Effective diffusion coefficient, D_{eff}; re-orientation rate (or characteristic re-orientation frequency), ν_{or}; characteristic diffusion frequency ν_D; diffusion coefficient in hexane, D_h, evaluated *via* eqn. (11)

	C/mol cm^{-3}	D_{eff}/cm^2 s^{-1}	D_h/cm^2 s^{-1}	ν_{or}/Hz	ν_D/Hz
$C_{10}E_5$ at w/h	1×10^{-7}	1.5×10^{-4}	4.9×10^{-5}	350	34
$C_{10}E_8$ at w/h	1×10^{-7}	3.0×10^{-5}	2.4×10^{-5}	5400	54
$C_{16}E_2$ (Brij52) at w/h	1×10^{-7}	2.5×10^{-5}	2.5×10^{-5}	70	2

The values of D_{eff}, are used together with the corresponding D_w and K_p, to estimate the values of D_h reported in Table 3. Thus the surface rheology also provides a way to estimate the diffusion coefficients of surfactants in oil, whose values are still largely unknown.

5.   Conclusions

In this work an experimental study on the surface properties of different non-ionic polyoxyethylenated surfactants is presented, covering equilibrium aspects, such as adsorption isotherms, partitioning, as well as kinetic aspects such as surface rheological properties related with internal processes of the adsorption layer and with the diffusional exchange with the bulk.

A drop shape tensiometer and a capillary pressure tensiometer were used to measure the dilational visco-elasticity of surfactant layers adsorbed at water/air interfaces as a function of the frequency of imposed oscillations in the range from 0.01 to 100 Hz.

In particular, the measurements with the CPT were performed according to the oscillating drop/bubble method. To this aim a specific treatment/methodology has been developed to get the dilational visco-elasticity from the measured pressure signal, which accounts for an intrinsic viscosity of the interfacial layer. In respect to previous approaches, the new treatment avoids the need for some critical calibration procedures, providing more accurate measurements and a stable performance of the apparatus.

The results show that models taking into account the re-orientation of the adsorbed molecules in the adsorption layer can be suitably applied to describe the observed features. Moreover, the investigation of the dilational properties provides a way to access the kinetics and transport features of these systems, such as the surfactant diffusion coefficients and in particular the characteristic time of the re-orientation process, which represents an important step forward in the understanding of the internal dynamics of adsorbed layers.

This result is made possible by the contemporaneous investigation of equilibrium properties, adsorption kinetics and surface rheology.

It is in fact particularly significant that in fitting eqn. (10) only the kinetic constants have been considered as variable, the other parameters being obtained by equilibrium measurements. This provides an important cross-checking for the adopted rheological model. In fact, in most of the surface rheological investigations published so far poor attention has been paid to the coherence between equilibrium and rheological parameters.

Being based on the investigation of the response to small perturbations of a well defined equilibrium state, the surface rheology is particularly useful to investigate the kinetic features of liquid–liquid systems. On the contrary, in the adsorption kinetic studies, for liquid–liquid systems interpretative problems can arise because the initial conditions of the surfactant partitioning are often ill defined.

The results reported here find a natural application in the field of emulsion/foams stability. The important role of surfactants in this field is well known. The reduction of interfacial tension is one of the main driving forces in the formation of such disperse systems, while their stability is mainly related to surface rheological properties of the adsorption layers.[40,41]

Appendix 1

Analysis of the bubble/cell system

A bubble or drop (phase 2), of volume V_d at the tip of a capillary of length L and inner radius α, is considered to oscillate inside a liquid phase (phase 1). The pressure P, measured by the pressure sensor, at distance r is given by the Navier–Stokes equation which for small oscillations, assuming spherical symmetry, is[29,31,42]

$$-\frac{\partial P}{\partial r} = \rho_1 \frac{\partial u}{\partial t} \tag{12}$$

where ρ_1 is the density of phase 1 and u is the flow velocity which is given by the continuity equation

$$\frac{\partial u}{\partial r} + \frac{2u}{r} = \frac{1}{B}\frac{\partial p}{\partial t} \tag{13}$$

where B is the compressibility of phase 1.

Phase 1 is assumed as incompressible liquid containing a gaseous phase of volume V_g whose variation is considered in the following for the calculation of the drop volume variation. This hypothesis allows vanishing of the second terms of eqn. (13) whose integration gives,

$$u = \left(\frac{R^0}{r}\right)^2 u_R = \left(\frac{R^0}{r}\right)^2 i\omega\Delta R \tag{14}$$

where $\omega = 2\pi\nu$ is the pulsation. u_R is the velocity for $r = R$, i.e. the drop radius which is considered oscillating at frequency ν around the reference value R^0. Introducing eqn. (14) into eqn. (12), one obtains

$$P = P_1 - \rho_1\omega^2 R^0\left(\frac{R^0}{r} - 1\right)\Delta R \tag{15}$$

where P_1 is the pressure close to the drop surface in phase 1.

The boundary condition at the drop surface for the pressure is

$$P_1 - P_2 = 2(\eta_1 - \eta_2)\frac{\partial u}{\partial r}\bigg|_{r=R} - \frac{2\gamma}{R} \tag{16}$$

where P_2 is the pressure close to the drop surface in phase 2, η_1 and η_2 are the kinematic viscosities in phase 1 and 2, respectively, and γ is the interfacial tension. Using again eqn. (14), eqn. (16) becomes

$$P_1 - P_2 = -4i\omega(\eta_1 - \eta_2)\frac{\Delta R}{R^0} - \frac{2\gamma}{R} \tag{17}$$

When the incompressible and quasi-stationary conditions are simultaneously satisfied[42] the pressure inside the capillary can be written in a simplified form where the Poiseuille term and the inertial term are considered,

$$P_2 - P_{\text{out}} = -\frac{8\eta_2 L}{\alpha^2}u - \frac{4\rho_2 L}{3}\frac{du}{dt} \tag{18}$$

where P_{out} is the pressure at the opposite side of the capillary, with respect to the drop.

The flow velocity inside the capillary is the drop volume variation,

$$u = \frac{1}{\pi\alpha^2}\frac{dV_d}{dt} \tag{19}$$

The drop volume variation is determined by the volume variation of the piezoelectric rod and of the effective gas bubble which is linked to the pressure by the perfect gas law. Then it can be written as

$$\frac{dV_d}{dt} = -\frac{dV_{pz}}{dt} + \frac{V_g^0}{P^0}\frac{dP}{dt} \tag{20}$$

Thus, assuming small amplitude sinusoidal variations for all the quantities involved, the pressure variation in the cell can be written using eqns. (15), (17) and (18) with some rearrangements,

$$\Delta P = \frac{2\gamma\Delta R}{R^{02}} - 2\frac{\Delta\gamma}{R^0} - 4i\omega(\eta_1 - \eta_2)\frac{\Delta R}{R^0} + \rho_1\omega^2 R^0\left(1 - \frac{R^0}{r}\right)\Delta R \\ + \left(i\omega K_1 - \omega^2 K_2\right)\left(\Delta V_{pz} - V_g^0\frac{\Delta P}{P^0}\right) \tag{21}$$

where

$$K_1 = \frac{8\eta_2 L}{\pi\alpha^4} \tag{22}$$

and

$$K_2 = \frac{4\rho_2 L}{3\pi\alpha^2} \tag{23}$$

Thus an expression for the dilational elasticity can be found by introducing its definition, eqn. (1), inside eqn. (21),

$$\frac{2}{R^0}\frac{\Delta A}{A^0}\varepsilon = \frac{2\gamma\Delta R}{R^{02}} - \Delta P - 4i\omega(\eta_1 - \eta_2)\frac{\Delta R}{R^0} + \rho_1\omega^2 R^0\left(1 - \frac{R^0}{r}\right)\Delta R$$
$$+ \left(i\omega K_1 - \omega^2 K_2\right)\left(\Delta V_{\mathrm{pz}} - V_{\mathrm{g}}^0\frac{\Delta P}{P^0}\right) \tag{24}$$

and by introducing the exponential formalism

$$\varepsilon = \frac{A^0 R^0}{2\tilde{A}}\left\{\left[\frac{2\gamma^0}{R^0} - 4i\omega(\eta_1 - \eta_2) + \rho_1\omega^2 R^{02}\right]\frac{\tilde{R}}{R^0}\right.$$
$$\left. - \left[\left(i\omega K_1 - \omega^2 K_2\right)\frac{V_g^0}{P^0} + 1\right]\tilde{P}e^{i(\varphi-\theta)} + \left(i\omega K_1 - \omega^2 K_2\right)\tilde{V}_{\mathrm{pz}}e^{-i\theta}\right\} \tag{25}$$

where $\tilde{P}$, and φ are measured quantities, while $\tilde{A}/A^0$ and $\tilde{R}$ are calculated from $\tilde{V}_{\mathrm{d}}$ and θ. These latter quantities are obtained by eqn. (20) according to the calculation reported in ref. 28,

$$\tilde{V}_{\mathrm{d}} = V_{\mathrm{g}}^0\sqrt{\left(\frac{\tilde{V}_{\mathrm{pz}}}{V_{\mathrm{g}}^0}\right)^2 + \left(\frac{\tilde{P}}{P^0}\right)^2 - 2\frac{\tilde{V}_{\mathrm{pz}}}{V_{\mathrm{g}}^0}\frac{\tilde{P}}{P^0}\cos\varphi} \tag{26}$$

$$\cos\theta = \frac{\frac{V_{\mathrm{g}}^0}{P^0}\tilde{P}\cos\varphi - \tilde{V}_{\mathrm{pz}}}{\tilde{V}_{\mathrm{d}}} \tag{27}$$

$\tilde{P}$ and φ are evaluated from the recorded pressure signal by extracting the component at the imposed frequency according to the Fast Fourier Transform algorithm, as explained previously.[28]

Acknowledgements

The work was partially supported by the European Space Agency under the MAP project "Fundamental and Applied Studies in Emulsion Stability-FASES" (AO-99-052).

References

1 L. Liggieri, F. Ravera and R. Miller, *Colloids Surf., A*, 2000, **175**, 51.
2 L. Liggieri, M. Ferrari, A. Massa and F. Ravera, *Colloids Surf., A*, 1999, **156**, 455.
3 D. Vollhardt, V. B. Fainerman and G. Emrich, *J. Phys. Chem. B*, 2000, **104**, 8536.
4 V. B. Fainerman, R. Miller and V. I. Kovalchuk, *J. Phys. Chem. B*, 2003, **107**, 6119.
5 F. Ravera, M. Ferrari and L. Liggieri, *Adv. Colloid Interface Sci.*, 2000, **88**, 129.
6 F. Ravera, M. Ferrari, R. Miller and L. Liggieri, *J. Phys. Chem. B*, 2001, **105**, 195.
7 R. Palazzolo, F. Ravera, M. Ferrari and L. Liggieri, *Langmuir*, 2002, **18**, 3592.
8 L. Liggieri, M. Ferrari and F. Ravera, *Langmuir*, 2003, **19**, 10233.
9 *Emulsion and emulsion stability*, ed. A. J. Sjoblom, Marcel Dekker, 1996, vol. 61.
10 J. R. Lu, M. Hromadova, R. K. Thomas and J. Penfold, *Langmuir*, 1993, **9**, 2417.
11 J. R. Lu, Z. X. Lee, R. K. Thomas, E. J. Staples, L. Thompson, I. Tucker and J. Penfold, *J. Phys. Chem.*, 1994, **98**, 6559.
12 M. Ferrari, L. Liggieri and F. Ravera, *J. Phys. Chem.*, 1998, **102**, 10521.

13 V. B. Fainerman, R. Miller, R. Wüstneck and A. V. Makievski, *J. Phys. Chem.*, 1996, **100**, 7669.
14 F. E. Bailey and J. V. Koleske, in *Non-ionic Surfactants*, ed. M. J. Schick, Surfactant Sci. Series, Marcel Dekker, New York, 1987, **vol. 23**, p. 927.
15 L. Liggieri and A. Passerone, *High Temp. Tech.*, 1989, **7**, 80.
16 G. Loglio, P. Pandolfini, R. Miller, A. V. Makieski, F. Ravera, M. Ferrari and L. Liggieri, in *Novel Methods to Study Interfacial Layers*, ed. D. Moebius and R. Miller, Studies in Interface Science Series, Elsevier, Amsterdam, 2001, **vol. 11**, p. 439.
17 C. Maze and G. Burnet, *Surf. Sci.*, 1969, **13**, 451.
18 L. Liggieri and F. Ravera, in *Drops and Bubbles in Interfacial Research*, ed. D. Mebius and R. Miller, Studies in Interface Science Series, Elsevier, Amsterdam, 1998, **vol. 6**, p. 239.
19 A. Passerone, L. Liggieri, N. Rando, F. Ravera and E. Ricci, *J. Colloid Interface Sci.*, 1991, **146**, 152.
20 R. Nagarajan and D. T. Wasan, *J. Colloid Interface Sci.*, 1993, **159**, 164.
21 G. A. MacLeod and C. J. Radke, *J. Colloid Interface Sci.*, 1993, **160**, 435.
22 K. Lunkenheimer and G. Kretzschmar, *Z. Phys. Chem. (Leipzig)*, 1975, **256**, 593.
23 H. D. Fruhner and K. D. Wantke, *Colloids Surf., A*, 1996, **114**, 53.
24 Y.-H. Kim, K. Koczo and D. T. Wasan, *J. Colloid Interface Sci.*, 1997, **187**, 29.
25 T. Horozov and L. Arnaudov, *J. Colloid Interface Sci.*, 1999, **219**, 99.
26 L. Liggieri, F. Ravera and A. Passerone, *J. Colloid Interface Sci.*, 1995, **169**, 226.
27 P. J. Breen, *Langmuir*, 1995, **11**, 885.
28 L. Liggieri, V. Attolini M. Ferrari and F. Ravera, *J. Colloid Interface Sci.*, 2002, **255**, 225.
29 V. I. Kovalchuk, E. K. Zholkovskij, J. Kraegel, R. Miller, V. B. Fainerman, R. Wüstneck, G. Loglio and S. S. Dukhin, *J. Colloid Interface Sci.*, 2000, **224**, 245.
30 V. I. Kovalchuk, J. Kraegel, E. V. Aksenenko, G. Loglio and L. Liggieri, in *Novel Methods to Study Interfacial Layers*, ed. D. Moebius and R. Miller, Studies in Interface Science Series vol. 11, Elsevier, Amsterdam, 2001, p. 485.
31 K. D. Wantke, H. Fruhner, J. Fang and K. Lunkenheimer, *J. Colloid Interface Sci.*, 1998, **208**, 34.
32 B. A. Noskov and G. Loglio, *Colloid Surf., A*, 1998, **143**, 167.
33 F. Ravera, M. Ferrari, L. Liggieri, R. Miller and A. Passerone, *Langmuir*, 1997, **13**, 4817.
34 L. Liggieri, F. Ravera, M. Ferrari, A. Passerone and R. Miller, *J. Colloid Interface Sci.*, 1997, **186**, 46.
35 K. D. Wantke and H. Fruhner, *J. Colloid Interface Sci.*, 2001, **237**, 185.
36 K. D. Wantke, H. Fruhner and J. Oertegren, *Colloids Surf., A*, 2003, **221**, 185.
37 S.-Y. Lin, R.-Y. Tsay, L.-W. Lin and S.-I. Chen, *Langmuir*, 1996, **12**, 6530.
38 H.-C. Chang, C.-T. Hsu and S.-Y. Lin, *Langmuir*, 1998, **14**, 2476.
39 R. Miller, G. Loglio and U. Tesei, *Colloid Polym. Sci.*, 1992, **270**, 598.
40 F. O. Opawale and D. J. Burgess, *J. Colloid Interface Sci.*, 1998, **197**, 142.
41 I. B. Ivanov and P. A. Kralchevsky, *Colloids Surf., A*, 1997, **128**, 155.
42 V. I. Kovalchuk, J. Kraegel, R. Miller, V. B. Fainerman, N. M. Kovalchuk, E. K. Zholkovskij, R. Wüstneck and S. S. Dukhin, *J. Colloid Interface Sci.*, 2000, **232**, 25.

Observation of interfacial tension minima in oil–water–surfactant systems with laser manipulation technique

Shujiro Mitani and Keiji Sakai

Institute of Industrial Science, University of Tokyo, 4-6-1 Komaba, Meguro-ku, Tokyo 153-8505, Japan

Received 14th April 2004, Accepted 5th July 2004
First published as an Advance Article on the web 21st September 2004

A laser beam passing through a liquid interface results in a deformation on the nanometer scale that is inversely proportional to the interfacial tension. Based on this principle, we developed a method to measure the liquid–liquid interfacial tension. The displacement excited by the pump laser is measured with another probe laser in a non-contact and non-destructive manner. The motion of the interface in response to the modulated intensity of the pump laser shows a characteristic spectrum yielding an accurate value of the interfacial tension. The new method is quite appropriate in the measurement of very low interfacial tension, and we applied it to the oil–surfactant–water system. First, we observed the change in the interfacial tension depending on the NaCl concentration in heptane and a water system containing AOT as a surfactant. The interfacial tension had a minimum value, $\sim 1\ \mu$N m^{-1}, for a certain concentration of NaCl. Secondly, the measurement of the ultra-low interfacial tension was carried out changing the temperature of the same system. In this experiment, the critical decrease in the tension was observed near temperature ranges where the microemulsion phase spontaneously appeared. The minimum of the interfacial tension was interpreted as the critical phenomenon close to the second order phase transition. The critical exponent of the interfacial tension obtained from the results is $\nu \sim 1.5$, which is the same as that expected from the mean field theory for the binary mixtures. The experiments successfully demonstrated that this method should be a new tool to study various interfacial phenomena. As another demonstration, we measured the response spectra on the colloidal liquid surface. The results show that the surface tension of a colloidal liquid is the same as that of pure water.

1. Introduction

Recently, the mixture systems of oil and water containing some surfactants have been in the spotlight and many studies have been performed. One of the remarkable properties of these systems is that the interface between oil and water has quite low tension less than 1 mN m^{-1}. This causes the appearance of various phases such as emulsion, microemulsion, or bicontinuous phase. They are quite important materials from the viewpoint of physical interest as well as industrial and medical applications. For instance, active studies have been performed on the relation between the microstructure and the intermolecular force in microemulsions, and the results are reflected in the development of new medicines. The size of microemulsion or emulsion droplet strongly depends on the interfacial energy, *i.e.* the interfacial tension, between oil and water. Therefore, precise measurement of the interfacial tension is a matter of great importance. Of the techniques established

DOI: 10.1039/b405493h

Faraday Discuss., 2005, **129**, 141–153   **141**

for measuring the interfacial tension, Wilhelmy's hanging plate method is a relatively easy technique and gives absolute values. However, it is hard to measure very low tensions because it is measured from the force that pulls the plate. In addition, the plate is in direct contact with the interface and may give the interface some undesirable effects, especially when the interface has some fragile structures such as adsorbed molecular layers. Another technique is the spinning-drop method: a narrow tube containing water with an oil droplet is rotated, and the change in the shape of the droplet gives the interfacial tension. Though useful for ultra-low tension, the spinning-drop method is not effective for systems including a microemulsion.[1,2]

We developed a new method for measuring interfacial tension that is characteristically non-contact yet useful over a wide range of tension, down to ultra-low tension.[3,4] Two focused laser beams are used in this method; one with a high intensity excites and deforms the interface[5] and the deformation is detected by the other laser with a lower intensity. This technique can be understood by an analogy to the laser trapping technique, which is used for a non-contact manipulation of very small particles. Therefore, we refer to our method as the laser interface manipulation (LIM) technique. There are three advantages in the LIM technique over other measuring methods. First, the interfacial wave with an arbitrary wavenumber is generated by modulating the pump laser beam and the spectroscopy of the wave provides various physical properties of the interface, such as the dynamic interfacial tension and the interfacial elasticity, or the adsorption rate of the amphipathic molecules. Secondly, the tension is obtained by measuring the interface deformation given by the external force. In Wilhelmy's method, for example, the tension is obtained as the force given by the interface to pull down the plate. Therefore, a lower tension means a weak force that is less accurate to be measured. In the LIM method, on the other hand, the lower tension makes a larger deformation and hence a higher accuracy. We can adjust the pump laser power to obtain the optimum deformation for measuring. The third advantage is the non-contact operation as mentioned above. Besides those problems of giving damage or too large deformation to the delicate molecular films, the contact probe is likely to pollute the interface seriously. The interfacial displacement in the LIM technique is of the order of 100 nm, at the maximum.

The LIM technique has also been used in other previous works.[6,7] These gave a large deformation in the liquid–liquid systems and studied the relation between the displacement of the interface and the pump laser intensity. Their purpose was to make a giant deformation, however, and not to study the interfacial properties. In this paper, we introduce a measurement system designed to give the surface or interfacial tensions more accurately, and then, we discuss the interfacial properties of the ionic surfactant system.

The basic function of ionic surfactants, such as Aerosol OT (AOT), is governed by the balance between their hydrophilic and hydrophobic nature, which depends on the counter ion concentration and the temperature. The interfacial tension between oil and water containing the ionic surfactant is also dependent on these conditions. For instance, Aveyard *et al.* observed that the interfacial tension between heptane and water with AOT had a minimum value, $\sim 1\ \mu\text{N m}^{-1}$, at a certain NaCl concentration.[8] They also studied the temperature dependence of the same system and found a minimum at a certain temperature.[9] Those studies showed that the interfacial tension undergoes a drastic change with the concentration of NaCl. We therefore decided to design experiments to measure interfacial tensions in heptane and water systems with AOT, changing the NaCl concentration and temperature. We have already reported on the relation between interfacial tension and NaCl concentration,[4] and we abstract the results in to this paper so that it is easy to understand the temperature dependence of the interfacial tension measurements. We discuss the dependence between tension and temperature in terms of critical phenomenon.

II.   Deformation of interface

The light propagating through the liquid–liquid interface or air–liquid interface causes a small deformation of the interface, which is brought about by the discontinuity of the light momentum in between. We assume that light with an intensity I per unit area travels from one medium with refractive index, n_1, and density, ρ_1, to the other with n_2 and ρ_2, as shown in Fig. 1. The light momentum is proportional to the refractive index of the medium, and the discontinuity at the interface generates the pressure p, which deforms the interface. The equation of the momentum

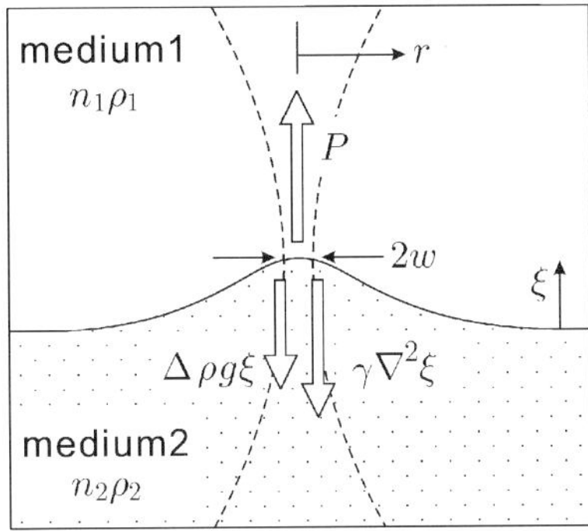

Fig. 1 Three kinds of force at the interface between two liquids with different refractive indexes and densities; P is the light pressure, $\gamma\nabla^2\xi$ is the Laplace force and $\Delta\rho g\xi$ is the gravity. The density of the medium 1 is less than that of medium 2, $\rho_1 < \rho_2$.

conservation is written as:

$$\frac{n_1 I}{c} = (1 - R)\frac{n_2 I}{c} - R\frac{n_1 I}{c} + p, \tag{2.1}$$

where p is defined as positive for the upward direction, R is the energy reflectivity of the light given by $R = (n_1 - n_2)^2/(n_1 + n_2)^2$ and c is the light velocity in a vacuum. The pressure applied to the interface is then given by:

$$p = \frac{2n_1 i}{c} \times \frac{n_1 - n_2}{n_1 + n_2}. \tag{2.2}$$

This equation shows that the light pressure p works from the optically dense matter to the dilute one, independently of the propagation direction. In the case of a laser beam with a Gaussian intensity profile, the pressure p has a dependence on the distance r from the center of the laser beam, and eqn. (2.2) is rewritten as:

$$p(r) = \frac{2n_1}{c} \times \frac{n_1 - n_2}{n_1 + n_2} \times \frac{2I_0}{\pi w^2}\exp\left(-\frac{2r^2}{w^2}\right), \tag{2.3}$$

where I_0 is the laser intensity and w is the half width of the beam at the relevant interface. This pressure should be balanced with the gravity and the Laplace force of the curved interface. The displacement of the interface $\xi(r)$ is given by:

$$p(r) - g\Delta\rho\xi(r) + \gamma\nabla^2\xi(r) = 0, \tag{2.4}$$

where $\Delta\rho$ represents the difference in density between the two media, g is the gravity constant, and γ is the interfacial tension. For the axially symmetric configuration, an arbitrary function can be expanded as the sum of the zero-th order Bessel function. In this case, the interface is deformed symmetrically on the interface plane, and the displacement $\xi(r)$ is written as

$$\xi(r) = P_0 \int_0^\infty \frac{kJ_0(kr)\exp\left(-w^2 k^2/8\right)}{\gamma k^2 + g\Delta\rho}\,\mathrm{d}k, \tag{2.5}$$

where J_0 is the zero-th order Bessel function and P_0 is given by

$$P_0 = \frac{I_0 n_1}{c\pi} \times \frac{n_1 - n_2}{n_1 + n_2}. \tag{2.6}$$

Eqn. (2.5) shows that the displacement of the liquid interface is proportional to the laser intensity and is, approximately, inversely proportional to the interfacial tension. The precise measurement of the interface deformation gives information on the properties of the liquid interface, such as the interfacial tension or the viscosity near the interface. The typical form of $\xi(r)$ calculated with the actual experimental parameters for the pure water surface is shown in Fig. 2. The surface is raised up by *ca.* 10 nm and the curvature at the area of laser beam irradiation is convex.

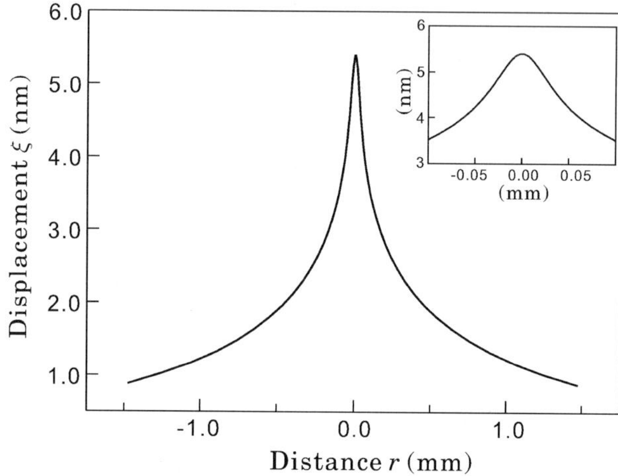

Fig. 2 Calculated curve for the displacement of the water surface under a pump laser with $\lambda = 532$ nm, $w = 30$ μm, and $I_0 = 500$ mW. The inset represents the shape around the center of the deformation in an expanded axis.

The deformation of the interface occurs while the pump laser irradiates the interface. If the laser intensity is modulated at the frequency ω, the displacement of the interface varies in time and excites the interfacial waves with various wavenumbers. The radiation pressure of the modulated laser is given by

$$p(r, t) = \frac{4P_0}{w^2}\exp\left(-\frac{2r^2}{w^2}\right)\exp(i\omega t). \qquad (2.7)$$

The displacement of the interface around $r = 0$ is represented as a function of time t as

$$\xi(r, t) = \frac{P_0}{\rho}\int_0^\infty \frac{k^2 J_0(kr)}{\omega^{*2} - \omega^2}\exp\left(-\frac{w^2 k^2}{8}\right)\exp(i\omega t)\mathrm{d}k. \qquad (2.8)$$

Here, ω^* is the characteristic frequency of the interfacial wave and given by the following dispersion relation with the wavenumber, k,

$$\left(i\rho_{\mathrm{tot}}\omega^* + 2\eta k^2\right)^2 + \rho_{\mathrm{tot}}\left(\gamma k^3 + \Delta\rho g k\right) = 4\eta^2 k^4\left(1 + \frac{i\rho_{\mathrm{tot}}\omega^*}{\eta k^2}\right)^{\frac{1}{2}}, \qquad (2.9)$$

where ρ_{tot} is the sum of the two densities, and η, that of the viscosity.[10] Eqn. (2.8) that gives the time-dependent response of the interface is obtained from the Navier–Stokes' equation with the boundary condition of the pressure at the interface, the incompressibility condition, and eqn. (2.7). While the displacement $\xi(r,t)$ takes the summed form of interfacial waves with various k, this ignores the effect of wave propagation and, therefore, has validity only at the region of laser illumination, $r < w$. Measuring $\xi(r,t)$ gives the interfacial tension and the total viscosity of the material. In the LIM technique, we measured the interfacial gradient at $r = w$ changing the laser modulation frequency ω, and taking the spectrum $S(\omega)$ as shown later,

$$S(\omega) \propto \int_0^\infty \frac{k^3 J_1(kr)}{\omega^{*2} - \omega^2}\exp\left(-\frac{w^2 k^2}{8}\right)\mathrm{d}k. \qquad (2.10)$$

Fig. 3 shows the calculated curve with the same parameters as Fig. 2. The spectrum gives us two characteristic values; one is the amplitude at the lowest frequency and the other is the characteristic frequency at which the curve suddenly decreases. The former represents the static interfacial displacement for the continuous pump laser. At low frequencies, the laser illuminates all the while the interface is deforming. Hence, the amplitude at the lower frequency limit, S_0, agrees with the static displacement given by eqn. (2.5) with $r = 0$, and yields the static interfacial tension.[4] The latter

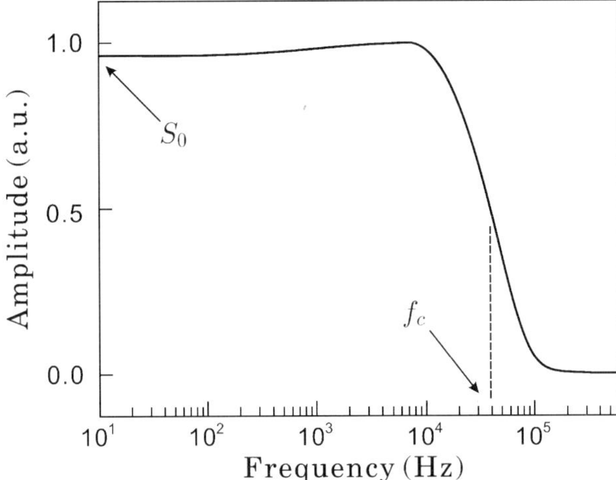

Fig. 3 Amplitude of the frequency spectrum $S(\omega)$ calculated for the water surface. The required parameters are the same as used in Fig. 2.

of the characteristic frequency indicates the typical time constant of the interface dynamics. As shown in Fig. 2, the deformed interface has a convex peak in the laser-illuminating region with a broad skirt of the meniscus. This shape is almost governed by the beam width of the pump laser and the excited waves have cut-off wavenumbers at around $k \sim \pi/2w$. The dispersion relation of the waves gives the characteristic frequency, $f_{\rm c}$, in the spectrum. The amplitude of the interface vibration decreases rapidly at around $f_{\rm c}$, which is roughly estimated to be $2\pi f_{\rm c} \sim (\gamma/\rho)^{1/2}(\pi/2w)^{3/2}$. The characteristic frequency provides us with a rough value of the interfacial tension.

III. Experiments

The typical experimental set-up of the LIM method for the oil–water interface is schematically shown in Fig. 4. Deformation of the interface is given by a frequency-doubled cw-Nd:YAG laser with a maximum power output of 1 W and a wavelength of 532 nm. We adjusted the pump laser intensity with an ND filter for fear of unexpectedly large deformation. Too large a deformation may not be represented by eqn. (2.6) nor (2.8), and is likely to destroy the interface.[11] The pump laser beam is focused onto the liquid interface by a lens with the focal length 150 mm upward through the window on the bottom of the sample cell. For the surface of the liquid impenetrable to the laser

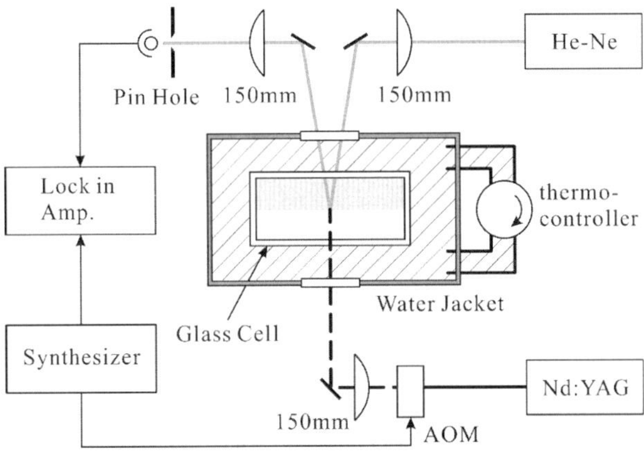

Fig. 4 Typical experimental set-up of the LIM method. The size of the glass cell is 10 mm height and 50 mm diameter.

light, such as the colloidal liquid, the pump laser beam goes downward from the air. The focused diameter of the pump beam is 65 µm. We switched on and off the YAG laser with the acousto-optic modulator (AOM). The first-order diffracted light through the AOM is blinking at the frequency ω. To detect the displacement of the excited interface, we used a He–Ne laser with an output power of 40 mW as a probe. The probe light is focused by a lens with the focal length 150 mm to a spot with 33 µm diameter on the excited area of the interface near the pump light, and reflected to the detector. To measure the maximum interface gradient, we swept the deformed part of the interface with the probe light and found the position where the maximum signal is obtained. The deformed interface works as a curved mirror. When the pump light is on and the interface is deformed, the reflected probe light goes to a different direction, and the probe light intensity detected through a pinhole decreased. The signal is detected by a lock-in amplifier to which reference is given by the modulation signal from the synthesizer.

This measuring system has two advantages, non-contact and a wide dynamic range of interfacial tension, over other interfacial tension measuring methods. As for the former advantage, the LIM method needs no foreign body besides the two laser beams to provide the interfacial properties. Hence, the interface is free from contamination and not broken from a mechanical viewpoint. Secondly, this method is advantageous in measuring interfacial tensions at extremely low levels. Eqn. (2.6) shows that the lower the interfacial tension is, the larger the deformation and the higher the sensitivity. For measuring a surface tension as high as water, we can increase the pump power to a level that yields a sufficient deformation. As mentioned in the following section, the LIM method has a dynamic range of interfacial tension from 100 mN m^{-1} to 1 µN m^{-1}.

We undertook three experiments with this technique to check the overall performance in measuring the interfacial tension, and showed that this system could provide a new approach to interfacial phenomena.

A.   Dependence of oil–water interfacial tension on salt concentration

The ionic surfactant, AOT for instance, changes its affinity to water or oil depending on the concentration of salt. Therefore, the interfacial tension in the system of oil and water with an ionic surfactant often makes a drastic change with a minimum value.[8] The range of the tension decreases from 1 mN m^{-1} down to 1 µN m^{-1}. This liquid system is suitable to check the usefulness of the present method in measuring low interfacial tensions. We performed a series of experiments and obtained γ in systems of heptane, water, and AOT, with different concentrations of NaCl ranging from 0.01–0.1%. The concentration of AOT to water was 2.47×10^{-3} mol l^{-1}, which is a little lower than the CMC of AOT at 25 °C, and there were no micelles in water. In our method, the interface is required to be flat as a mirror, so water with AOT and NaCl was first poured into the sample cell, and then heptane was poured onto the water gently. Fig. 5 shows typical examples of the frequency response spectra of the interface in systems with different salt concentrations. The solid lines are the spectra of eqn. (2.10) fitted with γ as the running parameter. Experimental points are in good agreement with the theoretical curves, and the frequency f_c corresponding to the limiting wavenumber, shown with dashed lines, depends on the salt concentration. The interfacial tensions obtained from the spectra are shown in Fig. 6, together with the previous results obtained with the spinning drop method.[8] This figure shows that the LIM method has an ability to measure the low interfacial tension close to 1 µN m^{-1} accurately. It is also found that the interfacial tension depends on the NaCl concentration and has a minimum at about 0.05 mol l^{-1}. This fact reflects the property of AOT as surfactant. At a salt concentration of 0.05 mol l^{-1}, the balance between the hydrophilic and hydrophobic nature of AOT gives the lowest interfacial tension. At other concentrations, the surfactant has either a hydrophilic or hydrophobic nature and, in both cases, the interfacial tension is larger than that at 0.05 mol l^{-1}. Such behavior of the interfacial tension is clearly observed with LIM measurement.

B.   Dependence of oil–water interfacial tension on temperature

In the previous section, we discussed the dependence of the interfacial tension on the counter-ion concentration in the oil–water–AOT system keeping aside the consideration of the effect of temperature. Aveyard et al. has already performed the experiments and discussed the dependence

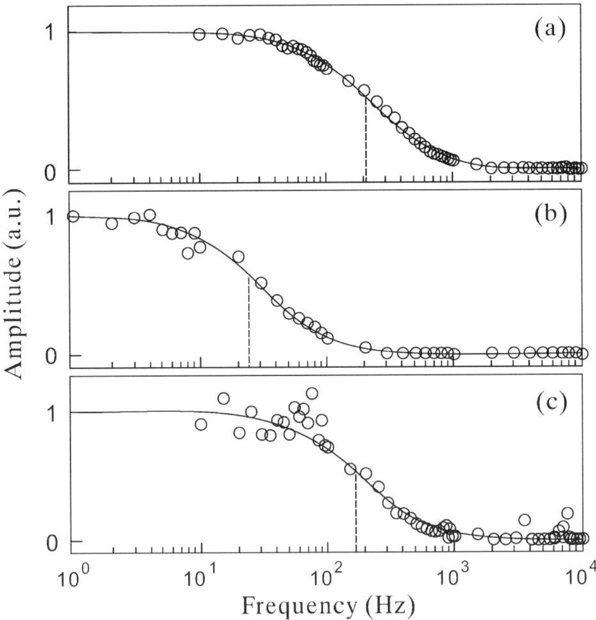

Fig. 5 Typical examples of the spectra obtained in the heptane–water–AOT systems with NaCl concentrations of (a) 0.03 mol l^{-1}, (b) 0.05 mol l^{-1} and (c) 0.1 mol l^{-1}. The dashed lines represent the characteristic frequency, f_c.

of γ on temperature. They introduced an equation of $d\gamma/dT$ with the entropy of the interface and the micelle, and assumed that γ has a minimum value when the micellisation entropy is balanced with the interface entropy. They showed the temperature dependence of the experimental results, without giving any reason why the interfacial tension changes its value with temperature. In this paper, we investigate the relation between the interfacial tension and the temperature and interpret the behavior in terms of the interface critical phenomenon. We designed a series of experiments to measure the interfacial tension between heptane and water with AOT and NaCl on changing the temperature. The temperature dependence expected in the ionic surfactant system is generally very small and hardly observable when the interfacial tension is larger than 1 mN m^{-1} as is the case without NaCl ions. Addition of NaCl to the AOT system decreases the interfacial tension as shown

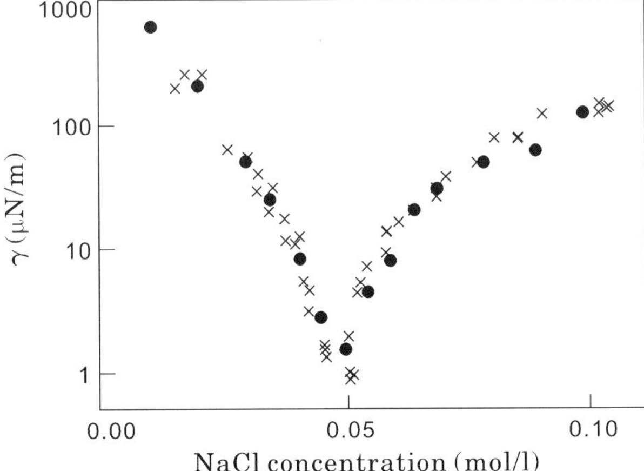

Fig. 6 Interfacial tension *vs.* NaCl concentration. The closed circles represent the experimental values of the present LIM method and the cross marks are the data of Aveyard *et al.* obtained from ref. 8.

in Fig. 6 and the temperature dependence of γ would clearly reveal itself. Therefore we measured the interfacial tensions of the system with 0.05 and 0.06 mol l^{-1} NaCl concentration changing the temperature, as shown respectively by the closed and the open points in Fig. 7. At certain regions of temperature, the microemulsion layer appeared on the interface. In these temperature regions, the interfacial tension was so low that self-emulsification occurred in the thermal fluctuation in a balance between entropy and interfacial energy. The emulsion layer scattered the pump and probe laser light, and the signal-to-noise ratio suddenly increased to prohibit further measurements of γ. We measured the interfacial tension at the temperatures at which the interface kept its flat plane. Either set of data in Fig. 7 is separated into two temperature regions (a) and (b) for 0.05 mol l^{-1} NaCl sample, and (c) and (d) for 0.06 mol l^{-1} sample. Both (b) and (d) regions are the temperature ranges where AOT is hydrophilic and there are more AOT molecules in water than in heptane. In the region (a) and (c), on the other hand, there are more AOT molecules in heptane for their hydrophobic property at higher temperatures.

Fig. 7 shows a marked feature in the relation between the interfacial tension and the temperature. In both 0.05 and 0.06 mol l^{-1} samples, the interfacial tension shows a drastic fall at a certain temperature region. The change rate of γ with respect to temperature increases to 20% per 1 °C. However, the absolute change in γ does not exceed 10 μN m^{-1}, an amount so small that only the LIM method can detect it. The decrease in γ at region (a) is almost the same as that at (c), and the increase at (b) is the same as that at (d). This fact suggests that the change in interfacial tension with temperature is independent of the salt concentration, and implies a possibility of critical behavior in the interfacial dynamics at each temperature region. In Fig. 8, the interfacial tensions are shown *versus* $|T - T_c|$ for the region of interest, where T_c is the critical temperature. The solid line in each figure represents the curve of the equation

$$\gamma = C|T - T_c|^{\nu}, \tag{3.11}$$

fitted to the observed points. Here ν is the critical exponent and C is a fitting constant. The fitted values of ν and T_c are shown in Table 1 for each temperature region. The difference in T_c between (a) and (b) is about 5 °C, and agrees with that between (c) and (d). In addition, ν has almost the same value in all temperature regions. This suggests that the critical behavior of the interfacial tension in respect to temperature seems independent not only of the counter-ion concentration, but also of the hydrophobic–hydrophilic balance of the surfactant.

The values of critical exponents are estimated in the following within the framework of the mean field approximation. In the steady-state two-phase system of oil and water, the concentration distribution $\psi(z)$ of one component is given so that the Helmholtz free energy has the minimum

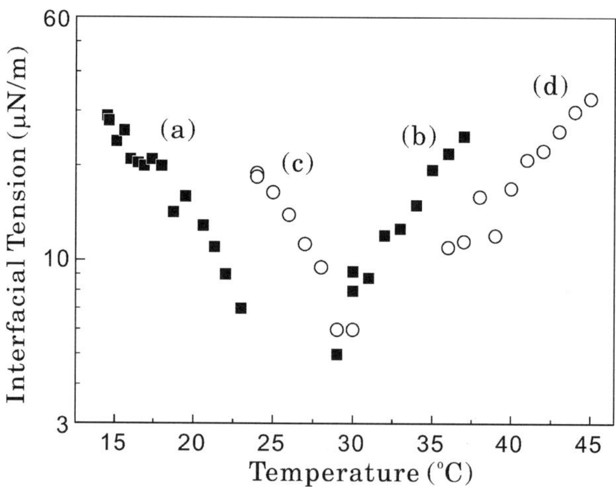

Fig. 7 Temperature dependence of interfacial tension for a NaCl concentration of 0.05 mol l^{-1} (■) and 0.06 mol l^{-1} (○).

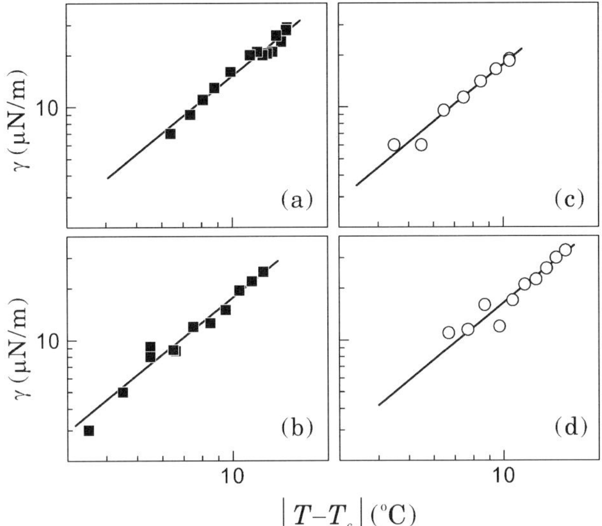

Fig. 8 Correlation between the interfacial tension and $|T - T_c|$. The lines represent eqn. (3.11) fitted to the data points.

value. Here z denotes the axis perpendicular to the interface. The Helmholtz free energy F is written as[12]

$$F = \int \left(-\frac{\varepsilon}{2}\psi^2 + \frac{c}{4}\psi^4 + \frac{B}{2}\left|\frac{\partial\psi}{\partial z}\right|^2 \right) \mathrm{d}z = \int \tilde{f}\,\mathrm{d}z, \qquad (3.12)$$

where

$$\varepsilon = \frac{4k_{\mathrm{b}}(T_c - T)}{a^3}, \quad c = \frac{16k_{\mathrm{b}}T}{3a^3}, \quad B = \frac{2k_{\mathrm{b}}T_c}{a},$$

and a and k_{b} are the mean distance between two molecules and Boltzmann constant, respectively. When the energy F is minimized, $\tilde{f}$ satisfies the following equation,

$$\frac{\partial\tilde{f}}{\partial\psi} - \frac{\partial}{\partial z}\frac{\partial\tilde{f}}{\partial(\partial\psi/\partial z)} = 0, \qquad (3.13)$$

and then $\psi(z)$ satisfies

$$-\varepsilon\psi + c\psi^3 - B\frac{\partial^2\psi}{\partial z^2} = 0. \qquad (3.14)$$

With the boundary condition of ψ, $\partial\psi/\partial z = 0$ at $z = \pm\infty$, the solution of eqn. (3.14) is written as

$$\psi(z) = \sqrt{\frac{\varepsilon}{c}}\tanh\left(\frac{z}{\xi}\right), \qquad (3.15)$$

where ξ is the correlation length and $\xi = a\sqrt{T_c/(T_c - T)}$. The length ξ represents roughly the increased thickness of the interface; as the temperature gets close to T_c, the interface thickness

Table 1 Critical temperature and exponent in the heptane–AOT–water system. Data sets (a) and (b) are for 0.05 mol l^{-1} NaCl concentration, and (c) and (d) are 0.06 mol l^{-1}

Data set	(a)	(b)	(c)	(d)
$T_c/°C$	29.4	24.5	34.5	29.3
ν	1.505	1.491	1.501	1.494

increases and the system changes over to consolution. The interfacial tension is given as the difference between the total free energy F and the free energy in the bulk, and

$$\gamma = \int_{-\infty}^{\infty} \left[\tilde{f} - \left(-\frac{\varepsilon}{2} \psi(\infty)^2 + \frac{c}{4} \psi(\infty)^4 \right) \right] \mathrm{d}z. \tag{3.16}$$

Ultimately, the interfacial tension is given as following with eqns. (3.15) and (3.16),

$$\gamma = \frac{2k_b \sqrt{T_c}}{a^2 T} (T_c - T)^{3/2}. \tag{3.17}$$

Hence, the value of the critical exponent is expected to be $\nu = 3/2$, which is in agreement with our experimental values. This agreement in ν between the mean field theory and the experiments suggests that this surfactant system has quite simple and ideal dependence on the temperature. In the general two-phase system, the correlation between the interfacial energy and the temperature is different from that predicted by eqn. (3.17) because of the fluctuation, and the value of the critical exponent is also different. In this case, however, AOT probably acts both as a molecule of heptane and as a molecule of water, and regulates the concentration of heptane and water molecules in the other phase. For example, when the number of heptane molecules in the water phase increases more than that in the equilibrium state owing to the density fluctuation, AOT shields the extra heptane molecules from water. As a result, the fluctuation is suppressed and the critical behavior of the interface in this system is described with the mean field theory.

In Table 1, the values of ν in (a) and (c) are slightly different with those in (b) and (d) by 0.01. This fact suggests that AOT has different degrees of suppressing effect on the fluctuation when AOT has hydrophilic or hydrophobic properties. The interaction between AOT and water depends more strongly on the temperature than that between AOT and heptane. With the LIM method, such a slight difference of surfactant activity can be detected because of its non-contact measuring.

C. Surface deformation in colloidal liquid

The surface deformation of pure liquids induced by the LIM method is described by eqn. (2.5) or eqn. (2.8). In this section, we discuss the surface behavior of colloidal liquids including small light scatterers. The colloidal particles exist underneath but not on the surface, and the forces working on the surface are only the light radiation pressure $p(r)$, the Laplace forces and the gravity when the pump light is incident on the surface downward from the air. The pump light penetrates the surface and is then scattered by the dispersed particles. In the scattering process, the light loses momentum and the loss works as the scattering force Φ on a dispersed particle, which is approximately given by

$$\Phi = \frac{nIC_{\text{scat}}}{c}, \tag{3.18}$$

where n is the refractive index of the particle and c the light velocity in a vacuum, I the light intensity per unit area and C_{scat} the total scattering cross section of light for one particle. The particles, dispersed homogeneously under the equilibrium state without the laser radiation, are forced to move slightly down by Φ when the light is incident. The new equilibrium of the colloidal particles is determined by the condition that the osmotic pressure Π due to the inhomogeneous distribution balances with the scattering force brought about by the laser radiation. By taking this new apparent pressure into account, the balance equation of eqn. (2.4) is modified into the form,

$$p - g\Delta\rho\xi + \gamma\nabla^2\xi - \Pi = 0. \tag{3.19}$$

The above equation shows that the force inducing the surface deformation decreases by Π, and the displacement of the surface of the colloidal solution is expected to be smaller than that of the pure water surface.

When the pump light is periodically modulated at frequency f, the particles in the subsurface region make an up-and-down motion. The amplitude of the motion is determined as the diffusion length of the particle allowed during half a period of the modulation, that is $\Delta z = \sqrt{k_b T / 3\pi\eta af}$, where k_b is the Boltzmann constant, T is the temperature, η is the viscosity of the surrounding liquid and a is the radius of the particle. In the colloidal solution, the light penetrates straight into the medium to a depth of the mean free path of the light,[13] which $l = 1/\varphi_0 C_{\text{scat}}$. The number of particles which contribute to the osmotic pressure is $\varphi_0 l$ for a unit surface area, φ_0 being the number density of particles in the equilibrium state without the laser radiation. The osmotic pressure on these particles is given by considering the concentration gradient of $\Delta\varphi = \varphi_0 \Delta z/l$, and the force applied to one particle is given as,

$$F = k_{\text{b}} T \left(\frac{\partial \log\varphi}{\partial z} \right)$$
$$= k_{\text{b}} T \, \frac{\Delta z}{\ell^2}.$$

(3.20)

The pressure Π is therefore given as

$$\Pi = F\varphi_0\ell = \sqrt{\frac{(k_{\text{b}}T)^3}{3\pi\eta f}} \times \frac{\varphi_0^2 C_{\text{scat}}}{a^{1/2}}.$$

(3.21)

To examine the effect of dispersed light-scattering particles and the apparent pressure Π, we observed the surface response spectra for polystyrene (PS) latex with diameters of 0.3, 0.6 and 0.8 μm for ten different volume fractions. In the experiments, the pump laser is incident downward to the opaque colloidal liquid. The highly concentrated colloidal liquid yields strong stray scattering of the pump and probe laser lights which would be a serious noise source for the light detector. Therefore, the experiments were carried out for dilute solutions with less than 0.25% of the volume fraction. The typical response spectra obtained for various concentrations are shown in Fig. 9. The amplitude at 70 Hz decreases with increasing volume fraction, while the characteristic frequency f_{c} is independent of the concentration. This fact suggests that the surface deformation decreases as the particle concentration increases, while the surface tension is independent of the concentration. Moreover, the suppression of the surface deformation is found to decrease with frequency. Fig. 10 represents the dependence of the surface tension and the surface displacement at 70 Hz on the volume fraction for various particle sizes. The surface tension approximately takes the same value as pure water for all concentrations and particle sizes. The reason for this strange phenomenon is that the PS particles are only in bulk but not on the surface, and the surface tensions in the dispersed systems are defined as the surface energy of the water. In other words, the surface displacement is determined by the macroscopic inhomogeneous structure near the surface, while the surface tension is given by the microscopic molecular origin. This agrees with eqn. (3.19) in which γ and Π are independent of each other. In addition, the displacement of the surface decreases with concentration, and the tendency is apparent for small particles. The relation between the suppression of the

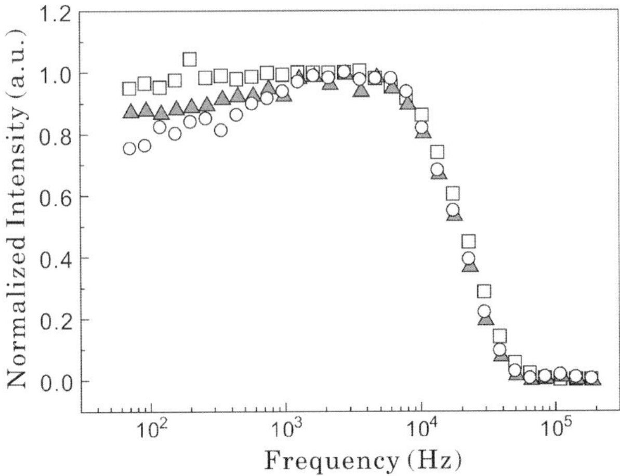

Fig. 9 Typical response spectra in polystyrene latex with volume fractions of 0.07% (□), 0.129% (▲) and 0.196% (○) at 25 °C. The ordinate is normalized so that the amplitude at 2.7 kHz is equal to 1.

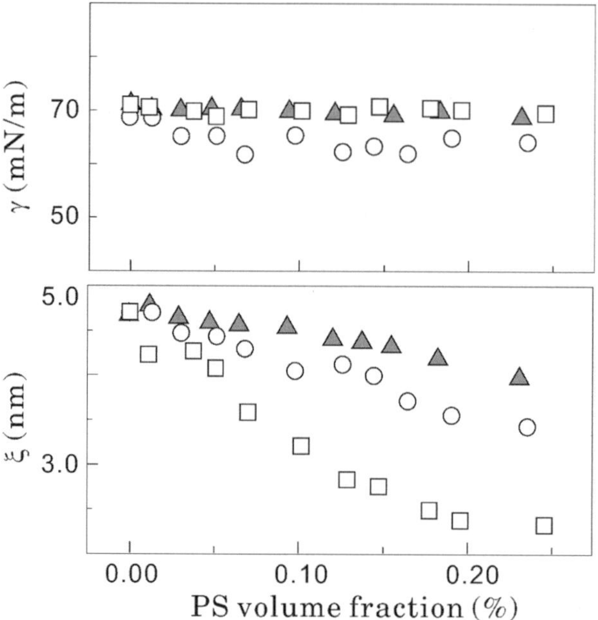

Fig. 10 Surface tension (upper) and surface displacement (lower) in each concentration of polystyrene latex with particle diameters of 0.3 μm (□), 0.6 μm (○) and 0.8 μm (▲). The surface displacements were given by comparing the amplitudes at 70 Hz with that of the pure water.

surface deformation and the concentration of dispersed particles is consistent with eqn. (3.21).

The dependence of the surface deformation at $f = 70$ Hz on Π is shown in Fig. 11, in which C_{scat} is calculated with the Mie scattering theory for spherical particles with the refractive index of polystyrene. Because all of the data points except the squares for 0.3 μm particles are on one straight line, we could conclude that eqn. (3.20) successfully explains the particle motion under the laser radiation. The abscissa is the value proportional to φ^2_0 for each particle diameter. The data points show discrepancy from the straight line for the 0.3 μm particle sample, which has a

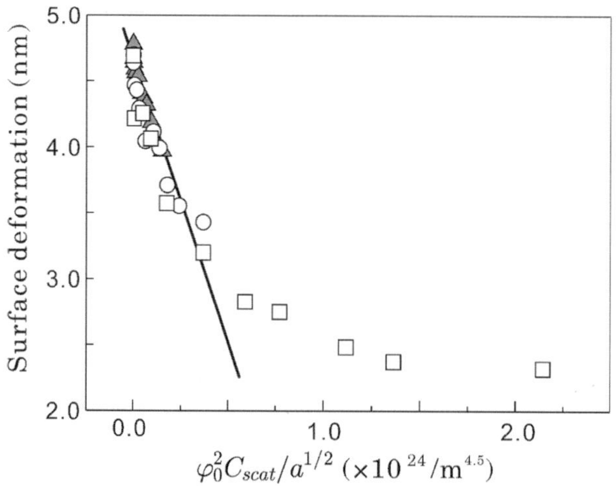

Fig. 11 Surface deformation *vs.* colloid localization pressure. The symbols are the same as in Fig. 10. The straight line is the eye guide.

concentration about ten times higher than the others. In high dispersion density, the particles might not move freely because of collision with each other. The colloid localization pressure, therefore, deviates from eqn. (3.20), and the data are not on the line.

The LIM method applied to the colloidal liquid surface provides many physical properties, including the surface tension and the viscosity of the pure medium, the size and the number density of the particles. The LIM method would be useful to study the dynamics of the colloid dispersion system near the surface.

We summarize the advantages of this laser interface manipulation method as a new technique for measuring the interfacial tension. It is useful over a wide dynamic range from 10^2 mN m^{-1} down to 10^{-3} mN m^{-1} providing a sufficient accuracy up to 0.2%. The key point is in the laser manipulation: we can adjust the driving power depending on the interfacial tension of the specimen under study, and make the ideal deformation for accurate measurements. Furthermore, the simple handling and the non-contact operation of this system have a great advantage in the measurements of very delicate interfaces appearing, for instance, in the critical consolute phenomenon, or fluids in medical production lines that strongly resist being contaminated. This technique has the potential to meet the demands of fundamental scientific studies, as well as industrial evaluation technology.

References

1 R. J. Good, J. T. Ho, G. E. Broers and X. Yang, *J. Colloid Interface Sci.*, 1985, **107**, 290.
2 R. J. Good, C. J. Vanoss, G. E. Broers and Y. Xin, *J. Colloid Interface Sci.*, 1986, **110**, 604.
3 K. Sakai, D. Mizuno and K. Takagi, *Phys. Rev. E*, 2001, **63**, 46302.
4 S. Mitani and K. Sakai, *Phys. Rev. E*, 2002, **66**, 31604.
5 A. Ashkin and J. M. Dziedzic, *Phys. Rev. Lett.*, 1973, **30**, 139.
6 A. Casner and J. P. Delville, *Phys. Rev. Lett.*, 2001, **87**, 54503.
7 A. Casner and J. P. Delville, *Opt. Lett.*, 2001, **26**, 1418.
8 R. Aveyard, B. P. Binks, T. A. Lawless and J. Mead, *Can. J. Chem.*, 1988, **66**, 3031.
9 R. Aveyard, B. P. Binks, S. Clark and J. Mead, *J. Chem. Soc., Faraday Trans.*, 1986, **82**, 125.
10 V. C. Levich, in *Physicochemical Hydrodynamics*, Prentice–Hall, Englewood Cliffs, NJ, 1962.
11 A. Canser and J. P. Delville, *Phys. Rev. Lett.*, 2003, **90**, 144503.
12 S. A. Safran, in *Statistical Thermodynamics of Surfaces, Interfaces, and Membranes*, Perseus Book, 1994.
13 S. Mitani, K. Sakai and K. Takagi, *Jpn. J. Appl. Phys.*, 2000, **39**, 146.

Structural studies of amphiphiles adsorbed at liquid–liquid interfaces using neutron reflectometry

Ali Zarbakhsh,*[a] Aránzazu Querol,[a] James Bowers[b] and John R. P. Webster[c]

[a] *Centre for Materials Research (Chemistry Department), Queen Mary, University of London, Mile End Road London, UK E1 4NS . E-mail: a.zarbakhsh@qmul.ac.uk; Fax: 020 7882 7427; Tel: 020 7882 3259*

[b] *Department of Chemistry, University of Exeter, Stoker Road, Exeter UK EX4 4QD*

[c] *CCLRC, ISIS Facility, Rutherford Appleton Laboratory, Chilton, Didcot, UK OX11 0QX*

Received 1st April 2004, Accepted 5th May 2004
First published as an Advance Article on the web 22nd September 2004

We report the application and refinement of a recently developed method for structural studies at a liquid–liquid interface using neutron reflectometry. The technique involves the entrapment of a thin oil layer between a silicon substrate and an aqueous subphase. The thin oil film is prepared by spin-coating an oil film on to an oleophilically treated silicon substrate. During the reflectivity measurement the sample is maintained in a horizontal position, and the angle of incidence of the neutron beam is varied using a supermirror. Attenuation of neutron reflectivity at the lowest angle of incidence is used to determine the oil-layer thickness. We report information regarding the structure at the interface between hexadecane and a 0.1% w/v aqueous solution of the triblock copolymer Pluronic L64 with $EO_{13}PO_{30}EO_{13}$ (EO = ethylene oxide; PO = propylene oxide) and the interface between hexadecane and a 3.7 mmol dm^{-3} ($\approx$critical micelle concentration) aqueous solution of the cationic surfactant tetradecyltrimethylammonium bromide ($C_{14}TAB$). For the $C_{14}TAB$ system, the reflectivity data unambiguously reveal the presence of a region highly concentrated in $C_{14}TAB$ on the oil side of the interface. For the Pluronic L64 system, the data suggest that the polymer adsorbs at the interface occupying both oil and water sides of the interface. Model scattering length density profiles that capture these features are presented and further models that better fit the data are discussed.

Introduction

Resolving structures at buried liquid–liquid interfaces contributes to our current understanding of many important physico-chemical and biological processes such as the transport properties of cell membranes and the stabilisation of emulsions used for drug delivery. Structure determination at liquid–liquid interfaces has proven in the past to be experimentally challenging, and it is only recently that significant progress has been made in enabling the structure at such buried interfaces to be routinely resolved. In particular, advances have been made in the application of non-linear spectroscopic and X-ray scattering methods to the study of liquid–liquid interfaces, and information regarding this progress can be found in the reviews by Richmond[1] and Schlossman.[2] However,

DOI: 10.1039/b404732j

despite the benefits provided by neutron reflectometry for the investigation of systems containing hydrogenous substances, progress in the application of neutron reflectometry to the study of liquid–liquid interfaces has been hampered by the lack of a suitable experimental design. The chief experimental problem is minimising the drastic attenuation of a neutron beam upon transmission through a liquid medium. To overcome this problem it is necessary to create and maintain a thin film of the upper liquid phase through which the incident and reflected neutron beams are transmitted. Pioneering experiments achieved this thin film either by reliance on spreading[3,4] or by condensation of a volatile oil on to an aqueous subphase.[5–9] Alternatively, the influence of oil on the interfacial structure can be examined by adsorbing the oil into an adsorbed monolayer.[10] Despite its success, the condensation method remains a very difficult experiment and possesses some limitations. This method relies on balancing the condensation and drainage rates of the oil film such that a sufficiently thick oil film wetting the lower liquid phase is maintained. However, the maintenance of a uniform condensed oil layer is a non-trivial and very time-consuming task. Furthermore, the technique is restricted to the study of volatile oils and therefore to the adsorption of oil-insoluble molecules. Consequently, we have developed a new, versatile and complementary approach for the investigation of the structure of adsorbed polymers and surfactants at oil–water interfaces.[11,12] This approach is suitable for the study of oil-soluble, water-soluble and insoluble amphiphiles as well as being amenable (and in the majority of cases, advantageous) to the study of non-volatile oils. In principle, the methodology is not restricted to the study of solely oil–water interfaces, however, it is upon this particular range of systems that we base our description. In our approach, a thin, uniform oil layer is trapped between an oleophilic silicon substrate and a bulk aqueous subphase. This oil film is created by spin coating a thin (~ 1 μm) layer of oil on the silicon substrate, freezing the oil, then sandwiching this layer between the silicon superphase and the aqueous subphase.

However, concerns remain from these early experiments regarding the behaviour of the oil film during the measurement. Since the sample is tilted by a small amount ($\leq 1.8°$) during the process of the reflectivity measurement, when an aqueous surfactant solution with low interfacial tension is being investigated potential drainage of the oil film becomes a real concern. Although one expects the long-range interactions to stabilise the film, it has been noticed in previous experiments that subtle shifts of total reflection edge position occur with time, suggesting film drainage. To circumvent the potential problems presented by film drainage and rupture, we have refined the experimental design by maintaining the sample in a horizontal geometry throughout the whole experiment. This is achieved by deflecting the neutron beam to the desired incidence angle by means of a supermirror.

A second change we have made to our previous methodology[12] involves the data analysis. Previously, we have measured the reflectivity from a Si–oil interface and have directly incorporated these data in the determination of the reflectivity from the buried oil–water interface. Although this is an elegant approach in principle, it is unfortunately flawed in practice. The disadvantage is that the measurement of the Si–oil reflectivity incurs a significantly higher background than the measurement from the Si–oil–water composite interface for many useful neutron refractive index contrasts. The consequence of this is a truncation of the useful momentum transfer range and ultimately a loss of detailed structural information. Accordingly, we have refined our data analysis approach.

In this paper we report the application of this refined approach to reflectivity measurements from liquid–liquid interfaces to determine the structure of the interface between hexadecane and aqueous solutions of amphiphiles. Our objectives were to establish confidence in the reproducibility of the measurements and to begin to obtain structural information regarding the organisation of adsorbed molecules at the liquid–liquid interface.

Experimental

Choice of system

As test systems we decided to examine systems of practical interest and for which a precedent in terms of structure determination using neutron reflection has been set. Adsorption of *n*-alkyltrimethylammonium halide surfactants has been extensively studied by neutron reflection. As an example of this class of surfactant we study here tetradecyltrimethylammonium bromide, $C_{14}TAB$.

For this system we are working sufficiently above its Krafft temperature, and the structure at the air–water interface of the surfactant is well documented.[13] Pluronic polymer surfactants are symmetric triblock copolymers of poly(ethylene oxide) (PEO) and polypropylene oxide (PPO) of general formula $(EO)_n(PO)_m(EO)_n$ and present a wide range of properties in solution depending on the specific values of n and m in each case. The copolymer studied here is the Pluronic L64 which has molar mass 2900 g mol^{-1}, with $n = 13$ and $m = 30$, that is, $(EO)_{13}(PO)_{30}(EO)_{13}$. A range of Pluronic surfactants has previously been studied at the hexane–water interface using the condensation method[9] and thus comparisons can be made between the results determined by the two methods. We use hexadecane as the oil, since it is involatile and possesses a convenient freezing temperature.

Sample environment

The general principles of the measurements have been discussed elsewhere,[11,12] as have the details of the sample cell, and preparation of the oleophilic silica surface by coupling of trimethylchlorosilane. The hexadecane film was frozen in place whilst the aqueous phase was syringed into the sample cavity. The sample cell was thermostatted at 298 ± 1 K during the reflectivity measurements, well above the bulk freezing temperature of $T_m \approx 289$ K for hexadecane.

Materials

Tetradecyltrimethylammonium bromide, $C_{14}TAB$, was obtained from Lancaster synthesis (98%), the L64 Pluronic from Fluka; hexadecane-h_{34} from Aldrich (99%), hexadecane-d_{34} from Cambridge Isotope Laboratories (>98 atom D%); D_2O from Fluorochem (99.9%). These substances were used without further purification for this study. A 3.7 mmol dm^{-3} solution of $C_{14}TAB$ in D_2O and a 0.35 mmol dm^{-3} (0.1% w/v) solution of Pluronic L64 in D_2O were prepared. The hexadecane used in the reflectivity measurements was a mixture of hexadecane-h_{34} and hexadecane-d_{34} with a scattering length density of nominally 4×10^{-6} Å^{-2}.

Neutron reflection measurements

Reflectivity measurements were conducted on the reflectometer SURF at the ISIS Spallation Neutron Source, Rutherford Appleton Laboratory, Didcot, UK. The neutron beam is polychromatic with wavelengths in the range $0.53 < \lambda < 6.9$ Å. To obtain the widest amenable momentum transfer, Q, range, reflectivity spectra are measured for a series of grazing incidence angles θ, where θ is the angle of incidence at the Si–oil interface, and $Q = (4\pi\sin\theta)/\lambda$ is the momentum transfer at the Si–oil interface. The nominal incidence angles used were 0.25, 0.32, 0.50, 1.18°. The actual angles of incidence are determined by performing detector angle scans in reflection geometry once the height alignment of the sample has been performed. Previously we have obtained these incidence angles by means of tilting the sample on a goniometer and utilising a θ–2θ scattering geometry. Here the angles are achieved through the use of a supermirror, allowing the sample cell to be maintained in a horizontal plane for the duration of the experiment, and a θ–θ scattering geometry is employed. The collimating slit settings are varied with incidence angle in order to measure all reflectivities with near-constant angular resolution. In these measurements the resolution, $\delta\theta/\theta$, was 4% and the sample was under-illuminated with an illuminated length of ~ 45 mm projected on the sample.

The measured reflectivity profiles are normalised relative to the incidence beam monitor spectrum and corrected for detector efficiency as per standard reflectivity measurement on a time-of-flight instrument,[14] and the data are subsequently corrected for the wavelength-dependent transmission through the silicon super-phase. Normalisation by the monitor spectrum also accounts for the supermirror efficiency. The data are truncated at an appropriate wavelength cut-off, which is determined by the angle of incidence of the neutron beam at the supermirror. Transmission measurements through samples of hexadecane with different H/D composition were measured in 2 mm Hellma spectrophotometric cuvettes. These measurements allowed the linear absorption coefficient, χ, to be determined for the oil phase. We discuss how the attenuation of the neutron beam upon transmission through the oil film is accounted for in the data analysis and also the process of normalising the reflectivity data on to an absolute scale in the following section. Reflectivity measurements were repeated both with the existing sample and with freshly prepared samples. Entirely reproducible reflectivity data were obtained. This provides confidence in stability of the oil layer and reliability of the measurements and is an essential factor for the further application of this method.

Data analysis

Neutron reflectivity is a technique sensitive to the average neutron refractive index, n, profile normal to an interface.[15] The dispersive refractive index can be written as

$$n(\lambda) \approx 1 - \frac{\lambda^2}{2\pi} Nb + i\frac{\lambda}{4\pi} N\sigma \tag{1}$$

where λ is the neutron wavelength, $Nb = \sum_i N_i b_i$ and $N\sigma = \sum_i N_i \sigma_i$, with N_i the number density, b_i the coherent scattering length, and σ_i the absorption and incoherent cross-section of nucleus i. The multiple Nb is known as the scattering length density of a medium with refractive index n. According to eqn. (1), the large difference in the scattering lengths of ^{1}H ($b = -3.7406$ fm) and ^{2}H ($b = 6.671$ fm) can be exploited in hydrogenous systems. For our present purpose it is convenient to replace σ with the linear absorption coefficient $\chi = \chi(\lambda)$ which embraces all loss processes and is determined from transmission measurements.

Since Nb is approximately linearly related to the volume fraction composition—$Nb \approx \sum_j \phi_j \overline{Nb_j}$, where ϕ_j is the volume fraction and $\overline{Nb_j}$ is the scattering length density of species j—a layer model with discrete strata (see Fig. 1) representing regions with different chemical composition can be constructed and the reflectivity from such a model can readily be calculated. Each layer, i, has a thickness, d_i, refractive index, n_i, scattering length density, Nb_i, and interfacial roughness, σ_i. The reflectivity can be calculated using, among other methods, the standard optical-matrix method[16] and the parameters of the proposed layer model can be optimised using non-linear least-squares fitting. The Gaussian interfacial roughness used in the modelling is the r.m.s. roughness of the interface. The incorporation of the roughness into the reflectivity calculation is based on the formalism proposed by Cowley and Ryan.[17]

For a relatively thin (typically < 10 μm) oil film, the reflections from both the Si–oil and oil–water interfaces contribute to the measured reflectivity. If the incidence angle and the Si–oil and oil–water scattering contrasts are appropriate, total reflection from both these interfaces occurs and two critical edges are observed; the locations of these edges correspond to total reflection from the Si–oil (high-λ edge) and Si–water (low-λ edge) interfaces. The intensity of the low-λ edge is reduced in magnitude due to the attenuation by the oil film. The reflectivity data for the lowest angles of incidence measured shown in Figs. 3 and 4 demonstrate this for a choice of $Nb_{oil} = 4.00 \times 10^{-6}$ Å^{-2} and $Nb_{water} = 6.35 \times 10^{-6}$ Å^{-2}. The decrease in intensity between the two critical edges is used to calculate the oil-film thickness using the Beer–Lambert law—the transmission, $T = T(\lambda)$, at a given wavelength is $T = \exp(-\chi l)$, where l is the pathlength—and this is incorporated in the reflectivity calculations.

Because of the nature of the measurements (using a white beam), the wavelength-dependence of the oil transmission means that the individual reflectivity spectra measured from the composite Si–oil–water interface cannot be directly overlapped prior to oil-transmission correction to create a standard reflectivity *vs.* Q data set. This presents a minor problem in terms of establishing scale (normalisation) factors for each spectrum, even if a total reflection edge is present. Thus scale factors for the individual spectra measured at different incidence angles are determined by calibration from a Si–D$_2$O interface for the same instrument geometry at each nominal angle. The measured data are in support of scale factors determined in this manner, with the agreement being good to within $\pm 2\%$. The reflectivity data shown have been normalised by these factors.

The starting point for the analysis of the reflectivity data is the thick film approximation for the calculation of the overall reflectivity, R_{tot}, from the Si–oil–water composite interface. It is assumed that the overall reflectivity arises from the reflections from two interfaces, with reflectivities R_1 from the Si–oil interface and R_2 from the oil–water interface, separated by a thick absorbing oil layer. For a sufficiently thick oil layer, the phase shift and hence the interference fringes arising from the presence of the layer are beyond the resolution of the experiment. In this approximation, the expression for R_{tot} is derived in the same manner as that for the reflectance from a thin film using classical optics except that the phase factor is replaced by an attenuation factor $A = A(\lambda)$ and the amplitudes (reflectance coefficients) are replaced by intensities (reflectivities). We assume from this point that all the measured spectra have been corrected to account for the attenuation due to transmission through the Si substrate and for supermirror efficiency.

In the following discussion we consider neutrons with a range of wavelengths λ (a 'white' beam) such as that used on SURF and incident with a single fixed incidence angle θ. The thick film reflectivity formula is:

$$R_{\text{tot}} = R_1 + \frac{A R_2 (1 - R_1)^2}{1 - A R_1 R_2} \tag{2}$$

where the attenuation factor incurred by crossing the oil film twice, A, is defined as

$$A = \exp\left(\frac{-2\chi d_{\text{oil}}}{\sin \theta_{\text{oil}}}\right) \tag{3}$$

where d_{oil} is the thickness of the oil film, and θ_{oil} is the incidence angle in the oil film. In earlier analyses, we transformed eqn. (2) to explicitly derive R_2 based on the measured R_{tot} and R_1. Here, to retain the desired Q-range, we perform optical-matrix calculations of a multiple-layer model for the complete interfacial system, but with the Si–oil and oil–water interfaces de-coupled to enable the thick film approximation to be applied to the calculation of R_{tot}. Fig. 1 shows the model used. R_1 is calculated using additional measurements (*e.g.*, from Si–D$_2$O reflectivity measurement) or assuming appropriate parameters for the silicon oxide layer and the trimethylsilane coupled layer. For the majority of cases the precise details of these parameters do not influence the outcome of the data

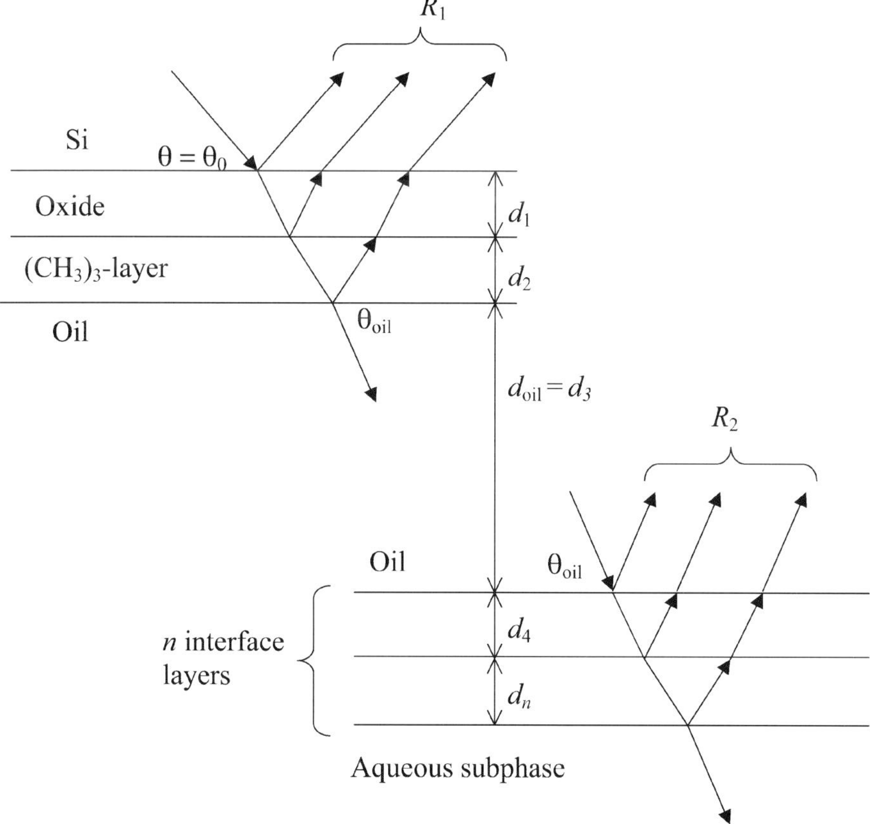

Fig. 1 Schematic representation of the model used for calculation of the structure at the buried oil–water interface. For clarity, only the primary reflections from each interface are shown.

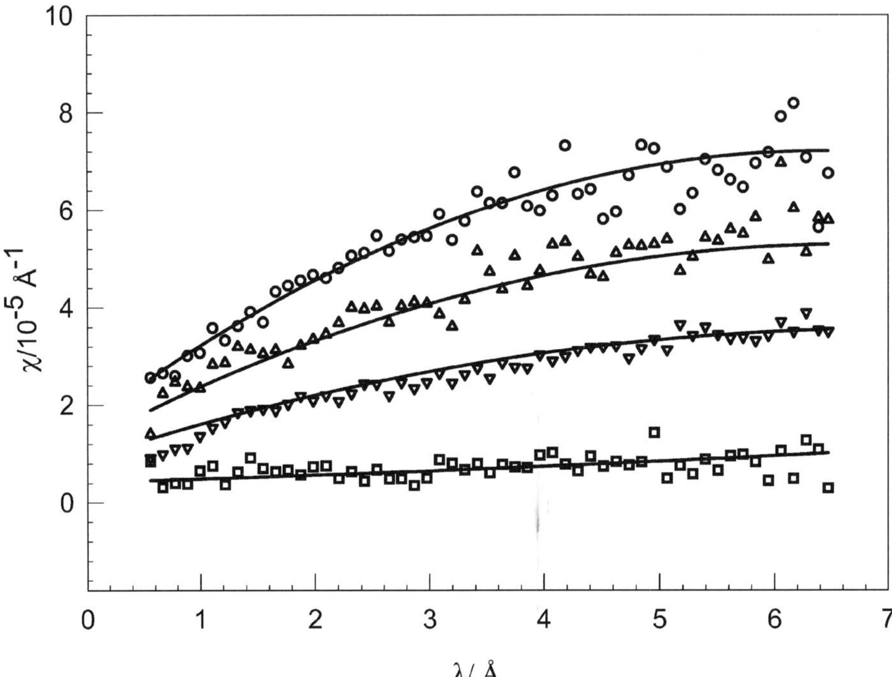

Fig. 2 Attenuation coefficient, χ, *versus* wavelength, λ, for hexadecane-h_{34} + hexadecane-d_{34} mixtures with scattering length densities, from top to bottom, of $Nb_{\mathrm{oil}} = 0.0$, 2.1, 4.2 and 6.8×10^{-6} Å^{-2}. The solid lines are polynomial fits to the data and are used to construct a global function, $\chi(Nb_{\mathrm{oil}}, \lambda)$, that enables the attenuation coefficient, A, to be determined for different Nb_{oil}.

analysis. The wavelength-dependent angles of incidence, θ_{oil}, of the neutrons propagating through the oil layer are retained as the incidence angles at the buried oil–water interface. With this spectrum of incidence angles, the calculation of R_2 is conducted for the layer structure at the buried oil–water interface; the analysis program caters for any number of layers at this interface. A macro, OILWATER.GCL,[18] is available in the OpenGENIE framework[19] for data analysis, and uses non-linear least-squares fitting employing either the Simplex algorithm or simulated annealing to determine the parameters describing the Si–oil–water composite interface.

The calculation of the attenuation factor, eqn. (3), involves the incorporation of $\chi(\lambda)$, which depends on the H/D composition of the oil. By fitting a series of polynomials to the transmission data for a range of H/D contrasts of oil and then interpolating between these polynomials, $\chi(\lambda)$ is determined for the specific contrast of oil used. The polynomial determined is:

$$\chi(Nb_{\mathrm{oil}}, \lambda) = \{(0.021\ Nb_{\mathrm{oil}} - 0.135)\ \lambda^2 + (-0.248\ Nb_{\mathrm{oil}} + 1.742)\ \lambda$$
$$+ (-0.177\ Nb_{\mathrm{oil}} + 1.617)\} \times 10^{-4} \tag{4}$$

where Nb_{oil} is the magnitude of the scattering length density of the oil expressed in Å^{-2} and λ is the magnitude of the wavelength expressed in Å. The experimental transmission data and polynomial fits are shown in Fig. 2. Before modelling the structure at the oil–water interface, the oil film thickness is determined from the R_{tot} data at the lowest incidence angle.

Results and discussion

The measured reflectivity data are shown in Figs. 3 and 4 for the C_{14}TAB and Pluronic L64 systems, respectively. The details of the specific incidence angles, θ, are given in the figure captions. The solid lines correspond to modelled reflectivity profiles based on the scattering length density profiles shown in Fig. 5. The neutron reflectivity data for all four angles were simultaneously fitted to a single model using the same values of layer thickness and scattering length density. In the modelling,

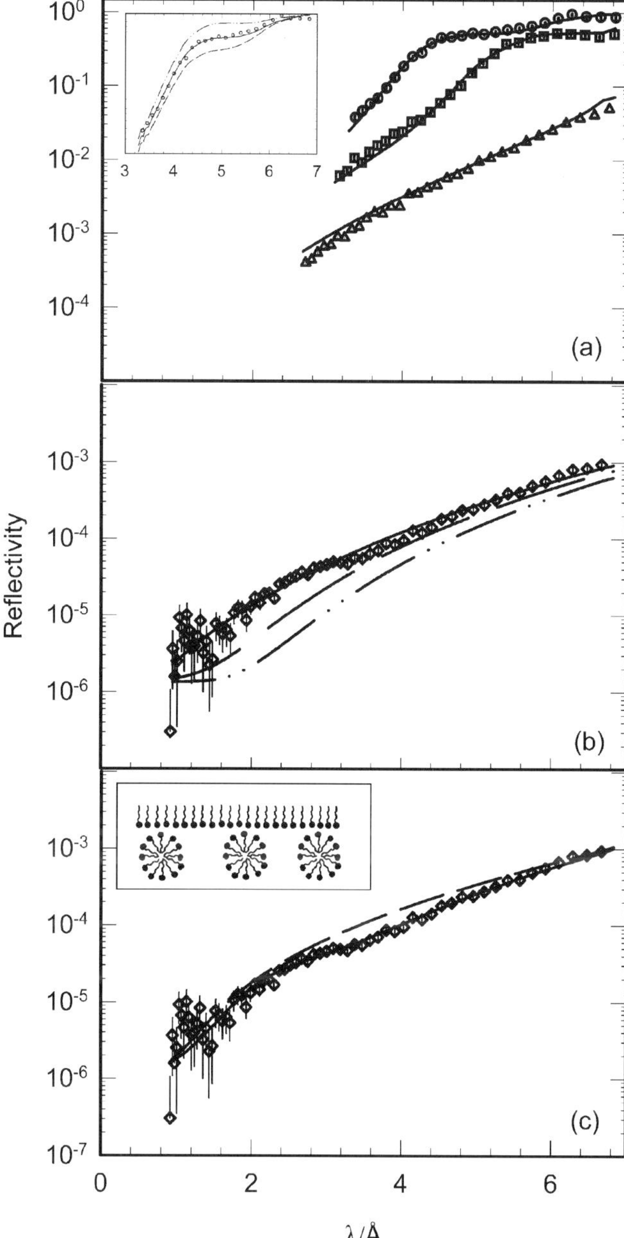

Fig. 3 Reflectivity data from the Si–hexadecane-solution interface for a solution of $C_{14}TAB$ in D_2O with concentration of 3.7 mmol dm^{-3}. The angles of incidence are (a) 0.29° ($\bigcirc$), 0.37° ($\square$), 0.52° ($\triangle$), and (b) and (c) 1.18° ($\diamond$). The solid lines in (a) and (b) are determined using a single-layer model. In (b) the calculated reflectivity for the bare oil–water interface in the absence of any roughness or adsorbed material is shown (--) as is the profile for a bare interface with roughness of 20 Å (-··-). In (c), the high angle data are shown again with the reflectivity from a two-layer model shown (—); also shown is the reflectivity corresponding to a one-layer model representing the adsorbed amount of material determined by Simister *et al.*[13] at the free surface of the solution (--). The scattering length density profiles corresponding to these models are shown in Fig. 5(a); see text for further details. The inset to (a) shows the sensitivity of the reflectivity to the oil film thickness: $d_{oil} = 1$ μm (-··-); $d_{oil} = 3$ μm (—); $d_{oil} = 5$ μm (--). The inset to (c) is a cartoon depicting a possible structure corresponding to the two-layer model used to calculate the reflectivity (—).

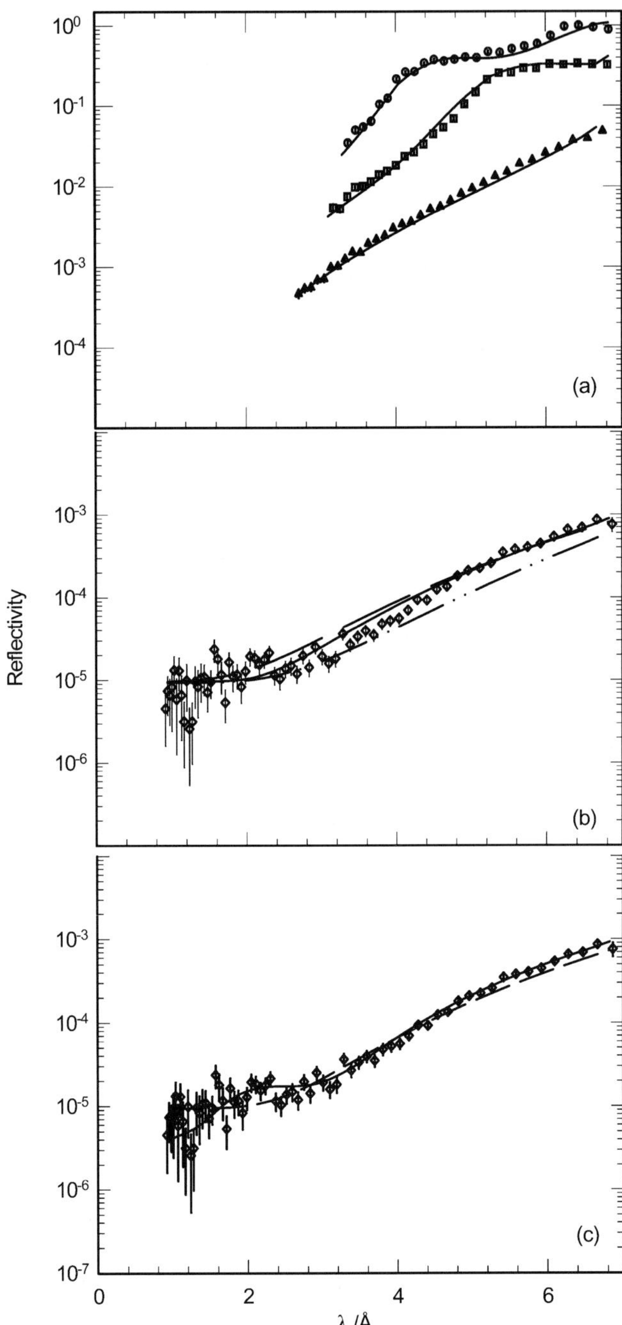

Fig. 4 Reflectivity data from the Si–hexadecane-solution interface for a solution of Pluronic L64 in D_2O with concentration of 0.35 mmol dm^{-3}. Data and modelled reflectivities for angles of incidence of (a) 0.26° ($\bigcirc$), 0.36° ($\square$), 0.51° ($\triangle$), and (b) and (c) 1.18° ($\diamond$). The solid lines in (a) and (b) are determined using a two-slab model. In (b) the calculated reflectivity for the bare oil–water interface in the absence of any roughness or adsorbed material is shown (--) as is the reflectivity profile for a bare interface with roughness of 20 Å (-··-). In (c) reflectivities calculated for two further models are shown: the two-slab model used in (b) with added roughness (--) and a more spatially extensive two-layer model with added diffuseness (—). The corresponding scattering length density profiles for these models are shown in Fig. 5(b); see text for details.

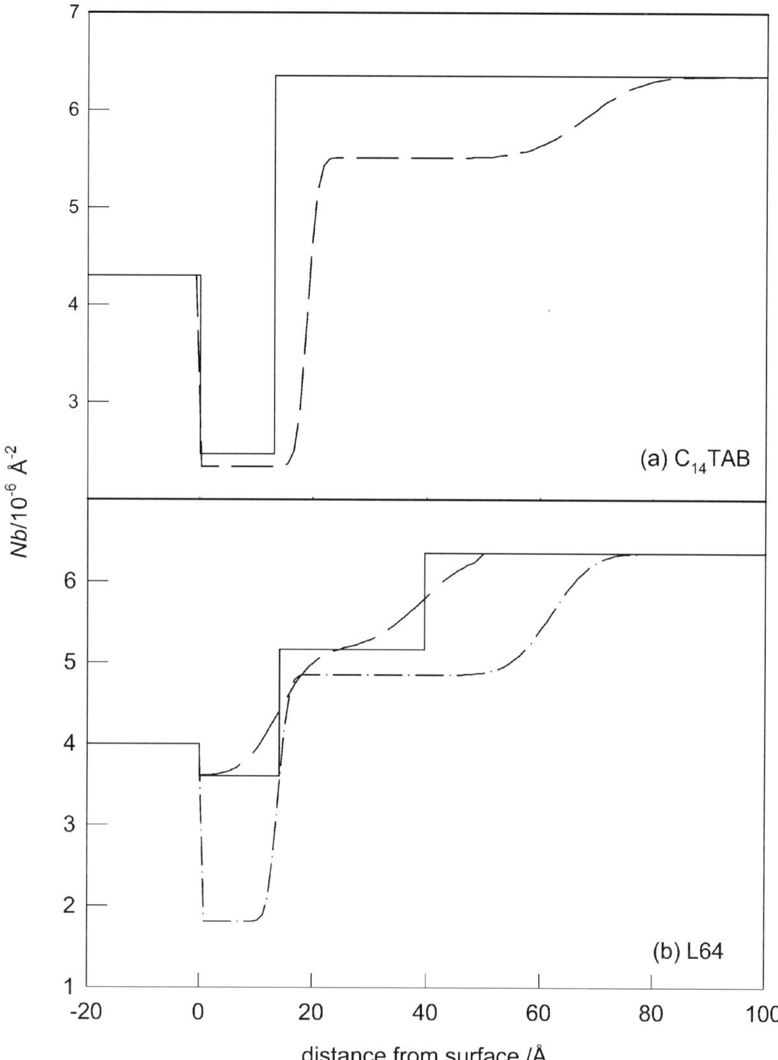

Fig. 5 Representative model scattering length density profiles used to model the hexadecane-solution interface for the two systems studied. (a) C_{14}TAB: the solid line in Fig. 3(b) (—); the solid line in Fig. 3(c) (--). (b) Pluronic L64: the solid line in Fig. 4(b) (—); the dashed line in Fig. 4(c) (--); the solid line in Fig. 4(c) (-··-).

the thickness and scattering length density of the oxide (oleophilic/trimethylsilane) film was $d = 8$ Å, $Nb = 3.4 \times 10^{-6}$ Å^{-2} ($d = 5$ Å, $Nb = 0.5 \times 10^{-6}$ Å^{-2}) with added interlayer roughness, each of 2 Å.

For the purposes of this paper, we have modelled the structure at the oil–solution interface using simple slab models with no added interlayer roughness. These models enable both the reflectivity data to be satisfactorily represented and the salient features of the interfacial structure to be captured without over parameterisation. Although this does not explicitly consider essential features of the interfacial structure, such as capillary-wave broadening and the specific oil, water and amphiphile distributions, such features may emerge from inspection of the scattering length density profiles.

To interpret the $Nb(z)$ profiles in terms of material composition, the amphiphiles have the following assumed Nb values: $Nb_{C_{14}TAB} = -0.23 \times 10^{-6}$ Å^{-2}; $Nb_{PEO} = 0.47 \times 10^{-6}$ Å^{-2} and $Nb_{PPO} = 0.29 \times 10^{-6}$ Å^{-2}, for the different blocks of the Pluronic L64.

For both systems, a thick oil film is present as demonstrated by the presence of a clear double reflection edge in the lowest-angle data in Figs. 3 and 4. The actual oil film thicknesses used are 3 μm and 6 μm for the C_{14}TAB and Pluronic L64 systems, respectively. The inset to Fig. 3 shows the sensitivity of the lowest-angle data to oil film thickness. Although we have fixed the angular resolution defined by the instrument geometry, it has been necessary to use a resolution of 8–10% in the data analysis to fit the data near the double critical edge for both systems. This may have implications for true oil film thickness determination, but for the present data this resolution only influences the fits to the data at the lowest two angles.

(a) C_{14}TAB at the hexadecane–water interface

It can be immediately seen from Fig. 3(b) that the reflectivities measured at the highest angle are significantly greater than the reflectivities calculated assuming a step-like oil–water interface with no roughening and no adsorbed material. With the contrast scheme used, *i.e.*, with protonated surfactant, this enhancement in reflectivity cannot be solely attributed to roughening or diffuseness of the interface, as illustrated in Fig. 3(b). The implication of this is that there is a minimum in $Nb(z)$. Although the significance is clear—the reflectivity data reveal the presence of a film of adsorbed protonated material at the interface—a clear interpretation is not. At the simplest level, this minimum in $Nb(z)$ may be interpreted as arising from the presence of the alkyl chains of the C_{14}TAB on the oil side of the interface, and in fact, the reflectivity data can reasonably be represented using a one-layer slab model in which the interfacial layer is rich in ^{1}H. The solid lines in Figs. 3(a) and (b) are the fits to such a model, the corresponding scattering length density profile is shown in Fig. 5(a). A layer of 13 Å with $Nb = 2.4 \times 10^{-6}$ Å^{-2} has been used in the fits shown. However, this pair of parameters is by no means unique and thus we have ascertained the correlation between fitted parameters that can be used to describe the reflectivity data. This is achieved by means of a contour plot of the weighted sum-of-squares deviations of two parameters, in this case the thickness and scattering length density of the interfacial slab. Such a contour plot is shown in Fig. 6. The minimum of this plot reveals that the acceptable parameters are: $d = 13 \pm 4$ Å and $Nb = (2.0 \pm 0.8) \times 10^{-6}$ Å^{-2}. The model we show here corresponds to the high-scattering-length-density limit of acceptable fits to the data and clearly demonstrates the presence of a protonated region. The thickness of the film is smaller than the length of a fully extended tetradecyl chain (~ 19 Å).

Unfortunately, the exact determination of volume fraction composition profiles is not possible here; with a single contrast it is not possible to determine a unique model of the interface structure in terms of the oil, water and amphiphile distributions. Accordingly, volume fraction profiles are not shown here and the adsorbed amounts presented are indicative. The adsorbed amount can be estimated using[20]

$$\Gamma = \frac{d(Nb_s - Nb_2)}{N_A V(Nb_s - Nb_a)} \tag{5}$$

where d is the adsorbed film thickness, N_A is Avogadro's constant, V is the molecular volume (which can be determined using Tanford's formula), and the Nb_i are scattering length densities of the solvent ($i = s$), the film ($i = 2$), and the adsorbate ($i = a$). Although the region containing the adsorbed surfactant probably contains both oil and water, eqn. (5) can nonetheless be used to estimate acceptable bounds for the adsorbed amount. Based on the presented model, these limits are 1.34×10^{-6} mol m^{-2} (1.96×10^{-6} mol m^{-2}) if oil (water) is assumed as the solvent. The adsorbed amounts correspond to an average area per molecule of 124 Å^2 (85 Å^2) for oil (water) as the solvent. At the air–solution interface[13] the area per molecule at the cmc is 45 ± 2 Å^2 and the film thickness is 19 ± 2 Å. The adsorbed amount at the oil–water interface therefore appears to be lower than found at the free surface. The dashed line in Fig. 3(c) represents the free surface model applied at the liquid–liquid interface, and clearly does not adequately represent the data. However, we should treat the estimated adsorbed amount with caution as we do not know the true surfactant distribution or the oil and water distributions underlying the adsorbed layer and the model of the interface has been purposely oversimplified. To illustrate this we show in Fig. 3(c) a further model that represents the data, with scattering length density profile also shown in Fig. 5(a). The modelled reflectivity is in

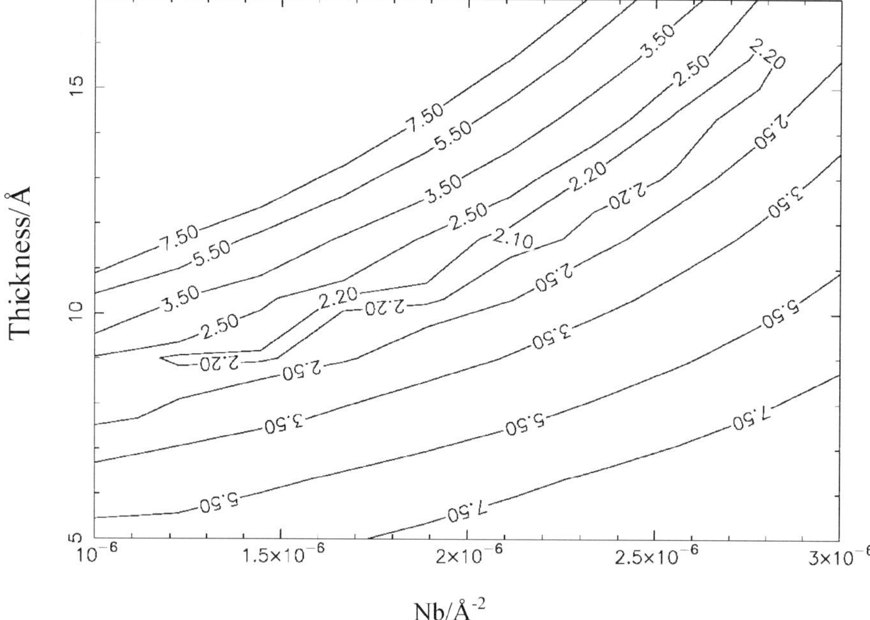

Fig. 6 Weighted sum-of-squares deviations as a function of scattering length density and layer thickness of the $C_{14}TAB$ layer at the oil–solution interface shown as a contour plot.

excellent agreement with the data. In this model, a primary adsorbed layer residing towards the oil-side of the interface contains the same adsorbed amount of material as reported for the free surface. However, to represent the data it is necessary to include an additional water-rich layer; whether this is a real or artificial feature of the interface remains open to further investigation. It is worth commenting on the fact that the thickness of the secondary layer is ~ 50 Å, which is of the order of twice the molecular length. It is plausible that this may suggest that a dilute layer of micelles forms beneath the monolayer at the oil–water interface. This is perhaps not a far-fetched suggestion. Since the water side of the monolayer is hydrophilic, further adsorption to the hydrophilic side of the interface will proceed as bilayer or fragmented bilayer/micelle adsorption, as sketched in the cartoon inset to Fig. 3(c), as has been found at hydrophilic solid surfaces.[21] Clearly, further work employing extensive isotopic substitution is in progress to address these issues.

(b)   Pluronic L64 at the hexadecane–water interface

A similar treatment of the data to that described above has been applied to this system, and the corresponding reflectivity data and model reflectivities are given in Figs. 4(a) and (b). Here the data generally lie below the reflectivity from the bare oil–solution interface, suggesting a diffuse or roughened interface. However, models incorporating a single interfacial roughness or a single interfacial slab with scattering length density intermediate to the two bulk phases do not satisfactorily model the data. It is necessary to use a two-slab model at the oil–solution interface to offer a reasonable representation of the data. Such a two-slab model is consistent with the structures found in earlier investigations[6,9] in which the PPO block is anchored at the oil side of the interface and the PEO blocks extend into the aqueous phase. In this model, the layer on the oil side of the interface has a thickness ~ 14 Å, which is $\sim 2R_g$ for the PPO block, where R_g is the radius of gyration. The layer on the aqueous side of the interface has been modelled here with a thickness of 25 Å, which is $\sim 50\%$ of the fully extended PEO chain. Clearly, the two-slab model is again an oversimplification of the interface structure: the oil and water distributions have not been treated, capillary wave roughening neglected, and the blocks have been assumed to be segregated when interpreting the slab widths in terms of molecular dimensions. However, it is clear that the simple model requires a thin region with scattering length density lower than that of the oil, and a thicker layer with scattering length density between that of the oil and the aqueous phase. The quality of the fit is improved by including some diffusion between the slabs and the sub- and super-phases, as shown in Fig. 4(c) (with scattering length density profile shown in Fig. 5(b)). Significantly improved fits to the data than those shown here can be obtained by allowing the thickness of the slab on the aqueous side of the interface to become greater than the length of a fully extended PEO chain (see Fig. 4(c) and Fig 5(b)). We have presented such a model here for completeness as it represents the best-quality fit to the reflectivity data. However, this model cannot be easily explained in terms of the molecular length scales present, and, as for the $C_{14}TAB$ system, further work is required to investigate whether the captured structural features are really present in the interfacial region. If indeed it is a real feature, the model suggests that the amphiphilic nature of the copolymer enables the less water-soluble PPO block to reside on the aqueous side of the interface and thus does not restrict the adsorption to monolayer formation. Such phenomena have been reported for other copolymers.[22]

In future data analysis and in conjunction with data obtained from other refractive index contrasts, it should, in principle, be possible to provide models in terms of capillary-wave broadening and explicit oil, water and amphiphile distributions. However, it is not expected that the capillary wave broadening will dominate the scattering length density profiles and complicate the interpretation of the scattering length density profiles presented here. This can be demonstrated by calculating the capillary wave roughness for the experimental conditions. The interfacial tension of the $C_{14}TAB$ solution–hexadecane interface is 7.0 mN m^{-1}, giving a capillary wave roughness at 298 K for the given instrument geometry of $\sim$9.5 Å. For the L64 Pluronic solution–hexadecane interface, interfacial tension is 11.7 mN m^{-1} and this corresponds to a capillary wave roughness of $\sim$7.5 Å; these roughness values are smaller than the dimensions of the molecules involved.

Summary

The main results of these experiments may be summarised as follows:

(i) We have demonstrated the ability to resolve details of the structure at oil–water interfaces using neutron reflectometry. The oil films required for these measurements can be reliably formed and maintained and the measured experimental data exhibit a high level of reproducibility. This level of reproducibility is essential for the employment of isotopic substitution in measurements permitting improved structural resolution. We have recently extended these studies to explore the structure of water-soluble synthetic peptides at a hexadecane–aqueous solution interface using further isotopic substitution.[23]

(ii) Even with the limited contrast variation employed in these measurements, a reasonably high level of structural information is attainable. For the $C_{14}TAB$ system there is evidence for surfactant enrichment at the interface and this structural feature is distinct from the capillary-wave broadening of the interface. The Pluronic L64 system also displays structure beyond the broadening of the interface; there is evidence for adsorption of copolymer on the oil side of the interface, and these results are consistent with the findings of other workers for related systems.

(iii) There still remain open questions regarding the nature of the oil, water and amphiphile distributions through the interface, the role of capillary waves, and determination of unambiguous adsorbed amounts. Detailed contrast variation studies are required.

Acknowledgements

The authors wish to thank the CCLRC for allocation of beamtime at ISIS. A. Q. acknowledges Queen Mary, University of London for the award of a studentship.

References

1   G. L. Richmond, *Chem. Rev.*, 2002, **102**, 2693.
2   M. L. Schlossmann, *Curr. Opin. Colloid Interface Sci.*, 2002, **7**, 235.
3   L. T. Lee, D. Langevin and B. Farnoux, *Phys. Rev. Lett.*, 1991, **67**, 2678.

4 U. Jeng, L. Eisbov, M. L. Crow and A. Steyerl, *Physica B (Amsterdam)*, 1996, **221**, 168.

5 T. Cosgrove, J. S. Phipps and R. M. Richardson, *Colloids Surf.*, 1992, **62**, 199.

6 J. S. Phipps, R. M. Richardson, T. Cosgrove and A. Eaglesham, *Langmuir*, 1993, **9**, 3530.

7 E. Dickinson, D. S. Horne, J. S. Phipps and R. M. Richardson, *Langmuir*, 1993, **9**, 242.

8 S. Calpin-Davies, B. J. Clifton, T. Cosgrove, R. M. Richardson, A. Zarbakhsh, C. Booth and G. Yu, in *Modern Aspects of Colloidal Dispersions*, ed. R. H. Ottewill and A. R. Rennie, Kluwer, Dortrecht, 1998, 159.

9 B. J. Clifton, T. Cosgrove, R. M. Richardson, A. Zarbakhsh and J. R. P. Webster, *Physica B (Amsterdam)*, 1998, **248**, 289.

10 B. P. Binks, D. Crichton, P. D. I. Fletcher, J. R. MacNab, Z. X. Li, R. K. Thomas and J. Penfold, *Colloids Surf., A*, 1999, **146**, 299.

11 A. Zarbakhsh, J. Bowers and J. R. P. Webster, *Meas. Sci. Technol.*, 1999, **10**, 738.

12 J. Bowers, A. Zarbakhsh, J. R. P. Webster, L. R. Hutchings and R. W. Richards, *Langmuir*, 2001, **1**, 140.

13 E. A. Simister, R. K. Thomas, J. Penfold, R. Aveyard, B. P. Binks, P. Cooper, P. D. I. Fletcher, J. R. Lu and A. Sokolowski, *J. Phys. Chem.*, 1992, **96**, 1383.

14 J. Penfold, *Physica B (Amsterdam)*, 1991, **173**, 1.

15 J. Penfold and R. K. Thomas, *J. Phys.: Condens. Matter*, 1990, **2**, 1369.

16 M. Born and E. Wolf, *Principles of Optics: Electromagnetic Theory of Propagation, Interference and Diffraction of Light*, Cambridge University Press, Cambridge, UK, 6th edn., 1980.

17 R. A. Cowley and T. W. Ryan, *J. Phys. D: Appl. Phys.*, 1987, **20**, 61.

18 J. R. P. Webster, *OILWATER.GCL*, Rutherford Appleton Laboratory, Didcot, UK, 2003.

19 F. A. Akeroyd, R. L. Ashworth, S. L. Campbell, S. D. Johnston, C. M. Moreton-Smith, R. G. Sergeant and D. S. Sivia, *OpenGENIE Reference Manual, Version 2.0, RAL-TR-1999-031*, 1999.

20 D. C. McDermott, D. Kanelleas, R. K. Thomas, A. R. Rennie and J. Penfold, *Langmuir*, 1993, **9**, 2404.

21 For example see J. Penfold, E. J. Staples, I. Tucker, L. J. Thompson and R. K. Thomas, *Physica B (Amsterdam)*, 1998, **223**, 248 and refs. 17–20 therein.

22 For example see J. Bowers, A. Zarbakhsh, J. R. P. Webster, R. W. Richards and L. R. Hutchings, *Langmuir*, 2001, **17**, 131; P. F. Dewhurst, M. R. Lovell, J. L. Jones, R. W. Richards and J. R. P. Webster, *Macromolecules*, 1998, **31**, 1261.

23 A. Zarbakhsh, J. Bowers, J. R. Lu, A. Querol and J. R. P. Webster, analysis in progress, unpublished.

Gibbs energies of transfer of chiral anions across the interface
water|chiral organic solvent determined with the help of
three-phase electrodes

Gibbs energies of transfer of chiral anions across the interface water|chiral organic solvent determined with the help of three-phase electrodes

Fritz Scholz* and Rubin Gulaboski

*Universität Greifswald, Institut für Chemie und Biochemie, Soldmannstrasse 23,
D-17489 Greifswald, Germany. E-mail: fscholz@uni-greifswald.de;
Fax: +49 3834 864 451; Tel: +49 3834 864 450*

Received 26th March 2004, Accepted 28th May 2004
First published as an Advance Article on the web 28th September 2004

For the first time the differences in free energies have been experimentally determined for the solvation of two enantiomeric ions in two enantiomeric solvents. Three-phase electrodes consisting of a droplet of a solution of decamethylferrocene in a chiral solvent (D- and L-2-octanol) attached to a graphite electrode and immersed in an aqueous solution containing chiral anions, allow the measurement of these data for the transfer of the enantiomeric ions across the water|chiral solvent interface. In all studied combinations the Gibbs energy to transfer an L-ion from water to the L-solvent was equal to the Gibbs energy of transfer of the D-ion from water to the D-solvent, and the same was found for the transfer of a L-ion from water to the D-solvent and for the D-ion from water to the L-solvent. In all cases, the combinations D-anion/D-solvent and L-anion/L-solvent have smaller standard Gibbs energies than the combinations D-anion/L-solvent and L-anion/ D-solvent. This can be explained by less favourable interactions between an anion and a solvent molecule in the latter cases.

Introduction

The interactions between chiral molecules or ions play a key role in biochemical systems. Therefore, the quantification of such interactions is important and appropriate methods have to be developed. Here we describe experiments aimed at determining the Gibbs energies of transfer of chiral ions from the achiral environment water to a chiral solvent. Fig. 1 depicts the possibilities to transfer a D- and L-ion from an aqueous solution to a D- and L-solvent. Because of the mirror symmetry of D- and L-forms of molecules, one must expect the Gibbs energy to transfer an L-ion from water to an L-solvent to be equal to the Gibbs energy of transfer of the D-ion from water to the D-solvent. The same should hold true for the L-ion from water to the D-solvent and for the D-ion from water to the L-solvent. To assess the role of the chiral centres in the interaction, one has to determine the differences in Gibbs energies between the couples D-ion/D-solvent and D-ion/L-solvent, and between the couples L-ion/D-solvent and L-ion/L-solvent. These data could not be determined until now because of a lack of appropriate techniques. In the standard 4-electrode technique[1] to study the ion transfer at the interface of immiscible liquids it is impossible to use solvents like octanol and no reports about the use of chiral solvents have been published so far. The reason that such solvents cannot be used in 4-electrode measurements is that they need electrolytes in the aqueous and in the

DOI: 10.1039/b404539d

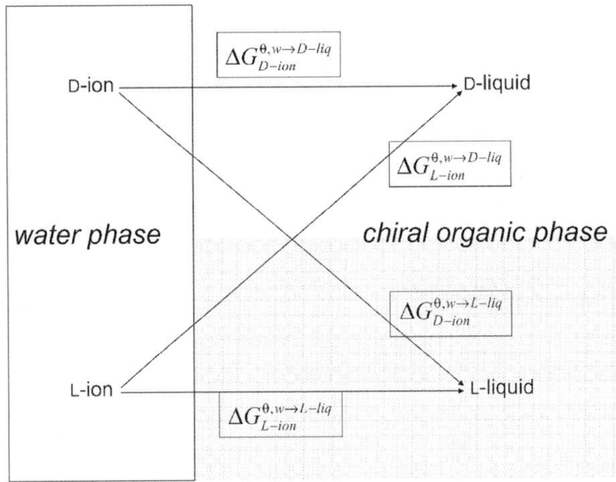

Fig. 1 Scheme of the possible processes of transfer of chiral ions across the water|chiral organic solvent interface and the associated standard Gibbs energies.

organic phase, and no electrolytes have been found for solvents like octanol allowing a polarisation of the water|organic interface. We have chosen the anions of three D- and L-amino acids and of D- and L-2-chloropropionic acid because they are simple systems typical of chiral molecules in biological systems. As solvents we have chosen D- and L-2-octanol because n-octanol is a standard solvent used to measure partition coefficients and quantify the lipophilicities of compounds.[2]

The experimental methodology

The determination of the differences in Gibbs energies of transfer of ionic enantiomers across water|chiral solvent interfaces is not trivial because, as it will be shown here, they are in the range of some kJ mol^{-1} only. In 2000, we introduced a new technique to determine Gibbs energies of ion transfer using a three-phase electrode and employing a 3-electrode potentiostat.[3] This approach has been used for the determination of transfer data of various anions[3–13] and cations[12,13] across the water|nitrobenzene,[3–7,11–13] water|n-octanol,[8,10,11] water|1,2-dichloroethane,[3] water|nitrophenyl-octyl ether,[11] water|D- and water|L-menthol interfaces.[9] The three-phase electrode experiments[14] make use of a working electrode modified by a droplet of an organic solution of a lipophilic neutral electroactive compound. The modified electrode, a counter and a reference electrode are immersed in the aqueous solution of a salt as in common 3-electrode voltammetric experiments. Provided that the neutral electroactive compound in the organic phase undergoes a 1-electron oxidation under anodic polarization, an anion transfer from water to the organic phase may accompany the oxidation. Generally, two ways of ion transfer are possible, either the transfer of the electrochemically produced cations from the organic phase to water, or the transfer of the anions from water toward the organic phase. What happens depends on the standard Gibbs energies of transfer of both pretenders. If the solvation of the electrochemically generated cations in the organic phase is stronger than the hydration of the anions present initially in the aqueous phase, then the transfer of the anions across the liquid|liquid interface will occur. In that case, one can determine the value of the standard Gibbs energy of transfer of the anions across the water|organic solvent interface.[3,11] By attaching a single droplet of a *chiral organic solvent* containing the electroactive compound decamethylferrocene to the surface of a paraffin impregnated graphite electrode, and immersing the modified electrode in an aqueous solution containing *chiral* anions, it is possible to measure the desired data. Fig. 2 depicts the situation at the three-phase electrode with an immobilized droplet of a chiral solvent. By applying an appropriate potential difference between the working and the reference electrode in such a set-up, dmfc is reversibly oxidized in a 1-electron step to dmfc$^+$ in both chiral 2-octanol solvents. This electron transfer at the graphite|D- or graphite|L-2-octanol interface

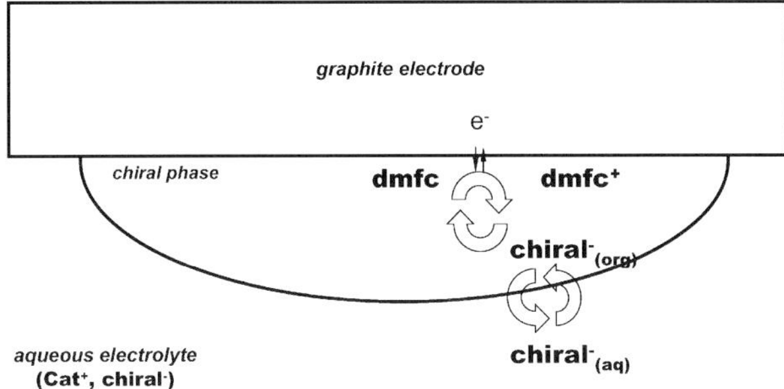

Fig. 2 Scheme of a three-phase electrode with an immobilized droplet of a chiral solvent containing decamethylferrocene. The electrode is immersed in a solution of a salt with chiral anions.

is coupled to a simultaneous transfer of chiral anions across the water|D- or water|L-2-octanol interface. The overall reaction at the three-phase electrode is as follows:

$$\text{dmfc}_{\text{(chiral 2-octanol)}} + \text{X}^-_{\text{(aq)}} \rightarrow \text{dmfc}^+_{\text{(chiral-2-octanol)}} + \text{X}^-_{\text{(chiral-2-octanol)}} + \text{e}^- \tag{I}$$

where X^- is the transferable chiral anion. If there are no kinetic constraints either for the electron or the ion transfer, the thermodynamic treatment of reaction (I) leads to the following form of the Nernst equation:[3,11]

$$E_c^{\ominus\prime} = E^{\ominus}_{\text{dmfc}^+_{(org)}|\text{dmfc}_{(org)}} - \frac{\Delta_W^{org}G_{X^-}^{\ominus}}{zF} - \frac{RT}{F}\ln\left(c_{X^-_{(aq)}}\right) + \frac{RT}{F}\ln\left(\frac{c_{\text{dmfc}_{(org)}}}{2}\right) \tag{1}$$

In eqn. (1) $E_c^{\ominus\prime}$ is the formal potential of the system portraying the coupled electron/ion transfer, $E^{\ominus}_{\text{dmfc}^+_{(org)}|\text{dmfc}_{(org)}}$ is the standard redox potential of dmfc/dmfc$^+$ in the organic phase (*i.e.*, D- or L-2-octanol), $\Delta_W^{org}G_{X^-}^{\ominus}$ is the standard Gibbs energy of transfer of the chiral anions X^- from water to the chiral 2-octanol, $c_{\text{dmfc}_{(org)}}$ is the initial concentration of dmfc in the organic phase, $c_{X^-_{(aq)}}$ is the concentration of the chiral anions in the aqueous phase, while R, T, F and z are the gas constant, absolute temperature, Faraday constant and ionic charge, respectively. Eqn. (1) allows determination of the values of $\Delta_W^{org}G_{X^-}^{\ominus}$ by measuring the formal redox potentials in the above discussed systems. As an indicator whether the entire reaction is electrochemically reversible one can use the dependence of the formal potential of the electrochemical system on the concentration of the anions in the aqueous phase. A slope of the dependence $E_c^{\ominus\prime}/\log\left(c_{X^-_{(W)}}\right)$ of about -60 mV, and stable voltammograms during consecutive cycling, are strong indications that the studied system behaves as described by reaction (I).

Experimental

Decamethylferrocene (dmfc) and 2-chloropropionic acid were products of ACRÖS, Germany; all the amino acids were products of BACHEM, Germany, while D- and L-2-octanol and all the other chemicals were purchased from Sigma–Aldrich, Germany. The square-wave and cyclic voltammetric experiments were performed with the aid of a conventional 3-electrode potentiostat (AUTOLAB PGSTAT 10, Eco-Chemie, Utrecht, Netherlands). An Ag/AgCl (sat. NaCl) served as reference electrode, while a Pt wire was the counter electrode. The working electrode was a paraffin impregnated graphite electrode (PIGE) with a radius of 2 mm. The procedure for preparing the PIGE is described elsewhere.[14,15] A droplet of 0.05 mol L^{-1} solution of dmfc dissolved in an organic solvent (D- or L-2-octanol) with a volume of 1 μL was attached to the surface of the working (PIGE) electrode and thereafter immersed in an aqueous electrolyte solution that contained one of the chiral forms of the investigated anions. The pH of the aqueous solution of the chiral amino acid anions of phenylalanine and lysine was 13 (1 mol L^{-1} NaOH), while pH of 9.80 (0.5 mol L^{-1} borax buffer) and 4.75 (1 mol L^{-1} acetate buffer) were explored for studying the transfer of chiral anions of tyrosine and 2-chloropropionate, respectively. The peak potentials of the redox reaction of dmfc coupled to the transfer of the chiral anions were determined by square-wave voltammetry (SWV).[16] Typical parameters for square-wave voltammetric measurements were: SW frequency $f = 10$ Hz, SW amplitude $E_{sw} = 50$ mV, and scan increment $dE = 1$ mV. In all cases additional measurements were performed with cyclic voltammetry at different scan rates (from 20 mV s^{-1} to 400 mV s^{-1}) in order to check the electrochemical reversibility of the system by measuring the anodic to cathodic peak separation and the peak current ratio, while the overall stability of the measurements was checked in repeated cycles (at least 10). In square-wave voltammetry the forward to backward peak current ratio is 1.0, the peak potential separation is 8 mV and the half-width is 140 mV at a frequency of 10 Hz. Concentration dependencies have been measured for all chiral anions in order to ensure that the entire system behaves according to reaction (I) and eqn. (1). The polarimetric measurements for determining the rotation angles of the solutions of the chiral anions were performed with the aid of a Perkin–Elmer 241 Polarimeter (Na 589) in order to check the stability of the compounds in the solutions.

Results and discussion

Fig. 3 depicts square-wave voltammograms of dmfc oxidation in the respective chiral octanol coupled to the transfer of anions of D-phenylalanine from water to D-2-octanol and to L-2-octanol, and the square-wave voltammograms of dmfc oxidation in the respective chiral octanol coupled to

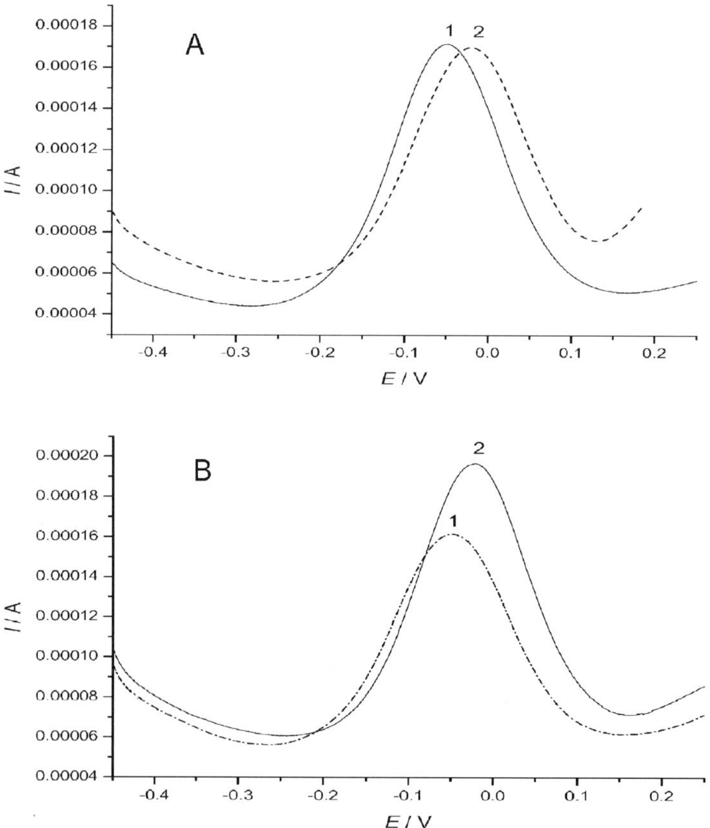

Fig. 3 (A) Square-wave voltammograms of dmfc oxidation in the respective chiral octanol coupled to the transfer of anions of D-phenylalanine from water to D-2-octanol (curve 1), and from water to L-2-octanol (curve 2). (B) Square-wave voltammograms of dmfc oxidation in the respective chiral octanol coupled to the transfer of anions of L-phenylalanine from water to L-2-octanol (curve 1) and from water to D-2-octanol (curve 2). For conditions see Experimental section.

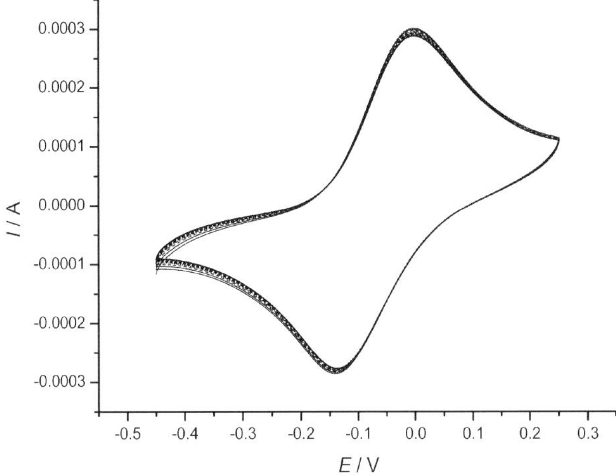

Fig. 4 Multiple scan cyclic voltammograms (2nd to 10th cycle) of dmfc oxidation in L-2-octanol coupled to the transfer of L-2-chloropropionate anions from water to L-2-octanol. Scan rate: 200 mV s^{-1}.

the transfer of anions of L-phenylalanine from water to L-2-octanol and to D-2-octanol. Distinct differences in the peak potentials can be determined. The standard deviation of peak potentials in repeated experiments was 1–3 mV only, which is equivalent to Gibbs energies of transfer of 0.1 to 0.3 kJ mol^{-1}. The slight mismatch of peak currents is due to small deviations in the droplet size in each experiment. These differences are without effect on the peak potentials. Fig. 4 shows the 2nd to 10th cycle of cyclic voltammograms of dmfc oxidation in L-2-octanol coupled to the transfer of L-2-chloropropionate anions from water to L-2-octanol. These curves illustrate the chemical reversibility of the system. The same stability was observed in all other cases of anion transfer reported in this paper. Fig. 5 depicts a typical plot of peak potentials of square-wave voltammograms of dmfc oxidation in L-2-octanol coupled to the transfer of L-2-chloropropionate anions from water to L-2-octanol *versus* concentration of L-2-chloropropionate in the aqueous phase. Clearly, the system follows the concentration dependence described with eqn. (1). From the peak potentials of the square-wave voltammograms the standard Gibbs energies of ion transfer have been calculated according to that equation. For this purpose the following standard potential of dmfc in the two chiral octanols was used: −0.317 V *vs* Ag/AgCl. The estimated standard Gibbs energies of transfer from water to D- or L-2-octanol of all chiral monoanionic forms considered in this study are given in Table 1. As can be seen from the data in Table 1, within the experimental error of peak potential measurements, in all cases the $\Delta_{W}^{org}G_{X^-}^{\ominus}$ values for the transfer of the D-ion to D-2-octanol are the same as the $\Delta_{W}^{org}G_{X^-}^{\ominus}$ values of transfer of the L-ion to L-2-octanol. The same holds true for the transfer of D-ions to L-2-octanol and for L-ions to D-2-octanol (*cf*. Fig. 6). This kind of symmetry of the results was expected; however, initial experiments gave very different results. It took some time to detect that these, obviously wrong results, were due to racemization reactions. Racemization could be detected in these cases by polarimetric measurements. Racemization and epimerization of amino acids are very slow processes at room temperature or below, and at around pH 7. Under these conditions amino acids have half life times in the ka (kilo years) range, which is utilized for age determinations of foraminifers, shells, bones, *etc*.[17] The racemization is strongly dependent on pH, and it is assumed that the racemization occurs *via* abstraction of the α-proton by OH$^-$. The rate constant for the proton abstraction is supposed to increase in the raw cationic, zwitterionic, anionic form, because of increasing electron-withdrawing and resonance stabilizing capacities of the substituents attached to the α-carbon atom.[18] Rather rapid racemizations have been observed in the case of both enantiomers of the amino acid leucine. Significant changes in the rotation angle were observed within the first few hours after dissolving an enantiomer of leucine in 1 M NaOH. The racemization processes of the other studied amino acids were much slower, *i.e.*, changes in the rotation angles were observable only after a period of one day or more. In order to prevent the undesirable racemization, the electrochemical measurements were performed immediately after

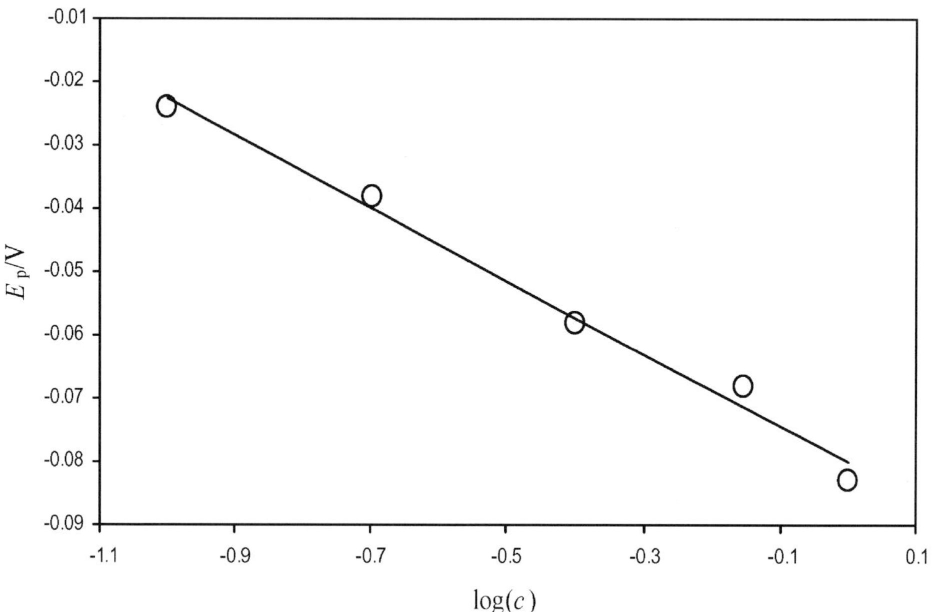

Fig. 5 Peak potentials of square-wave voltammograms of dmfc oxidation in L-2-octanol coupled to the transfer of L-2-chloropropionate from water to L-2-octanol *versus* concentration of L-2-chloropropionate in the aqueous phase.

preparing the fresh solutions of the amino acids in basic media. For the applicability of the three-phase electrode measurements it is of utmost importance to perform parallel experiments to control the stability of the chiral ions towards racemization. Especially in strongly alkaline solutions there is the danger of rather rapid racemization reactions. In our previous communication on the transfer of the anions of D- and L-tryptophan from water to D- and L-menthol an asymmetry of Gibbs transfer energies was observed.[9] In the course of the experiments reported here, we have made the observation that a precipitate forms at the liquid|liquid interface with these solvents. It is rather difficult to detect these precipitates in the experiments with the three-phase microdroplet modified electrodes. However, by simple shaking of the organic phase together with the aqueous solution, it was obvious that a precipitation of tryptophan occurs at the water|D- and water|L-menthol interfaces. A similar precipitation of tryptophan was observed at the water|D-2-octanol and water|L-2-octanol interfaces. Therefore no reliable data could be obtained for the anion of tryptophan.

Table 1 Standard Gibbs energies of transfer of the chiral anions from water to D-2-octanol ($\Delta_{\mathrm{W}}^{\mathrm{D\text{-}2\text{-}oct}}G_{\mathrm{X}^-}^{\ominus}$), from water to L-2-octanol ($\Delta_{\mathrm{w}}^{\mathrm{L\text{-}2\text{-}oct}}G_{\mathrm{X}^-}^{\ominus}$), and from D-2-octanol to L-2-octanol ($\Delta_{\mathrm{D\text{-}2\text{-}oct}}^{\mathrm{L\text{-}2\text{-}oct}}G_{\mathrm{X}^-}^{\ominus}$) The values are estimated by using the value of the standard redox potential of dmfc/dmfc$^+$ in *n*-octanol from ref. 8

Monoanion of:	$\Delta_{\mathrm{W}}^{\mathrm{D\text{-}2\text{-}oct}}G_{\mathrm{X}^-}^{\ominus}$/kJ mol^{-1}	$\Delta_{\mathrm{w}}^{\mathrm{L\text{-}2\text{-}oct}}G_{\mathrm{X}^-}^{\ominus}$/kJ mol^{-1}	$\Delta_{\mathrm{D\text{-}2\text{-}oct}}^{\mathrm{L\text{-}2\text{-}oct}}G_{\mathrm{X}^-}^{\ominus}$/kJ mol^{-1}
D-Lysine	30.80	32.45	1.65
L-Lysine	32.40	30.70	−1.70
D-Tyrosine	27.80	29.90	2.10
L-Tyrosine	29.80	27.70	−2.10
D-Phenylalanine	25.80	28.90	3.10
L-Phenylalanine	28.95	25.85	−3.10
D-2-Cl-propionic acid	22.60	23.40	0.80
L-2-Cl-propionic acid	23.50	22.65	−0.85

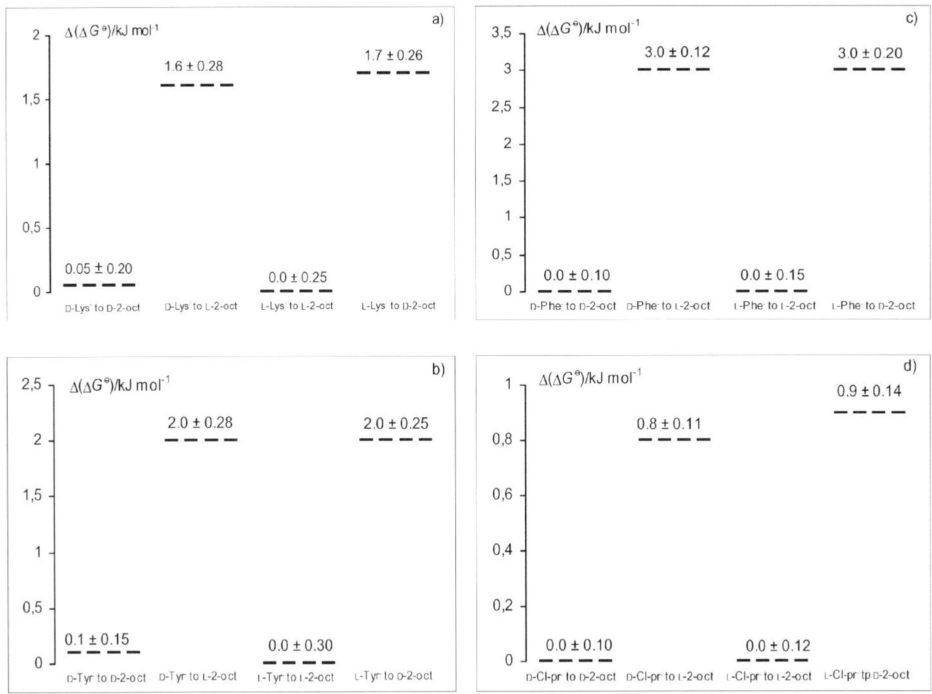

Fig. 6 Diagrams showing the differences in the standard Gibbs energies of transfer of the studied chiral anions. The symbols Lys$^-$, Phe$^-$, Tyr$^-$, and Cl-propionate$^-$ denote the monoanionic forms of lysine, phenylalanine, tyrosine and 2-chloropropionic acid, respectively, while the symbol 2-oct stands for 2-octanol.

Conclusions

To the best of our knowledge this is the first report of the differences of Gibbs energies of solvation of chiral ions in chiral solvents. There are no calorimetric studies of these systems, probably because of the low solubilities of salts in chiral solvents like D- and L-2-octanol. Given that the differences in Gibbs energies are rather small this will make it even more difficult to obtain similar data by calorimetry.

The differences in standard Gibbs energies of ion transfer suggest that the anions probably form diastereomeric complexes with one solvent molecule. If the anions where to be housed in a sphere of solvent molecules without such kind of specific interaction, *i.e.*, if the anions would see only a randomized solvation sphere, the differences would be negligible. Fig. 7 shows intuitively drawn complexes between D- and L-2-octanol and D- and L-anions of amino acids. Obviously, one has to assume one strong hydrogen bond between the OH-group of octanol and the COO$^-$-group of the amino acid anion. The hexyl group of octanol is itself rather large and at room temperature it will need even more space due to its rotation, so that the NH$_2$ group is more repelled in the D/L and L/D than in the D/D and L/L combinations, making the latter two arrangements energetically slightly more favourable. The experimentally determined energetic differences between the D/D and L/L pairs on one side and the D/L and L/D on the other side are in the range of typical van-der-Waals' forces. The steric differences may be responsible for the different van-der-Waals' interactions between the anions and solvent molecules.

Future efforts must be focussed on molecular dynamic calculations of interaction energies of chiral anions with chiral solvent molecules. It is hoped that the experimental data reported here may serve for comparison with theoretical data.

Stereoselective interactions between chiral molecules have tremendous importance both for analytical and synthetic separations. Chiral phases are used in chromatography to separate enantiomers,[19] and chiral reagents are used in electrophoretic separations.[20] However, the

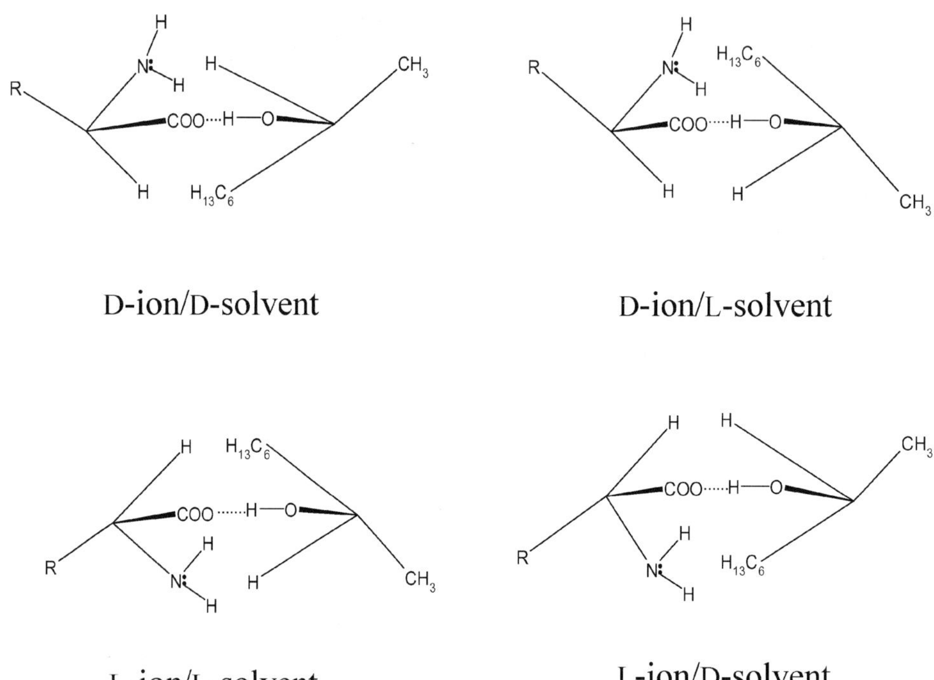

Fig. 7 Interactions of chiral anions of amino acids with the chiral solvents D- and L-2-octanol.

conditions for both techniques are much more complicated than for simple solvation experiments. Thus it is believed that the data produced in transfer experiments using three-phase electrodes may find an easier theoretical explanation than data derived from interactions with chirally derivatized solid surfaces, or data from electrophoretic measurements, *i.e.*, with the involvement of a strong electric field and under dynamic conditions.

Acknowledgements

F. Scholz acknowledges support by Deutsche Forschungemeinschaft (DFG) and Fonds der Chemischen Industrie (FCI), R. Gulaboski thanks Deutscher Akademischer Austauschdienst (DAAD) for the provision of a PhD scholarship.

References

1 H. H. J. Girault and D. J. Schiffrin, *Electrochemistry of Liquid–Liquid Interfaces*, in *Electroanalytical Chemistry, A Series of Advances*, ed. A. J. Bard, vol. 15, 1989, Dekker, New York, pp. 1–141.
2 A. Leo, C. Hansch and D. Elkins, *Chem. Rev.*, 1971, **71**, 525.
3 F. Scholz, Š. Komorsky-Lovrić and M. Lovrić, *Electrochem. Commun.*, 2000, **2**, 112.
4 (*a*) Š. Komorsky-Lovrić, K. Riedl, R. Gulaboski, V. Mirčeski and F. Scholz, *Langmuir*, 2002, **18**, 8000; (*b*) Š. Komorsky-Lovrić, K. Riedl, R. Gulaboski, V. Mirčeski and F. Scholz, *Langmuir*, 2003, **19**, 3090.
5 R. Gulaboski, K. Riedl and F. Scholz, *Phys. Chem. Chem. Phys.*, 2003, **5**, 1284.
6 R. Gulaboski and F. Scholz, *J. Phys. Chem. B*, 2003, **107**, 5650.
7 R. Gulaboski, V. Mirčeski and F. Scholz, *Amino Acids*, 2003, **24**, 149.
8 R. Gulaboski, V. Mirčeski and F. Scholz, *Electrochem. Commun.*, 2002, **4**, 277.
9 F. Scholz, R. Gulaboski, V. Mirčeski and P. Langer, *Electrochem. Commun.*, 2002, **4**, 659.
10 G. Bouchard, A. Galland, P.-A. Carrupt, R. Gulaboski, V. Mirčeski, F. Scholz and H. H. Girault, *Phys. Chem. Chem. Phys.*, 2003, **5**, 3748.
11 R. Gulaboski, A. Galland, G. Bouchard, K. Caban, A. Kretchmer, P.-A. Carrupt, Z. Stojek, H. H. Girault and F. Scholz, *J. Phys. Chem. B*, 2004, **108**, 4565.

12   V. Mirčeski, R. Gulaboski and F. Scholz, *Electrochem. Commun.*, 2002, **4**, 813.
13   F. Scholz, R. Gulaboski and K. Caban, *Electrochem. Commun.*, 2003, **5**, 929.
14   F. Scholz, U. Schröder and R. Gulaboski, *The Electrochemistry of Particles and Droplets Immobilized on Electrode Surfaces*, Springer–Verlag, Berlin, 2004.
15   F. Scholz and B. Meyer, in *Electroanalytical Chemistry. A Series of Advances*, ed. A. J. Bard and I. Rubinstein, Marcel–Dekker, New York, 1998, vol. 20, p. 1.
16   M. Lovrić, in *Electroanalytical Methods, Guide to Experiments and Applications*, ed. F. Scholz, Springer–Verlag, Berlin, 2002, p. 111.
17   M. A. Geyh and H. Schleicher, *Absolute Age Determinations*, Springer, Berlin, 1990.
18   J. L. Bada, *J. Am. Chem. Soc.*, 1972, **94**, 1371.
19   M. E. Y. Cabusas, *Chiral Separations on HPLC Derivatized Polysaccharide CSPs: Temperature, Mobile Phase and Chiral Recognition Mechanism Studies*, PhD Thesis, Virginia Polytechnic Institute and State University, 1998.
20   T. de Boer, *Selectivity Enhancement in Capillary Electrokinetic Separations via Chiral and Molecular Recognition*, PhD Thesis, Rijksuniversiteit Groningen, 2001.

General discussion

Prof. Scholz opened the discussion of Dr Dagastine's paper: When you change the solution composition you will, most probably, also change the probe–solution interface. Since your organic compounds are certainly slightly soluble in the aqueous phase, there may be adsorption competition among these organics and the ions on the probe surface. Thus it may happen that you get a layer of adsorbed organic molecules on the tip surface. In that case an attractive interaction between the tip and the droplet will not be a surprise. Can you really exclude such tip surface alterations?

Dr Dagastine responded: All the experiments are undertaken in saturated oil solutions. Butyl acetate is slightly water soluble whereas the octyl acetate is sparingly soluble in aqueous solution. The solubility of the butyl acetate in solution is the smallest with the sodium perchlorate. If there is adsorption to the hydrophilic surface, it is more likely to occur the higher the solubility, but the attraction is observed for the case with the lowest butyl acetate solubility. Furthermore, if adsorption of the organic solute onto the silica were an issue it would be expected that it would be more significant for octyl acetate than butyl acetate, yet no difference was seen.

Prof. Unwin said: I have several points and questions on your paper.
1. Following on from the comments of Prof. Scholz, could you describe how you treat the quartz spheres prior to your experiments and comment on how the surface chemistry of the quartz influences the force curves you measure? For example, how would the force curves differ for the extremes of a hydrophilic sphere, pre-treated with piranha solution, and a silanised hydrophobic sphere?
2. As a related point, could you comment on whether the force curves change when you make sequential measurements?
3. It is usual to find a statistical analysis of force curves derived from AFM measurements. Could you comment on the level of uncertainty in the force curves you measure and the various parameters you derive?

Dr Dagastine replied: (1) and (2): The attraction observed in the system is attributed to a specific ion effect, which seems unrelated to the hydrophobicity of the silica. In this study the silica spheres were exposed to a generated ozone atmosphere for 30 min just prior to use. One could expect that if the oil was coating the silica sphere then the observed behaviour could change after the initial measurement. This may be likely for hydrophobic surfaces, but not for this case of a hydrophilic silica sphere as discussed in the response to the previous question. Each force measurement is taken independently from the rest, sequential measurements do not exhibit large variations in force behaviour.

Although unrelated to the measurements presented in our paper we have included a review of the body of previous force measurements between rigid spherical particles and *either* oil droplets *or* bubbles where the hydrophobicity of the probe was modified. The first measurement of a particle between a deformable air–water interface with an AFM by Ducker *et al.* in 1994[1] varied the hydrophobicity of the silica probe, along with several other subsequent studies by Fielden *et al.*[2] and Presuss *et al.*[3,4] In Ducker's study the hydrophilic surface was exposed to water plasma and hydrophobic particles were created by chemisorbing a self-assembled monolayer of octadecyltrichorosilane resulting with a water contact angle near 115°. For the hydrophilic case, the force behaviour exhibited a short-range attraction attributed to an attractive electrostatic double layer (EDL) force with a constant surface potential. The force behaviour with the hydrophobic silica exhibited a traumatic attractive jump-in before any EDL forces were observed. Also, upon addition

DOI: 10.1039/b416301j

of anionic surfactant, these long-range attractions were no longer observed and EDL repulsion was observed. The hydrophobic interactions in these systems have led to estimations of the contact angle at the three phase contact line of bubble with the particle. Furthermore, a study by Yakubov *et al.*[5] used hydrophobic polystyrene microspheres interaction with an air bubble to probe the effects of line tension on the wettability with water.

The bubble particle interactions have repulsive van der Waals, but similar behaviour has been observed between rigid spherical particles and oil droplets. Mulvaney *et al.*[6] and Hartley *et al.*[7] performed measurements between hydrophilic silica and alkanes in inorganic electrolyte solutions. A DLVO force behaviour was observed with an EDL repulsion followed by a short-range attraction from van der Waals forces. Upon adsorption of surfactant onto the alkane–solution interface attraction was no longer observed. Aston and Berg have also observed long range attractions, attributed to hydrophobic force for the interaction between either polystyrene beads or silanated glass spheres and hexadecane droplets in aqueous solutions.[8,9] Again with the addition of surfactant these long range attractions were mediated or removed.

(3) The question of statistically significant atomic force microscopy measurements can sometimes have an ambiguous interpretation. Most studies, including this one, employ hundreds of individual force curve measurements. While this is a large number of measurements, it is by no means statistically significant if one takes into account the actual surface area probed on a given sample. The interaction area probed is several hundred square nanometers on a single sample that is a square centimetre. Perhaps several different samples are repeated, but this is still a small population to examine. To add to this complication is that force measurements with an AFM do not necessarily measure large ensemble averages of molecules. As an example, AFM is commonly used to measure the interactions of single molecules.[10–12] It is crucial that AFM investigations report all force behaviours observed, not just the one the investigators may deem statistically significant.

In this study, each type of experiment was repeated on a number of different occasions with different spheres, substrates and new solutions. The variability in the force data is within 10–20%, which is comparable to the error in determination of the cantilever spring constant.

1   W. A. Ducker, Z. G. Xu and J. N. Israelachvili, *Langmuir*, 1994, **10**, 3279.
2   M. L. Fielden, R. A. Hayes and J. Ralston, *Langmuir*, 1996, **12**, 3721.
3   M. Preuss and H.-J. Butt, *Langmuir*, 1998, **14**, 3164.
4   M. Preuss and H.-J. Butt, *J. Colloid Interface Sci.*, 1998, **208**, 468.
5   G. E. Yakubov, O. I. Vinogradova and H.-J. Butt, *J. Adv. Sci. Tech.*, 2000, **14**, 1783.
6   P. Mulvaney, J. M. Perera, S. Biggs, F. Grieser and G. W. Stevens, *J. Colloid Interface Sci.*, 1996, **183**, 614.
7   P. G. Hartley, F. Grieser, P. Mulvaney and G. W. Stevens, *Langmuir*, 1999, **15**, 7282.
8   D. E. Aston and J. C. Berg, *J. Colloid Interface Sci.*, 2001, **235**, 162.
9   D. E. Aston and J. C. Berg, *Ind. Eng. Chem. Res.*, 2002, **41**, 389.
10 V. T. Moy, E.-L. Florin and H. E. Gaub, *Science*, 1994, **266**, 257.
11 M. Rief, F. Oesterhelt, B. Heymann and H. E. Gaub, *Science*, 1997, **275**, 1295.
12 C. Friedsam, M. Seitz and H. E. Gaub, *J. Phys: Condens. Matter*, 2004, **16**, S2369.

Prof. Hampe asked: When considering the interaction between two liquid phases and a solid phase as has been done in the AFM experiments of Dagastine *et al.*, the question of wettability and contact angle springs into mind. *E.g.*, the phenomenon named 'jump-in' should have a correspondence in the complete wettability of the solid by one of the liquid phases. Have the contact angles of the systems been measured? And if so, what is the interpretation of the contact angle data in view of the AFM experiments?

Dr Dagastine answered: The question of wettability is certainly relevant to the behaviour of force curves as mentioned in the responses to Prof. Unwin, but this question actually brings two separate points together. The first is what causes "jump-in" and the second is what happens once "jump-in" has occurred and how does this relate to wettability. The force measurement up to the point of "jump-in" is an equilibrium measurement where a description of the interaction force between the drop and the particle is sufficient to understand the force behaviour. The origin of jump-in is from an attractive surface force such that the gradient of the force is equal to the spring constant of the cantilever. This attractive force may be a macroscopic van der Waals force, which is attractive in this system; it could also be the result of a charge reversal or charge regulation in the electrostatic double layer force. Both of these behaviours may not correspond to the wettability of the particle. For the hydrophobic force, where the mechanism and the origins of the force are not completely understood, this may correspond to the wettability behaviour of the system.[1] Thus, jump-in is not necessarily related to wetting. Also, the state after jump-in may not completely wet the particle. As mentioned above, the results of Ducker,[2] Fielden,[3] Preuss[4,5] and Yakubov[6] show that in some cases complete engulfment occurs and in others finite contact angles are measured from the formation of three phase contact lines. Another possibility pointed out from modelling by Bhatt *et al.*[7] and Dagastine and White[8] shows that if there are primary and secondary minima in the disjoining pressure, jump-in can take place between these minima, where the probe jumps from one attractive region to an inner repulsive one, where the two surfaces are not in contact.

1   J. N. Israelachvili, *Intermolecular and Surface Forces*, Academic Press, New York, 1992.
2   W. A. Ducker, Z. G. Xu and J. N. Israelachvili, *Langmuir*, 1994, **10**, 3279.
3   M. L. Fielden, R. A. Hayes and J. Ralston, *Langmuir*, 1996, **12**, 3721.
4   M. Preuss and H.-J. Butt, *Langmuir*, 1998, **14**, 3164.
5   M. Preuss and H.-J. Butt, *J. Colloid Interface Sci.*, 1998, **208**, 468.
6   G. E. Yakubov, O. I. Vinogradova and H.-J. Butt, *J. Adv. Sci. Tech.*, 2000, **14**, 1783.
7   D. Bhatt, J. Newman and C. J. Radke, *Langmuir*, 2001, **17**, 116.
8   R. R. Dagastine and L. R. White, *J. Colloid Interface Sci.*, 2002, **247**, 310.

Prof. Schlossman asked: Have you considered the relevance of elastic properties of the interface (*e.g.*, bending moduli, viscosity) on the description of the jump-in process?

Dr Dagastine replied: The question raises a good point about the dynamic process during jump-in, however this is irrelevant to whether jump-in is observed *or* the origins of the non-DLVO attraction observed in the paper. As mentioned in a previous response: "The force measurement up to the point of "jump-in" is an equilibrium measurement where a description of the interaction force between the drop and the particle is sufficient to understand the force behaviour. The origin of jump-in is from an attractive surface force such that the gradient of the force is equal to the spring constant of the cantilever."

The description of the force measurement stops at the point of jump-in. The conditions to achieve jump-in are determined by the equilibrium interaction forces and the cantilever spring constant, but once the mechanical instability in the cantilever is reached this becomes a dynamic process. One would speculate that any time dependent phenomenon of the interface (*i.e.* interfacial rheological response, bulk viscosity, *etc.*) may contribute to the behaviour during the jump-in process.

Prof. Rusling asked: Is it possible to change the size of your particle probe into the nanometer range, and if so what would you expect the size effects to show?

Dr Dagastine replied: It is possible to change the probe radius to one-micron with standard particle attachment methods, but controlling the position of attachment with smaller particles can be difficult since most methods employ optical microscopy. It is likely that these difficulties could be overcome through novel attachment methods or new methods to micromanipulate particles. Using smaller particles may raise other issues not as relevant for larger particles. The spherical probe is used to provide well-defined geometry and a surface that is easy to characterize in contrast to an AFM tip, which is a much smaller dimension but the tip geometry can be irregular and difficult to characterize. The larger interaction area also provides an increased sensitivity if one scales by the interaction area or sphere radius. Another issue is that probes on the micron scale typically have surface roughness on the order of less than a nm for silica or 10 to 15 nm for polymer microspheres.[1] As the probe size scales down, the surface roughness does not necessarily scale with radius, thus for small particles the geometry may not be as well defined.

To speculate on how the force behaviour would change with size, one could employ the Derjaguin approximation[2] for a sphere and plate

$$F(D) = 2\pi a E(D)$$

where *a* is the radius of the probe and *E* is the interaction between flat surfaces. The overall magnitude and range of the force decrease with the interaction area. To respond to the comment by

Dr Liggieri, the above expression shows how the surface energy scales with radius for a solid liquid interface.

One can speculate on aspects unique to the measurement of liquid–liquid systems as well. The radial component of the force for this interaction scales by roughly the $\sqrt{aD_o}$ where D_o is the intersurface separation, thus the interaction area is dependent on both intersurface separation and the probe geometry.[3] The closest approach of the probe will change as well. As mentioned in the paper, wrapping occurs for a number of these systems where only repulsion between the oil drop and the particle is observed. The pressure at which wrapping occurs is given by

$$\Pi(D_\mathrm{w}) = 2\gamma\left(\frac{1}{a} + \frac{1}{R_o}\right)$$

where there is an inverse relation to the probe radius. With very small probe geometries, say 200 nm in radius, the wrapping pressure would occur at a much smaller separation. As discussed in the paper, if the wrapping thickness is smaller than the length scale of the ever-present van der Waals attraction, then it is likely that jump-in instead of wrapping would be observed.[4]

1   R. F. Considine, R. A. Hayes and R. G. Horn, *Langmuir*, 1999, **15**, 1657.
2   R. J. Hunter, *Foundations of Colloid Science*, Clarendon Press, Oxford, 1995.
3   D. Y. C. Chan, R. R. Dagastine and L. R. White, *J. Colloid Interface Sci.*, 2001, **236**, 141.
4   R. R. Dagastine and L. R. White, *J. Colloid Interface Sci.*, 2002, **247**, 310.

Dr Horozov asked: The jump-in or wrapping cases are related to the stability of the water film between the probe and oil drop.

Did you measure the stability of water films in a similar system *e.g.*, silica plates in water and approaching oil drops?

Dr Dagastine replied: The question is correct in that the wrapping or jump-in behaviour is dependent on the stability of the water film. The governing forces behind the film stability are primarily surface forces and interfacial tension. As discussed above, the conditions for wrapping (given in previous responses) are based on the presence of a sufficiently strong repulsion coupled with a sufficiently pliable interface, such that the liquid–liquid interface begins to reflect the curvature of the rigid probe. The wrapping separation or film thickness must be larger than the length scale of the attractive forces in the system, otherwise the probe and liquid–liquid interface jump-in and the stable film is not formed. Theoretical studies have shown that if the interaction forces are constant only changes in the interfacial tension can cause a transition from jump-in to wrapping as the interfacial tension is significantly decreased.[1]

The suggested film stability experiments are a considerably different physical situation where the radial length scale is orders of magnitude larger. For example, scanning angle ellipsometry measurements of oil droplets approaching a flat silica surface have a radial length scale on the order of tens of microns.[2–4] Studies using a modified SFA and a mercury drop approaching a flat silica plate also have a much larger radial length scale.[5–8] More classical experiments including the Scelkudo–Exerowa cell to examine thin film stability for foams have radial films on the order of millimetres.[9] The radial length scale of the "wrapping films" in the AFM measurement is on the order of $\sqrt{aD_\mathrm{w}}$ where D_w is the wrapping thickness. For these experiments this is on the order of 100 nm in the radial direction.

All of the film stability experiments mentioned above have hydrodynamic components and formation of barrier rings during drainage due to the larger radial length scale. The AFM measurements are equilibrium force measurements, thus hydrodynamic drainage effects are not observed and the force behaviour is independent of approach speed.

A comparison between the AFM measurements and the film stability experiments can be drawn for equilibrium behaviour. The equilibrium film in the stability measurements provides a measure of the disjoining pressure curve between a silica surface and liquid–liquid interface at a *single* separation. The thickness is measured very accurately and the disjoining pressure is estimated or calculated.[3–4] This is in contrast to the AFM experiment, where the applied force is varied and the *entire* disjoining pressure curve can be measured. There is the opportunity to compare this one point disjoining pressure measurement to the AFM results which we are currently carrying out in parallel studies using AFM and scanning angle ellipsometry on the same oil system.

1   R. R. Dagastine and L. R. White, *J. Colloid Interface Sci.*, 2002, **247**, 310.
2   D. G. Goodall, M. L. Gee and G. W. Stevens, *Langmuir*, 2001, **17**, 3784.
3   D. G. Goodall, G. W. Stevens, D. Beaglehole and M. L. Gee, *Langmuir*, 1999, **15**, 4579.
4   D. G. Goodall, M. L. Gee, G. Stevens, J. Perera and D. Beaglehole, *Colloids Surf., A*, 1998, **143**, 41.
5   J. N. Connor and R. G. Horn, *Langmuir*, 2001, **17**, 7194.
6   J. N. Connor and R. G. Horn, *Rev. Sci. Instrum.*, 2003, **74**, 4601.
7   J. N. Connor and R. G. Horn, *Faraday Discuss.*, 2003, **123**, 193.
8   J. N. Connor, R. G. Horn and S. J. Miklavcic, *Uzbek J. Phys.*, 1999, **1**, 99.
9   A. Scheludko and D. Exerowa, *Kolloid Z.*, 1959, **165**, 148.

Dr Sakoguchi asked: I'm very interested in complex formation reaction at the liquid–liquid interface. What do you think about the possibility of using your technique in order to obtain information about the molecular structure at the liquid–liquid interface?

Dr Dagastine replied: A number of contributions in this discussion have shown excellent examples of how spectroscopic and novel physical chemistry methods can probe the orientation and structure of molecules at the liquid–liquid interface. The measurements of interaction forces with colloidal probe microscopy are very sensitive to the presence and structure of molecules at the liquid–liquid interface, but do not probe the molecular structure directly compared to spectroscopic methods. Colloidal probe microscopy at the liquid–liquid interface provides a method to develop quantitative correlations between the observed force behaviour and the molecules absorbed at the interface or present in solutions. This type of structure-function relation can then be used effectively in understanding and controlling coalescence phenomena. The AFM measurement coupled with spectroscopic techniques may be an approach to probe both the interaction forces and molecular structure simultaneously.

Prof. Unwin asked: Have you considered using AFM to probe the potential distribution at liquid/liquid interfaces under electrochemical control, using either partitioning ions or an external source to control the potential across the liquid/liquid interface?

Dr Dagastine replied: We have not considered this, but it seems likely it could provide some insight into the behaviour observed in these systems. For the paper presented in this discussion the technique might create an added complexity since the observed force behaviour is dependent on the type of ion present. This type of approach has been used in the modified SFA with a mercury droplet,[1,2] but this has not been adapted to AFM for liquid–liquid interfaces.

1   J. N. Connor and R. G. Horn, *Langmuir*, 2001, **17**, 7194.
2   J. N. Connor and R. G. Horn, *Rev. Sci. Instrum.*, 2003, **74**, 4601.

Prof. Girault asked: $NaClO_4$ will distribute to the organic more than $NaNO_3$. Therefore, the liquid/liquid interface will be changed according to the distribution.

Did you take that into account in your model?

Dr Dagastine responded: The model used in the paper presented for the electrostatic double layer force was the full numerical solution to the Poisson–Boltzmann equation between flat plats in aqueous solution using the numerical methods of Chan *et al.*[1] This model does not take into account ion size, the width of the interface, or the distribution of ions in the organic phase. The motivation behind using the less complicated model is based on several reasons. First, this modelling approach has proved successful with the force measurements between alkane with and without the addition of surfactant,[2–4] where all of the model input parameters were measured independently.[5] The separation distances encountered in most of the force measurements are in the region where the EDL force is expected to follow an exponential behaviour.

Since the force behaviour observed is dependent on the type of ion present, it seems plausible that some additional information may need to be included in the treatment of the electrostatic double layer forces. It is also important to note, that varying the surface potential or charge by twenty to thirty percent did change the qualitative trends in the model predictions. This would imply that if there were a deviation form the expected EDL force behaviour it must be significant and not just a small perturbation to the surface charge or potential.

1   D. Y. C. Chan, R. M. Pashley and L. R. White, *J. Colloid Interface Sci.*, 1980, **77**, 283.
2   P. G. Hartley, F. Grieser, P. Mulvaney and G. W. Stevens, *Langmuir*, 1999, **15**, 7282.
3   S. A. Nespolo, D. Y. C. Chan, F. Grieser, P. G. Hartley and G. W. Stevens, *Langmuir*, 2003, **19**, 2124.
4   R. R. Dagastine, D. C. Prieve and L. R. White, *J. Colloid Interface Sci.*, 2004, **269**, 84.
5   S. A. Nespolo, M. A. Bevan, D. Y. C. Chan, F. Grieser and G. W. Stevens, *Langmuir*, 2001, **17**, 7210.

Prof. Girault asked: Why so few experiments with scanning probes?

Dr Dagastine replied: Over the last eight years scanning probe microscopy has been effective at measuring the interaction forces at liquid–liquid interfaces with colloidal probe AFM[1–9] as well as measuring the interaction forces between two oil droplets in aqueous solution.[10] One of the main complications using a scanning probe method is interfacial deformation creates difficulties in analysis and interpretation. There are a number of studies dedicated to just the analysis of the colloidal probe–liquid droplet experiment.[8,9,11–16] As mentioned in a response previously, colloidal probe microscopy at the liquid–liquid interface has been a useful method to develop quantitative correlations between the observed force behaviour and the molecules absorbed at the interface or present in solution. There have also been a few imaging studies, but again droplet deformation gives rise to interpretation difficulties.[17,18] As the technological aspects of the scanning probe microscope improve, such as closed loop feedback in piezoelectric scanners and faster imaging times incorporated into commercial instruments, studies may become more common. Also, near field scanning optical microscopy (NSOM) can provide methods to perform spectroscopy on smaller length scales compared to traditional techniques.[19] NSOM can operate in a non-contact mode using light as the feedback signal, thus it may circumvent some of the difficulties encountered in AFM from interfacial deformation. The study of living cellular systems is an area with ever-increasing use of scanning probe methods.[20] This has also led to integration of AFM with a variety of methods including fluorescent microscopy, laser scanning confocal microscopy and some spectroscopy methods.[20] These studies range from very soft mammalian cells to more rigid bacteria and plant cells. If one expands the definition of a liquid–liquid interface to include a lipid bilayer, then there are a large number of experiments using scanning probe methods. It was unfortunate that there was virtually no representation at this discussion from the biophysics or physical chemistry communities using scanning probe methods with biological systems.

1   P. Mulvaney, J. M. Perera, S. Biggs, F. Grieser and G. W. Stevens, *J. Colloid Interface Sci.*, 1996, **183**, 614.
2   B. A. Snyder, D. E. Aston and J. C. Berg, *Langmuir*, 1997, **13**, 590.
3   P. G. Hartley, F. Grieser, P. Mulvaney and G. W. Stevens, *Langmuir*, 1999, **15**, 7282.
4   D. E. Aston and J. C. Berg, *Ind. Eng. Chem. Res.*, 2002, **41**, 389.
5   D. E. Aston and J. C. Berg, *J. Colloid Interface Sci.*, 2001, 235, 162.
6   G. Gillies, C. A. Prestidge and P. Attard, *Langmuir*, 2002, **18**, 1674.
7   G. Gillies and C. A. Prestidge, *Adv. Colloid Interface Sci.*, 2004, **108–109**, 197.
8   S. A. Nespolo, D. Y. C. Chan, F. Grieser, P.G. Hartley and G. W. Stevens, *Langmuir*, 2003, **19**, 2124.
9   R. R. Dagastine, D. C. Prieve and L. R. White, *J. Colloid Interface Sci.*, 2004, **269**, 84.
10   R. R. Dagastine, G. W. Stevens, D. Y. C. Chan and F. Grieser, *J. Colloid Interface Sci.*, 2004, **273**, 339.
11   D. Y. C. Chan, R. R. Dagastine and L. R. White, *J. Colloid Interface Sci.*, 2001, **236**, 141.
12   D. Bhatt, J. Newman and C. J. Radke, *Langmuir*, 2001, **17**, 116.
13   R. R. Dagastine and L. R. White, *J. Colloid Interface Sci.*, 2002, **247**, 310.
14   P. Attard and G. Gillies, *Aust. J. Chem.*, 2001, **54**, 477.
15   G. Gillies, C. A. Prestidge and P. Attard, *Langmuir*, 2001, **17**, 7955.
16   P. Attard and S. J. Miklavcic, *J. Colloid Interface Sci.*, 2002, **247**, 255.
17   S. D. A. Connell, S. Allen, C. J. Roberts, J. Davies, M. C. Davies, S. J. B. Tendler and P. M. Williams, *Langmuir*, 2002, **18**, 1719.
18   A. P. Gunning, A. R. Mackie, P. J. Wilde and V. J. Morris, *Langmuir*, 2004, **20**, 116.
19   A. Lewis, H. Taha, A. Strinkovski, A. Manevitch, A. Khatchatouriants, R. Dekhter and E. Ammann, *Nature Biotechnol.*, 2003, **21**, 1378.
20   R. M. Henderson, *J. Cell Sci.*, 2003, **116**, 761.

Dr Bain opened the discussion of Dr Liggieri's paper: Maldarelli and Liu have argued for a phase transition model in more hydrophobic members of the C_iE_j surfactants. You have used a re-orientation model. How would a phase transition manifest itself in your rheological measurements?

Dr Liggieri replied: Maldarelli and Liu argued for the formation of small aggregates (dimers, trimers, *etc.*).

The characteristic time associated with such transitions should be very short. Thus, in that case we would expect a system behaving according to the Lucassen–Van der Tempel (diffusion controlled relaxation) model. The phase transition should, however, influence the isotherm, and thus ε_0.

In a previous paper we have modelled the rheological behaviour of an adsorption layer presenting a surface phase transition from small to large aggregates, coupled to diffusional exchange.[1] Large aggregates were assumed to form above a critical surface pressure. The model predicts, in particular, the appearance of a discontinuity in the module of the surface visco-elasticity, $|\varepsilon|$, at the critical surface pressure, due to the different packing of the molecules inside the aggregates.

The model has been successfully applied to explain/interpret the surface rheological behaviour of n-dodecanol adsorbed at water–air,[2] allowing kinetic and equilibrium parameters to be estimated.

1 R. Palazzolo, F. Ravera, M. Ferrari and L. Liggeri, *Langmuir*, 2002, **18**, 3592.
2 L. Liggieri, F. Ravera and M. Ferrari, *Langmuir*, 2003, **19**, 10233.

Dr Cicuta asked: Two experimental techniques are used, to give results over a wide frequency range. Could you comment on how you match the experimental conditions in the two set-ups, for example is the strain deformation the same? I see that in most cases the matching is very good, but in Fig. 9 of your paper, it seems less satisfactory.

Dr Liggieri replied: Both utilised methods rely on the linear response of the system for small area perturbations. Thus in the two set-ups we apply area variations of relative amplitude smaller than 0.2. In this range, in fact, the measured visco-elasticity is nearly independent of the relative perturbation amplitude. The two methods are however affected by different experimental errors which may cause some mismatch. For the elasticity measured by the capillary pressure tensiometer the error can be evaluated on the order of mN m^{-1}, which is compatible with the data reported in Fig. 9 of our paper.

Prof. Unwin asked: Eqn. (9) of your paper, which you used to fit your dynamic data, contains many variables, some of which you can obtain from equilibrium measurements. Nonetheless, you appear to derive three parameters from your dynamic measurements: the diffusion coefficient in solution; the reorientation rate; and the intrinsic surface viscosity. Could you comment on the precision with which these model-dependent parameters can be obtained using this method? I ask this partly because some of the diffusion coefficients appear high. For example, the value for the diffusion coefficient of $C_{12}E_{23}$ in water of 9.0×10^{-6} cm^2 s^{-1} is what one might expect for a small molecule rather than a surfactant.

Dr Liggieri replied: In Table 2 the values of diffusion coefficients obtained from the elasticity data (D), are compared with those (D_{adk}) obtained by applying the Ward–Tordai model to experiments of adsorption kinetics at fresh interfaces (see Fig. 8 of our paper). For $C_{10}E_5$ and $C_{10}E_8$ the values agree within about 2×10^{-6} cm^2 s^{-1}. Assuming D to scale as the square root of the molecular weight, these values are also consistent with those measured by PFGSE-NMR (see for example, B. Faucompre *et al.*[1]

The reported value for $C_{12}E_{23}$ is actually larger than those of $C_{10}E_8$ and $C_{10}E_5$, which is in contrast with its larger molecular weight. As stressed in the text, commercial $C_{12}E_{23}$ samples where utilised, which may contain smaller CiEj's and low molecular weight impurities, justifying the observation of a larger value of D. For this surfactant the value of D_{adk} is missing since the quality of results from adsorption kinetics experiments is very poor due to the strong surface activity of this surfactant.

Error estimation on parameters obtained by best-fit of multi-parametric systems is very difficult to give. The method provides anyway at least the order of magnitude of the re-orientation rate. The estimated value is probably much more accurate than when it is found independently of the surfactant concentration, as for example for $C_{10}E_5$.

1   B. Faucompre and B. Lindman, *J. Phys. Chem.*, 1987, **91**, 383.

Dr Bain asked: The rate constants, ν_{or}, for reorientation of the surfactants are $O(10^2\,s^{-1})$, which is 6 orders of magnitude slower than equilibration times for surfactant monolayers in MD and MC simulations, for low surface coverages. What is the nature of the collective motions of the surfactant that gives rise to such (relatively) slow dynamics?

Dr Liggieri replied: A possible explanation for a collective behaviour could arise from the circumstance that, according to the physical view of the two-state model, the adsorbed molecules are highly packed already at rather low surface pressures. In fact, at small adsorption, molecules are allowed to adsorb in a status with high molar area. Thus, assuming for example the isotherm parameters for $C_{10}E_5$ in Table 1, at the surface pressure of 5 mN m^{-1}, corresponding to a bulk concentration of about 10^{-9} mol cm^{-3}, one finds already a surface coverage, $\omega\Gamma$, of about 0.8.

According to this view one should find larger values of the rate constant, ν_{or}, in the range of very low surface pressures (say below 5 mN m^{-1}). This range involves however very low bulk concentration and a vanishing rheological response, which makes these investigations nearly impossible.

Dr Bain commented: The reorganisation processes observed by Dr Liggieri and by other workers occur on timescales of milliseconds. Molecular simulations of surfactant layers typically equilibrate in nanoseconds, at least for systems that are not too dense. The nature of the collective processes that give rise to the slow reorganisation kinetics in surfactant monolayers remain a mystery and are a challenge for both simulators and experimentalists, since the timescales are long even for coarse-grained simulations and short for most spectroscopic or structural probes

Dr Liggieri responded: A possible justification for the occurrence of a collective behaviour assuming a two-state like model has been given in the previous answer to Dr Bain.

I agree that efforts should be made, in particular as far as simulation is concerned, to clarify these mechanisms.

Moreover, some "direct" optical techniques for adsorption investigation (ellipsometry, SHG) are probably mature enough to be applied to the highly dynamic situations related to surface rheology studies.

Prof. Rathman opened the discussion of Dr Mitani's paper: The method involves deforming the interface, so that the interfacial area in the measurement region increases and decreases with the oscillations of the laser. When surfactants are adsorbed at the interface, the oscillation is probably so rapid that the surfactant does not have time to desorb and re-adsorb during each oscillation cycle; thus, the interfacial tension would not be constant. Even though oscillations are relatively small (< 10 nm amplitude) the change in surfactant surface concentration (molecules area^{-1}) could be significant, and so the changes in interfacial tension would be significant.

Dr Mitani responded: In our experiments, we disregard the effect of the change of the surface density of the surfactant on the interfacial tension when the interface is deformed. The reason is that the density of the surfactant on the interface is low such as a gas, and the surfactant molecules can move freely on the surface, while they are limited to coming and going between the surface and the bulk. Therefore the small deformation of the interface does not change the interfacial tension, even when the modulation frequency of the interface is ~ 1 kHz. If the motion of the surfactant on the surface is limited, we can get the surface tension safely from the experimental spectrum by adding the elasticity term to the dispersion relation (eqn. (2.9) of our paper).

Prof. Kornyshev asked: Does your effect depend on polarization of incident light?

Dr Mitani replied: Because we used the optically isotropic liquids as the specimens and the pump laser penetrates the interface perpendicularly, the change of the light momentum at the interface is independent of the polarization of the pump light. If you measure the surface of a non-isotropic liquid such as a liquid crystal, the result would depend on the polarization of the pump light.

Dr Liggieri asked: The laser beam causes locally a heating of the interface. Has any effect been observed due to that local temperature increase?

Dr Mitani replied: Actually we observed that the interface deformed larger than the expected value in the LIM method when we used a pump laser of large intensity. We believe the cause of this to be the thermal expansion of the liquid on absorbing the pump light. Therefore we keep the intensity as low as possible in the measuring process. In addition, we measure the spectrum of the interface response, that is to say the time-dependent deformation. The liquid interface spends more time on expanding thermally than on deforming by light force, and we can divide the deformation caused by changing the light momentum and by heating. Accordingly we ignore the heating effect of the laser on the interfacial tension.

Dr Grayeff asked: Does Dr Mitani consider that the type of the initial standard water utilised in electrochemical experimental work is significant: different types of deionised water may contain different ionic ratios?
Comment:
(1) When an aqueous medium grade vermiculite solution is electrometrically measured the pH of the water (in this case tap water) can vary due to the ionic content of the water.

1   See Mitani's paper, Section A.
2   I. Lin, M. V. Smalley and S. G. Grayeff, *Overview of Vermiculite Modifications*, The Civil Engineering, Dept., The Technion.

Dr Mitani responded: In our experiments, we used distilled water without additional ions. The section A in our paper is the summary of our previous work, ref. 4 (of our paper), and we did not mention in detail the materials used in the experiments. As shown in section B, the concentration of ionic molecules in water such as surfactant or NaCl may easily change the surface or interfacial tension.

Prof. Penfold opened the discussion of Dr Zarbakhsh's paper: The results for the adsorption of $C_{14}TAB$ at the hexadecane–water interface is surprising for 3 reasons:
(i) We know from neutron reflectivity studies[1] that at the c.m.c. common small molecule surfactants adsorb at the air–water interface as a monolayer, and we have demonstrated this specifically for $C_{14}TAB$.[2] Sub-surface ordering of micelles is observed, but at much larger concentrations $\gg$ 10 times the c.m.c.[3] Furthermore, in this regime of concentration micellar ordering has been predicted from computer simulations by Smit *et al.*,[4] at the oil–water interface.
(ii) The total amount of $C_{14}TAB$ at the hexadecane/water interface (85 to 124 Å^2) is low compared to what is measured at the air–water interface (~ 45 Å^2 at the c.m.c.[2]). We have directly compared the adsorption of SDS at the air–water interface by neutron reflectivity[5] and at the hexadecane–water interface by SANS from SDS stabilised hexadecane in water emulsion.[7] From these measurements we observe rather similar amounts of adsorption at saturation at the two interfaces.
(iii) The thickness of the adsorbed layer is low compared to the surfactant chain length, and we have observed similar thicknesses for SDS.[6] However, this is in contrast to what is observed in slightly different circumstances. Lu *et al.*,[7] have measured the adsorption and structure of a cationic surfactant monolayer at the air–water interface to which dodecane is added (adsorbed) within the monolayer to saturation. (This can be considered as a nascent liquid–liquid interface). In this case the addition of the dodecane resulted in a larger alkyl chain thickness being measured for the cationic surfactant.

1   J. R. Lu, R. K. Thomas and J. Penfold, *Adv. Colloid Interface Sci.*, 2000, **84**, 143.

2   (*a*) E. A. Simister, E. M. Lee, R. K. Thomas and J. Penfold, *J. Phys. Chem.*, 1992, **96**, 1373; (*b*) E. A. Simister,
    R. K. Thomas, J. Penfold, R. Avegard, B. P. Binks, P. Cooper, P. D. I. Fletcher, J. R. Lu and A. Sokolowski,
    J. Phys. Chem., 1992, **96**, 1383.
3   (*a*) J. R. Lu, E. A. Simister, R. K. Thomas and J. Penfold, *J. Phys. Chem.*, 1993, **97**, 13907; (*b*) E. M. Lee
    et al., *J. Phys.*, 1989, **7**, 75; (*c*) E. M. Lee *et al.*, *Colloid Polym. Sci.*, 1990, **82**, 99.
4   (*a*) B. Smit, P. A. J. Hilbers, K. Esselink, L. A. M. Rupert, N. M. van Os and A. G. Schlijper, *J. Phys. Chem.*,
    1991, **95**, 6361; (*b*) B. Smit, P. A. J. Hilbers, K. Esselink, L. A. M. Rupert, N. M. van Os and A. G. Schlijper,
    Nature, 1990, **348**, 624.
5   J. R. Lu, A. Marrocco, T. J. Su, R. K. Thomas and J. Penfold, *J. Colloid Interface Sci.*, 1993, **158**, 303.
6   E. Staples, J. Penfold and I. Tucker, *J. Phys. Chem. B*, 2000, **104**, 606.
7   (a) J. R. Lu, R. K. Thomas, R. Aveyard, B. P. Binks, P. Cooper, P. D. I. Fletcher, A. Sokolowski and
    J. Penfold, *J. Phys. Chem.*, 1992, **96**, 10971; (*b*) J. R. Lu, R. K. Thomas, B. P. Binks, P. D. I. Fletcher and
    J. Penfold, *J. Phys. Chem*, 1995, **99**, 4113; (*c*) J. R. Lu, Z. X. Li, R. K. Thomas, B. P. Binks, D. Crichton,
    P. D. I. Fletcher, J. R. McNab and J. Penfold, *J. Phys. Chem.*, 1998, **102**, 5785.

Dr Zarbakhsh replied: We believe a single layer model does not fully describe our data. A single layer model does indeed give a low adsorbed amount, however, a two layer model yields a comparable adsorbed amount to that found at the air–water interface.

It may be possible that the monolayer oil experiment may not represent a true bulk oil/water interface.

Dr Leermakers asked: When equilibrating the system, surfactants may travel through the oil to the SiO_2 surface. Does this occur and is the surfactant pool in the interface sufficiently large to account for this loss of surfactant? In other words is the surfactant adsorption at the oil–water interface still in equilibrium with c.m.c.-level surfactant concentrations? Again in other words: What is the surfactant concentration in the water phase after equilibration?

Dr Zarbakhsh replied: The interface is in contact with c.m.c.-level concentrations of the bulk (relatively large reservoir ~ 20 cm^3) at all times.

On the matter of the surfactant travelling to the SiO_2–oil interface, we are not sensitive to this although we plan to determine this in the future.

Prof. Schlossman asked: Your current models do not include the effect of interfacial roughness. Please comment on the effect of neglecting the roughness on the other parameters in your models. The ability to correctly identify the interfacial roughness depends critically on the background measurements. Please also comment on the method used to measure the background scattering.

Dr Zarbakhsh replied: As Prof. Schlossman suggests, the background levels can limit the ability to determine the true interfacial roughness. Currently we fit a constant background level to our reflectivity and this is found to be adequate for air–water interfaces. We currently do not subtract the background. We will, however, in the future by measuring the background either side of the specular peak. It is also important to reduce the true background, the dominant part of which originates from the sample/sample environment. We hope to achieve this by reducing the volume of the bulk solution and having a roughened silicon trough.

Prof. Schlossman asked: It would be helpful to identify a reference system that has been (or can be) studied by other techniques. For example, independent experiments and analysis of a common system by both X-ray and neutron reflectivity that provided complementary, but consistent, information on that system would provide a baseline to judge the capabilities of these techniques. Can you suggest appropriate systems for this type of study?

Dr Zarbakhsh replied: A well-defined water-soluble surfactant such as CTAB should also be characterisable using your X-ray technique. Currently we cannot use oil soluble surfactants, which would partition at solid–liquid interfaces.

Dr Caruana commented: I was very encouraged to hear that you have some preliminary analysis of the clean liquid/liquid interface (hexadecane/water) which gives the interfacial thickness to be 6 Å. We have used a different approach to stabilise a liquid/liquid interface in the form of an emulsion and studied it with small angle neutron scattering. We determined the water/DCE interface to be very similar in thickness to what you quote.

Dr Zarbakhsh responded: Good. We are investigating the slight discrepancy for the bare interfacial roughness determined using neutrons and those reported using X-rays. The role of the oleophilic/trimethylsilane layer will also be investigated.

Dr Caruana asked: Would it be possible to use your technique to study the water/DCE interface? This would require the water layer to be frozen on the silicon block as the DCE is more dense.

Dr Zarbakhsh answered: Yes. There may be two possible ways to approach this problem:
(i) To spin coat a silicon block with 1,2-dichloroethane (DCE) and then fill the cell up upside down with water acting as a bulk phase. Then bending the neutron beam up to obtain the desired Q-range.
(ii) A thin water (D_2O) layer could be deposited on a hydrophilic silicon block (evaporation, spinning or dipping) and frozen. DCE would then be spun on this layer (higher freezing point, so as not to damage the water film when bulk DCE is introduced) and then sandwich it by bulk DCE.

Prof. Deutsch asked: The method of preparation of the thin oil layer between the silicon and water is very clever. I would like, however, to ask about the fits. Von Neumann reportedly said once that "with four parameters I can fit an elephant, and with five I can make him wiggle his trunk". To assess the quality and significance of a fit it is important to know which parameters were varied in the fit and which were kept fixed, and at what values. Could Dr Zarbakhsh please expand on this for the fits shown in Figs. 3 and 4 of his paper?

Dr Zarbakhsh replied: The thickness and scattering length density of the oxide (oleophilic/trimethylsilane) film with added interlayer roughness were fixed. The oil layer thickness deduced from the attenuation from the $0.26°$ angle measurement was subsequently fixed.
The layer thickness and the scattering length density of the layer at the interfaces were fitted only. The uniqueness of this pair of parameters were ascertained by means of a contour plot of the weighted sum-of-squares deviations of two parameters, in this case the thickness and scattering length density of the interfacial slab.
These values were then used in our calculations (Figs. 3 and 4 of our paper) and reflectivities for all four angles were then calculated simultaneously.

Prof. Samec opened the discussion of Prof. Scholz's paper: The three-phase electrode that you used is actually a membrane system (I),

$$C|M|W \tag{I}$$

where C represents the graphite electrode, M is the organic phase containing ferrocene species Fc, and W is the aqueous electrolyte solution containing anion X^-. The approximate thickness of the membrane can be estimated from the volume of the droplet as *ca.* 80 μm. The electric current I is associated with the overall reaction (II),

$$Fc(M) + X^-(W) \rightarrow Fc^+(M) + X^-(M) + e^-(C) \tag{II}$$

The theory of amperometry for such a system should be based on a balance of the electric current, rather than on the electroneutrality condition, which of course applies to any finite volume element of space. Following our theoretical approach to such a system,[1] the applied potential E splits into two contributions plus a constant involving terms associated with the reference electrode,

$$E = \Delta_M^C \phi_{Fc^+/Fc}^{0'} + (RT/F)\ln[c_{Fc^+}(M)/c_{Fc}(M)] - \Delta_M^W \phi_{X^-}^{0'} + (RT/F)\ln[c_{X^-}(M)/c_{X^-}(W)] + \text{const} \tag{1}$$

Eqn. (1) is practically identical with eqn. (2) of your previous paper.[2] Assuming that the transport both inside the membrane and in the aqueous solution is controlled by the linear diffusion, the interfacial concentrations c_i's in eqn. (1) can be expressed using the convolution theorem as

$$
\begin{aligned}
c_{X^-}(M) &= m/\kappa_{X^-}(M) \\
c_{X^-}(W) &= (m_{l,X^-} - m)/\kappa_{X^-}(W) \\
c_{Fc}(M) &= (m_{l,Fc} - m)/\kappa_{Fc}(M) \\
c_{Fc^+}(M) &= m/\kappa_{Fc^+}(M)
\end{aligned}
\tag{2}
$$

where m is the convolution current,

$$
m = \pi^{-1/2} \int_0^t I(\tau)(t - \tau)^{-1/2} d\tau
\tag{3}
$$

The current *vs.* potential relationship then takes the form of an integral equation[1]

$$
E = \Delta_M^C \phi_{Fc^+/Fc}^{0'} - \Delta_M^W \phi_{X^-}^{0'} + (RT/F) \ln \frac{m^2 \kappa_{Fc}(M) \kappa_{X^-}(W)}{\kappa_{Fc^+}(M) \kappa_{X^-}(M)(m_{l,Fc} - m)(m_{l,X^-} - m)} + \text{const}
\tag{4}
$$

where $m_{l,i} = \kappa_i c_i^0$ is the limiting convolution current, and κ_i's are the mass transport coefficients comprising as the case may be the migration effect.[1] The equation for the reversible half-wave potential $E_{1/2}^{rev}$ has the form depending on whether the transport is controlled by diffusion of the anion X^- in the aqueous phase (when $m_{l,Fc} \gtrsim m_{l,X^-} \geq m$), or by ferrocene species in the organic phase (when $m_{l,X^-} \gg m_{l,X^-} \geq m$). In the former case, the substitution $m = m_{l,X^-}/2$ into eqn. (4) yields

$$
E_{1/2}^{rev} \approx \Delta_M^C \phi_{Fc^+/Fc}^{0'} - \Delta_M^W \phi_{X^-}^{0'} + (RT/F) \ln \frac{\kappa_{X^-}(W)^2}{2\kappa_{Fc^+}(M)\kappa_{X^-}(M)} + (RT/F) \ln \frac{c_{X^-}^0(W)}{c_{Fc}^0(M)} + \text{const}
\tag{5}
$$

In the latter case, which apparently corresponds to your experimental conditions, the substitution $m = m_{l,Fc}/2$ into eqn. (4) yields

$$
E_{1/2}^{rev} = \Delta_M^C \phi_{Fc^+/Fc}^{0'} - \Delta_M^W \phi_{X^-}^{0'} + (RT/F) \ln \frac{\kappa_{Fc}(M)^2}{2\kappa_{Fc^+}(M)\kappa_{X^-}(M)} + (RT/F) \ln \frac{c_{Fc}^0(M)}{c_{X^-}^0(W)} + \text{const}
\tag{6}
$$

Apart from the contribution from the mass transfer coefficients, eqn. (6) agrees with eqn. (1) of your paper.

1 Z. Samec, A. Trojánek, J. Langmaier and E. Samcová, *J. Electroanal. Chem.*, 2000, **481**, 1.
2 G. Bouchard, A. Galland, P.-A. Carrupt, R. Gulaboski, V. Mirčeski, F. Scholz and H. H. Girault, *Phys. Chem. Chem. Phys*, 2003, **5**, 3748.

Prof. Scholz responded: We have published a theoretical derivation and the experimental proof that the dependence of the formal potential on the logarithm of the anion concentration has an opposite sign depending on the ratio of the activities of the anions in water to that of dmfc in the organic phase.[1] In that publication the diffusion coefficients have been taken into account as well.

1 Š. Komorsky-Lovrić, M. Lovrić and F. Scholz, *Collect. Czech. Chem. Commun.*, 2001, **66**, 434.

Prof. Janata asked: Since the graphite electrode is in contact with both the aqueous and organic phase a mixed potential is formed. Has it been taken into account and/or eliminated?

Prof. Scholz replied: In all experiments performed so far, the potential was controlled externally. Since all electrode processes that, in principle, could occur at the graphite aqueous interface have much larger overpotentials than the studied process at the three-phase junction (and its extension), no effect of the graphite aqueous interface was detectable.

Prof. Janata asked: It appears that this "triple point" has a special significance in your experiment. Can you elaborate on this?

Prof. Scholz answered: Yes, indeed, the three-phase junction line is of special importance. Since the organic phase does not contain deliberately added electrolyte, its initial conductivity is very low. Only at the three-phase junction it will acquire by free partition an ionic conductivity that is sufficient to drive the coupled electron and ion transfer across the two interfaces graphite | organic phase and aqueous phase | organic phase. Since the electrode process creates dmfc$^+$ and anions in the organic phase, the electrode reaction extends along these interfaces.

Prof. Janata asked: If the process is initiated at the triple point and then progresses to the rest of the electrode surface then there should be a transient current. Has this been observed and can you show such data?

Prof. Scholz replied: Until now, no potential step experiments monitoring current time transients have been performed by us.

Prof. Karyakin asked: By definition this system is a combination of thin layer–three-phase electrode. A three-phase electrode with thick organic layer would not have standard potential dependent on counter anion hydrophilicity.

Prof. Scholz replied: The term thin-layer electrode is misleading as the electrode consists of one or several immobilized droplets. The main feature is the three-phase arrangement. When a thick layer would cover the entire graphite surface, no electrochemical response can be measured because of the high resistance of the organic phase. However, in the three-phase arrangement, when the droplets are rather "thick" the response remains because of the unique feature of the three-phase junction that enables both the electron and ion transfer.

Dr Walker asked: Can you comment on whether or not the kinetics are different for the L(solute) → L(solv)/D(solute) → D(solv) *vs.* L(solute) → D(solv)/D(solute) → L(solv) systems?
If the kinetics are similar would that imply similar mechanisms of ion transfer?
Could different kinetics be used to differentiate different mechanisms of ion transfer?

Prof. Scholz replied: We could not see any differences for which different kinetics could be accounted for.

Prof. Unwin asked: Your analysis requires that the experimental response you measure is truly reversible. Mass transport in this system would appear to be complicated, with the reaction starting at the triple phase contact, where it will be characterised by high mass transport and resistive effects within the drop. Could you comment on the likely implications of such effects for your analysis, and whether resistive effects, in particular, have any influence on your data? The CV in Fig. 4 (of your paper) shows a peak to peak separation which appears to be close to 150 mV. What separation is expected for a reversible process at a droplet electrode of this type?

Prof. Scholz responded: The effect of the ohmic drop was studied in a theoretical simulation assuming, for reasons of simplicity, a conical form of the droplet.[1] Indeed it was found that the theoretical peak separation is between 125 and 150 mV, depending on the experimental conditions. The mid-peak potential may deviate by 7 mV from the thermodynamically expected value.

1   M. Lovrić and F. Scholz, *J. Electroanal. Chem.*, 2003, **540**, 89.

Prof. Samec addressed Prof. Unwin: Eqn. (4) in my comment above to Prof. Scholz's paper indicates that the reversible voltammetric peak potential difference cannot take the value of 59 mV as for a univalent ion transfer across ITIES. Numerical calculation shows that the peak separation in this case is 91 mV.[1]

1   Z. Samec, A. Trojánek, J. Langmaier and E. Samcová, *J. Electroanal. Chem.*, 2000, **481**, 1.

Prof. Kornyshev asked: What is the nature of chiral effects in the free energy of transfer? Is it just the effect on van der Waals interactions?

Prof. Scholz replied: Yes, we suppose that the differences in van der Waals interactions are responsible for the different Gibbs energies of transfer.

Dr. Wakisaka asked: When you use a chiral organic phase, and a D-, L-mixture aqueous phase, is it possible to extract one optical isomer from the aqueous phase into the chiral organic phase, by controlling electric potential?

Prof. Scholz replied: In principle, such extractions should be feasible. However, since the energetic differences are very small, *i.e.*, the signals of mixtures are overlapping, a multi-step process will be necessary to achieve interesting separation yields.

Dr Bain asked: You have demonstrated that your voltammograms are not affected by transport rates across the liquid–liquid interface. Could one change the experimental parameters in order to probe chiral effects on transport rates across the liquid–liquid interface, or is diffusion/convection so complicated that it would be difficult to disentangle different kinetic processes?

Prof. Scholz replied: So far, we have not searched for effects of chirality on the kinetics of the process. Such effects can be expected and it would be interesting to study them.

Dr Liggieri commented: Interfacial tensiometric/rheometric methodologies can be very effective to investigate some aspects of the general items addressed in the discussion. Examples of these investigations are:

(1) Transfer across the interface of amphiphiles or of other species interacting with the adsorbed molecules can be investigated by the dynamic interfacial tension of freshly formed interfaces.[1]

(2) Nature and characteristics of the barriers to transport to/across the interface have been investigated by dynamic interfacial tension in well stirred bulks[2] and in drop shape tensiometers equipped with multiple feed systems.[3]

(3) Information about the interfacial layer structure and dynamic behaviour and about the relationships between bulk and interfacial properties are obtained by combining measurements of the Π–c or Π–A isotherms, of the dynamic interfacial tension (adsorption kinetics) and of the dilational and shear dynamic surface visco-elasticity (see our paper or ref. 4).

Technological advancement makes continuous improvements of these techniques possible and their full potential still needs to be explored. Significant progress will arise from coupling "fast" tensiometric/rheometric techniques (drop shape, oscillating bubble/drop, capillary pressure tensiometry) with optical ones (BAM, ellipsometry, SHG *etc.*); which is now at an initial stage.

1   M. Ferrari, L. Liggeri, F. Ravera, C. Amodio and R. Miller, *J. Colloid Interface Sci.*, 1997, **186** 40.
2   T. F. Svitova, M. J. Wetherbee and C. J. Radke, *J. Colloid Interface Sci.*, 2003, **261**, 170.
3   Miller *et al.*, *Food Hydrocolloids*, 2004, submitted.
4   L. Liggieri, F. Ravera and M. Ferrari, *Langmuir*, 2003, **19**, 10233.

Dr Liggieri added: At present we do not have reliable techniques for the measurement of dynamic interfacial tension (IT) and surface visco-elasticity working in the range of very low IT (<1 mN m^{-1}). This range is very important for emulsion science and technology. It is then worth exploring the potential offered in this respect by the technique presented in Mitani's paper.

Biocomposite films synthesized at a fluid/fluid interface

James F. Rathman* and Pei Sun

Department of Chemical and Biomolecular Engineering, The Ohio State University, 140 W. 19th Ave., Columbus OH 43210-1180, USA. E-mail: rathman.1@osu.edu; Tel: (614) 292 3760

Received 16th July 2004, Accepted 20th July 2004
First published as an Advance Article on the web 22nd September 2004

In the synthesis of mesostructured particles and films, the cooperative self-organization of amphiphilic molecules in the presence of reactive species is a key factor in the reaction mechanism. This paper presents a method for preparing structured collagen films synthesized at fluid/fluid interfaces. This work is an extension of previous efforts in our group to synthesize structured silica films in a reaction system confined at the interface between two immiscible fluid phases, providing an additional level of control over the structural evolution that occurs during reaction. Synthesis at a liquid/liquid interface was shown to provide excellent control over the mesostructure of the final product, avoiding a major problem encountered in many film synthesis techniques in which the reaction occurs at a liquid/solid interface: namely, the undesired effect of the solid surface on the film structure. The focus of this paper is the synthesis of structured composite films containing amphiphilic phospholipids and collagen. These films provide a way to pattern cell growth on biocompatible surfaces and a model system for studying the self-assembly mechanism of lipids and collagen. Self-assembled monolayers and bilayers composed of phosphatidylethanolamine (PE) lipids and collagen were investigated to determine how regularly patterned films can be prepared in a manner that preserves the bioactive properties of the collagen. Mixtures of PEs and acid-soluble collagen were spread on an aqueous subphase in a Langmuir trough. Surface pressure–area compression isotherms for the composite lipid/collagen monolayers provide information about interactions between these components. Langmuir–Blodgett (LB) techniques were utilized to transfer the composite films onto freshly cleaved mica. The mica-supported films were characterized by atomic force microscopy. The lipid/collagen ratio in the composite films was found to be the most important factor in determining how the collagen is assembled and distributed. The temperature and pH of the aqueous subphase, the process for spreading collagen on the subphase, the deposition speed, and the deposition pressure are factors that can be used to selectively control the film patterning. For most of these experimental factors, the range over which a highly structured uniform film can be fabricated over significant length scales is generally very narrow. Based on the experimental results and understanding of the fundamental interactions involved, a mechanism for the co-self-assembly of phospholipids and collagen is suggested. The adhesion and growth of Chinese hamster ovary (CHO) cells on the patterned film surfaces demonstrates the biocompatibility of these composite films.

DOI: 10.1039/b410919h

Introduction

Phosphatidylethanolamines (PE) are among the important components found in biological membranes. These zwitterionic phospholipid molecules self-assemble into well organized bilayer structures. When adsorbed at fluid/fluid interfaces, monolayers composed of mixtures of different PEs containing saturated or unsaturated acyl chains with various lengths often exhibit interesting two-dimensional patterns.[1–3] For example, mixtures of DOPE (1,2-dioleoyl-*sn*-glycero-3-phospho-ethanolamine) and DPPE (1,2-dipalmitoyl-*sn*-glycero-3-phosphoethanolamine) form hetero-geneous monolayers that exhibit distinct DOPE-rich and DPPE-rich domains.[2]

Langmuir–Blodgett (LB) technology provides a way to transfer a self-assembled monolayer from an air/liquid interface onto a solid surface. Multilayered structures can be fabricated by repeating the deposition process. LB films supported by a solid substrate have many applications, including microelectronics, sensors, catalysis, and coatings. In comparison to evaporation or spin-coating methods, the LB technique provides an additional level of control over the molecular organization and therefore a more sensitive way to control surface pattern formation by adjusting the surface molecular density and the temperature and pH of the aqueous subphase. Since weak interactions are typically involved in the self-assembly process, self-assembled structures are sensitive to small changes in these and other variables, such as traces of contamination. Film defects incurred during the transfer process are often difficult to avoid. Bottom-up procedures to fabricate ultrathin films have had limitations of realizing the reproducibility and long-range uniformity that are critical for commercial applications.

Collagen is a triple stranded peptide that is the most abundant structural protein in the extracellular matrix (ECM) of connective tissues. Uneven distribution of hydrophobic regions and charged residues along the surface of a collagen monomer leads to a unique and highly ordered way for collagen to self-assemble forming complex fibrous network structures *in vivo*.[4–6] Collagen plays a crucial role in specific binding of many different kinds of proteins and cells, such as fibronectin and platelets.[7,8] Collagen thin films supported by solid substrates with well-defined surface patterns and highly-ordered inner structures are of major importance in drug delivery, biosensors and biomedical implant materials due to its biocompatibility.

Supported ultrathin films of collagen have previously been fabricated by introducing acetic solutions of collagen monomers onto a solid substrate through adsorption or coating methods.[9–14] When the substrate surface properties, the concentration, pH and temperature of the collagen solution and the method used to dry the supported film are properly selected, collagen fibril or network coatings can be fabricated evenly distributed on the surface of a solid. These conventional techniques are primarily based on the interactions between collagen and the solid surface of the substrate, and are also dependent on the dewetting mechanism, which generally limits the ability to control both the lateral distribution of collagen molecules at the nano-scale and the degree of collagen polymerization.[15]

There have been very few prior studies of pure collagen or collagen/phospholipid LB films at vapor/liquid interfaces or on a solid support.[16,17] At 25 °C, a 0.1 wt% acetic solution of type I collagen (pH 3.0) was spread onto the surface of ultrapure water in a LB trough and successfully transferred onto silicon-coated glass by horizontal deposition.[16] Ghannam *et al.* used a different method to incorporate collagen at the air/water interface by injecting a collagen solution in 0.5M acetic acid into the aqueous subphase with or without a phospholipid spread monolayer on the surface.[17] After 30 min waiting, the collagen spread at the interface yielded a compression isotherm similar to water-insoluble monolayers, indicating that collagen is a good surface-active molecule that can penetrate the air/water interface and interact with the lipid monolayer spontaneously. The previous studies of collagen at interfaces with or without lipids have not addressed the problem of accurately controlling the amount of collagen delivered to the interface. Although amphiphilic proteins may be spread as a monolayer alone or with lipids, acid-soluble collagen molecules dissolve into the subphase if spread by themselves on an aqueous subphase. It is even more difficult to evaluate the actual amount of collagen incorporated in the floating lipid monolayer after the equilibrium between the interface incorporation and subphase solubility is reached. A new method that effectively prevents partitioning of collagen into the subphase is presented in this paper. The ability to spread soluble macromolecules such as collagen at an interface in this manner provides a useful technique for the preparation of ultrathin composite films.

The domains observed in heterogeneous phospholipid monolayers or bilayers floating at air/water interfaces can serve as templates for patterning macromolecules at the nanometer scale. This approach can potentially be used to form various patterned collagen features in a well-controlled way by incorporating collagen in a properly selected lipid matrix using LB techniques. Patterning collagen assemblies by reaction at fluid interfaces in the presence of phospholipids has not been reported. The current study focuses on developing a new procedure to fabricate collagen-containing ultrathin films with well-defined 2-D patterns and various surface morphologies, in which DOPE/DPPE monolayers serve as templates to localize collagen features and control the degree of collagen self-assembly. The novel procedure successfully realizes the long-range uniformity of the co-self-assembled composite in a reproducible manner.

Experimental section

Materials

DOPE and DPPE were obtained from Avanti Lipids (Alabaster, AL) and used without further purification. DPPE was dissolved in chloroform, DOPE and the mixtures of DOPE/DPPE were soluble in chloroform/methanol (4/1 vol/vol). The total lipids concentration in the spreading solutions was 0.7 mg ml^{-1}.

Type I collagen extracted from calf skin was supplied by Sigma-Aldrich (St. Louis, MO). In 1.0 M acetic acid, collagen solution at a concentration of 0.1 mg ml^{-1} was stirred for at least 4 h at ambient temperature until the solution was transparent.

Measurements

Surface pressure isotherms for lipid-only and lipid/protein mixtures with various molar ratios were measured at 25 °C using a Langmuir–Blodgett trough from Nima Technology (Coventry, England). Surface pressure was measured using the Wilhelmy plate method. The subphase was an aqueous buffer solution of potassium dihydrogen phosphate, sodium hydroxide, and sodium bicarbonate at a pH of 7.2. Highly purified water (Milli Q, electrical resistivity >18 MΩ cm) was used in all experiments.

To obtain a lipid LB film on a solid support, a volume of 40 μl mixed DOPE/DPPE solution was first spread dropwise on the subphase using a microsyringe with the trough barriers set at the maximum area of 280 cm^2. After spontaneous spreading on the subphase, the monolayer was kept for 10 min to allow complete evaporation of the volatile solvents in the spreading solution. The monolayer was then compressed in a stepwise quasi-static mode using area increments of 1 cm^2, a wait interval of 16 s, and a barrier speed of 25 cm^2 min^{-1}. Upon the surface pressure reaching a target value (typically 25 mN m^{-1} or higher), the surface pressure was maintained at a constant value by the LB trough by automatic adjustment of the barrier position as necessary. Typically, the surface area continued to slowly decrease during this phase, compressing the film to maintain constant surface pressure, indicating that the film was not in a steady state and subject to molecular re-organization within the film. When the barrier stopped moving, the LB films were transferred onto freshly cleaved mica using vertical up deposition with a lifting speed of 1 mm min^{-1}. The surface pressure was kept constant throughout the deposition step.

Attempts to prepare a pure collagen monolayer or to simultaneously spread both lipids and collagen were unsuccessful due to the dissolution of collagen into the subphase. It was observed that collagen could be deposited at the interface and prevented from partitioning into the subphase if a sequential deposition process was used. The lipid monolayer was first spread and then compressed to non-zero, but still very low, surface pressure. At this point the collagen spreading solution was delivered. In each of the experiments reported here, collagen was delivered after the DOPE/DPEE monolayer had been compressed to a low surface pressure, Π_S. The collagen solution in acetic acid at a concentration of 0.1 mg ml^{-1} was spread drop by drop using a microsyringe on top of the lipid monolayer. During the spreading process, the surface pressure was maintained constant at Π_S by moving the barrier and increasing the surface area. The composite film was allowed to age for 15 min after spreading the collagen and was then compressed until reaching the desired deposition surface pressure, Π_D, generally 25 mN m^{-1} or higher. The LB film was held at Π_D until the barrier stopped moving and then transferred onto mica by vertical up deposition.

The LB films were air-dried for 24 h before the observation using tapping mode atomic force microscopy (AFM). A Nanoscope III AFM from Digital Instruments (Santa Barbara, CA) was used for investigating the surface morphology of the film in air and at ambient temperature. Tapping mode imaging using silicon cantilevers was performed at a low spring constant of 0.3 N m^{-1}.

Results and discussions

DOPE/DPPE monolayer

The behavior of DPPE/DOPE monolayers was observed to be very sensitive to the pH of the subphase due to NH_3^+-containing phosphoethanolamine (PE) lipid head groups; the protonation (charge) of this group is determined by a pH-sensitive dissocation equilibrium, so the monolayer structure and the resulting film organizations are also expected to be sensitive to the pH of the subphase. For studies at neutral pH, the subphase for this system was buffered at pH 7.2.

The surface pressure (Π) *vs.* area per molecule (A) isotherms for the DPPE/DOPE system at 25 °C were investigated. This is the 2-dimensional analogy of pressure–volume isotherms of compressible 3-dimensional bulk phases. As shown in Fig. 1, there is no apparent phase transition for single-lipid isotherms for either DPPE or DOPE. The DPPE isotherm is typical of a highly condensed film while the DOPE isotherm suggests a more loosely packed film. The isotherms of their mixtures with various molar ratios are intermediate between those of the two pure components and the area at which Π starts to increase above zero decreases with the proportion of DPPE in the mixture. The carbon double bond existing in the DOPE tails enhances the steric repulsions between adjacent molecules and increases the average area occupied by an individual molecule in the mixed monolayer. The collapse pressures do not show dependence on the molar proportions in the mixed films, implying at least partial immiscibility of the two lipids in all proportions studied.

To further investigate the miscibility of DOPE and DPPE in the binary film, the area per molecule at surface pressures of 25 mN m^{-1} and 40 mN m^{-1} are plotted against the molar fraction of DOPE in the mixture (Fig. 2). The dotted lines shown in Fig. 2 represent the ideal additive relation:

$$A = X^* A_{\text{DOPE}} + (1 - X)^* A_{\text{DPPE}}$$

where A, A_{DOPE} and A_{DPPE} are the average molecular areas for the mixture and the pure components and X is the molar fraction of DOPE in the mixture. For both surface pressures the area curves exhibit positive deviations from the ideal relation at DOPE molar ratios below 0.8, indicating stronger repulsive interactions between DPPE and DOPE in their mixtures than between pure lipid molecules. Since DOPE and DPPE have the same head groups, the additional repulsive

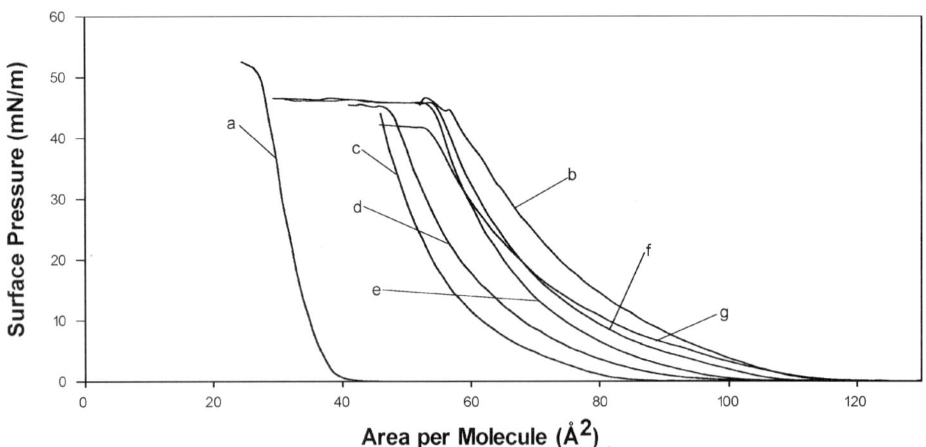

Fig. 1 Π–A isotherms for (a) DPPE, (b) DOPE, and their mixtures at various molar ratios (DOPE : DPPE): (c) 1 : 4, (d) 2 : 3, (e) 1 : 1, (f) 3 : 2 and (g) 4 : 1 at 25 °C on aqueous buffer solution (pH = 7.2).

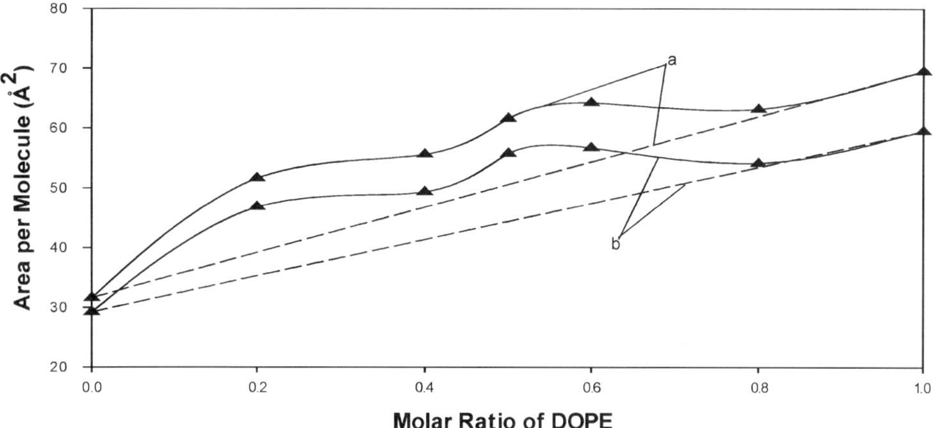

Fig. 2 Comparison of ideal additive (dashed line) and actual (solid line) relationship between average area and molar ratio of DOPE to DPPE: (a) $\Pi = 25$ mN m^{-1}; (b) $\Pi = 40$ mN m^{-1}.

interactions can be mainly attributed to the steric effect of the kinked DOPE tails. At slightly lower pH, in which pure water (pH = 6.8) was used as the subphase, there is additional electrostatic repulsions between PE head groups, which results in larger average areas of 0.72 nm^2 and 0.42 nm^2 occupied by DOPE and DPPE, respectively, at surface pressure of 40 mN m^{-1}.[2] Ideal additivity was observed for all DOPE proportions.[2] Significant deviation from area additivity suggests possible miscibility in the mixture; such deviation was not observed in this study, so DOPE and DPPE are assumed to be totally immiscible, which will be demonstrated further by the atomic force microscope (AFM) topographic study.

The films consisting of DOPE/DPPE mixtures with various molar compositions were deposited onto mica at surface pressure $\Pi = 25$ mN m^{-1}. The morphology of these binary monolayers as characterized by AFM in air is illustrated in Fig. 3. Phase separation is clearly observed at all molar ratios, with topographically higher closed domains embedded in a lower continuous phase. A previous study reported similar domain formation for the supported DOPE/DPPE monolayer transferred from the water subphase at 40 mN m^{-1}, where the domains were assigned to pure DPPE.[2] The area fraction occupied by DPPE calculated from isotherms at 25 mN m^{-1} is in good agreement with the area fraction shown in Fig. 3 for all molar ratios. The relative height difference of 1 nm between the domains and the background is in good agreement with the expected heights of DOPE and DPPE monolayers (1.8 nm and 2.8 nm, respectively) leading to the conclusion that the domains are composed of nearly pure DPPE and that DPPE is excluded from the continuous matrix consisting of DOPE.

The shape and the size of the DPPE domains can be adjusted by varying the molar ratio of DOPE to DPPE (Fig. 3). With increasing proportions of DOPE in the mixture, the DPPE domains vary from well-defined circular shapes to more elongated snow flake-like shapes. In Fig. 3a, the film with the lowest DOPE ratio at 0.2 exhibits well-defined circular domains with diameters ranging from a couple of microns to more than 10 microns. Upon further addition of DOPE in the mixture, the domain shape becomes less rounded, changing to multi-lobed patches (Fig. 3b and 3c) and finally rosette patterns (Fig. 3d) as the DOPE ratio increases to 0.8. Though at various molar ratios, the DPPE domains vary distinctly in shape, they are relatively homogeneous in average size. However, the size distribution and the domain localization appear to be more uniform by increasing the fraction of DOPE, which could be ascribed to more scattered DPPE domains and weaker interactions between separated DPPE domains. The surfaces were re-examined after air-drying for up to 10 days and no visible difference in morphology was observed, indicating the stability of the supported lipid film in an ambient environment.

For binary lipid mixtures with complete phase separation, the interactions along the boundary give rise to a 2-D line tension. A balance between interfacial line tension and repulsive electrostatic/ steric interactions, which induce the formation of circular domains and elongated domains,

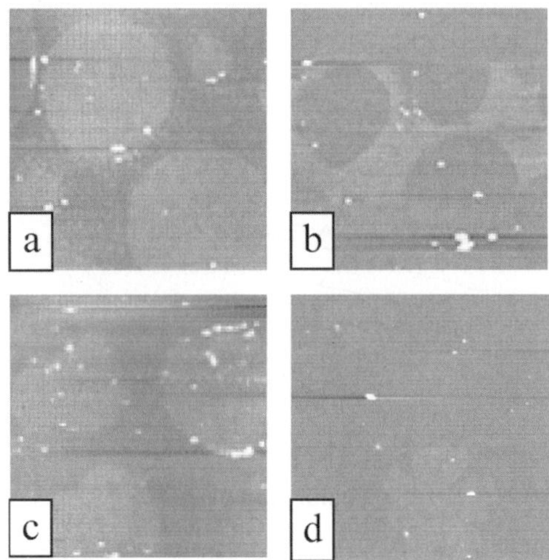

Fig. 3 AFM images of DOPE/DPPE LB monolayer on mica deposited at 25 mN m^{-1} at various molar ratios of (DOPE : DPPE): (a) 1 : 4, (b) 2 : 3, (c) 1 : 1, (d) 4 : 1. Each image is a 20 × 20 μm scan.

respectively, mainly determines the shape and the size of the 2-D lipid domains. The negatively-charged mica surface should exhibit similar binding to the positively charged NH_3^+ head group of DPPE and DOPE, which should not be an essential factor influencing the domain formation. At lower proportions of DOPE and therefore less steric repulsion in the mixture, the dominating line tension causes the formation of circular domains. As the DOPE fraction is increased, steric repulsion coming from the kinked hydrocarbon tails results in more elongated domain shapes.

Collagen/DOPE/DPPE LB film

The synthesis of collagen assemblies in the form of natural fibrils has been reported on micro-fabricated surfaces with defined micron-sized features, achieved by adsorption of collagen from monomeric solutions.[18–20] This method provides no control over the degree of collagen association and results in irregular spatial localization of collagen with features typically on the scale of 100 μm. Based on self-assembly behavior and LB technique, collagen assemblies with well-defined 2-D patterns at nano-scale templated by lipids have not been previously reported.

Collagen LB film

To accomplish collagen polymerization in the fluid environment provided by a self-assembled lipid matrix at a vapor/liquid interface, the first challenge is to deliver collagen molecules at the interface and prevent their dissolution into the aqueous subphase, which is an obvious prerequisite for reproducible and systematic formation of films by the LB deposition. One prior study reported that collagen behaved as an insoluble monolayer at an air/water interface based on only one compression Π–A isotherm;[17] however, this is not sufficient enough to demonstrate that collagen molecules remain at the interface. In this study, collagen monolayer compression Π–A isotherms were obtained when different amounts of collagen were delivered at the air/water interface (Fig. 4). As the amount delivered was increased, the isotherm shifted to the left, indicating a lower average area per molecule. At a specific surface pressure, the actual average area per molecule should be the same regardless of the delivery amount if all the collagen remains at the interface. The reduced average area calculated based on the total delivery amount indicates a significant amount of collagen molecules penetrating into the aqueous subphase.

The actual surface concentration of collagen is difficult to determine and to control due to the partial solubility of collagen in the aqueous subphase. As discussed previously, spreading collagen onto a slightly compressed lipid monolayer effectively prevented dissolution of collagen monomer

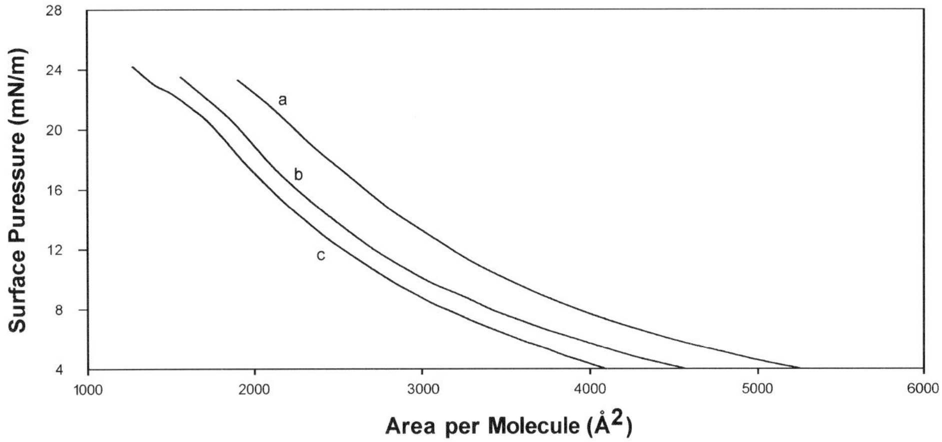

Fig. 4   *Π–A* isotherms for collagen with a delivery amount of: (a) 1.5 mg, (b) 2.0 mg and (c) 3.0 mg at 25 °C on ultrapure water (pH = 6.8).

into the subphase, thereby providing the key for experimentally patterning polymerized collagen at air/water interfaces in a well-controlled way.

Collagen/PE binary LB film

Collagen displayed a completely different mixing behavior with DOPE and DPPE. At molar DOPE : collagen ratios ranging from 80 : 1 to 100 : 1, collagen can be spread on top of DOPE monolayers at an air/water interface and polymerize to form fibril aggregates (Fig. 5a) very similar to collagen structures observed in nature. However, when incorporated with DPPE at the same conditions, no collagen self-assembly and polymerization was observed at any lipid : collagen molar ratio. The loosely packed monolayer of DOPE acts as an effective 2-dimensional solvent for collagen, while DPPE monolayers, more densely packed due to their straight hydrocarbon tails, exclude collagen. The kinked DOPE tails have larger cross sectional area for collagen molecules to localize and aggregate in between. The hydrophobic clusters along collagen molecules are expected to interact with the hydrophobic tails of DOPE and to be exposed to the air phase at the surface of the supported film. The tightly packed DPPE tails have no space to incorporate collagen molecules within the matrix. Therefore, the phase-separated DOPE/DPPE binary system allows us to exploit this incompatibility between lipid components to pattern collagen assemblies at 2-D.

Collagen/DOPE/DPPE ternary LB film

The patterning procedure developed in this study is based on the co-self-assembly of collagen and PE lipids by introducing collagen in a previously spread lipid monolayer on an aqueous surface. Restricting collagen molecules to the interface region is the first key step in this method. If significant amounts of monomeric collagen dissolve into the subphase, the reproducibility of polymerized collagen patterns and the long-range uniformity of the sample are impossible to realize, a major challenge for all surface patterning methods based on self-assembly.

The fabrication of the composite collagen/DOPE/DPPE monolayer was performed in a LB trough. A phosphate buffer solution at pH 7.2 at constant temperature 25 °C was used as the subphase, having been found to be suitable for the formation of DOPE/DPPE patterns and for the assembly of collagen. The compression isotherm for a collagen/DOPE/DPPE composite film (DOPE : DPPE : collagen = 80 : 20 : 1) at 25 °C was investigated. As shown in Fig. 6, the collagen/DOPE/DPPE isotherm is positioned to the right of the isotherms for single lipids and lipid mixtures, suggesting a more loosely packed film due to the incorporation of bulky collagen molecules.

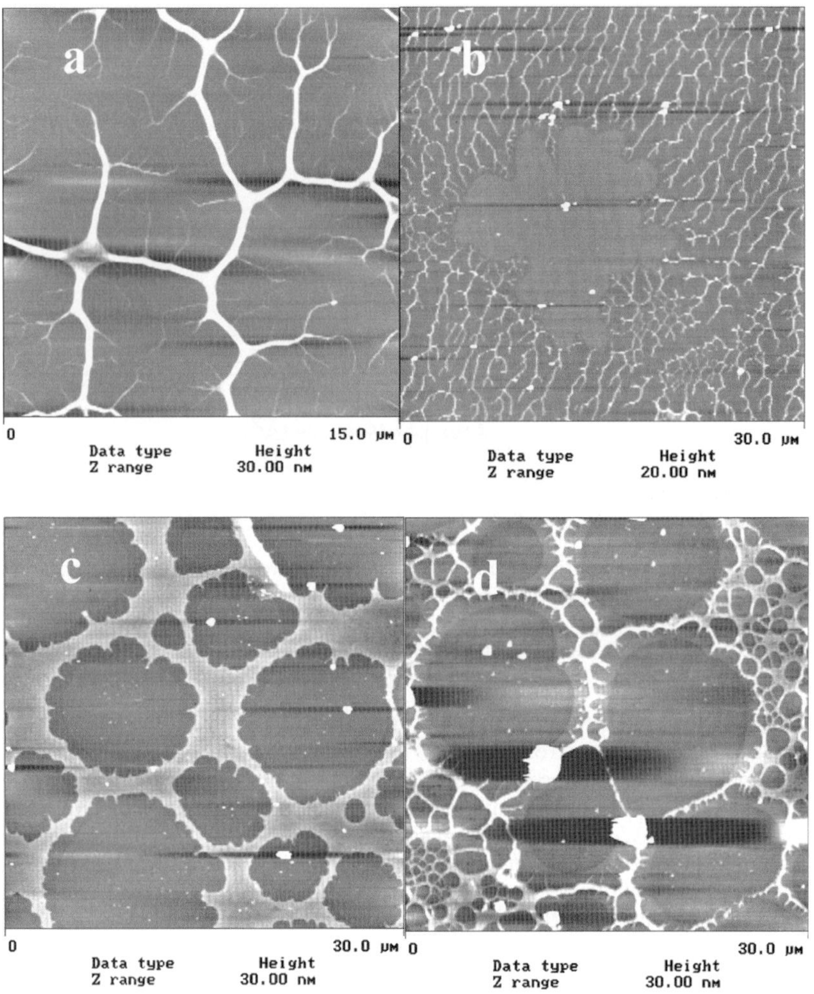

Fig. 5 AFM images of composite LB monolayers deposited at 25 mN m^{-1} on mica: (a) DOPE/collagen film at a molar ratio (DOPE:collagen) of 80:1 and DOPE/DPPE/collagen film at molar ratios (DOPE:DPPE:collagen) of: (b) 80:20:1, (c) 70:70:1, (d) 4:16:1.

Based on the general procedure described in the Experimental section, many controllable factors can be adjusted to explore their effects on the collagen pattern formation. The collagen/lipid molar ratios and the procedure to deliver collagen solution at the interface proved to be key factors that determine the successful incorporation of collagen into the lipid film and the formation of collagen patterns. The deposition pressure, the deposition speed and the method of preparing monomeric collagen acidic solution also have important effects on the long-range uniformity of the sample and the reproducibility of the fabrication procedure. After a series of comparative investigations, fabrication of uniform collagen/lipid composite films with well-defined collagen patterns was realized by optimizing these key factors.

The collagen lateral localization and the type of collagen aggregation are very sensitive to the molar ratio of DOPE:DPPE:collagen in the monolayer. At a proper molar ratio, collagen is effectively integrated into DOPE-rich regions and is totally excluded from the more condensed DPPE domains. The shape and the size of the DPPE domains from which collagen is excluded are mainly determined by the molar ratio between the lipids, indicating that the introduction of collagen does not result in a considerable rearrangement of the lipid lateral organization. The shape

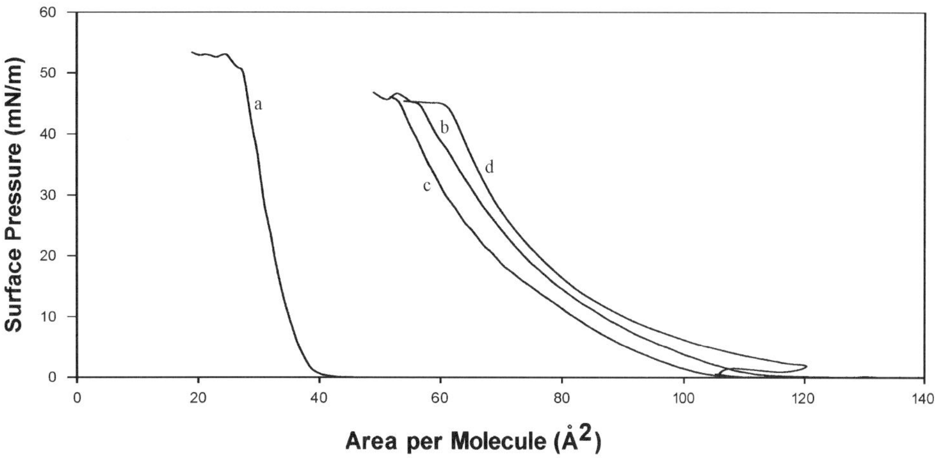

Fig. 6 *Π–A* isotherms for (a) DPPE, (b) DOPE, (c) lipid mixture with DOPE:DPPE = 4:1 and (d) lipid/collagen mixture with DOPE:DPPE:collagen = 80:20:1 at 25 °C on buffer solution (pH = 7.2).

of the domains appears a bit more irregular along the domain interface than in the absence of collagen, suggesting some effect of collagen on the line tension along this boundary. A wide variety of polymerized collagen structures are observed between the DPPE domains, from fine branching bundles evenly dispersed in the DOPE matrix to highly condensed and interwoven sheets that completely cover the DOPE phase (Fig. 5b–d). The interwoven structure of collagen is unique in terms of its possible applications as porous membrane or molecular filter. The degree of collagen interconnection or polymerization and the resulting type of aggregation can be partially controlled by adjusting the molar ratio between the lipids and collagen. A wide range of lipids and collagen compositions has been investigated with the lipid molar ratio fixed at 1:4, 1:1 or 4:1 (DOPE:DPPE). The range of the molar collagen/lipid ratio in which intricate collagen pattern formation is observed is fairly narrow and sometimes even specific for various compositions of binary lipids. The range of molar collagen/DOPE ratios at which the patterns form decreases as the relative proportion of DPPE increases, which could be ascribed to the compatibility of collagen with loosely packed DOPE phases. Since the DOPE provides the matrix to disperse the collagen structures, collagen exhibits less condensed assemblies and more even distribution on the surface as the DOPE content increases. The collagen features protrude from the background showing an average height of 6 nm for the branching bundles and 20 nm for the highly polymerized sheet. Since the collagen monomer is 300 nm in length and 1.5 nm in diameter, collagen molecules should be aligned with their long axis parallel to the surface in all kinds of the assemblies.

Regardless of the collagen/lipid composition, collagen patterns could not be obtained on mica when the collagen is introduced on top of a lipid monolayer that is either initially fully expanded or fully compressed (to 25 mN m^{-1}). In the first case, AFM images exhibit only the regular lipid domains and no trace of collagen monomer or polymerized aggregates. The absence of collagen on the supported film has two possible explanations. The widely spread lipids give collagen enough space to disperse into the aqueous subphase due to the solubility of collagen in the aqueous phase and weak interactions between lipids and collagen. This explanation is supported by the fact that, during the spreading of collagen, the surface pressure did not increase. Another possibility is that vertical deposition can not be applied to a collagen film with highly disordered collagen molecules. The previously reported successful transfer of a film of pure collagen was performed by horizontal lift.[16] In the work reported here, collagen spreading is conducted after a lipid film has been compressed to a target pressure in the second case. A multilayer structure is produced in which collagen fibril structures are observed on top of the lipid monolayer with lipid domains underneath. The branching collagen bundles spread over the sample surface without selective attachment to the lipid template. The compressed lipid monolayer with closer packing prevents the collagen structure from moving through the air/water interface during the collagen delivery step. The restricted motion of the lipid alkyl chains reduces the fluidity of both DPPE domains and DOPE matrix and therefore reduces the affinity of collagen to the lipid monolayer. It is reasonable to expect that the spread collagen is squeezed out and assembles on top of the lipid monolayer with their long axis parallel to the surface, which is confirmed by the height analysis from AFM images. Inspired by the above results, the surface pressure at which the collagen is spread on the floating lipid film (denoted as spreading pressure, Π_S) was tested for a range of values between 0 to 25 mN m^{-1}. During the addition of collagen solution, the surface pressure is maintained constant at a specific Π_S resulting in a necessary increase in surface area. Π_S values in the range of 0.2–1.4 mN m^{-1} were found to effectively confine collagen at the interface.

When a collagen feature is obtained, enhanced reproducibility and long-range uniformity are additional challenges. Although the reproducibility and long-range uniformity have not been realized completely, the current investigation demonstrates the importance of a few factors that need to be further addressed. In the current study, the collagen solution is prepared by dissolving type I collagen extracted from calf skin in 1 M acetic acid and stirring for at least 4 h at ambient temperature. A collagen concentration of approximately 0.1 mg ml^{-1} gives the highest quality films when other parameters are fixed. Concentrated collagen solutions are not likely to be monomeric solutions and may therefore be undesired because of the possible deposition of aggregates onto the interface, while over-diluted spreading solutions are also not optimum since the necessary volume to be delivered is too high and may change the subphase pH. During the spreading of collagen solution by a microsyringe at the interface, the collagen solution is introduced dropwise at many widely dispersed locations. The deposition pressure (Π_D) and deposition speed show moderate influence on sample uniformity and reproducibility. The lowest deposition speed allowed by the LB trough, 1 mm min^{-1}, was utilized for vertical lifting to provide enough time for the attachment of the composite film to the mica surface. The combination of Π_D above 25 mN m^{-1} and Π_S in the range of 0.2–1.4 mN m^{-1} exhibits a significant effect on selective control over pattern fabrication. Generally, lower Π_S combined with lower Π_D results in less aggregated and more dispersed collagen features, such as fine branches, while higher Π_S with higher Π_D leads to more condensed collagen structures such as continuous collagen sheets fully covering the DOPE region. Systematic studies are needed to optimize the spreading and deposition pressures.

Successful adhesion and growth of Chinese hamster ovary (CHO) cells on the patterned film surfaces demonstrate the biocompatibility of these composite films. Fig. 7 illustrates the spreading morphology of CHO cells grown on a collagen/lipid composite film, as observed by optical microscopy. The composite film is sufficiently stable to survive the sterilization and cell culture procedures.

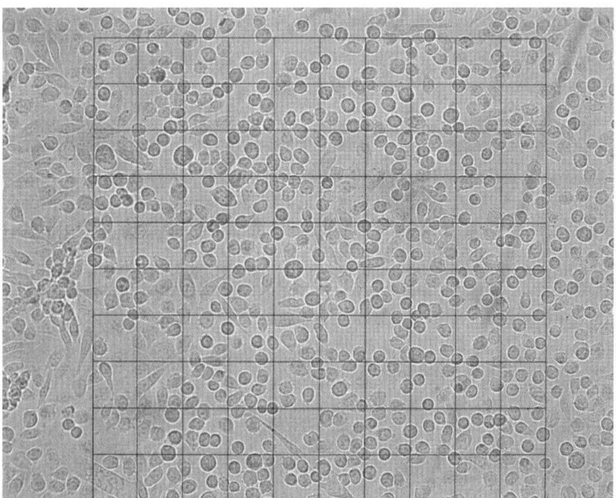

Fig. 7 Optical micrograph of Chinese hamster ovary (CHO) cells cultured on a DOPE/DPPE/collagen (molar ratio 80:20:1) film supported by mica after 24 h incubation (100×, each grid is 50 × 50 μm).

Conclusions

The studies reported here form the basis for examination and fabrication of more complex collagen-containing structures of biological importance in a more controllable way. Monolayers of phospholipids can be used to pattern ECM proteins two-dimensionally. Phospholipids orientate themselves at an air/water interface with chains pointing towards the air and heads towards the aqueous subphase with long-range order. The specific alignment and possible microdomain formation of phospholipids can be exploited to incorporate proteins at an interface and to induce a wide variety of two-dimensional protein configurations through hydrophobic and electrostatic interactions between lipids and proteins. The stability, biocompatibility and long-range uniformity of the fabricated composite film and the reproducibility of the fabricating procedure are of great importance for future applications.

Future work will focus on optimizing and simplifying the current fabricating procedure, extending the general procedure to other phospholipid systems and developing this promising technology to its full potential.

Acknowledgements

This work was funded by the U.S. National Science Foundation (Grant # CTS-9986488). The authors wish to thank Dr Yoon-Seob Lee and Poonam Nigam for intellectual contributions to this work.

References

1 Y. F. Dufrene, W. R. Barger, Green, D. John-Bruce and G. U. Lee, *Langmuir*, 1997, **13**, 4779.
2 J. M. Solletti, M. Botreau, F. Sommer, Brunat, T. M. Duc and M. R. Celio, *J. Vac. Sci. Technol. B*, 1996, **14**(2), 1492.
3 E. Györvary, W. M. Albers and J. Peltonen, *Langmuir*, 1999, **15**, 2516.
4 K. Okuyama, N. Go and M. Takayanagi, *Chem. Lett.*, 1978, **509**.
5 B. Brodsky and J. A. M. Ramshaw, *Matrix Biol.*, 1997, **15**, 545.
6 D. P. Yurchenco, D. E. Birk and R. P. Mecham, *Extracellular Matrix Assembly and Structure*, Academic Press, San Diego, 1994.
7 E. D. Hay, *Cell Biology of Extracellular Matrix*, Plenum Press, New York, 2nd edn., 1991.
8 *Protein Profile*, Academic Press, London, 1994, vol. **1**.
9 G. C. Wood and M. K. Keech, *Biochem. J.*, 1960, **75**, 588.
10 M. Gale, M. S. Pollanen, P. Markiewicz and M. C. Goh, *Biophys. J.*, 1995, **68**, 2124.
11 M. Mertig, U. Thiele, J. Brakt, G. Leibiger, W. Pompe and H. Wendrock, *Surf. Interface Anal.*, 1997, **25**, 514.
12 E. A. G. Chernoff and D. A. Chernoff, *J. Vac. Sci. Technol. A*, 1992, **10**(4), 596.
13 Y. F. Dufrêne, T. G. Marchal and P. G. Rouxhet, *Appl. Surf. Sci.*, 1999, **144**, 638.
14 C. C. Dupont-Gillan, B. Nysten and P. G. Rouxhet, *Polym. Int.*, 1999, **48**, 271.
15 C. C. Dupont-Gillain and I. Jacquemart, *Surf. Sci.*, 2003, **539**, 145.
16 A. Higuchi, M. Yoshida, T. Ohno, T. Asakura and M. Hara, *Cytotechnology*, 2000, **34**, 165.
17 M. M. Ghannam, M. M. Mady and W. A. Khalil, *Biophys. Chem.*, 1999, **80**, 31.
18 R. G. Thakar, F. Ho, N. F. Huang, D. Liepmann and S. Li, *Biochem. Biophys. Res. Commun.*, 2003, **307**, 883.
19 A. Folch and M. Toner, *Biotechnol. Prog.*, 1998, **14**, 388.
20 S. Miyamoto, A. Ohashi and J Kimura, *Sens. Actuators, B*, 1993, **13–14**, 196.

Using the dynamic, expanding liquid–liquid interface in a Hele–Shaw cell in crystal growth and nanoparticle assembly

Debabrata Rautaray, Ritwik Kavathekar and Murali Sastry*

Materials Chemistry Division, National Chemical Laboratory, Pune 411008, India. E-mail: sastry@ems.ncl.res.in; Fax: +91 20 25893952/25893044; Tel: +91 20 25893044

Received 14th April 2004, Accepted 7th May 2004
First published as an Advance Article on the web 16th September 2004

The liquid–liquid interface has been used with considerable success in the synthesis of advanced materials ranging from (bio)minerals to inorganic membranes to nanoparticles. In almost all such cases, the interface is static. The Hele–Shaw cell in which a viscous fluid is displaced by a less viscous one in a constrained manner has been invaluable in the study of dynamic instabilities at interfaces and in the study of viscous fingering pattern formation. However, the potential of the Hele–Shaw cell in carrying out reactions at the interface between the two fluids leading to the formation of inorganic materials has been largely unrecognized and underexploited. Realizing that the dynamic liquid–liquid interface in a Hele–Shaw cell would provide opportunities to control a variety of time-scales associated with material formation, we have started a program on the use of the Hele–Shaw cell in materials synthesis. In this discussion paper, we present some of our recent results on the growth of calcium carbonate crystals in the Hele–Shaw cell by the reaction of Ca^{2+} ions electrostatically complexed with carboxylate ions pinned to the interface with carbonate ions present in the aqueous part of the biphasic reaction medium. We show that both polymorph selectivity and the morphology of the crystals may be modulated by varying the experimental conditions in the cell. We also discuss the possibility of using the dynamic interface in the Hele–Shaw cell to cross-link gold nanoparticles in water through bifunctional linkers present in the oil phase and investigate the nature of the structures formed.

Introduction

Synthesis of advanced inorganic materials with control over structure, size and morphology is often driven by commercial requirements in areas as diverse as electronics, pigments and cosmetics, ceramics and medical industries.[1,2] Insofar as materials engineering is concerned, much of the research has centered on the use of biomimetic templates such as Langmuir monolayers at the air–water[3–9] and liquid–liquid interface,[10,11] self-assembled monolayers (SAMs),[12–15] lipid bilayer stacks[16–18] and functionalized polymer surfaces[19–21] to achieve such control.

Liquid–liquid interfaces play a critical role in many chemical, physical, and biological processes. The past decade has witnessed a huge increase of research interest in the study of liquid–liquid interfaces, fueled in part by new experimental and theoretical methods. Interfaces play a key role in modern science and technology and the unique features and phenomena induced in them have attracted the interest of researchers in many different fields. In particular, liquid–liquid interfaces

DOI: 10.1039/b405599n

have drawn much attention; not only do these interfaces have structural and dynamical nonlinear properties which are not observed in bulk liquids, but also they are integral to many chemical and biological systems.[22]

The liquid–liquid interface is important in various chemical processes such as phase transfer catalysis,[23–29] solvent extraction,[26] and ion-selective electrode operation.[25,30] Phase transfer catalysis at the liquid–liquid interface has attracted much attention because reaction rates are considerably enhanced by transfer across the liquid–liquid interface to yield high product and selectivity,[31–33] separation and recovery[29] and reuse of the catalyst[23] while maintaining high levels of catalytic activity. The liquid–liquid interface also offers a fertile medium for the assembly of nanoparticles,[34–40] and for the chemical manipulation of nanoparticles.[36] For example, interfaces between immiscible fluids, *i.e.*, on the surface of droplets, have been shown to be ideal for the assembly of elastic, semipermeable capsules composed of micrometer-sized colloidal particles.[41,42] Self-assembly of chemically functionalized nanoparticles at the toluene–water interface, coupled with chemical crosslinking of the attached ligands, provides a simple and flexible route for the fabrication of ultrathin, composite organic–inorganic membranes.[42] Efrima and co-workers first demonstrated that metal liquid-like films (MELLFs) of silver could, under stringent conditions, be synthesized at the interface between an organic solvent such as dichloromethane and water.[37,38] Subsequent reports have developed on this approach and the organization of negatively charged colloidal gold particles *via* electrostatic interactions with cationic surfactant molecules,[39] and the assembly of gold nanoparticles into one-dimensional superstructures[40] at the liquid–liquid interface have been shown. The controlled assembly of nanoparticles both in solution and on suitable surfaces is a problem of current interest with direct relevance to commercial applications of nanoscale matter. Interparticle interactions have also been studied at liquid–liquid interfaces.[36,43] For example oxides such as TiO_2 and metals such as Au nanoparticles can be assembled at the interface upon applying a potential bias[44,45] between two immiscible electrolyte solutions.[46] The transferring of monolayers from liquid–liquid interfaces using the Langmuir–Blodgett technique has also been studied in great detail.[47–49]

The liquid–liquid interface is of considerable importance in biological systems, particularly in biomedical engineering, pharmacology, and food processing. For example, the hydrocarbon–water interface provides an excellent medium to cellular interfaces where many important biological functions and/or processes occur.[50] These include receptor–ligand interactions, energy transduction, electron transfer, enzyme coupling, and triggering of cells by hormones and neurotransmitters.[50] The liquid–liquid interface provides an excellent model for interactions of macromolecules at relatively mobile interfaces as well.[50] The liquid–liquid interface is also crucial to food systems and food processing. For example, the distribution and/or interactions of proteins and surfactants at this interface greatly influences the properties of food emulsions and microemulsions.[51]

The important biomineral calcium carbonate has been studied in much detail due to its abundance in nature and also its vast industrial application. Calcium carbonate is a relatively complex mineral system due to the existence of three stable polymorphs calcite, aragonite and vaterite.[52] Calcium carbonate crystallization has been achieved under Langmuir monolayers,[53] within lipid bilayer stacks,[17] using polymers,[19] bicontinuous emulsions,[54] using mixed solutions of surfactants and block copolymers,[55] on functionalized gold nanoparticles,[56] monolayer films of gold nanoparticles,[57] as well as on free nanoparticles in solution.[58] The synthesis of gold nanoparticles has also received considerable attention due to their interesting applications in single electron tunneling[59] and non-linear optical devices.[60] There is a growing interest in the organization of nanoparticles in two- and three-dimensional structures. The main challenge in this area is to develop protocols for the organization of crystalline arrays of nanoparticles wherein both the size and separation between the nanoparticles in the arrays can be tailored. Applications based on the collective properties of the organized particles require flexibility in controlling the nanoarchitecture of the film.[61] Attempts have been made to assemble nanoparticles in two-dimensional structures by a variety of methods that include self-assembly of the particles during solvent evaporation,[62] immobilization by covalent attachment at the surface of the self-assembled monolayers,[14,15] or surface modified polymers,[63] electrophoretic assembly onto suitable substrates,[64] electrostatic attachment to Langmuir monolayers at the air–water interface[7–9] and air–organic solvent interface,[11] and by diffusion into ionizable fatty lipid films.[18] Attempts have also been made to synthesize oxides such as zinc oxide,[65] gallium oxide,[66] and ceramics such as $BaCrO_4$[67] and $BaSO_4$[68–70] particularly in microemulsions where interactions between surfactant molecules coating the crystallites were implicated in the assembly process.[67,68–70]

From the above discussions, it is clear that the liquid–liquid interface has been extensively investigated in the synthesis of advanced materials. However, in all these studies, the interface (often charged due to the presence of ionizable surfactants) was *static* and other than providing a scaffold on which material growth could take place, was more or less passive. The idea of generating *expanding* liquid–liquid interfaces and investigating their role in modulating the morphology/crystallography of inorganic materials grown at these interfaces is a relatively new one developed in this laboratory and in terms of understanding, is still at an elementary stage. In order to achieve controllable expanding liquid–liquid interfaces for growth of materials, we have used a Hele–Shaw cell[71] commonly used by physicists to study the phenomenon of viscous fingering.[71] The schematic of the process is given in Fig. 1A and consists of injecting a less viscous immiscible liquid into another that is constrained between two parallel plates. In our first set of experiments, we have studied the growth of the minerals $BaSO_4$[72] and $CaCO_3$[73] in a radial Hele–Shaw cell (Fig. 1A) as model systems. This was accomplished by taking the fatty acids such as stearic acid/aerosol OT in the organic phase and carrying out the reaction of the appropriate metal ions (Ba^{2+}/Ca^{2+}) with the corresponding counterions (SO_4^{2-}/CO_3^{2-}) present in the aqueous phase (injected phase). The reaction leading to mineral formation occurred at the interface due to strong complexation of the metal cations with the charged fatty acid/AOT leading to unusual morphologies of the crystals.[72,73] In the case of $BaSO_4$ crystals, it was observed that the crystals were organized into highly linear superstructures over large length-scales[72] whereas in the case of $CaCO_3$ crystallization, we have observed the room temperature growth of almost phase-pure aragonite needles.[73] By judicious choice of the experimental parameters in the Hele–Shaw cell, we have achieved control over the degree of recrystallization of aragonite to calcite and thus, have obtained crystals with hollow, cylindrical morphology.[73]

In this Faraday discussion paper, we address the issue of expanding liquid–liquid interfaces in materials synthesis in greater depth and present results of our investigation of two problems. The first concerns the role of viscosity in determining the morphology and assembly of $CaCO_3$ crystals

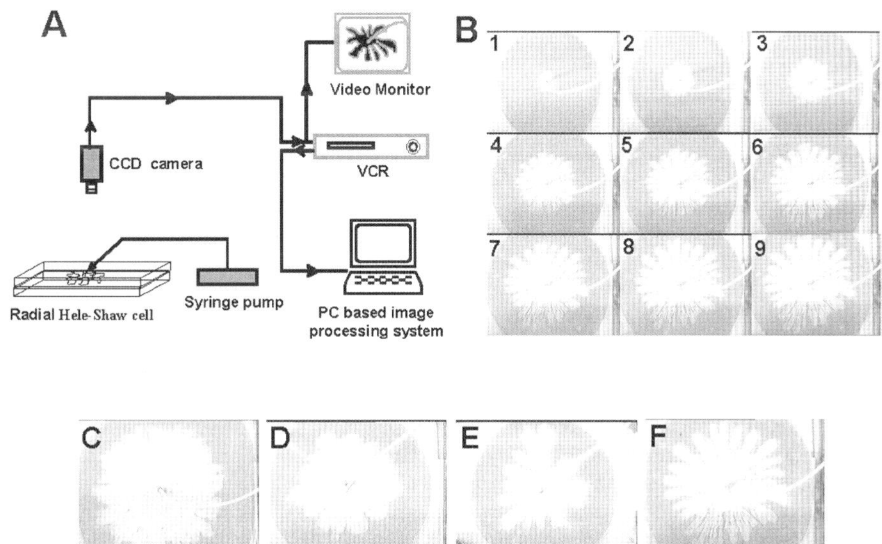

Fig. 1 (A) Schematic showing the various elements in the radial Hele–Shaw experiment. (B) Images of the patterns recorded at different times during injection of aqueous solution of $CaCl_2$ and Na_2CO_3 (transparent phase) into the AOT–chloroform–paraffin solution (colored phase). (C–F) Images of the patterns formed at the end of injection of aqueous solutions of $CaCl_2$ and Na_2CO_3 (transparent phase) into the AOT–chloroform–paraffin phase of different viscosities: C, 0.002678 Pa s; D, 0.003434 Pa s; E, 0.00488 Pa s and F, 0.046 Pa s (see text for details).

grown in a radial Hele–Shaw cell. We have also studied, for the first time, the assembly of gold nanoparticles using bifunctional linker molecules at an expanding liquid–liquid interface. Presented below are the details of this study.

Experimental

The radial Hele–Shaw cell used in this study was composed of two 1 cm thick, 30×30 cm^2 float-glass plates. Spacers of thickness $b = 300$ μm were used between the top and bottom glass plates. The viscous fluid [mixture of a solution of chloroform containing AOT ($C_{20}H_{37}NaO_7S$, MW $= 444.56$, 1×10^{-3} M) and paraffin] was taken in the cell gap and thereafter, an aqueous solution containing $CaCl_2$ and Na_2CO_3 at salt solution supersaturation (S_R) of 30 (10 mL of 1×10^{-2} M aqueous $CaCl_2$ and 10 mL of 1.36×10^{-3} M aqueous Na_2CO_3) was injected through a hole (0.5 mm diameter) drilled at the center of the top glass plate, using an automated fluid delivery system (Fig. 1A). Experiments were performed at five different solution viscosities by varying the AOT–chloroform solution : paraffin ratios of 4 : 1 (viscosity, $\eta = 0.002678$ Pa s), 1.5 : 1 ($\eta = 0.003434$ Pa s), 1 : 1 ($\eta = 0.00412$ Pa s), 1 : 1.5 ($\eta = 0.00488$ Pa s) and 1 : 4 ($\eta = 0.046$ Pa s) as the displaced fluid at a constant plate separation (S_P) of 300 μm. A small amount of dye was added to the chloroform–paraffin mixture to yield sufficient contrast between the aqueous and organic solutions for easy capture of the images with a CCD (charge-coupled device) camera (Fig. 1A). The experiments were performed at a constant volumetric flow rate (VFR) of 0.5 ml min^{-1}. The evolution of the finger pattern with time was followed for the above cases using a CCD camera connected to a video recorder at an image capture rate of 25 images s^{-1} (Fig. 1A). On completion of the injection process (typically 15–20 min), the organic solution was carefully removed and the remaining aqueous $CaCO_3$ precipitate was washed with copious amounts of doubly distilled water and placed on Si(111) wafers and on glass substrates for scanning electron microscopy (SEM)/energy dispersive analysis of X-rays (EDAX), Fourier transform infrared spectroscopy (FTIR) and X-ray diffraction (XRD) measurements, respectively.

Samples for SEM and EDAX measurements were prepared by drop-coating films of the $CaCO_3$ crystals on Si(111) wafers. These measurements were carried out on a Leica Stereoscan-440 scanning electron microscope equipped with a Phoenix EDAX attachment. EDAX spectra were recorded in the spot-profile mode by focusing the electron beam onto specific regions of the film. Fourier transform infrared spectroscopy (FTIR) measurements were carried out in the diffuse reflectance mode at a resolution of 2 cm^{-1} on a Perkin–Elmer FTIR-Spectrum One instrument. XRD analysis of the $CaCO_3$ drop-coated films on glass substrates were carried out on a Phillips PW 1830 instrument operating at a voltage of 40 kV and a current of 30 mA with Cu Kα radiation.

The Hele–Shaw set up was also used for assembling gold nanoparticles at the expanding liquid–liquid interface. In this experiment, a viscous fluid [mixture of toluene containing ethylenediamine (NH_2–CH_2–CH_2–NH_2) and paraffin] was taken in the cell gap and thereafter, an aqueous solution of sodium borohydride reduced gold nanoparticles was injected into the Hele–Shaw cell. This experiment was performed with a viscous solution of toluene and paraffin in the ratio of 1 : 4 (viscosity, $\eta = 0.043$ Pa s) as the displaced fluid with a constant plate separation of 300 μm and a steady VFR of 0.5 ml min^{-1}. In another experiment, the assembly of gold nanoparticles was carried out using the radial Hele–Shaw cell where an aqueous solution of sodium borohydride reduced aqueous gold nanoparticles was taken in the cell gap and thereafter, chloroform containing ethylenediamine was injected into the cell at a constant S_P of 300 μm and a constant VFR of 0.5 ml min^{-1}. On completion of the injection process, the organic solutions were carefully removed in both experiments and the remaining aqueous solutions were characterized by UV-visible spectroscopy (UV-vis) and transmission electron microscopy (TEM) measurements.

UV-vis spectroscopy measurements were carried out on a Jasco-V-570 UV/VIS/NIR spectro-photometer operated at a resolution of 2 nm. Samples for TEM analysis were prepared by drop-coating films of the Au nanoparticle solutions onto carbon-coated copper TEM grids, allowing the grids to stand for 2 min following which the extra solution was removed using a blotting paper. TEM analysis of the gold nanoparticle samples was carried out on a JEOL model 1200EX instrument operated at an accelerating voltage of 120 kV.

Results and discussion

The displacement of a viscous fluid by a less viscous one leads to the formation of finger-like patterns known as 'viscous fingering'.[71] The morphology of the viscous-finger patterns generated is a function of many parameters such as the flow rate, difference in viscosities of the two fluids and the interfacial tension. In the first part of this study, we investigate the role of variation in viscosity of the displaced phase in modulating the morphology of $CaCO_3$ crystals grown at the liquid–liquid interface. This represents an important advance on our preliminary investigation into $CaCO_3$ growth in a Hele–Shaw cell presented earlier.[73] We recollect that the $CaCO_3$ crystals are grown by reaction of an injected aqueous solution of $CaCO_3$ into a viscous chloroform–paraffin mixture containing the surfactant AOT (scheme in Fig. 1A). Fig. 1B shows pictures of the Hele–Shaw cell showing the evolution of the finger patterns with time recorded during the injection of an aqueous solution of $CaCl_2$ and Na_2CO_3 into the AOT–chloroform–paraffin mixture ($\eta = 0.046$ Pa s). As can be seen in Fig. 1B, the viscous fingering pattern evolves as the injection of the aqueous phase into the organic phase proceeds (time of injection increasing from 1–9). At completion of the injection, the highly branched, viscous fingering pattern is clearly seen. The viscous fingering patterns obtained in separate experiments where the viscosity of the organic phase was varied is shown in Fig. 1C–F [$\eta = 0.002678$ Pa s (C); 0.003434 Pa s (D); 0.00488 Pa s (E) and 0.046 Pa s (F)]. One can clearly notice the significant variation in the morphologies of the branched patterns for the different viscosities of the displaced fluid used. As the viscosity increases, the morphology of the patterns become progressively more branched (Fig. 1C–F). The fractal dimensions of the patterns have been determined for the different viscosities used and are listed in Table 1. It is observed that there is an increase in the fractal dimensions (Table 1) on increasing the viscosity of the displaced organic phase supporting the increase in the complexity of the patterns observed (Fig. 1C–F).

It should be pointed out that the flow rate was decided based upon the induction time observed for initiation of crystal nucleation. The slow VFR coupled with the difference in viscosities of the two fluids[74] is responsible for the complex, branched interface and thus provides a model system for understanding crystal growth processes at such dynamic liquid–liquid interfaces.

Fig. 2A and B show representative SEM images at different magnifications of the $CaCO_3$ crystals grown in the Hele–Shaw cell during displacement of the AOT–chloroform–paraffin organic phase ($\eta = 0.002678$ Pa s) by an aqueous solution of $CaCl_2$ and Na_2CO_3 at an S_R of 30. The substrate was densely populated with flower-like $CaCO_3$ crystals (Fig. 2A). The inset of Fig. 2A shows a magnified view of a few of the crystallites in greater detail showing that most of the flower-like $CaCO_3$ crystals consist of either four or six petals and that the petals originate from a central point. Fig. 1B shows the flower-like $CaCO_3$ crystals consisting of six petals in finer detail. The petals of the $CaCO_3$ crystals appear to be extremely thin and flat. The inset of Fig. 2B clearly shows the flatness of the $CaCO_3$ crystals. The length of the $CaCO_3$ petals in the flower-like assemblies is in the range of 300–600 nm while the widths are typically in the range of 200–300 nm. From the SEM images, we are unable to estimate the thickness of the $CaCO_3$ petals. EDAX analysis of the petals in the $CaCO_3$ flowers (curve 1, Fig. 3A) yielded a $Ca:C:O$ atomic ratio close to the expected stoichiometry. Along with expected Ca, C and O, strong signals of Na and S components from the surfactant AOT are seen indicating surface binding of the AOT molecules to the $CaCO_3$ petals. The FTIR spectrum

Table 1 Details of the fractal dimensions of the finger patterns, viscosity of the organic phase and observed morphology of $CaCO_3$ crystals in the Hele–Shaw cell experiments

Obs No.	Mixture of chloroform : paraffin	Viscosity/ Pa s	Fractal dimension (FD)	Morphology of $CaCO_3$ crystals formed
1	4:1	0.002678	1.88 ($\pm$ 0.03)	Flower like calcite with two/four petals
2	1.5:1	0.003434	1.91 ($\pm$ 0.01)	Flower like calcite as well as patterned structures
3	1:1	0.00412	1.94 ($\pm$ 0.01)	Patterned calcite with regular steps
4	1:1.15	0.00488	1.97 ($\pm$ 0.02)	Patterned calcite with smaller and sharper crystallites
5	1:4	0.046	1.99 ($\pm$ 0.01)	Patterned calcite assemblies

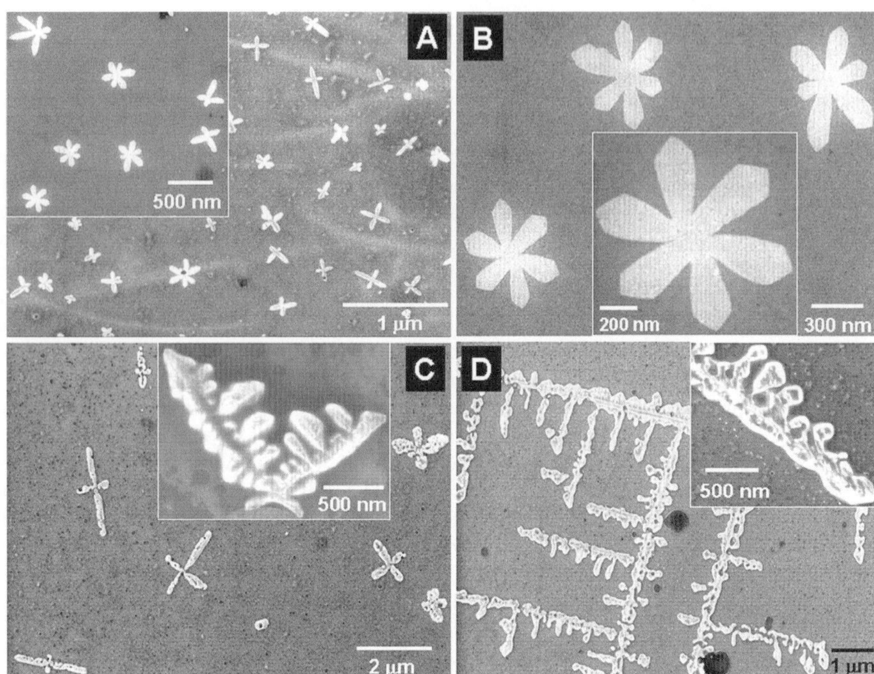

Fig. 2 (A and B)—Low and high magnification SEM images, respectively, of $CaCO_3$ crystals grown in the Hele–Shaw cell with an organic phase viscosity of 0.002678 Pa s. (C and D)—Low and high magnification SEM images, respectively, of $CaCO_3$ crystals grown in the Hele–Shaw cell with an organic phase viscosity of 0.003434 Pa s. The insets in all the figures show magnified views of representative $CaCO_3$ crystals in the main part of the figure (see text for details).

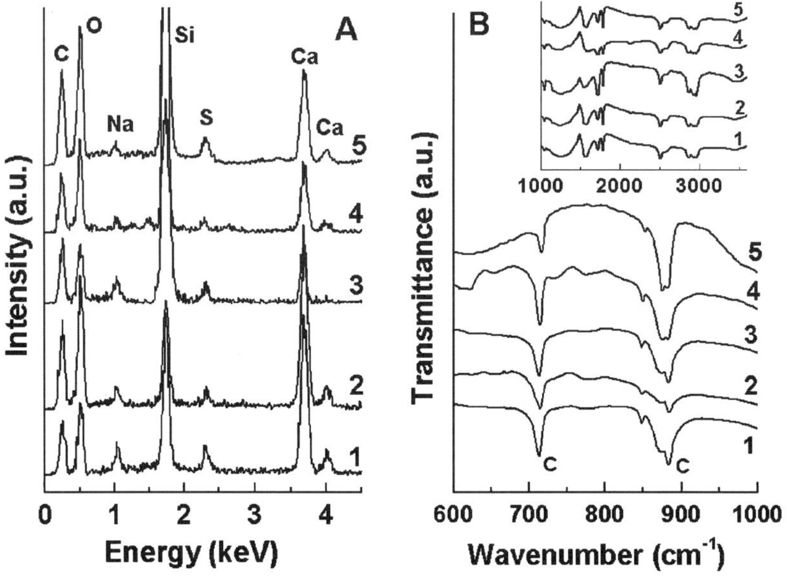

Fig. 3 (A) Spot-profile EDAX spectra recorded from the films of $CaCO_3$ crystals grown in the Hele–Shaw cell with organic phase viscosities of 0.002678 Pa s (curve 1); 0.003434 Pa s (curve 2); 0.00412 Pa s (curve 3); 0.00488 Pa s (curve 4) and 0.046 Pa s (curve 5). (B and inset) FTIR spectra in different spectral windows recorded from: $CaCO_3$ crystals grown in the Hele–Shaw cell with organic phase viscosities of 0.002678 Pa s (curve 1); 0.003434 Pa s (curve 2); 0.00412 Pa s (curve 3); 0.00488 Pa s (curve 4) and 0.046 Pa s (curve 5).

recorded from the $CaCO_3$ flowers is shown as curve 1 in Fig. 3B. The absorption bands at 712 and 872 cm^{-1} are characteristic of calcite.[75] Prominent absorption bands are also seen at 1056, 1700, 2850 and 2950 cm^{-1} (curve 1 in the inset of Fig. 3B). The band at 1056 cm^{-1} is assigned to the S=O stretching vibration of the sulfonate group present in the surface-bound AOT molecules.[76] The band at 1700 cm^{-1} is due to carbonyl stretch vibrations in the AOT molecules and the two bands at 2850 and 2950 cm^{-1} have been assigned to the methylene symmetric and antisymmetric stretching vibrations in the hydrocarbon chains, respectively. The XRD pattern recorded from the flower-like $CaCO_3$ crystals shown in Fig. 1A and B is displayed as curve 1 in Fig. 4. It is observed that the $CaCO_3$ crystals formed in this experiment consist of predominantly the calcite polymorph.[77]

Fig. 2C and D show SEM images at different magnifications of $CaCO_3$ crystals grown in the Hele–Shaw cell during the displacement of AOT–chloroform–paraffin organic solution ($\eta = 0.00412$ Pa s) by an aqueous solution of $CaCl_2$ and Na_2CO_3 at an S_R of 30. In this experiment we observe both $CaCO_3$ crystals with flower-like morphology (Fig. 2C) as well as more extended, quasi-linear superstructures of $CaCO_3$ crystals (Fig. 2D). The $CaCO_3$ flowers shown at the lower magnification (Fig. 2C) appear on a gross scale to be quite similar to those obtained in the lower viscosity experiment (Fig. 2A). At higher magnification, however (inset of Fig. 2C) differences are observed. The petals in the $CaCO_3$ flowers are not smooth and flat as in the earlier experiment. Furthermore, the presence of the highly branched linear structures of the crystals appears to be related to an increase in viscosity of the displaced fluid which, as seen in the images of the patterns formed for different viscosities of the organic phase (Fig. 1C–F), can be correlated with increasing complexity of the viscous-fingering patterns during the injection process. The inset in Fig. 2D shows a high magnification SEM image of a particular region of the quasi-linear $CaCO_3$ crystals clearly revealing the branched nature of the $CaCO_3$ structures. EDAX analysis of the $CaCO_3$ crystals shown in Fig. 2C and D yielded a Ca : C : O atomic ratio close to the expected stoichiometry along with Na and S components from the AOT capping layer (curve 2 in Fig. 3A). The FTIR spectrum (curve 2 in Fig. 3B) and XRD pattern (curve 2 in Fig. 4) recorded from this sample revealed the formation of the polymorph, calcite.

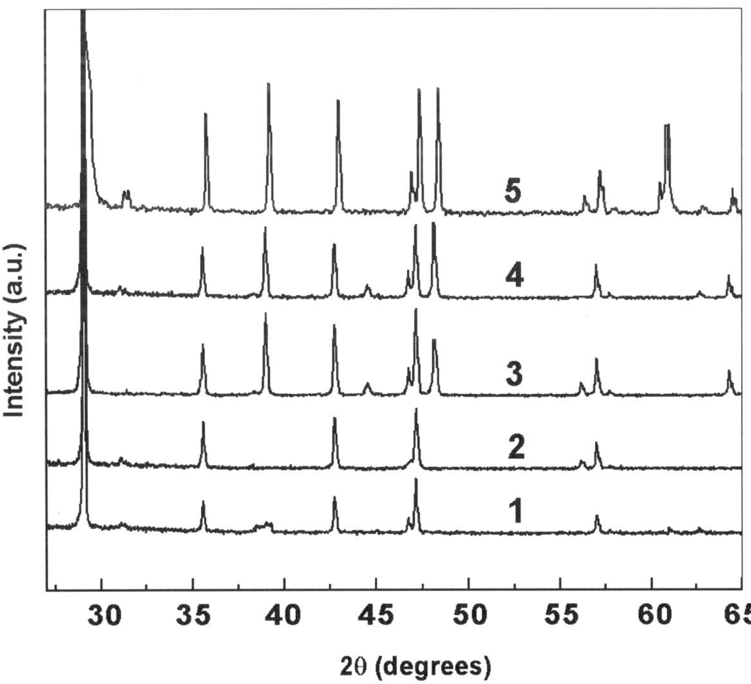

Fig. 4 XRD patterns recorded from $CaCO_3$ crystals synthesized in the radial Hele–Shaw cell with organic phase viscosities of 0.002678 Pa s (curve 1); 0.003434 Pa s (curve 2); 0.00412 Pa s (curve 3); 0.00488 Pa s (curve 4) and 0.046 Pa s (curve 5).

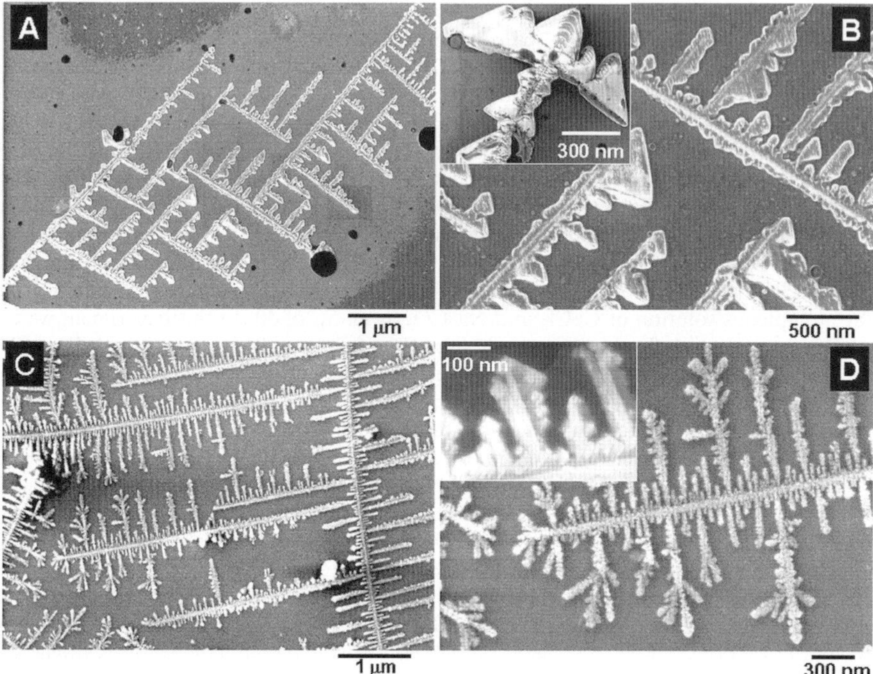

Fig. 5 (A and B)—Low and high magnification SEM images, respectively, of CaCO$_3$ crystals grown in the Hele–Shaw cell with an organic phase viscosity of 0.00412 Pa s. (C and D)—Low and high magnification SEM images, respectively, of CaCO$_3$ crystals grown in the Hele–Shaw cell with an organic phase viscosity of 0.00488 Pa s. The insets in B and D show a magnified view of representative crystals shown in the main part of the figure.

Fig. 5A and B show SEM images at different magnification of CaCO$_3$ crystals grown in the Hele–Shaw cell during the displacement of AOT–chloroform–paraffin ($\eta = 0.003434$ Pa s) by an aqueous solution of CaCl$_2$ and Na$_2$CO$_3$ at an S_R of 30. In gross detail, the CaCO$_3$ structures that are formed are similar to those synthesized in the Hele–Shaw cell at an organic solution viscosity of 0.002678 Pa s. The CaCO$_3$ crystals in this experiment exhibit a highly branched morphology, with evidence of secondary nucleation and growth around a linear 'core' (Fig. 5A). The higher magnification SEM image (Fig. 5B) shows that the edges of the CaCO$_3$ crystals are very sharp with steps observed regularly on the CaCO$_3$ super structures (inset of Fig. 5B). Similar steps have been observed previously on calcite crystals.[78,79] Orme *et al.* have observed such steps in calcite crystals where they have found that site-specific binding of amino acid residues to surface steps changes the step-edge free energies, thereby giving rise to modifications that propagate from atomic length scales to macroscopic length scales.[78] Sugawara *et al.* have succeeded in preparing periodically patterned CaCO$_3$ films with regular surface relief structures on a thin matrix of cholesterol-modified pullulan from an aqueous solution containing poly(acrylic acid).[79] That similar structures may be obtained in an experiment involving crystal growth at a dynamic liquid–liquid interface is an interesting outcome of this work.

The EDAX measurements from the patterned CaCO$_3$ structures obtained above (curve 3 in Fig. 3A) indicated the presence of Ca, C and O along with Na and S arising from surface-bound AOT molecules. The formation of the calcite polymorph is confirmed by FTIR (curve 3 in Fig. 3B) and XRD analysis (curve 3 in Fig. 4). Fig. 5C and D show SEM images of CaCO$_3$ crystals grown in the Hele–Shaw cell during the displacement of AOT–chloroform–paraffin ($\eta = 0.00488$ Pa s) by an aqueous solution of CaCl$_2$ and Na$_2$CO$_3$ under conditions identical to the other experiments. In keeping with the trend, an increase in viscosity leads to an enhancement of the branched calcite structures, as observed in Fig. 5C. The individual crystallites in the branched assemblies are thinner and finer than those observed in the lower viscosity experiments (Fig. 5D). There also appears to be more secondary crystal growth on both sides of the linear axis. The higher magnification SEM image (inset of Fig. 5D) clearly shows that the branches are in turn made up of smaller crystallites with sharp edges. EDAX analysis confirms the presence of Ca, C and O in the right stoichiometric ratio along with Na and S from surface-bound AOT molecules (curve 4 in Fig. 3A). The FTIR (curve 4 in Fig. 3B) and XRD (curve 4 in Fig. 4) data confirm that the structures observed are indeed calcite crystals.

At the highest viscosity studied ($\eta = 0.046$ Pa s), the fingers formed in the radial Hele–Shaw cell are at their narrowest (Fig. 1E) and the corresponding fractal dimension is the highest. Fig. 6A–D show SEM images at different magnification of $CaCO_3$ crystals grown in the Hele–Shaw cell involving the experiment in the organic phase being of this viscosity. The increase in complexity of the finger patterns seen in Fig. 1E has clearly translated into the formation of quasi-linear, patterned assemblies of $CaCO_3$ crystals of very large length scales (Fig. 6A and B) densely populating the substrate surface. In some regions, the growth of the calcite crystals is very irregular (Fig. 6C). The higher magnification SEM image of one of the ordered assemblies (Fig. 6D) clearly shows that the $CaCO_3$ crystals are in fact assemblies of smaller $CaCO_3$ crystallites and also that in this case, growth of the secondary crystals proceeds out of the plane of the substrate as well (inset, Fig. 6D). The EDAX spectrum from the $CaCO_3$ crystals assemblies observed in Fig. 6 showed the expected Ca, C and O signals at stoichiometric ratios along with strong signals of Na and S from the AOT surface monolayer (curve 5 in Fig. 3A). FTIR analysis from the $CaCO_3$ crystal assemblies show characteristic strong absorption bands at 712 and 872 cm^{-1} which are attributed to the calcite polymorph (curve 5 in Fig. 3B). The absorption band at 1056 cm^{-1} arises from S=O stretching vibration of the sulfonate group present in the AOT molecules (curve 5 in the inset of Fig. 3B). The XRD pattern recorded from the patterned $CaCO_3$ assemblies shown in Fig. 6 is displayed as curve 5 in Fig. 4 indicating the formation of calcite polymorph in this case as well. One interesting observation in all the XRD spectra (curves 1–5 in Fig. 4) is the gradual appearance of new Bragg peaks as the organic phase viscosity increases.

To demonstrate the generality of this method for not only growing but assembling inorganic materials as well, we have examined the possibility of cross-linking gold nanoparticles using bifunctional molecules such as ethylenediamine at the expanding liquid–liquid interface. The assembly of Au nanoparticles was achieved using the Hele–Shaw cell by injecting an aqueous solution of sodium borohydride reduced gold nanoparticles into a solution of toluene containing diamine and paraffin having a viscosity of 0.046 Pa s. During injection of the nanoparticle solution, it is expected that the diamine molecules present at the interface would complex with gold nanoparticles leading to their aggregation at the interface. It is known that amine groups bind strongly with gold nanoparticles,[80–84] and therefore, this is expected to be a new strategy for nanoparticle assembly. Fig. 7A shows a representative viscous fingering pattern recorded during the injection of borohydride reduced gold nanoparticles (colored phase) into a mixed solution of toluene containing diamine and paraffin (transparent phase) showing the complex fingering pattern formed (Fig. 7A).

Fig. 7B shows the UV-vis spectra recorded from pure paraffin solution (curve 1) and from the sodium borohydride reduced gold nanoparticle solution before injection into the Hele–Shaw cell (curve 2). It is observed that there is no evidence of absorption in the 400–800 nm window region of the electromagnetic spectrum from pure paraffin solution (curve 1). The gold nanoparticles show a prominent and sharp absorption band at 520 nm. This absorption band is characteristic of gold nanoparticles and arises due to excitation of surface plasmon vibrations in the nanoparticles. Curve 3 in Fig. 7B shows the UV-vis spectrum of gold nanoparticles assembled in the Hele–Shaw cell during the injection of an aqueous solution of sodium borohydride reduced gold nanoparticles into the toluene solution of the diamine and paraffin having a viscosity of 0.046 Pa s. Two absorption bands at 526 and 640 nm are observed in the UV-vis spectrum from these gold nanoparticles. The absorption band at 526 nm is due to out-of-plane plasmon vibrations in the particles while the longer wavelength absorption band (640 nm) arises due to excitation of the in-plane plasmon vibration from aggregates of the gold nanoparticles.[85] It is well known that highly dispersed Au nanoparticles in solution exhibit only a single peak at 520 nm while crosslinked particles show two absorption maxima. As the aspect ratio of the aggregates increases, the in-plane plasmon peak intensifies and shifts to longer wavelengths.[85]

Fig. 8A–C show representative TEM images at different magnifications of Au nanoparticle assemblies formed in the Hele–Shaw cell experiment discussed above. As was indirectly inferred

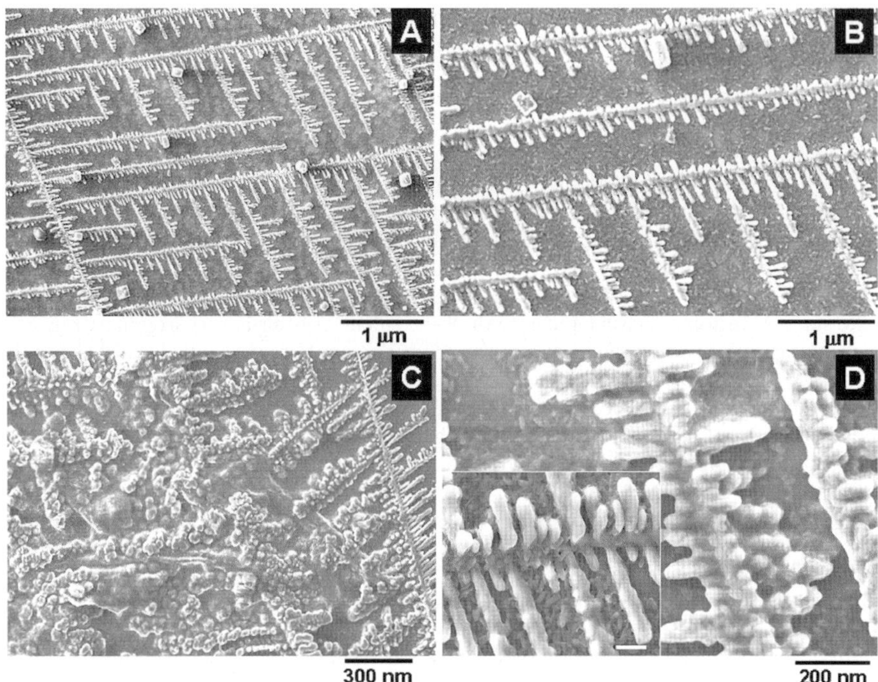

Fig. 6 (A–D) SEM images of CaCO$_3$ crystals at different magnifications, grown in the Hele–Shaw cell with an organic phase viscosity of 0.046 Pa s. The inset in D shows a magnified view of the main image with a scale bar of 100 nm.

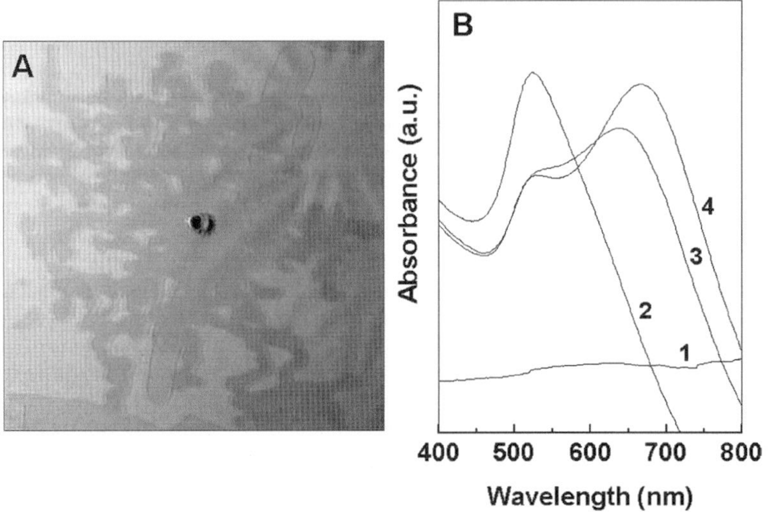

Fig. 7 (A) Image of the viscous-finger pattern formed after injection of an aqueous solution of borohydride reduced gold nanoparticles (colored phase) into the ethylenediamine–toluene–paraffin solution ($\eta = 0.046$ Pa s; transparent phase). (B) UV-vis spectra recorded from: curve 1—pure paraffin solution; curve 2—borohydride reduced aqueous gold nanoparticle solution; curve 3—gold nanoparticles assembled in the Hele–Shaw cell during the injection of an aqueous solution of borohydride reduced gold nanoparticles into the diamine–toluene–paraffin mixture ($\eta = 0.046$ Pa s); curve 4—gold nanoparticles assembled in the Hele–Shaw cell during the injection of chloroform containing diamine into borohydride reduced aqueous gold nanoparticles.

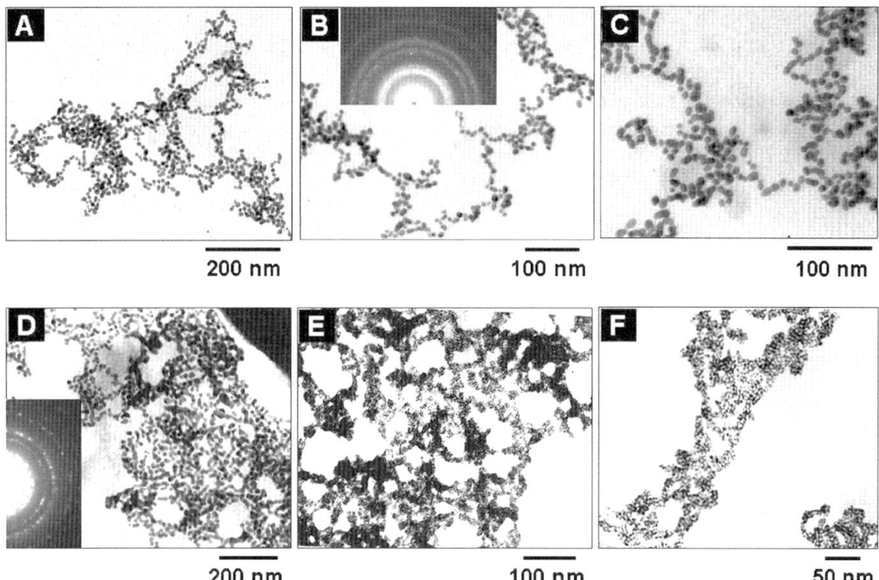

Fig. 8 (A–C) TEM images at different magnifications of gold nanoparticles assembled in the Hele–Shaw cell during the injection of an aqueous solution of borohydride reduced gold nanoparticles into the ethylenediamine–toluene–paraffin solution ($\eta = 0.046$ Pa s). (D–F) TEM images at different magnifications of gold nanoparticles grown in the Hele–Shaw cell during the injection of chloroform containing diamine into borohydride reduced aqueous gold nanoparticles. The insets in B and D show the SAED patterns recorded from the gold nanoparticles formed in both the experiments.

from the UV-vis measurements, the gold nanoparticles are assembled into quasi-linear superstructures (Fig. 8A and B). The selected area electron diffraction (SAED) analysis of the nanoparticles shown in Fig. 8B yielded a sharp diffraction pattern characteristic of fcc gold (inset, Fig. 8B). The highly open, quasi-linear structure of the gold nanoparticle assemblies is very clearly seen in Fig. 8C. The fact that the particles do not aggregate into dense structures normally observed during destabilization of nanoparticles in solution indicates that the crosslinking agent coupled with the assembly at the interface is responsible for the superstructures observed in the Hele–Shaw cell.

In order to understand the role of the solution viscosity and whether the nature of the solvent has any role to play in the nanoparticle assembly, an experiment was performed wherein chloroform containing diamine was injected into an aqueous solution of borohydride reduced gold nanoparticles (taken as the viscous fluid) without any attempts at increasing the viscosity by the addition of paraffin. The UV-vis spectrum recorded from the gold nanoparticle solution collected from this experiment is shown as curve 4 in Fig. 7B. Two absorption bands at 520 and 668 nm are observed in the UV-vis spectrum. The main difference with respect to the previous experiment is the shift in the in-plane plasmon vibration to a longer wavelength of 668 nm indicating the formation of more extended open nanoparticle assemblies. Fig. 8D–F show TEM images recorded at different magnification of Au nanoparticle assemblies observed in the Hele–Shaw cell experiment during the displacement of aqueous solution of gold nanoparticles by the diamine–chloroform solution. Dense networks of Au nanoparticles are observed in the TEM images (Fig. 8D and E) and in terms of extent and complexity of packing differ greatly from that observed in the earlier experiment. The magnified TEM image (Fig. 8F) reveals that the Au nanoparticle networks consist of smaller clusters of well separated Au nanoparticles. The formation of clusters of nanoparticles and their subsequent assembly into networks may be responsible for the shift towards longer wavelengths (668 nm) in the UV-vis spectrum from this sample (curve 4 in Fig. 7B).

In conclusion, we have shown that a dynamic liquid–liquid interface is a highly versatile medium for growing and assembling minerals as well as in the assembly of metal nanoparticles. The viscosity of the organic phase and thereby the nature of the viscous fingering interfacial structures obtained could be varied leading to differences in the morphology of the calcite crystals grown at the interface. These effects also influence the assembly of gold nanoparticles *via* bifunctional cross-linkers at the interface. These results demonstrate the versatility of the expanding liquid–liquid interface in a Hele–Shaw cell in designing new materials and suggests great potential for development. It also brings up a number of fundamental issues related to the interface and how they may influence the inorganic structures formed. While the complexity of the interface does play an important role, it is not yet clear how hydrodynamic flows at the interface affect the reaction rate of ions at the interface (calcium and carbonate ions in this study) or how they influence the superstructure of gold nanoparticle assemblies. Rational design of the interface together with a better understanding of hydrodynamic flow issues and diffusion time scales of reacting species need to be developed in order to add value to this method for materials design and synthesis.

Acknowledgements

D. R. thanks the Department of Science and Technology (DST), Government of India for a research fellowship. We acknowledge the SEM/EDAX, TEM and XRD facilities of the Center for Materials Characterization, NCL Pune.

References

1 S. Mann and G. A. Ozin, *Nature (London)*, 1996, **382**, 313.
2 E. Matijevic, *Curr. Opin. Colloid Interface Sci.*, 1996, **1**, 176.
3 B. R. Heywood and S. Mann, *Langmuir*, 1992, **8**, 1492.
4 B. R. Heywood and S. Mann, *J. Am. Chem. Soc.*, 1992, **114**, 4681.
5 A. L. Litvin, S. Valiyaveettil, D. L. Kaplan and S. Mann, *Adv. Mater.*, 1997, **9**, 124.
6 D. Rautaray, S. R. Sainkar, N. R. Pawaskar and M. Sastry, *CrystEngComm.*, 2002, **4**, 626.
7 M. Sastry, K. S. Mayya, V. Patil, D. V. Paranjape and S. G. Hegde, *J. Phys. Chem. B*, 1997, **101**, 4954.
8 V. Patil, K. S. Mayya, S. D. Pradhan and M. Sastry, *J. Am. Chem. Soc.*, 1997, **119**, 9281.
9 M. Sastry, K. S. Mayya and V. Patil, *Langmuir*, 1998, **14**, 5921.
10 D. Rautaray, A. Kumar, S. Reddy, S. R. Sainkar, N. R. Pawaskar and M. Sastry, *CrystEngComm*, 2001, 45.
11 K. S. Mayya and M. Sastry, *Langmuir*, 1999, **15**, 1902.
12 J. Kuther, G. Nelles, R. Seshadri, M. Schaub, H. J. Butt and W. Tremel, *Chem. Eur. J.*, 1998, **4**, 1834.
13 J. Aizenberg, A. J. Black and G. M. Whitesides, *J. Am. Chem. Soc.*, 1999, **121**, 4500.
14 V. L. Colvin, A. N. Goldstein and A. P. Alivisatos, *J. Am. Chem. Soc.*, 1992, **114**, 522.
15 K. Bandyopadhyay, V. Patil, K. Vijayamohanan and M. Sastry, *Langmuir*, 1997, **13**, 5244.
16 D. Rautaray, A. Kumar, S. Reddy, S. R. Sainkar and M. Sastry, *Cryst. Growth Des.*, 2002, **2**, 197.
17 C. Damle, A. Kumar, M. Bhagwat, S. R. Sainkar and M. Sastry, *Langmuir*, 2002, **18**, 6075.
18 S. Mandal, S. Phadtare, P. R. Selvakannan, R. Pasricha and M. Sastry, *Nanotechnology*, 2003, **14**, 878.
19 S. Feng and T. Bein, *Science (Washington, D. C.)*, 1994, **265**, 1839.
20 G. Falini, M. Gazzano and A. Ripamonti, *Adv. Mater.*, 1994, **6**, 46.
21 J. J. J. M. Donners, R. J. M. Nolte and N. A. J. M. Sommerdijk, *J. Am. Chem. Soc.*, 2002, **124**, 9700.
22 T. Takahashi, H. Yui and T. Sawada, *J. Phys. Chem. B*, 2002, **106**, 2314.
23 J. J. Krueger, M. D. Amiridis and H. J. Ploehn, *Ind. Eng. Chem. Res.*, 2001, **40**, 3158.
24 Y. Selzer and D. Mandler, *J. Phys. Chem. B*, 2000, **104**, 4903.
25 L. X. Dang, *J. Phys. Chem. B*, 1999, **103**, 39.
26 Y. Uchiyama, I. Tsuyumoto, T. Kitamori and T. Sawada, *J. Phys. Chem. B*, 1999, **103**, 4663.
27 G. D. Yadav and C. A. Reddy, *Ind. Eng. Chem. Res.*, 1999, **38**, 2245.
28 H. S. Wu, *Chem. Eng. Sci.*, 1996, **51**, 827.
29 A. Bhattacharya, *Ind. Eng. Chem. Res.*, 1996, **35**, 645.
30 Y. Chen, Z. Gao, F. Li, L. Ge, M. Zhang, D. Zhan and Y. Shao, *Anal. Chem.*, 2003, **75**, 6593.
31 C. M. Starks, *J. Am. Chem. Soc.*, 1971, **93**, 195.
32 Z. Zhang, I. Tsuyumoto, S. Takahashi, T. Kitamori and T. Sawada, *J. Phys. Chem. A*, 1997, **101**, 4163.
33 Z. Zhang, I. Tsuyumoto, S. Takahashi, T. Kitamori and T. Sawada, *J. Phys. Chem. B*, 1998, **102**, 10 284.
34 P. Pieranski, *Phys. Rev. Lett.*, 1980, **45**, 569.
35 O. D. Velev, K. Furusawa and K. Nagayama, *Langmuir*, 1996, **12**, 2374.
36 Y. Lin, H. Skaff, T. Emrick, A. D. Dinsmore and T. P. Russell, *Science (Washington, D. C.)*, 2003, **299**, 226.
37 D. Yogev and S. Efrima, *J. Phys. Chem.*, 1988, **92**, 5754.
38 H. Schwartz, Y. Harel and S. Efrima, *Langmuir*, 2001, **17**, 3884.
39 K. S. Mayya and M. Sastry, *Langmuir*, 1999, **15**, 1902.
40 D. Wyrwa, N. Beyer and G. Schmid, *Nano Lett.*, 2002, **2**, 419.
41 A. D. Dinsmore, M. F. Hsu, M. G. Nikolaides, M. Marquez, A. R. Bausch and D. A. Weitz, *Science (Washington, D. C.)*, 2002, **298**, 1006.

42 Y. Lin, H. Skaff, A. Boker, A. D. Dinsmore, T. Emrick and T. P. Russell, *J. Am. Chem. Soc.*, 2003, **125**, 12 690.

43 M. G. Nikolaides, A. R. Bausch, M. F. Hsu, A. D. Dinsmore, M. P. Brenner and D. A. Weitz, *Nature (London)*, 2002, **420**, 299.

44 H. Jensen, D. J. Fermyn, J. E. Moser and H. H. Girault, *J. Phys. Chem. B*, 2002, **106**, 10 908.

45 D. J. Fermyn, H. Jensen, J. E. Moser and H. H. Girault, *Chem. Phys. Chem.*, 2003, **4**, 85.

46 T. Shioya, S. Nishizawa and N. Teramae, *J. Am. Chem. Soc.*, 1998, **120**, 11 534.

47 P. Liljeroth, A. Malkia, V. J. Cunnane, A. K. Kontturi and K. Kontturi, *Langmuir*, 2000, **16**, 6667.

48 E. Squitieri and I. Benjamin, *J. Phys. Chem. B*, 2001, **105**, 6412.

49 C. Fradin, D. Luzet, A. Braslau, M. Alba, F. Muller and J. Daillant, *Langmuir*, 1998, **14**, 26.

50 A. Gajraj and R. Y. Ofoli, *Langmuir*, 2000, **16**, 4279.

51 E. Dickinson, S. R. Euston and C. M. Woskett, *Prog. Colloid Polym. Sci.*, 1990, **82**, 65.

52 K. M. McGrath, *Adv. Mater.*, 2001, **13**, 989.

53 B. R. Heywood and S. Mann, *Adv. Mater.*, 1994, **6**, 9.

54 D. Walsh and S. Mann, *Adv. Mater.*, 1997, **9**, 658.

55 L. Qi, J. Li and J. Ma, *Adv. Mater.*, 2002, **14**, 300.

56 I. Lee, S. W. Han, H. J. Choi and K. Kim, *Adv. Mater.*, 2001, **13**, 1617.

57 J. Kunther, R. Seshadri, G. Nelles, H. J. Butt, W. Knoll and W. Tremel, *Adv. Mater.*, 1998, **10**, 401.

58 J. Kuther, R. Seshadri, G. Nelles, W. Assenmacher, H. J. Butt, W. Mader and W. Tremel, *Chem. Mater.*, 1999, **11**, 1317.

59 R. P. Andres, T. Bein, M. Dorogi, S. Feng, J. J. Henderson, C. P. Kubiak, W. Mahoney, R. G. Osifchin and R. Reifenberger, *Science (Washington, D. C.)*, 1996, **272**, 1323.

60 P. Galletto, P. F. Brevet, H. H. Girault, R. Antoine and M. Broyer, *J. Phys. Chem. B*, 1999, **103**, 8706.

61 C. P. Collier, R. J. Saykally, J. J. Shiang, S. E. Henriches and J. R. Heath, *Science (Washington, D. C.)*, 1997, **277**, 1978.

62 Z. L. Wang, *Adv. Mater.*, 1998, **10**, 13.

63 F. G. Freeman, K. C. Grabar, K. J. Allison, R. M. Bright, G. A. Davis, A. P. Guthrrie, M. B. Hommer, M. A. Jackson, P. C. Smith, D. G. Walter and M. J. Natan, *Science (Washington, D. C.)*, 1995, **267**, 1629.

64 M. Giersig and P. Mulvaney, *J. Phys. Chem.*, 1993, **97**, 6334.

65 J. Y. Lao, J. Y. Huang, D. Z. Wang, J. G. Wen and Z. F. Ren, *Nano Lett.*, 2003, **3**, 235.

66 S. Sharma and M. K. Sunkara, *J. Am. Chem. Soc.*, 2002, **124**, 12 288.

67 M. Li, H. Schnablegger and S. Mann, *Nature (London)*, 1999, **402**, 393.

68 J. D. Hopwood and S. Mann, *Chem. Mater.*, 1997, **9**, 1819.

69 M. Li and S. Mann, *Langmuir*, 2000, **16**, 7088.

70 M. Summers, J. Eastoe and S. Davis, *Langmuir*, 2002, **18**, 5023.

71 The radial Hele–Shaw cell is a simple set-up to study the phenomenon of viscous fingering and the factors that control the finger patterns. Such 'fingers' are formed during displacement of a viscous fluid by a less viscous one and this phenomenon has important implications in secondary oil recovery, electrochemical deposition, combustion and fluid flow in porous media. E. Ben-Jacob and P. Garik, *Nature (London)*, 1990, **343**, 523 and references therein. While the Hele–Shaw cell has been used extensively in understanding the underlying physics of dynamic instabilities at liquid–liquid interfaces, its use in chemistry in probing surfactant-mediated interfacial recognition events has only recently been attempted M. Sastry, A. Gole, A. G. Banpurkar, A. V. Limaye and S. B. Ogale, *Curr. Sci.*, 2001, **81**, 191. We are not aware of any report on the synthesis of materials in a Hele–Shaw cell.

72 D. Rautaray, A. Banpurkar, S. R. Sainkar, A. V. Limaye, S. Ogale and M. Sastry, *Cryst. Growth Des.*, 2003, **3**, 449.

73 D. Rautaray, A. Banpurkar, S. R. Sainkar, A. V. Limaye, N. R. Pavaskar, S. Ogale and M. Sastry, *Adv. Mater.*, 2003, **15**, 1273.

74 D. Bonn, H. Kellay, B. M. Amar and J. Meunier, *Phys. Rev. Lett.*, 1995, **75**, 2132.

75 G. Falini, S. Albeck, S. Weiner and L. Addadi, *Science (Washington, D. C.)*, 1996, **271**, 67.

76 *Spectrometric Identification of Organic Compounds*, ed. R. M. Silverstein and F. X. Webster, John Wiley and Sons, Inc., New York, 6th edn., p. 107.

77 The XRD patterns were indexed with reference to the unit cell of the calcite structure ($a = b = 4.989$ Å, $c = 17.062$ Å, space group D_{3D}^6-$R3c$, ASTM, chart card no. 5-0586) from ASTM chart.

78 C. A. Orme, A. Noy, A. Wierzbicki, M. T. McBride, M. Grantham, H. H. Teng, P. M. Dove and J. J. DeYoreo, *Nature (London)*, 2001, **411**, 775.

79 A. Sugawara, T. Ishii and T. Kato, *Angew. Chem., Int. Ed. Engl.*, 2003, **42**, 5299.

80 A. Kumar, S. Mandal, P. R. Selvakannan, R. Pasricha, A. B. Mandale and M. Sastry, *Langmuir*, 2003, **19**, 6277.

81 L. O. Brown and J. E. Hutchison, *J. Am. Chem. Soc.*, 1999, **121**, 882.

82 L. O. Brown and J. E. Hutchison, *J. Phys. Chem. B*, 2001, **105**, 8911.

83 D. V. Leff, L. Brandt and J. R. Heath, *Langmuir*, 1996, **12**, 4723.

84 M. Green and P. O'Brien, *Chem. Commun.*, 2000, 183.

85 T. Jensen, L. Lelly, A. Lazarides and G. C. Schatz, *J. Cluster. Sci.*, 1999, **10**, 295.

Frank Marken,[*a] **Katy J. McKenzie,**[a] **Galyna Shul**[b] **and Marcin Opallo**[b]

[a] *Department of Chemistry, Loughborough University, Loughborough, Leicestershire, UK LE11 3TU. E-mail: f.marken@lboro.ac.uk*
[b] *Institute of Physical Chemistry, Polish Academy of Sciences, ul. Kasprzaka 44/52, 01-224 Warszawa, Poland*

Received 14th April 2004, Accepted 26th May 2004
First published as an Advance Article on the web 22nd September 2004

Biphasic electrode systems allow electrochemical reactions to be driven in a microphase of organic liquid (typically 1–100 nL), which is coupled *via* ion transfer processes to the surrounding aqueous electrolyte medium. Microdroplet deposits on basal plane pyrolytic graphite as well as thin film deposits of the organic phase within a mesoporous titanium oxide host film are investigated. Cobalt tetraphenylporphyrin (CoTPP) is dissolved in the organic liquid 4-(3-phenylpropyl)-pyridine (PPP) and deposited in the form of microphases at suitable electrode surfaces. The electrode is immersed in aqueous electrolyte environments. It is shown that two stable and highly reversible one-electron metal-centred redox processes occur consistent with $Co(III/II)TPP$ and $Co(II/I)TPP$ in the presence of axial pyridine ligands. The electrochemical characteristics for both processes are strongly affected by the liquid | liquid ion exchange accompanying the redox processes. The potential for both the $Co(III/II)TPP$ and the $Co(II/I)TPP$ redox processes can be adjusted independently by the choice of the nature and concentration of the aqueous electrolyte. The reversible potential observed for the CoTPP metal complex is dominated by the Gibbs energy of transfer for the 'spectator ions'. Conditions can be chosen to eliminate ion transfer effects on the potential scale for biphasic oxidation and reduction processes.

1. Introduction

Biphasic electrode systems are of considerable interest for processes involving ion accumulation and detection,[1] as modified electrode systems,[2] and for mediated interfacial electron transfer processes.[3] In order to bring two immiscible phases in contact with an electrode surface, static (*e.g.* immobilised thin films,[4] pastes,[5] or droplets[6] of organic solvents) and dynamic (*e.g.* the confluence reactor,[7] a flow through microreactor,[8] biphasic sonoelectrochemistry,[9] or the dropping liquid electrode[10]) methods have been proposed. The deposition of microdroplets of an immiscible liquid on a graphite electrode and immersion of this electrode in aqueous electrolyte media has emerged recently as a very convenient and powerful approach for (i) measuring Gibbs energies of transfer for anions[11] and for cations,[12] (ii) for investigating biphasic electrocatalytic processes,[13] and (iii) for studying biphasic photoelectrochemical processes.[14] The unique advantage of the microdroplet approach is the use of the triple phase boundary effect: electron transfer processes can occur at the electrode |

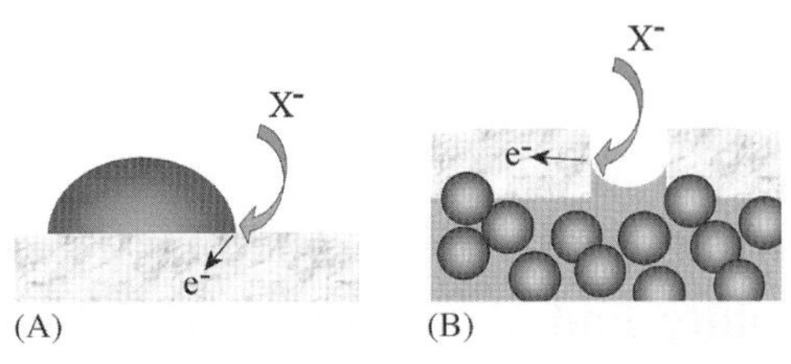

(A)   (B)

Scheme 1

organic | aqueous electrolyte triple interface and no intentionally added electrolyte is required in the organic phase. Therefore, for microdroplet modified electrodes the solvent window for ion transfer processes is not limited by competing background ion transfer processes. The transfer of sulfate, chromate,[15] phosphate,[16] as well as for a range of drugs[17] and peptides[18] has been investigated at triple phase boundary electrode systems.

In this report, electrochemical processes are studied for a novel system, microdroplets of 4-(3-phenylpropyl)pyridine (PPP) immobilised at graphite and in mesoporous ceramic electrodes. The organic solvent PPP is non-volatile, highly water immiscible, and a suitable medium for water insoluble transition metal complexes such as those of porphyrins or phthalocyanines. Cobalt tetraphenylporphyrin (CoTPP) is a very well known redox system[19] related to the naturally occurring vitamin B_{12} or cobalamin system[20,21] with applications predominantly in electrocatalysis.[22] CoTPP is only sparsely soluble in many organic solvents and entirely insoluble in aqueous media. Immobilisation of CoTPP at electrode surfaces has been previously achieved by adsorption *e.g.* onto suitable carbon supports,[23] by coordination binding,[24] and by dissolution into a thin film of benzonitrile.[25] The substantially increased solubility of CoTPP in pyridine containing solvents is due to the complexation of the two axial positions in the ligand sphere giving a $Co(TPP)L_2$ metal complex with L = pyridine derivative.[26] Here, CoTPP is dissolved in PPP, immobilised in the form of microdroplets directly onto suitable electrode surfaces, and studied when immersed into aqueous electrolyte solution.

Two types of biphasic electrode systems are compared for the electrochemical oxidation and reduction of CoTPP in PPP solution. First, microdroplets deposited onto basal plane pyrolytic graphite electrodes are investigated (see Scheme 1A). It is shown that only for a microdroplet deposit well-defined voltammetric responses are detected. Second, a novel 'porotrode' geometry[27] is investigated (see Scheme 1B). It has recently been shown that porous silicon oxide/graphite composite materials allow biphasic processes to occur efficiently.[28] Here, the electrode is based on a mesoporous TiO_2 (anatase) nanoparticle deposit, which has been sputter-coated with a thin and porous gold film. The organic liquid film is 'buried' in the mesoporous structure and the triple phase boundary electrode | organic solution | water (electrolyte) is located in between organic and aqueous phases. Both types of biphasic electrodes show characteristic voltammetric features consistent with a coupled biphasic process with ion transfer coupled to electron transfer.[29]

2. Experimental

2.1. Instrumentation

Voltammetric measurements were performed with a computer controlled Eco Chemie PGSTAT20 Autolab potentiostat system. Experiments were conducted in staircase voltammetry mode with a platinum gauze counter and saturated calomel reference electrode (SCE, REF401, Radiometer). The working electrode was a 300 µm diameter glassy carbon disk electrode, a 4.9 mm diameter basal plane pyrolytic graphite electrode (Pyrocarbon, Le Carbone UK), or a tin-doped indium oxide (ITO) coated glass (10 mm × 60 mm, resistivity 20 Ω square^{-1}) obtained from Image Optics Components Ltd. (Basildon, Essex). The ITO electrode surface was modified with a porous metal oxide film (see below) and gold coated in a Polaron sputter coating unit. An Elite tube furnace system was employed for cleaning ITO electrode surfaces (at 500 °C in air) and for calcining TiO_2 phytate films (at 500 °C in air). Prior to conducting electrochemical experiments, all solutions were purged with argon (BOC, UK). All experiments were carried out at a temperature of 22 ± 2 °C. The dielectric constant of 4-(3-phenylpropyl)pyridine, $\varepsilon = 6.6$, was determined with a liquid dielectric bridge (Calvert Designs). Scanning electron microscopy images were obtained with a Leo 1530 Field Emission Gun Scanning Electron Microscope (FEGSEM) system. Prior to FEGSEM imaging, the sample surface was scratched with a scalpel blade.

2.2. Chemical reagents

Titania sol (anatase, *ca.* 6 nm diameter, 30–35% in aqueous HNO_3, pH 0–3) was obtained from Tayca Corp., Osaka, Japan and diluted 100-fold with deionised water. 5,10,15,20-Tetraphenyl-21H,23H-porphine cobalt(II), 4-(3-phenylpropyl)pyridine, $LiNO_3$, $NaNO_3$, KNO_3, NMe_4NO_3, NBu_4Cl, NaSCN, $NaClO_4$, $NaIO_3$, NaN_3, NaF, NaOH, $HClO_4$ (60%), KCl, and phytic acid dodecasodium salt hydrate [*myo*-inositol hexakis (dihydrogen phosphate) dodecasodium salt] (all Aldrich) were obtained commercially in analytical or the highest purity grade available. NBu_4PF_6 was obtained in electrochemical grade (Fluka). Demineralised and filtered water was taken from an Elga water purification system (Elga, High Wycombe, Bucks, UK) with a resistivity of not less than 18 MΩ cm.

2.3. Working electrode preparation

Most experiments were conducted with a 4.9 mm diameter basal plane pyrolytic graphite electrode, which was renewed by polishing on a fine (P1000) grade carborundum paper (see Fig. 1A). For experiments with tin-doped indium oxide electrodes (ITO), mesoporous films of TiO_2 phytate were deposited following a layer-by-layer dip coating strategy.[30] A clean glass or ITO surface (washed with ethanol and water, dried, and 30 min heat treated at 500 °C in air) is dipped into a solution of TiO_2 nanoparticles followed by rinsing. The surface charge of the resulting nanoparticle deposit is reversed by dipping into a solution of phytic acid (40 mM in pH 3 aqueous solution) followed by rinsing. The process is carried out by a robotic Nima dip coating carousel (DSG–Carousel, Nima

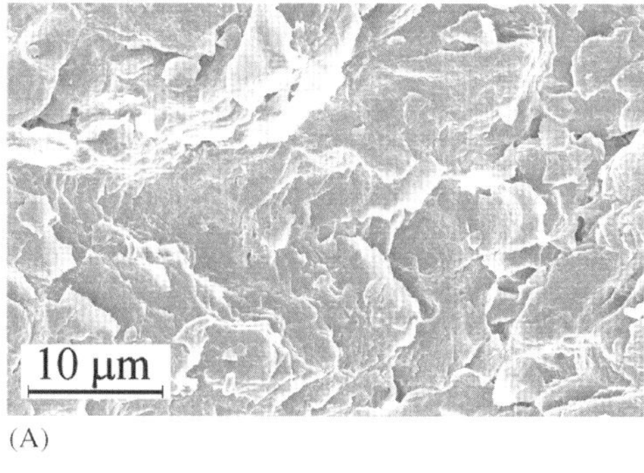

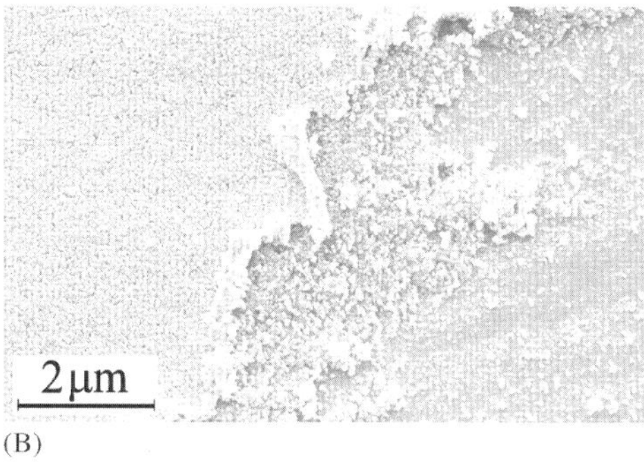

Fig. 1 FEGSEM images of (A) the surface of a basal plane pyrolytic graphite electrode and (B) the surface of a ITO electrode with a 15 layer deposit of TiO_2 nanoparticles sputter-coated with 20 nm porous gold (the surface was scratched to expose the film cross section).

Technology, Coventry, UK) and repeated to give multi-layer deposits of TiO_2 phytate with approximately 30 nm thickness per layer. After the deposition of 15 layers, the TiO_2 phytate film is dried and calcined prior to gold sputter coating with 20 nm porous gold (see Fig. 1B). The porous nature of the gold coating is due to the topography of the underlying oxide film.

3. Results and discussion

3.1. The oxidation and reduction of CoTPP dissolved in PPP

A solution with a concentration of 15 mM CoTPP and 0.1 M NBu_4PF_6 in PPP was prepared by gentle heating. The supporting electrolyte remained in solution only at slightly elevated temperature and therefore all voltammetric experiments directly in PPP solution were conducted at 80 °C. A 300 μm diameter glassy carbon electrode was employed to study the voltammetric behaviour of CoTPP under these conditions. Two redox processes, an oxidation (quasi reversible) and a reduction (reversible), were detected at $E_{mid} = -0.16$ V $vs.$ SCE and at $E_{mid} = -1.08$ V $vs.$ SCE, respectively (not shown). These processes are readily identified as metal-centered and one electron in nature by comparison with the electrochemical characteristics of CoTPP in pyridine containing solvents[26] (eqns. (1) and (2)).

$$Co(II)TPP \rightarrow Co(III)TPP^+ + e^- \tag{1}$$

$$Co(II)TPP + e^- \rightarrow Co(I)TPP^- \tag{2}$$

3.2. The oxidation and reduction of CoTPP immobilised in PPP microdroplets at a graphite electrode

A solution of 15 mM CoTPP in PPP was deposited onto the surface of a basal plane pyrolytic graphite electrode (see Fig. 1A) by solvent evaporation from acetonitrile (50-fold dilution). Fig. 2 shows typical voltammograms obtained for a volume of (A) 100 nL and (B) 400 nL PPP solution of CoTPP deposited onto basal plane pyrolytic graphite and immersed in aqueous 0.1 M $NaClO_4$. Voltammograms obtained under these conditions are highly reproducible and stable as long as proton donors (*e.g.* dihydrogen phosphate anions) or reactive molecules (*e.g.* CO_2) are absent from the electrolyte medium. Again two electrochemical responses are detected and identified at -0.296 V $vs.$ SCE (Co(III/II)) and at -1.184 V $vs.$ SCE (Co(II/I)). However, the biphasic nature of the electrode process requires ion exchange between aqueous and organic phase to be taken into consideration in the overall mechanism. Based on the neutral state of the starting material, Co(II)TPP, the following two overall processes can be proposed (eqns. (3) and (4)).

$$Co(II)TPP(organic) + X^-(aq) \rightarrow Co(III)TPP^+(organic) + X^-(organic) + e^- \tag{3}$$

$$Co(II)TPP(organic) + M^+(aq) + e^- \rightarrow Co(I)TPP^-(organic) + M^+(organic) \tag{4}$$

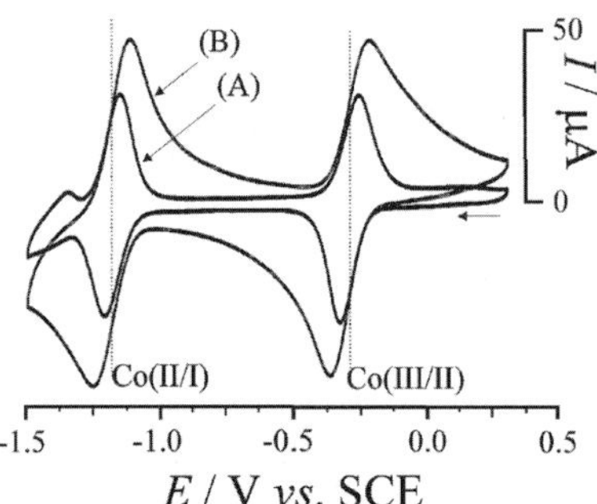

Fig. 2 Cyclic voltammograms (scan rate 50 mV s^{-1}, third cycle shown) for the oxidation and the reduction of (A) 100 nL and (B) 400 nL of a solution of 15 mM CoTPP in PPP deposited in the form of microdroplets onto a 4.9 mm diameter basal plane pyrolytic graphite electrode and immersed in aqueous 0.1 M $NaClO_4$.

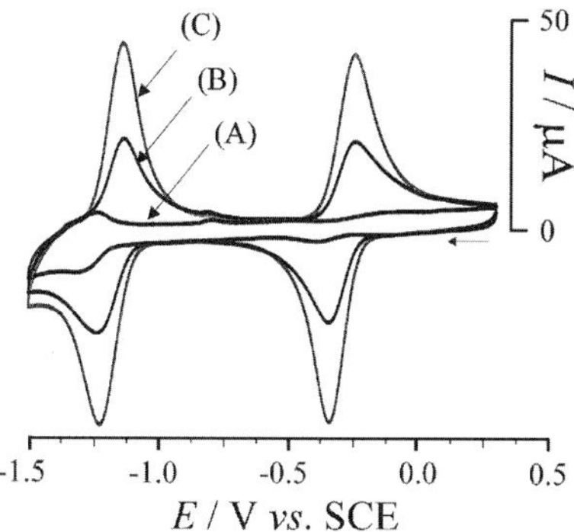

Fig. 3 Cyclic voltammograms (scan rate 50 mV s^{-1}, third cycle shown) for the oxidation and the reduction of 100 nL of a solution of (A) 0 mM, (B) 7.5 mM, and (C) 15 mM CoTPP in PPP deposited in the form of microdroplets onto a 4.9 mm diameter basal plane pyrolytic graphite electrode and immersed in aqueous 0.1 M NaClO$_4$.

The oxidation process, Co(III/II), is proposed to be accompanied by the transfer of an anion from the aqueous phase into the organic phase. In contrast, the charge balance during reduction, Co(II/I), is achieved *via* transfer of a metal cation from the aqueous phase into the organic phase (*vide infra*).

In Fig. 2 it can be seen that increasing the amount of deposit at the electrode surface from 100 nL to 400 nL causes only a modest increase in peak current and a considerable broadening of the voltammetric signals. This perhaps surprising behaviour can be explained with the triple phase boundary reaction zone present at the electrode surface.[29] Scheme A depicts a microdroplet deposit and the reaction zone for electron and ion transfer is indicated. By increasing the volume of deposit, the average microdroplet size is increased and the triple phase boundary zone decreased. The presence of bigger droplets also causes mass transport effects to impact on the shape of voltammograms as can be seen from the broadened and asymmetric peaks (see Fig. 2B).

The concentration of CoTPP in PPP has a considerable effect on the voltammetric characteristics. Fig. 3 shows cyclic voltammograms obtained for the oxidation and reduction of (A) 0 mM, (B) 7.5 mM, and (C) 15 mM CoTPP in 100 nL PPP deposited on a basal plane pyrolytic graphite electrode. Under these conditions, both the peak currents and the charge under the peaks are proportional to the concentration of CoTPP within the microdroplet deposit. (The origin of the reversible impurity process at -1.2 V *vs.* SCE has not been identified; see Fig. 3A). Fig. 4 demonstrates the effect of scan rate on the voltammetric signals. Well-defined voltammograms for both oxidation and reduction are observed over a range of scan rates (from 2 mV s^{-1} to 200 mV s^{-1}) and the linear increase in the

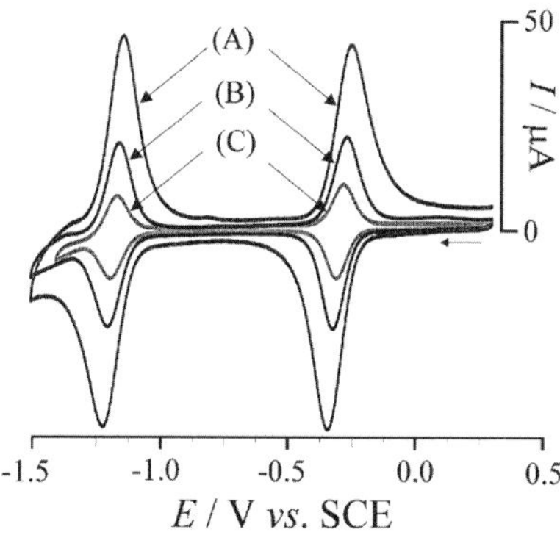

Fig. 4 Cyclic voltammograms (scan rates (A) 50 mV s^{-1} (B) 20 mV s^{-1} (C) 10 mV s^{-1}, third cycle shown) for the oxidation and the reduction of 100 nL of a solution of 15 mM CoTPP in PPP deposited in the form of microdroplets onto a 4.9 mm diameter basal plane pyrolytic graphite electrode and immersed in aqueous 0.1 M NaClO$_4$.

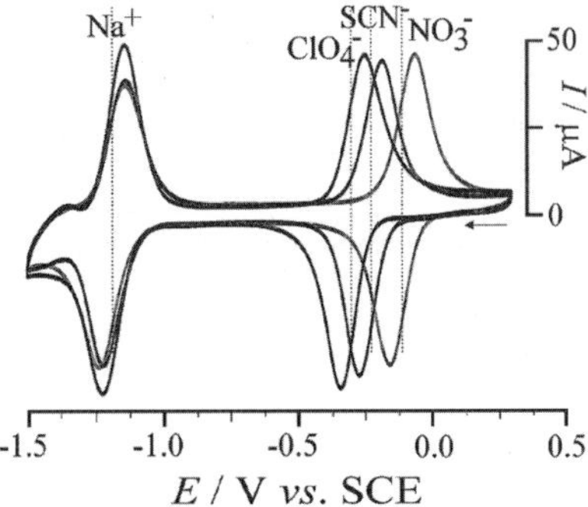

Fig. 5 Cyclic voltammograms (scan rate 50 mV s^{-1}, third cycle shown) for the oxidation and the reduction of 100 nL of a solution of 15 mM CoTPP in PPP deposited in the form of microdroplets onto a 4.9 mm diameter basal plane pyrolytic graphite electrode and immersed in aqueous solutions of 0.1 M NaClO$_4$, 0.1 M NaSCN, and 0.1 M NaNO$_3$.

peak current is consistent with the bulk electrolysis of the content of microdroplets deposited on the basal plane pyrolytic graphite electrode. The peak-to-peak separation between oxidation and reduction responses for the Co(III/II) system decreases from 97 mV to 54 mV and to 32 mV for scan rates of 50 mV s^{-1}, 20 mV s^{-1}, and 10 mV s^{-1}, respectively. Integration of the current under the peaks gives a charge of *ca.* 150 µC approximately consistent with the amount of charge anticipated for a 100 nL deposit of 15 mM CoTPP in PPP.

In order to further investigate the effect of the liquid | liquid ion exchange process coupled to the electron transfer, experiments in various aqueous electrolyte systems were conducted. Fig. 5 shows the effect of the anion on the oxidation of 15 mM CoTPP in 100 nL PPP. In the presence of a more hydrophobic anion, ClO$_4{}^-$, the oxidation process becomes more facile and the oxidation is detected at less positive potentials. Changing the anion but keeping the sodium cation constant causes only the oxidation response to change in potential. This behaviour is consistent with the process described by eqn. (3). In contrast, changing the cation in the aqueous electrolyte solution with a constant anion, nitrate, causes (in first approximation) only the reduction response to shift (see Fig. 6). This observation is consistent with the process described by eqn. (4).

The transfer of anions and cations from the aqueous into the organic phase is dominated by and correlated to the Gibbs energy of transfer. An expression for the Gibbs energy of transfer (here expressed as potential) can be based on Born's theory[31,32] (eqn. (5)).

$$\Delta\Phi^{O \to W} = \frac{N_A e^2}{8\pi\varepsilon_0 rF}\left(\frac{1}{\varepsilon^O} - \frac{1}{\varepsilon^W}\right) \tag{5}$$

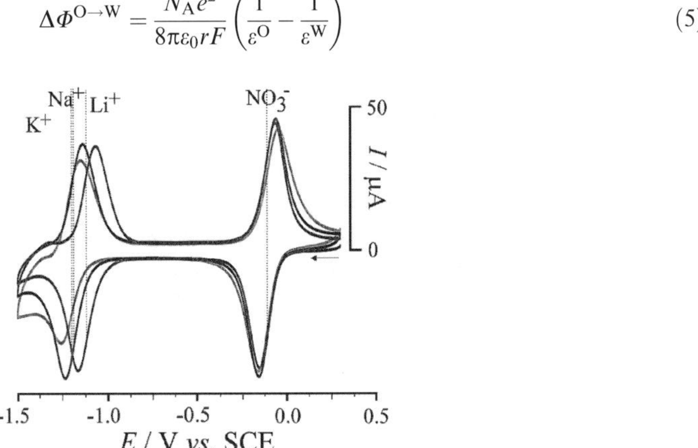

Fig. 6 Cyclic voltammograms (scan rate 50 mV s^{-1}, third cycle shown) for the oxidation and the reduction of 100 nL of a solution of 15 mM CoTPP in PPP deposited in the form of microdroplets onto a 4.9 mm diameter basal plane pyrolytic graphite electrode and immersed in aqueous solutions of 0.1 M NaClO$_4$, 0.1 M KClO$_4$, and 0.1 M LiClO$_4$.

Table 1 Voltammetric data (scan rate 0.05 V s^{-1}) for the oxidation and reduction of 15 mM CoTPP in PPP (volume 100 nL) deposited onto a basal plane pyrolytic graphite and immersed in aqueous electrolyte solution

Aqueous electrolyte	Conc./ mol L^{-1}	Process Co(II/I)			Process Co(III/II)		
		E_p^{red}/V vs. SCE	E_p^{ox}/V vs. SCE	E_{mid}/V vs. SCE	E_p^{red}/V vs. SCE	E_p^{ox}/V vs. SCE	E_{mid}/V vs. SCE
NaClO$_4$	0.1	−1.220	−1.143	−1.181	−0.338	−0.250	−0.294
NaSCN	0.1	−1.218	−1.123	−1.170	−0.273	−0.184	−0.228
NaF	0.1	−1.240	−1.175	−1.207	0.110	0.182	0.146
NaOH	0.1	−1.227	−1.168	−1.197	0.130	0.178	0.154
NaIO$_3$	0.1	−1.246	−1.149	−1.197	−0.019	0.110	0.045
NaN$_3$	0.1	−1.221	−1.156	−1.188	−0.112	−0.021	−0.066
LiNO$_3$	0.1	−1.169	−1.065	−1.117	−0.156	−0.065	−0.110
NaNO$_3$	0.1	−1.220	−1.143	−1.181	−0.149	−0.078	−0.113
KNO$_3$	0.1	−1.246	−1.182	−1.214	−0.143	−0.071	−0.107
NaNO$_3$	2.0	−-1.158	−1.076	−1.117	−0.201	−0.115	−0.158
NaNO$_3$	0.2	−1.208	−1.106	−1.157	−0.173	−0.065	−0.119
NaNO$_3$	0.02	−1.244	−1.153	−1.198	−0.153	0.0	−0.076
NMe$_4$NO$_3$	0.1	−1.253	−1.156	−1.204	−0.152	−0.058	−0.105
NBu$_4$Cl	0.1	−1.052	−0.931	−0.992	−0.067	0.032	−0.018
KCl	0.1	−1.247	−1.162	−1.207	−0.039	0.029	−0.005

In this equation the potential difference required for the transfer of a mono-anion, $\Delta\Phi^{O\rightarrow W}$, is related to Avogadro's constant, N_A, the elementary charge, e, the vacuum permittivity, ε_0, the radius of the anion, r, Faraday's constant, F, and the dielectric constants for the organic and aqueous medium, ε^O and ε^W. Voltammetric data for ion transfer processes involving a wide range of anions and cations are summarised in Table 1. Data for aqueous 0.1 M electrolyte solution are also plotted in Fig. 7. The plot shows the inverse of the ionic radii as a function of the midpoint potentials for both oxidation and reduction processes. From eqn. (5) it can be seen that a simple linear relationship is expected to hold for the potential *versus* the inverse ionic radius. The dotted lines in Fig. 7 indicate the theoretically predicted slope (assuming a dielectric constant of $\varepsilon^O = 4.8$) for the transfer of cations and anions. The agreement is only approximate and the dielectric constant deviates from the value determined for pure PPP with a dielectric bridge, $\varepsilon^O = 6.6$. Furthermore, cations strongly solvated by polar solvent molecules (Li$^+$, Na$^+$, and K$^+$) deviate strongly from the behaviour predicted by eqn. (5). However, for bulky tetraalkylammonium cations and for all anions reasonable agreement between simple Born theory prediction and experiment is observed. However, the limitations of this approach and the effects of coextraction of water molecules have been pointed out by Osakai *et al.*[33] More complex expressions for calculating the Gibbs energy of transfer have been proposed.[34]

It is interesting to compare biphasic voltammetric data with data obtained in PPP solution in the absence of water. Two horizontal bars in the plot shown in Fig. 7 indicate the halfwave potentials for the oxidation and the reduction of CoTPP dissolved in PPP (0.1 M NBu$_4$PF$_6$) (*vide supra*). Eliminating the potential gap between the two bars allows cation and anion transfer scales to be compared directly. Also, by choosing an electrolyte system such as LiNO$_3$, conditions can be found under which the effects of ion transfer on voltammetric responses at biphasic electrodes are eliminated and the potential scale matches that in 'simple' organic solution environments. Essentially, this type of experiment allows voltammetric experiments to be conducted in microphases of organic solvents in the absence of intentionally added electrolyte.

Finally, changing the concentration of the aqueous electrolyte also causes a tell-tale shift in the voltammetric responses consistent with the trend predicted by the Nernst expressions for eqns. (3) and (4). Fig. 8 shows voltammograms obtained for the oxidation and reduction of 15 mM CoTPP immobilised in 100 nL PPP microdroplets deposited onto basal plane pyrolytic graphite and immersed in aqueous electrolyte with concentrations of (A) 2.0 M, (B) 0.2 M, and (C) 0.02 M NaNO$_3$. Voltammetric responses for oxidation and for reduction processes shift in opposite

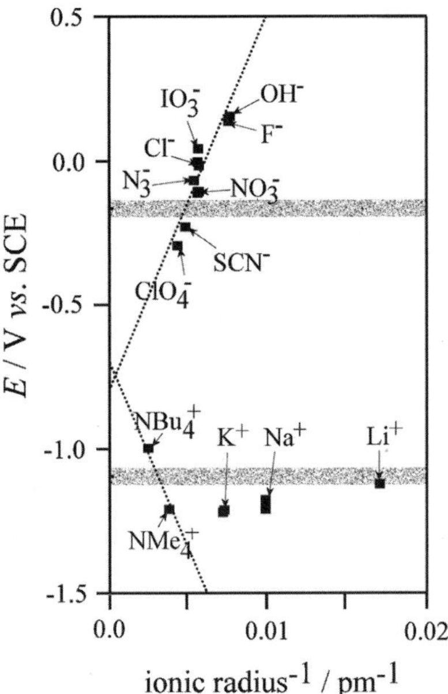

Fig. 7 Plot of the inverse ionic radius for different ions (taken from Lowenschuss data[37]) *versus* the midpoint potentials for ion transfer processes accompanying the oxidation and reduction of CoTPP in PPP at biphasic electrodes immersed in aqueous 0.1 M electrolyte solution.

direction upon changing the concentration of the electrolyte and the shift is consistent with 41 ± 2 mV per decade change in $NaNO_3$ concentration (see Table 1).

3.3. The oxidation and reduction of CoTPP immobilised in a PPP microphase at a mesoporous oxide electrode

The formation of random arrays of microdroplets is possible at basal plane pyrolytic graphite electrodes due to the microstructure and wetting properties of this type of electrode surface (see Fig. 1A). However, alternative electrode systems with a less random interface are desirable. In order to form a biphasic interfacial structure at electrode surfaces, mesoporous oxide films can be employed.[35] We have demonstrated recently the use of TiO_2 (anatase) nanoparticles for the layer-

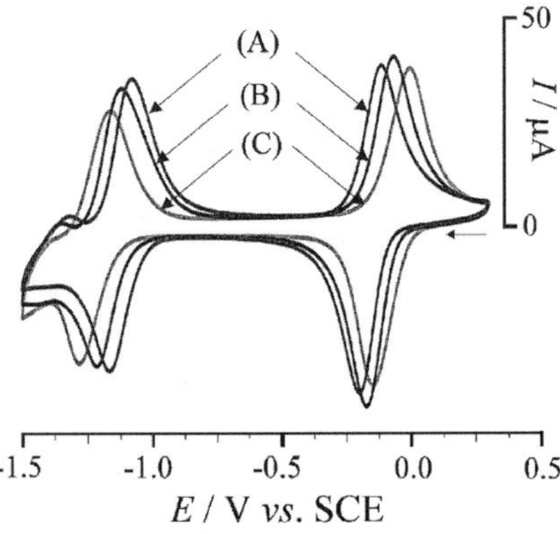

Fig. 8 Cyclic voltammograms (scan rate 50 mV s^{-1}, third cycle shown) for the oxidation and the reduction of 100 nL of a solution of 15 mM CoTPP in PPP deposited in the form of microdroplets onto a 4.9 mm diameter basal plane pyrolytic graphite electrode and immersed in aqueous (A) 2.0 M, (B) 0.2 M, and (C) 0.02 M $NaNO_3$.

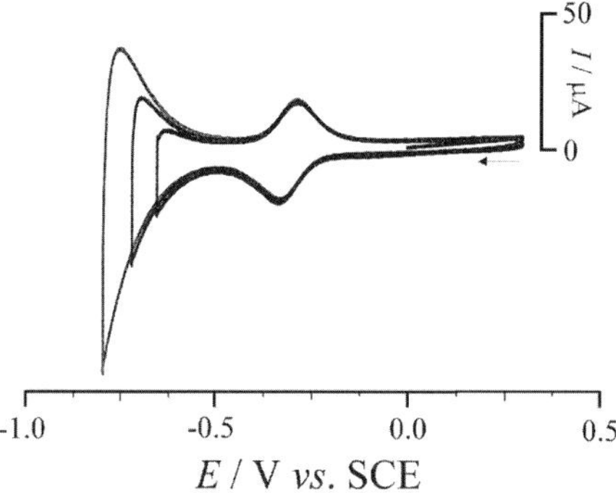

Fig. 9 Cyclic voltammograms (scan rate 50 mV s^{-1}, third cycle shown) for the oxidation and the reduction of 100 nL of a solution of 15 mM CoTPP in PPP deposited onto a porous 1 cm × 1 cm ITO/TiO$_2$/Au electrode (see text) and immersed in aqueous 0.1 M NaClO$_4$.

by-layer growth of a mesoporous oxide films at tin-doped indium oxide (ITO) electrodes and the immobilisation of redox proteins.[36] These TiO$_2$ films allow diffusion within the pores and have been demonstrated to conduct electrical signals.[27] Here it is demonstrated in a proof-of-principle experiment, that mesoporous oxide films can also be employed beneficially as a host for an organic phase in biphasic voltammetry.

The oxide film grown layer-by-layer from TiO$_2$ nanoparticles of approximately 6 nm diameter is shown in Fig. 1B. An even film deposit with approximately 300–400 nm thickness is produced. In order to avoid resistivity problems in the presence of the organic film deposit, a porous gold film of *ca.* 20 nm thickness is sputter-coated onto the surface of the electrode. This gold film allows an extended triple phase boundary to be formed (see Scheme B) with triple phase boundary regions in direct contact with the aqueous solution phase. Fig. 9 shows three consecutive voltammograms obtained for the oxidation of 15 mM CoTPP in 100 nL PPP deposited onto the surface of a 1 cm^2 ITO electrode coated with TiO$_2$ and a gold layer and immersed in aqueous 0.1 M NaClO$_4$. The oxidation response is detected with a midpoint potential of -0.30 V *vs.* SCE consistent with results obtained on a basal plane pyrolytic graphite. The voltammetric response for the oxidation of CoTPP at the ITO/TiO$_2$/Au electrode is well-defined and consistent with substantial electrolysis of the 100 nL deposit. The free volume within the oxide host can be estimated as approximately 10–20 nL and therefore increasing the amount of deposit may not lead to a further increase in the voltammetric signal. Experiments conducted in aqueous solutions of 0.1 M NaSCN and 0.1 M NaNO$_3$ resulted in well-defined oxidation responses detected at $E_{mid} = -0.22$ V *vs.* SCE and $E_{mid} = -0.10$ V *vs.* SCE consistent with results obtained at basal plane pyrolytic graphite (see Table 1).

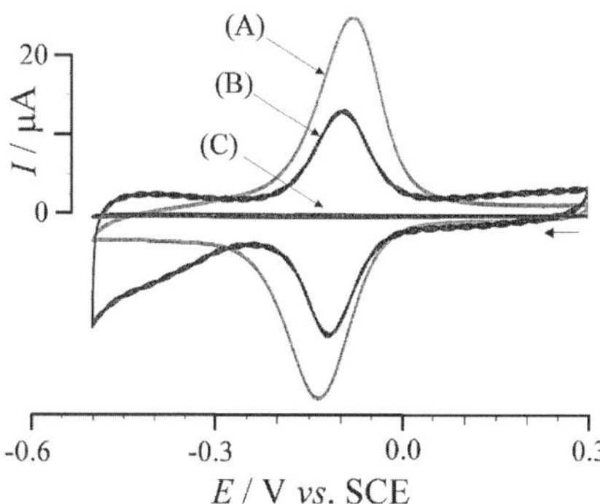

Fig. 10 Cyclic voltammograms (scan rate 50 mV s^{-1}) for the oxidation of 50 nL of a solution of 15 mM CoTPP in PPP deposited onto (A) a 4.9 mm diameter basal plane pyrolytic graphite electrode, (B) a porous 1 cm × 1 cm ITO/TiO$_2$/Au electrode, and (C) a porous 1 cm × 1 cm ITO/TiO$_2$ electrode and immersed in aqueous 0.1 M NaNO$_3$.

Similar experiments with a plane uncoated ITO electrode or a TiO_2 coated ITO electrode resulted only in very poorly defined voltammetric responses. In Fig. 10 voltammograms are shown for the oxidation of 50 nL of 15 mM CoTPP in PPP deposited onto (A) a basal plane pyrolytic graphite electrode, (B) a gold coated TiO_2 thin film electrode, and (C) a TiO_2 thin film electrode without gold coating. The gold contact appears to be essential for the process to occur.

Upon scanning the potential applied to a TiO_2 coated electrode into a more negative range, a new strong reduction response is detected (see Fig. 9) commencing at *ca.* -0.5 V *vs.* SCE. This reduction response is fully reversible and capacitive in nature. This voltammetric response is detected in the absence and in the presence of PPP solution. In the absence of TiO_2 this signal is not detected and therefore direct and reversible charging of TiO_2 nanoparticles is proposed as an explanation. Unfortunately, this TiO_2 based voltammetric signal is preventing the detection of the Co(II/I) reduction response. Future work will focus on introducing other types of nanoparticulate oxides (such as Al_2O_3 and SiO_2) in order to widen the potential window for this type of novel biphasic electrode system.

4. Conclusions

Two types of biphasic electrode system have been employed for the oxidation and reduction of CoTPP in 4-(3-phenylpropyl)pyridine as the unsupported electrolyte medium. A random array of microdroplets on a basal plane pyrolytic graphite electrode gives well-defined voltammetric signals consistent with (i) CoTPP oxidation and simultaneous anion transfer and (ii) CoTPP reduction and simultaneous cation transfer. Conditions have been identified under which potential shifts due to ion transfer are minimised. The second type of biphasic electrode is based on a thin film of mesoporous oxide coated with a porous gold layer. The organic phase is immobilised within the mesoporous host and a triple phase boundary formed at the organic solution | gold | aqueous electrolyte interface.

Acknowledgements

F. M. thanks the Royal Society for a University Research Fellowship. K. J. M. thanks the EPSRC for a studentship. We thank Tayca Corp., Osaka, Japan for generous support with titania sol. The partial support from the Polish–British Partnership Programme sponsored by the British Council and Committee for Scientific Research (Project WAR/314/248) is gratefully acknowledged.

References

1 F. Marken, A. Blythe, R. G. Compton, S. D. Bull and S. G. Davies, *Chem. Commun.*, 1999, 1823.
2 C. E. Banks, T. J. Davies, R. G. Evans, G. Hignett, A. J. Wain, N. S. Lawrence, J. D. Wadhawan, F. Marken and R. G. Compton, *Phys. Chem. Chem. Phys.*, 2003, **5**, 4053.
3 See, for example: C. N. Shi and F. C. Anson, *J. Phys. Chem. B*, 2001, **105**, 8963.
4 C. N. Shi and F. C. Anson, *J. Phys. Chem. B*, 2001, **105**, 1047.
5 See, for example: P. Tomcik, C. E. Banks, T. J. Davies and R. G. Compton, *Anal. Chem.*, 2004, **76**, 161.
6 J. C. Ball, F. Marken, F. L. Qiu, J. D. Wadhawan, A. N. Blythe, U. Schröder, R. G. Compton, S. D. Bull and S. G. Davies, *Electroanalysis*, 2000, **12**, 1017.
7 K. Yunus, C. B. Marks, A. C. Fisher, D. W. E. Allsopp, T. J. Ryan, R. A. W. Dryfe, S. S. Hill, E. P. L. Roberts and C. M. Brennan, *Electrochem. Commun.*, 2002, **4**, 579.
8 S. Sawada, M. Taguma, T. Kimoto, H. Hotta and T. Osakai, *Anal. Chem.*, 2002, **74**, 1177.
9 J. D. Wadhawan, F. Marken and R. G. Compton, *Pure Appl. Chem.*, 2001, **73**, 1947.
10 See, for example: C. J. Slevin and P. R. Unwin, *Langmuir*, 1997, **13**, 4799.
11 U. Schröder, R. G. Compton, F. Marken, S. D. Bull, S. G. Davies and S. Gilmour, *J. Phys. Chem. B*, 2001, **105**, 1344.
12 F. Scholz, R. Gulaboski and K. Caban, *Electrochem. Commun.*, 2003, **5**, 929.
13 A. J. Wain, J. D. Wadhawan and R. G. Compton, *Chem. Phys. Chem.*, 2003, **4**, 974.
14 J. D. Wadhawan, R. G. Compton, F. Marken, S. D. Bull and S. G. Davies, *J. Solid State Electrochem.*, 2001, **5**, 301.
15 F. Marken, C. M. Hayman and P. C. B. Page, *Electroanalysis*, 2002, **14**, 172.
16 F. Marken, C. M. Hayman and P. C. B. Page, *Electrochem. Commun.*, 2002, **4**, 462.

17 G. Bouchard, A. Galland, P. A. Carrupt, R. Gulaboski, V. Mirceski, F. Scholz and H. H. Girault, *Phys. Chem. Chem. Phys.*, 2003, **5**, 3748.
18 R. Gulaboski and F. Scholz, *J. Phys. Chem. B*, 2003, **107**, 5650.
19 See, for example: D. G. Davis in *The Porphyrins*, ed. D. Dolphin, Academic Press, London, 1978, vol. 5, p. 127.
20 R. Banerjee and S. W. Ragsdale, *Ann. Rev. Biochem.*, 2003, **72**, 209.
21 Y. Hisaeda, T. Nishioka, Y. Inoue, K. Asada and T. Hayashi, *Coord. Chem. Rev.*, 2000, **198**, 21.
22 See, for example: C. Shi and F. C. Anson, *Inorg. Chem.*, 2001, **40**, 5829.
23 See, for example: E. H. Song, C. N. Shi and F. C. Anson, *Langmuir*, 1988, **14**, 4315.
24 See, for example: S. Z. Zou, R. S. Clegg and F. C. Anson, *Langmuir*, 2002, **18**, 3241.
25 T. D. Chung and F. C. Anson, *J. Electroanal. Chem.*, 2001, **508**, 115.
26 L. A. Truxillo and D. G. Davis, *Anal. Chem.*, 1975, **47**, 2260.
27 K. J. McKenzie, F. Marken and M. Opallo, *Bioelectrochemistry*, 2004, **64**, DOI: 10.1016/j.bioelechem.2004.03.008.
28 (*a*) M. Opallo and M. Saczek-Maj, *Chem. Commun*, 2002, 448; (*b*) M. Saczek-Maj and M. Opallo, *Electroanalysis*, 2003, **15**, 566; (*c*) M. Opallo, M. Saczek-Maj, G. Shul, C. M. Hayman, P. C. Bulman Page, F. Marken, *Electrochim. Acta*, in print.
29 F. Marken, R. D. Webster, S. D. Bull and S. G. Davies, *J. Electroanal. Chem.*, 1997, **437**, 209.
30 K. J. McKenzie, F. Marken, M. Hyde and R. G. Compton, *New J. Chem.*, 2002, **26**, 625.
31 M. Born, *Z. Phys.*, 1920, **1**, 45.
32 U. Schröder, R. G. Evans, R. G. Compton, B. Wood, D. J. Walton, R. R. France, F. Marken, P. C. B. Page and C. M. Hayman, *J. Phys. Chem. B*, 2002, **106**, 8697.
33 T. Osakai, A. Ogata and K. Ebina, *J. Phys. Chem. B*, 1997, **101**, 8341.
34 H. H. J. Girault and D. J. Schiffrin, *Electroanal. Chem.*, 1989, **15**, 1.
35 J. Niedziolka and M. Opallo, *Electrochem. Commun.*, 2003, **5**, 924.
36 K. J. McKenzie and F. Marken, *Langmuir*, 2003, **19**, 4327.
37 Y. Marcus, *Ion Properties*, Marcel Dekker, New York, 1997.

Phase separation of water–alcohol binary mixtures induced by the microheterogeneity

Akihiro Wakisaka* and Takahiro Ohki

National Institute of Advanced Industrial Science and Technology, Onogawa 16-1, Tsukuba 305-8569, Japan. E-mail: akihiro-wakisaka@aist.go.jp; Fax: +81 29 861 8252; Tel: +81 29 861 8271

Received 15th April 2004, Accepted 20th May 2004
First published as an Advance Article on the web 21st September 2004

The relationship between liquid–liquid phase separation and microheterogeneity in water–primary alcohol mixtures was examined by analysing the mass spectra of clusters generated through the fragmentation of liquid droplets. By comparing the cluster structures of water–ethanol, –1-propanol, and –1-butanol binary mixtures at various alcohol concentrations, we discovered differences in the molecular clusters that control phase separation. We also studied the role of water in alcohol self-association. Alcohol self-association is promoted in the presence of a small amount of water (*ca.* 10 ∼ 20 wt%), in which the water–water hydrogen-bonding network is weak and does not contribute to alcohol self-association. We have demonstrated that alcohol self-association is also promoted by non-ideal mixing with other alcohols. The self-association of alcohol molecules complements the loss of stabilization energy caused by the relatively weak coexisting interactions. This complementary relationship among intermolecular interactions is an inherent property of solutions, and plays a key role in the phase separation process.

Introduction

Primary alcohols such as methanol, ethanol and 1-propanol are miscible with water at any mixing ratio. Beyond this, however, primary alcohols are not completely miscible with water. Specifically, 1-butanol shows liquid–liquid phase separation from 6.4 to 80.2 wt% 1-butanol concentrations at 20 °C.[1,2] This means that in the two-phase region, the 1-butanol concentrations in the water rich phase and in the 1-butanol rich phase correspond to 6.4 and 80.2 wt%, respectively. To explain why the liquid–liquid interface is generated in 6.4 to 80.2 wt% water–1-butanol mixtures, information regarding the microscopic structure of the liquids is required. There have been a lot of reports on the microscopic structures of water–alcohol binary mixtures *via* experimental and theoretical approaches such as mass spectrometry,[3–8] neutron diffraction,[9] X-ray diffraction,[10,11] molecular dynamic simulations,[12] Monte Carlo simulations,[13,14] *etc.* Most of them report that there are microheterogeneous structures in water–alcohol binary mixtures. Even the most miscible water–methanol mixtures are known to have some microheterogeneous structures.[3,4,9] Therefore, the formation of a liquid–liquid interface should be dependent on the microheterogeneous structures formed by intermolecular interactions in these solutions. To address this, this study focused on the microheterogeneous structures of water–ethanol, water–1-propanol and water–1-butanol binary mixtures through the analysis of mass spectra of clusters isolated from liquid droplets.

DOI: 10.1039/b405391e

On a related topic, we would like to focus on why the self-association of alcohols is increased by mixing with water. The molecular self-association of organic molecules in water is generally assumed to be the result of hydrophobic interactions. Using properties unique to water, the fine tuning of organic reactions in aqueous media have been attempted.[15,16] However, we report here that the molecular self-association of an alcohol can also be promoted by mixing with another alcohol, in the case that one alcohol is barely soluble in the other. It is noted here that solvent-assisted molecular self-association is an inherent property of solutions, both aqueous and non-aqueous. Since molecules in the liquid phase are surrounded by several other molecules, several simultaneous intermolecular interactions must be involved in molecular clustering. It has been suggested that molecular self-association in the liquid phase is controlled by the *relative* values of the intermolecular interaction energies rather than by the *absolute* ones.[17,18] In binary mixtures, particularly, molecular self-association can be promoted by the coexistence of different kinds of intermolecular interactions, in comparison with that in pure liquids. The possibility that the phase separation of water–alcohol mixtures is also induced by this molecular self-association inherent in the liquid phase is discussed.

When two kinds of liquids are mixed, the resulting binary mixtures usually deviate from ideal mixing. The physicochemical properties of binary mixtures, such as the excess enthalpy of mixing,[19] partial vapour pressure,[20] viscosity,[21,22] self-diffusion constant,[23] *etc.*, are quite different from those calculated based on ideal mixing. We attempt to demonstrate here that the physicochemical properties of binary liquid mixtures are closely related to the microheterogeneous structures at the cluster level.

Experimental

Mass spectrometry for clusters

The microscopic structures of binary mixtures were analyzed through the mass spectra of clusters isolated from liquid droplets. The principle of this specially designed mass spectrometry (Fig. 1) has been reported previously.[24] The mass spectrometer consists of a four-stage differentially pumped vacuum chamber, a heated sample injection nozzle and a quadrupole mass filter (Extrel C50). Sample solutions are continuously injected into the vacuum chamber at a flow rate of $0.1 \sim 0.18$ cm^3 min^{-1}. When part of the nozzle is heated to $110–160\,°C$, a constant flow of liquid droplets is generated due to the increase in the pressure inside the nozzle caused by the vaporization of part of the solution. Due to the strong surface tension of water, higher temperatures are needed for the samples with higher water content. Owing to the pressure gradient, the resulting liquid droplets are led to the second and then the third chamber, which leads to fragmentation of the liquid droplets into clusters *via* adiabatic expansion. During the fragmentation process, the weakly interacting molecules are vaporised as monomeric molecules. The resulting clusters are ionized by an electron impact at 30 eV, and then analyzed by a quadrupole mass filter.

To minimize vaporization of molecules from the clusters, the pressure gradient from the nozzle to the quadrupole mass filter is kept as small as possible by use of the four-stage differential pumping system. To reduce collision-based interactions, the nozzle is situated coaxial to the skimmers and the mass filter. This lowers the resolution of the mass analyzer compared to a conventional apparatus, however, it is necessary to insure clusters are formed and maintained.

Coefficient of viscosity

The viscosity coefficients for water–alcohol binary mixtures were measured by a vibration method viscometer (Sine-wave Vibro Viscometer (SV-10), A&D Co. Ltd.) at $20\,°C$.

Chemicals

Methanol, ethanol, 1-propanol, 1-butanol and 1-pentanol with special grade ($>99\%$, Wako) were used without further purification. Water was purified using Milli-RX 12 and MILLI-Q SP. TOC (MILLIPORE) filters.

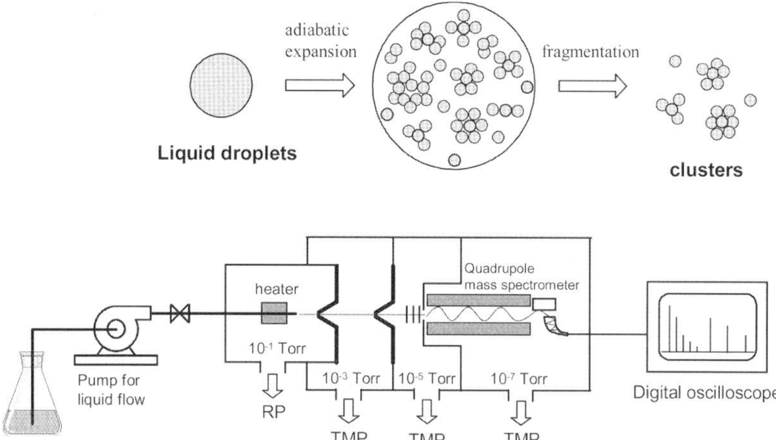

Fig. 1 Schematic picture of a mass spectrometer specially designed for the analysis of clusters. RP and TMP correspond to a rotary pump and a turbo molecular pump for vacuum creation. A liquid droplet beam is formed between the nozzle and the first skimmer. The cluster formation process through fragmentation of the liquid droplets is also shown here.

Results and discussion

1. Microheterogeneity in water–alcohol binary mixtures

1.1. Mass spectrometric analysis of clusters. In order to demonstrate how microscopic structures are connected with macroscopic phase separation, molecular clustering in the water–ethanol, water–1-propanol and water–1-butanol binary mixtures was studied by mass spectrometry.

(i) Water–ethanol binary mixtures. Water and ethanol are miscible at any mixing ratio. Macroscopically, these mixtures appear homogeneous. However, it has already been reported that the microscopic structures at the cluster level are not homogeneous.[7,8] Through our specially designed mass spectrometry, this microheterogeneity was also observed. To examine the dependence of this microheterogeneity on the water–ethanol mixing ratio, the mass spectra of the water–ethanol binary mixtures with ethanol concentrations: [EtOH] = 0, 2.5, 5.0, 11.9, 13.5, 22.1, 31.1, 39.0, 52.3, 63.0, 79.3, 91.1, 95.8 and 100 wt% were observed. The mass spectra observed at 0, 5.0, 52.3, 79.3 and 100 wt% are shown here as representative of the concentration dependence.

0 ≤ [EtOH] ≤ 13.5 wt%. Fig. 2 shows a mass spectrum of pure water. A series of water clusters, $H^+(H_2O)_n$, is observed as a result of the fragmentation of the hydrogen-bonding network of water molecules. Some peaks are observed as prominent peaks compared with other peaks. The largest peak is that of $H^+(H_2O)_{21}$, which is called the 'magic-number species'.[25,26] This magic-number property represents the existence of relatively stable water clusters.

At low ethanol concentrations (Fig. 3(a)), the observed clusters are mainly composed of hydrogen-bonded water clusters, expressed as $H^+(C_2H_5OH)_m(H_2O)_n$: $m = 0 \sim 6$, $n = 0, 1, 2, 3, \ldots$. In Fig. 3(a), a series of clusters with the same number of ethanol molecules (m) is connected by one line. Each series of clusters has a similar mass distribution to that of pure water. This indicates that the hydrogen-bonding network of water is not influenced by the presence of ethanol at lower ethanol concentrations. Furthermore, the clusters $H^+(C_2H_5OH)_m(H_2O)_n$ represented by m–n again possess the magic-number property for each series of clusters at 0–21, 1–20, 2–19, 3–18, 4–17 and 5–16. These magic-number clusters are composed of the same total number of molecules: $m + n = 21$. Ethanol molecules are capable of substitutional interaction with hydrogen-bonded water clusters intrinsic to pure water. This is likely the reason clustering structures of pure water were observed in ethanol concentrations lower than 13.5 wt%.

13.5 wt% ≤ [EtOH]. At ethanol concentrations above 13.5 wt%, the cluster structures based on the hydrogen-bonding network intrinsic to pure water are disrupted, and the self-association of

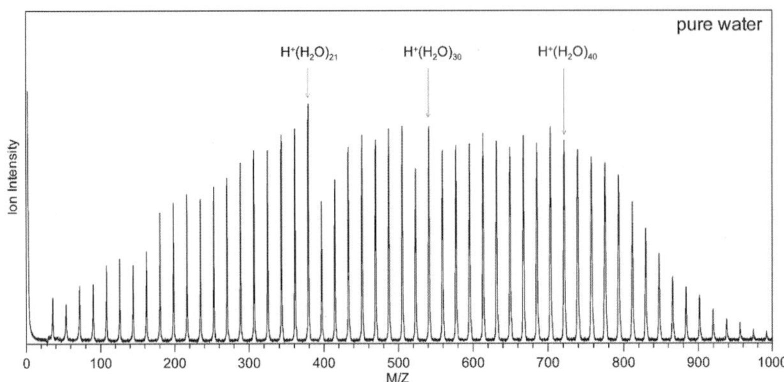

Fig. 2 A mass spectrum of clusters isolated from pure water. The peaks observed at the same interval of 18 u correspond to the water clusters, $H^+(H_2O)_n$: $n = 2, 3, 4, \ldots$. The peak position of $n = 1$ is out of the effective sensitivity. Some peak assignments are shown in the spectrum.

ethanol became prominent. At 52.3 wt%, shown in Fig. 3(b), ethanol self-association clusters interacting with water molecules (hydrated ethanol clusters) are observed. The hydration number (n) in the observed $H^+(C_2H_5OH)_m(H_2O)_n$ clusters increases with an increase in the number of ethanol molecules (m), which can be seen by the series of clusters having the same number of ethanol molecules (m), connected by the line in Fig. 3(b). This indicates two kinds of characteristics: (1) the self-association of ethanol is clearly dominant and forms the major structures in the binary mixture, (2) with an increase in the number of ethanol molecules (m) in the $H^+(C_2H_5OH)_m(H_2O)_n$ clusters, the hydrated cluster structure is more favourable than the self-association of ethanol molecules, especially at $m \geq 8$. As described in the following sections, these clustering structures are dependent on the alcohol, which suggests a correlation with macroscopic phase separation.

With a further increase in the ethanol concentration, the hydrated clusters with a large number of ethanol molecules ($m \geq ca.$ 8) decreases, as shown in the spectrum of [EtOH] = 79.3 wt% (Fig. 3(c)). The ethanol self-association clusters larger than $H^+(C_2H_5OH)_4$ were also markedly reduced at 95.8 wt% ethanol concentration. At 100 wt% ethanol, ethanol clusters larger than $H^+(C_2H_5OH)_4$ are hardly observed, as shown in Fig. 3(d). In ethanol–water binary mixtures, ethanol self-association is heavily promoted by the presence of water.

(ii) Water–1-propanol binary mixtures. Water and 1-propanol are also miscible at any mixing ratio. To examine the microheterogeneity's dependence on the 1-propanol concentration, the mass spectra of binary mixtures with 1-propanol concentrations: [PrOH] = 3.26, 6.37, 15.9, 27.1, 45.5, 58.8, 69.0, 83.3, 93.0, 96.7, 98.4 and 100 wt% were observed. The fundamental properties of the cluster structures in the water–1-propanol mixtures were similar to those observed in the water–ethanol mixtures. At lower 1-propanol concentrations, the molecular clustering is controlled by the hydrogen-bonding network intrinsic to pure water. Since the mass number of $(C_3H_7OH)_3$ equals that of $(H_2O)_{10}$, many peaks are overlapped in the higher mass region. Therefore, the analyses of the mass spectra of 3.26 and 6.37 wt% 1-propanol were difficult. However, with increasing 1-propanol concentration, the clustering structures changed to 1-propanol self-association structures instead of the intrinsic water structure. The intrinsic water structure disintegrated at lower alcohol concentrations compared with the water–ethanol mixture. As shown in Fig. 4, at higher 1-propanol concentrations, the major cluster structures in the water–1-propanol mixtures are controlled by 1-propanol self-association. For larger clusters, the hydrated cluster structure is more favourable than non-hydrated ones. With decreasing water concentration, the self-association of 1-propanol decreases, as observed for the water–ethanol mixtures. It should be noted that the hydration for the 1-propanol clusters is weaker than that for the ethanol clusters, based on a comparison between Figs. 3(b) and (c), and Figs. 4(a) and (b).

(iii) Water–1-butanol binary mixtures. Water–1-butanol binary mixtures have a two-phase region at 1-butanol concentrations $6.4 \leq [BuOH] \leq 80.2$ wt% at 20 °C. In this region, the 1-butanol

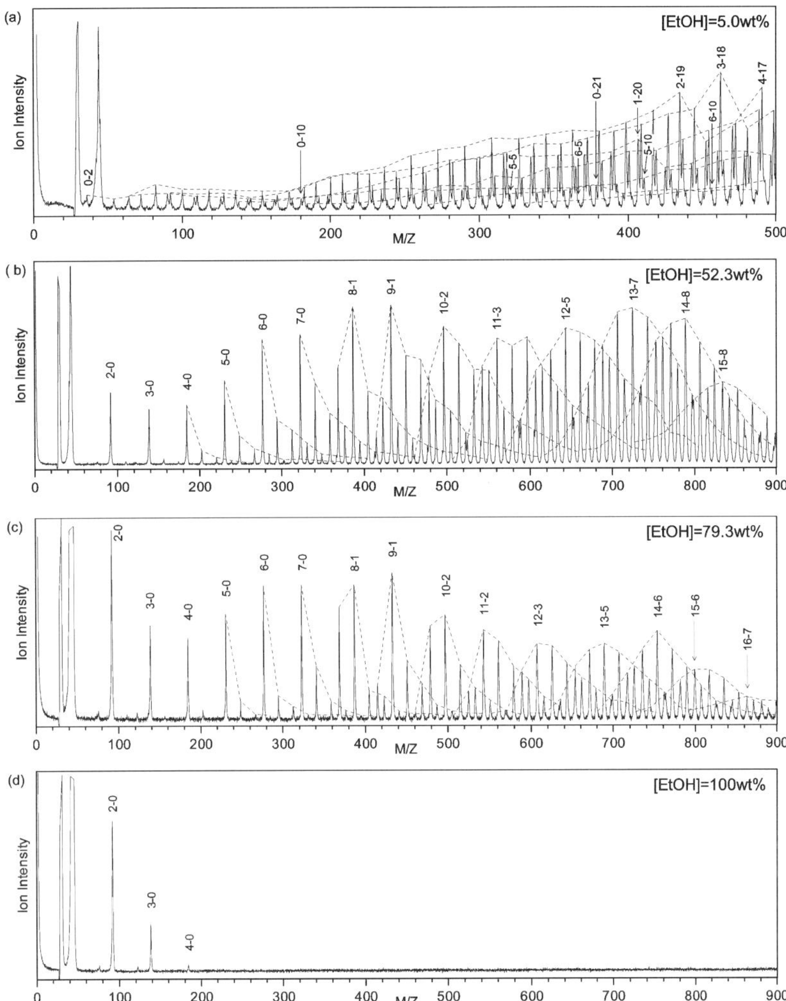

Fig. 3 Mass spectra of clusters isolated from water–ethanol binary mixtures at various ethanol concentrations: (a) 5.0, (b) 52.3, (c) 79.3, and (d) 100 wt% of ethanol. The clusters having the same number of ethanol molecules (m) in the clusters $H^+(C_2H_5OH)_m(H_2O)_n$ are connected by lines. The paired numbers represent m–n of $H^+(C_2H_5OH)_m(H_2O)_n$. The peaks observed at <50 u correspond to monomeric ethanol and its fragmented peaks.

concentrations for the water rich and the 1-butanol rich phase are fixed at 6.4 and 80.2 wt%, respectively. Accordingly, we have measured mass spectra at concentrations close to 6.4 and 80.2 wt% 1-butanol. Figs. 5(a) and (b) show the mass spectra of [1-BuOH] = 6.0 and 80.2 wt%, respectively.

At 6.0 wt%, two kinds of clusters are observed. One is hydrated 1-butanol, and the other is a 1-butanol self-association cluster. The clusters including the same number of 1-butanol molecules (m) in $H^+(C_4H_9OH)_m(H_2O)_n$ are connected by one line. A series of hydrated 1-butanol clusters show a mass distribution similar to that of intrinsic pure water. The magic-number property is also obvious at the peaks m–n representing $H^+(C_4H_9OH)_m(H_2O)_n$: 0–21, 1–20, 2–19, 3–18, 4–17, and 5–16 in Fig. 5(a). This indicates that the hydrogen-bonding network intrinsic to pure water is sustained at 6.0 wt%. On the other hand, the 1-butanol self-association clusters, such as 2–0, 3–0, 4–0,…, 9–0 are also predominant in Fig. 5(a). This implies that, at the cluster level, there has already been microscopic phase separation between the hydrogen-bonded water clusters and the self-associated 1-butanol clusters.

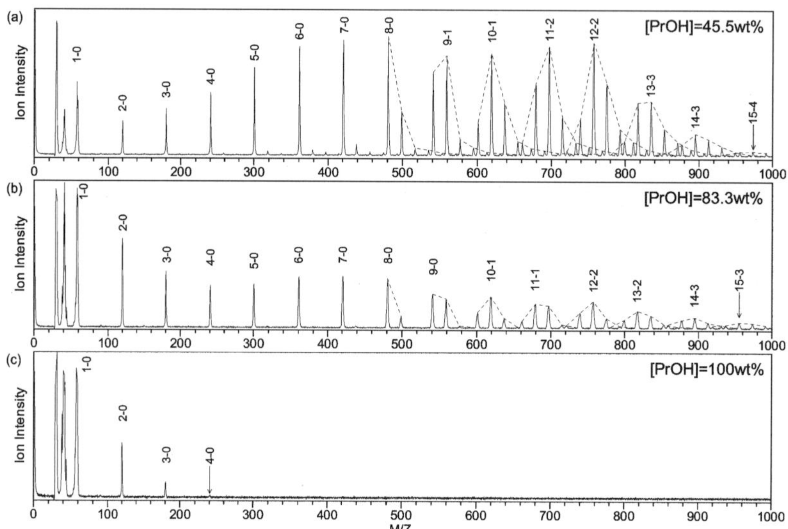

Fig. 4 Mass spectra of clusters isolated from water–1-propanol binary mixtures at various 1-propanol concentrations: (a) 45.5, (b) 83.3, and (c) 100 wt% of 1-propanol. The clusters including the same number of 1-propanol molecules (m) in the clusters $H^+(C_3H_7OH)_m(H_2O)_n$ are connected by lines. The paired numbers represent m–n of $H^+(C_3H_7OH)_m(H_2O)_n$. The peaks observed at <50 u correspond to the fragmented 1-propanol.

For water–ethanol and water–1-propanol binary mixtures, the hydrogen-bonded water clusters and the alcohol self-association clusters do not coexist. When the ethanol and 1-propanol self-association clusters were formed, predominantly at higher alcohol concentrations, the hydrogen-bonding network intrinsic to pure water disintegrated, as shown in Figs. 3 and 4.

At [BuOH] = 80.2 wt%, which is close to the concentration of the 1-butanol rich phase in the two-phase region, most of the observed clusters are formed through 1-butanol self-association, $H^+(C_4H_9OH)_m$, as shown in Fig. 5(b). With the increase in the size of the $H^+(C_4H_9OH)_m$ clusters, the interaction with water molecules also increased. However, the hydration for the 1-butanol self-association clusters is much weaker than that for the ethanol and the 1-propanol self-association clusters.

As shown in Figs. 5(c) and (d), with a further increase in the 1-butanol concentration, the number of 1-butanol self-association clusters is markedly reduced. In comparing Figs. 5(b), (c) and (d), 1-butanol self-association is clearly promoted by the presence of water. However, the hydrogen-bonding network of water is not present as in pure water when the 1-butanol self-association occurs, as shown in Fig. 5(b). This suggests that the alcohol self-association induced by the presence of water is not related to the formation of the hydrogen-bonding network of water molecules.

1.2. Mechanism of the phase separation based on the cluster structures. In comparing the observed cluster structures for water–ethanol, water–1-propanol and water–1-butanol binary mixtures, we can point out the following two characteristics, of which (i) and (ii) correlate with the phase separation of 1-butanol.

(i) Change in the type of clustering with varying alcohol concentration. All of the observed water–alcohol binary mixtures show a similar change in the type of clustering with varying alcohol concentrations. At lower alcohol concentrations, the observed clusters are composed of a hydrogen-bonding network intrinsic to pure water. With an increase in the alcohol concentration, the self-association of alcohol molecules is markedly promoted. With a further increase in the alcohol concentration above *ca.* 80 wt%, the self-association of alcohol molecules is decreased.

In the water–ethanol and water–1-propanol binary mixtures, the hydrogen-bonding network of water is disrupted when the alcohol self-association clusters are generated (Figs. 3(b) and 4(a)). This makes it possible to maintain a single phase even when the alcohol self-association clusters form.

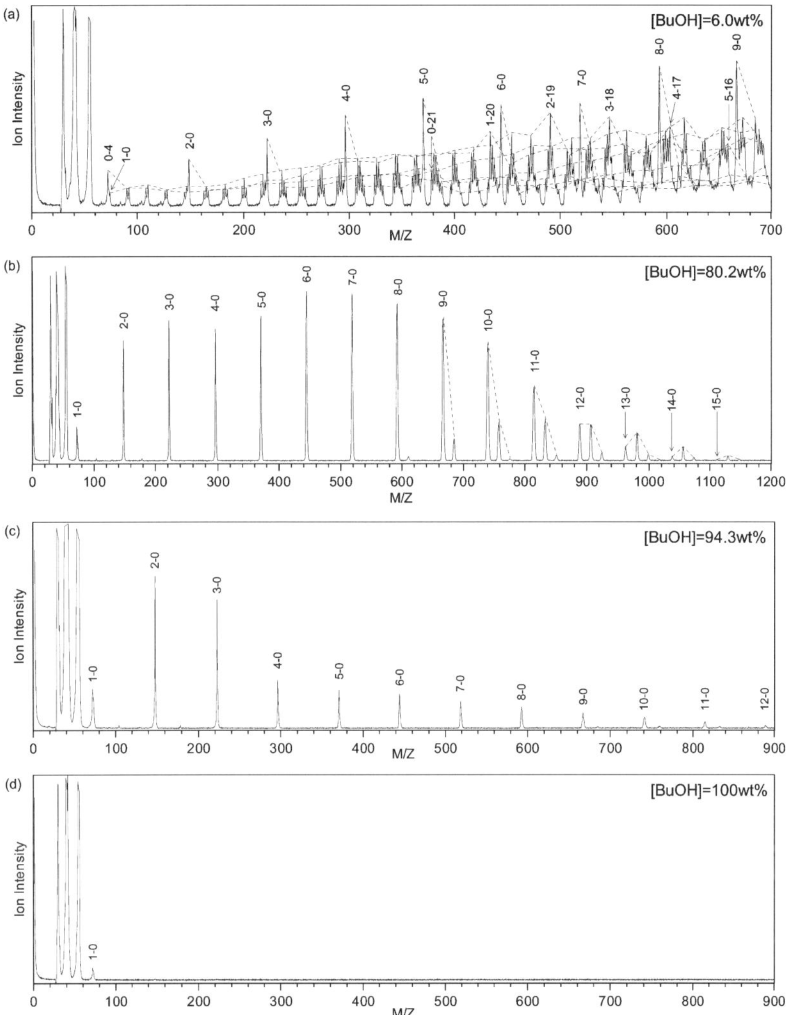

Fig. 5 Mass spectra of clusters isolated from water–1-butanol binary mixtures at various 1-butanol concentrations: (a) 6.0, (b) 80.2, (c) 94.3, and (d) 100 wt% of 1-butanol. The clusters including the same number of 1-butanol molecules (m) in the clusters $H^+(C_4H_9OH)_m(H_2O)_n$ are connected by lines. The paired numbers represent m–n of $H^+(C_4H_9OH)_m(H_2O)_n$. The peaks observed at <60 u correspond to the fragments of 1-butanol.

On the other hand, in the water–1-butanol binary mixtures, 1-butanol self-association clusters are formed without disintegrating the hydrogen-bonding network of water even at low 1-butanol concentrations (Fig. 5(a)). In this binary mixture, phase separation is induced by the coexistence of the 1-butanol self-association clusters and the hydrogen-bonded water clusters. Since 1-butanol self-association clusters are observed in the 1-butanol rich phase in greater numbers (Fig. 5(b)), the 1-butanol rich phase is thought to be generated from the 1-butanol self-associating clusters formed in the water rich phase. In the cases of the water–ethanol and water–1-propanol binary mixtures, the self-associated alcohol clusters do not coexist with the hydrogen-bonded water clusters, because the hydrogen-bonding network of water disintegrates to hydrate the alcohol self-association clusters.

(ii) Difference in the hydration for alcohol self-association clusters. As described above, the induction of a phase separation is dependent on the hydration structures of the alcohol self-association clusters. To see the difference in the hydration of the alcohol self-association clusters, the mass distributions of $H^+(ROH)_m(H_2O)_n$: $ROH = C_2H_5OH$, C_3H_7OH and C_4H_9OH, for each m value were compared. The difference in hydration for the alcohol self-association clusters is notable for $H^+(ROH)_m(H_2O)_n$ with $m \geq 8$ at higher alcohol concentrations (around 80 wt%). The mass distribution of the hydrated clusters $H^+(ROH)_m(H_2O)_n$ for each (m) value ($8 \leq m \leq 15$) as a function of the number of water molecules (n) is shown in Fig. 6. The mass distributions for $ROH = C_2H_5OH$, C_3H_7OH and C_4H_9OH are reproduced from the observed mass spectra in Figs. 3(c), 4(b) and 5(b), respectively.

For a series of alcohol self-association clusters, the hydration number (n) obviously increases with an increase in the number of alcohol molecules (m) in the $H^+(ROH)_m(H_2O)_n$ clusters. Furthermore, the hydration number (n) obviously decreases with an increase in the molecular size of the alcohol from ethanol to 1-butanol. In the $H^+(C_2H_5OH)_m(H_2O)_n$ clusters, the clusters with a large number of ethanol molecules, such as $m = 14$ or 15, show a maximum distribution around $n = 6$ or 7. On the other hand, in the $H^+(C_4H_9OH)_m(H_2O)_n$ clusters, the clusters with $m = 14$ or 15 show a sharp maximum at $n = 1$. This indicates that the 1-butanol clusters are less hydrated than the ethanol or 1-propanol clusters, and that the resulting clusters in the water–1-butanol binary mixtures are more heterogeneous than those in the water–ethanol or in the water–1-propanol binary mixtures.

When we look at the change in the molecular clustering structures starting from pure alcohol and begin increasing the water concentration, the three alcohols observed here have a common characteristic in that the alcohol self-association is heavily promoted by the addition of water. However, the difference in the level of hydration for the alcohol clusters, as shown in Fig. 6, controls the generation of a macroscopic liquid–liquid interface.

1.3. Viscosity coefficient from a microscopic point of view. To show a correlation between the observed cluster structures and the properties of solutions, the viscosity coefficients (η) of the water–ethanol, water–1-propanol and water–1-butanol mixtures were measured at various alcohol concentrations. Fig. 7 shows the η values for these three binary mixtures plotted as functions of the alcohol wt%. In the case of the water–ethanol binary mixtures, η is at a maximum around 40–50 wt%. This viscosity coefficient correlates well with the formation of the ethanol self-association clusters observed in Fig. 3. As shown in Fig. 3, the ethanol self-association clusters grow through mixing with water, and their formation reaches a maximum around 40–50 wt% ethanol. If ethanol self-association clusters are surrounded by water molecules, the lifetime of the ethanol self-association clusters will increase due to a decrease in the rate of molecular exchange. Based on this, the increase in viscosity due to the mixing of water and ethanol can be explained by the promotion of ethanol self-association.

On the other hand, at lower ethanol concentrations, the size of the ethanol self-association clusters decreases and instead the hydrogen-bonding network of water dominates, as observed for the 5.0 wt% ethanol mixture (Fig. 3(a)). When hydrogen-bonding water clusters are formed at higher water concentrations, the resulting water clusters should be surrounded by other water molecules. The fact that the η value for pure water is the smallest in the plots of Fig. 7 indicates that the clustering water molecules are easily exchanged with neighbouring water molecules. Water molecules easily form clusters, but the structures of the water clusters are easily modified in contact with neighbouring water molecules. This is the reason for the small η values even when larger hydrogen-bonding water clusters exist.

The η values for the water–1-propanol and water–1-butanol binary mixtures plotted in Fig. 7 are also larger in the mixture than in each pure liquid. If no clusters were formed by the mixing of water and alcohol, the viscosity coefficient would change in a more linear fashion with the alcohol concentration from 0 to 100 wt%. Using η' to denote an ideal value of the viscosity coefficient, the difference between the observed η and η', $\eta - \eta'$, represents the excess viscosity coefficient that results from alcohol self-association. Assuming that the η' values are estimated from a linear approximation between the value in pure water and that in pure alcohol, $\eta - \eta'$ values were plotted as functions of alcohol concentration in Fig. 8(a). This suggests that alcohol self-association would be promoted around $40 \sim 55$ wt%.

To confirm the relationship between the viscosity and the self-association of alcohol, the peak intensity ratio of the relatively large self-association clusters of alcohol observed in the mass spectra to the relatively small self-association clusters, R_c defined by eqn. (1), is plotted as a function of

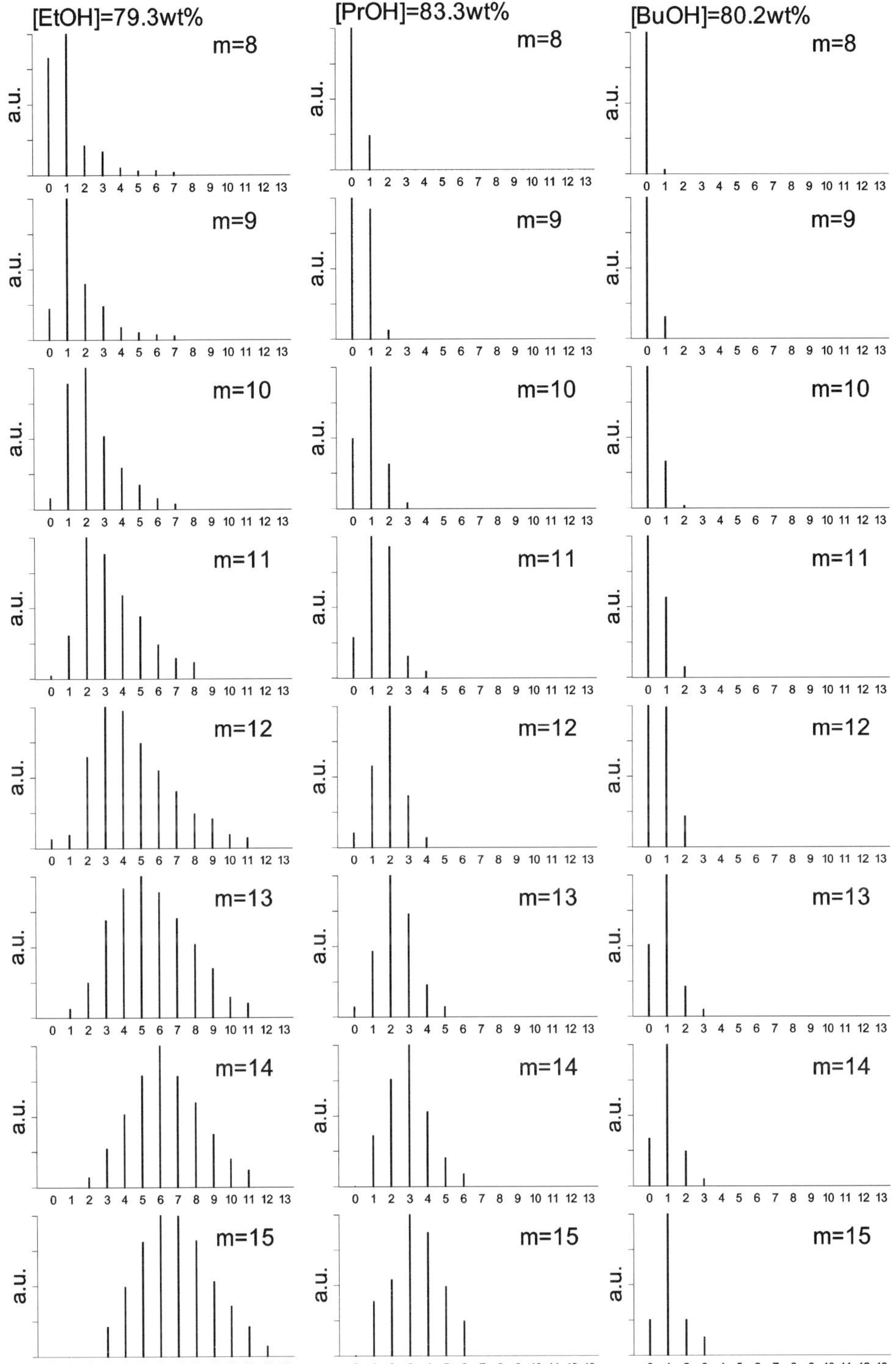

Fig. 6 Mass distributions of the hydrated clusters $H^+(ROH)_m(H_2O)_n$ for m values of $8 \leq m \leq 15$ as a function of the number of water molecules (n). The mass distributions for $ROH = C_2H_5OH$, C_3H_7OH and C_4H_9OH are reproduced from Figs. 3(c), 4(b) and 5(b), respectively.

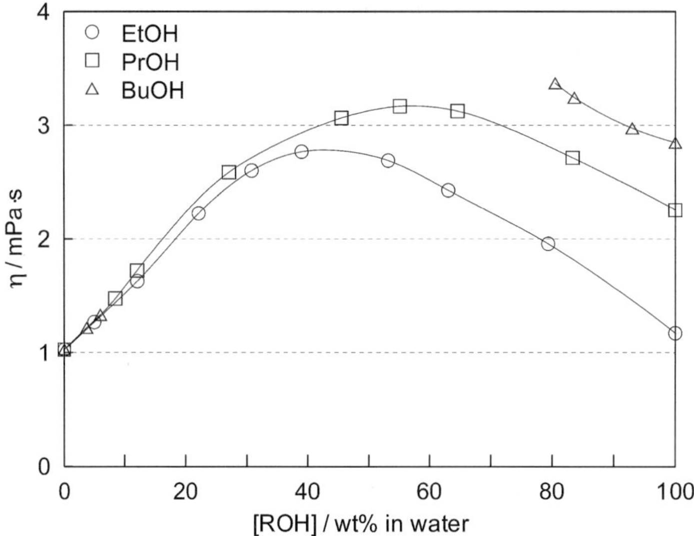

Fig. 7 Viscosity coefficients for water–ethanol, –1-propanol and –1-butanol binary mixtures as functions of alcohol wt%.

alcohol concentration in Fig. 8(b). $I[\mathrm{H}^+(\mathrm{ROH})_m(\mathrm{H_2O})_n]$ represents the peak intensity.

$$R_c = \frac{\sum\limits_{m=5}^{10}\sum\limits_{n=0}^{3} I\left[\mathrm{H}^+(\mathrm{ROH})_m(\mathrm{H_2O})_n\right]}{\sum\limits_{m=2}^{10}\sum\limits_{n=0}^{3} I\left[\mathrm{H}^+(\mathrm{ROH})_m(\mathrm{H_2O})_n\right]} \tag{1}$$

In order to count the clusters with an alcohol rich composition, the $\mathrm{H}^+(\mathrm{ROH})_m(\mathrm{H_2O})_n$ clusters for $0 \leq n \leq 3$ were counted. Since clusters with $m \geq 5$ are hardly formed in pure ethanol and pure 1-propanol, we assumed that the clusters with $m \geq 5$ are relatively large clusters. Furthermore, the clusters with $m > 10$ are mainly formed as hydrated clusters with $n > 3$. Therefore, we counted the clusters with $m \leq 10$ as self-association clusters.

Figs. 8(a) and (b) show that the viscosity of the water–alcohol binary mixtures increases with an increase in the formation of the relatively large alcohol self-association clusters. This indicates that the cluster structures observed in the mass spectra directly reflect the intermolecular interaction in the binary liquid mixtures.

2.   Molecular self-association promoted in binary mixtures rather than in pure liquids

The molecular self-association of alcohol is strongly promoted by mixing with water, as shown in Figs. 3–5. In general, it is believed that this kind of molecular self-association of organic molecules in water is due to hydrophobic effects, and that the hydrogen-bonding network of water coexists with the molecular self-association clusters of organic molecules. The hydrophobic interaction may have provided the assumption that the strong hydrogen-bonding network of water induces the self-association of hydrophobic molecules. Accordingly, it has been thought that water has a special ability to promote molecular self-association.

However, when water is added to pure alcohol, the number of alcohol self-association clusters markedly increases before the formation of the hydrogen-bonding network of water, as shown in Figs. 3–5. Clearly, the large hydrogen-bonding network of water is not indispensable to the molecular self-association of the alcohol. This suggests that the molecular self-association of the alcohol might be promoted by the coexistence of relatively weak alcohol–water intermolecular interactions, which cannot be observed as a cluster. If this is true, the molecular self-association should also be promoted by the presence of other molecules that interact weakly with the alcohol.

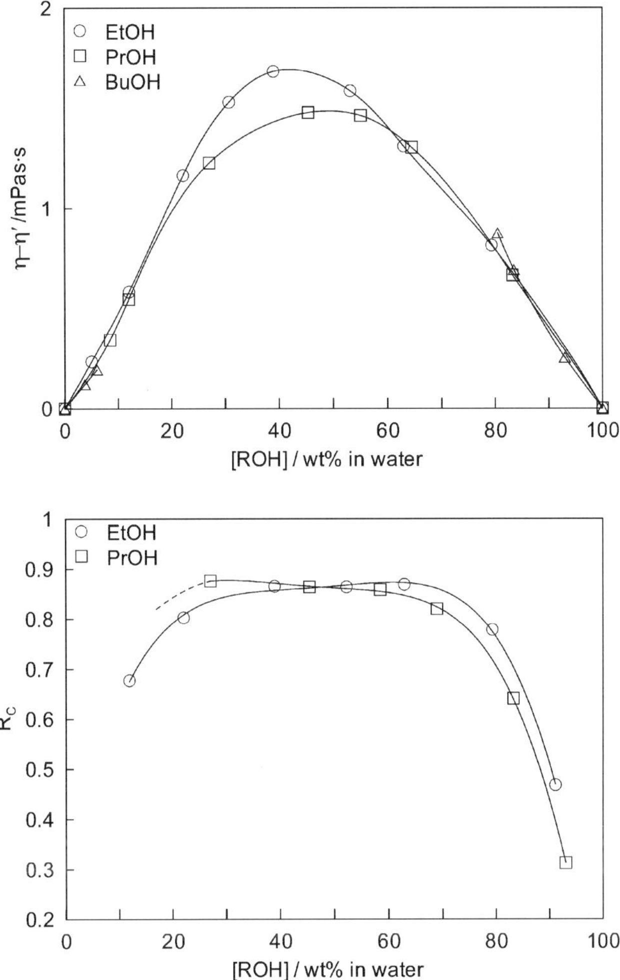

Fig. 8 (a) Excess viscosity coefficient, $\eta - \eta'$, as functions of the alcohol wt%. (b) Ratios of relatively large self-association clusters of ethanol and 1-propanol to all of the self-association clusters of the alcohols, estimated by eqn. (1), as functions of the alcohol wt%.

For a series of binary mixtures between primary alcohols, this idea was demonstrated experimentally.

2.1. Mass spectrometric analysis of clusters in binary alcohol mixtures. As a series of binary alcohol mixtures, we focused on mixtures of methanol (MeOH) with another alcohol (ethanol (EtOH), 1-propanol (PrOH), 1-butanol (BuOH), or 1-pentanol (PeOH)), and mixtures of EtOH with another alcohol (PrOH, BuOH, or PeOH).

Molecular self-association in the binary alcohol mixtures was investigated through mass spectrometric analyses of clusters using the mass spectrometer shown in Fig. 1. The mass spectra for a series of binary mixtures of methanol with other alcohols are shown in Fig. 9. For the MeOH–EtOH mixture, the clusters are equally composed of MeOH and EtOH molecules (Fig. 9(a)). In the cluster $H^+(MeOH)_m(EtOH)_n$ represented by m–n, the peak intensity ratio of the two-molecule clusters 2–0, 1–1 and 0–2, seems to be related to the probability distribution (1 : 2 : 1). Similarly, the distribution of the three-molecule clusters 3–0, 2–1, 1–2 and 0–3, also seems to be controlled by the probability distribution. This ensures that the MeOH–EtOH mixture is nearly equivalent to complete mixing at the molecular level.

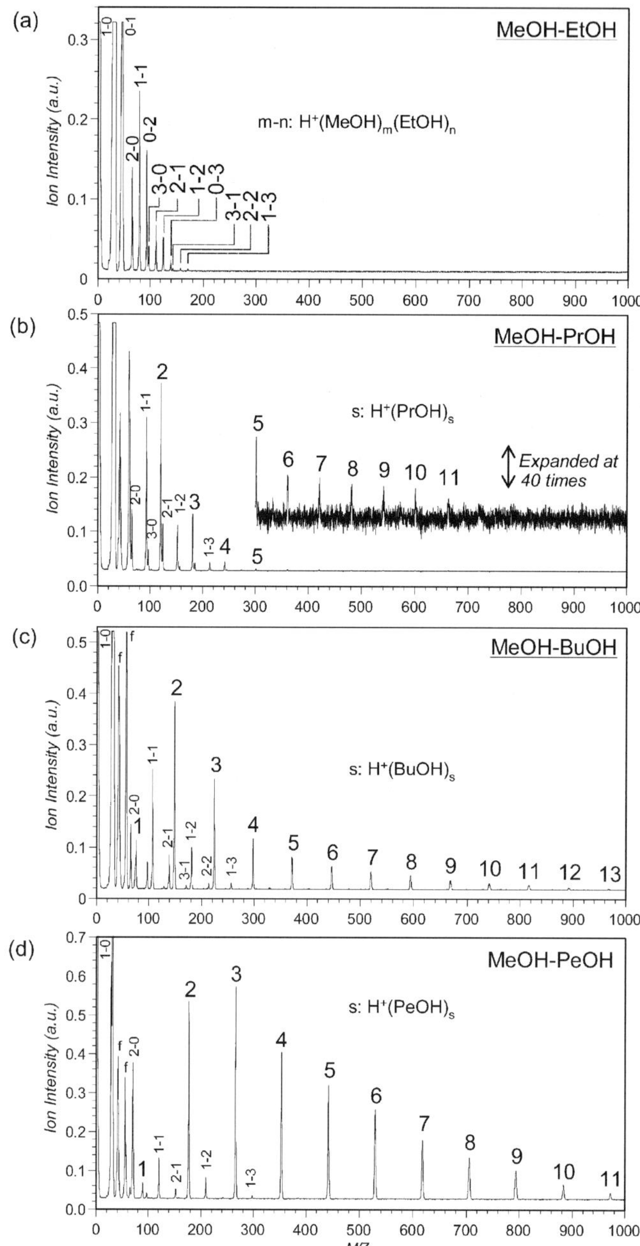

Fig. 9 Mass spectra of clusters generated from methanol–another alcohol (MeOH–ROH: 70 ml + 30 ml) binary mixtures: (a) MeOH–EtOH, (b) MeOH–PrOH, (c) MeOH–BuOH, and (d) MeOH–PeOH. The paired numbers m–n and the single numbers written above the peaks represent m–n for $H^+(MeOH)_m(ROH)_n$ and s for $H^+(ROH)_s$, respectively; f represents fragmented species of alcohol molecules.

When the molecular size of the alcohol mixed with methanol was increased from PrOH to PeOH, the self-association clusters composed of only the larger alcohol molecules became more prominent (Figs. 9(b)–(d)). The MeOH molecules were hardly present in the observed BuOH- and PeOH-self-association clusters. This indicates that their interactions with MeOH were relatively weak.

The same kind of molecular self-association promoted in the binary mixtures was also clearly demonstrated in the binary mixtures of ethanol with other alcohols (PrOH, BuOH, or PeOH). For the EtOH–PrOH mixture, the resulting clusters were equally composed of EtOH and PrOH molecules. The self-association clusters composed of only the larger alcohol molecules became more prominent with an increase in the size of the alcohol molecules in the ethanol mixture.

When the difference in the molecular size of the alcohols is small, *e.g.*, MeOH–EtOH and EtOH–PrOH, one component can be solvated by the other component, which leads to nearly complete mixing (mutual solvation) at the molecular level. On the other hand, when the difference in the molecular size of the two alcohols is relatively large, *e.g.*, MeOH–PeOH, MeOH–BuOH and EtOH–PeOH, the self-association of the larger alcohol molecules becomes more favorable to mutual solvation. The smaller alcohol (MeOH or EtOH) surrounds the self-association clusters. In the absence of MeOH or EtOH, the self-association clusters of the larger alcohol molecules were hardly formed, as shown in Fig. 5(d).

2.2. Solvent-assisted molecular self-association and 'complementary relationship'. As shown in Fig. 9, the molecular self-association of organic molecules is obviously promoted not only by mixing with water but also by mixing with non-aqueous solvents. We have already reported that the size of the solvent molecules is a factor in controlling the molecular self-association and solvation structures.[24,27–29] In order to explain this solvent-assisted molecular self-association behaviour, we would like to propose a type of interaction we call 'complementary relationship' as part of the intermolecular interactions in binary mixtures.

In pure alcohol, each intermolecular interaction is equivalent, and the neighbouring molecules are exchangeable with one another. This leads to a random structure (complete mixing) rather than a self-association cluster structure. Binary alcohol mixtures where the size of the alcohol molecules is similar, *e.g.*, MeOH–EtOH and EtOH–PrOH, also belong to this category.

On the other hand, in binary mixtures where the size of the component alcohol molecules is considerably different, the intermolecular interactions are also considerably different. For example, in the MeOH–PeOH binary mixtures, the MeOH–PeOH interaction is less stable than the PeOH–PeOH interaction due to the lack of inter-alkyl interactions. In this case, in order to complement the loss of stabilization of the PeOH molecules caused by the contacts with the MeOH molecules through PeOH–PeOH interactions, the self-association of the PeOH molecules would be promoted in the binary mixture.

The cluster structure is thus determined as a result of the balance of interactions working at the same time in a solution. Therefore, it is possible for molecular self-association to be more favourable in mixtures than in pure liquids. The complementary relationship indicates that molecular self-association is promoted by the presence of other relatively weakly interacting molecules. Molecular self-association is controlled not only by the *absolute* interaction energy, but also by the *relative* interaction energy in a solution. The various cluster formation processes are likely to be controlled by this factor as an inherent property of solutions.

We have already reported on the clustering in electrolyte solutions that also followed the complementary relationship as defined here.[17,18] In a salt solution, each ion interacts with the solvent molecules and the counterions. According to the complementary relationship, the solvation of each ion is influenced by the interactions with the counterions. For example, in a methanol solution of LiCl, solvated Li^+ and Cl^- are present. When we added 18-crown-6 (crown ether) to this methanol solution of LiCl, the solvation of Cl^- by the methanol molecules were markedly increased. This can be explained by the complementary relationship. The 18-crown-6 forms a stable complex with Li^+, which reduces Li^+–Cl^- interactions. As a result of the complementary relation, the reduction of the electrostatic interactions experienced by Cl^- was complemented by the increase in the solvation of Cl^- by the methanol molecules.[17]

2.3 Non-ideal mixing and cluster structures. The deviation from an ideal solution is seen in partial vapor pressures as functions of mole fractions of the components of the solution. According to Raoult's law, the partial vapour pressure of one component in an ideal solution is proportional to the mole fraction of that component.[30] For the binary mixtures of methanol with the larger alcohols (EtOH, PrOH, BuOH or PeOH), the partial vapour pressures of methanol (P_M) as functions of the

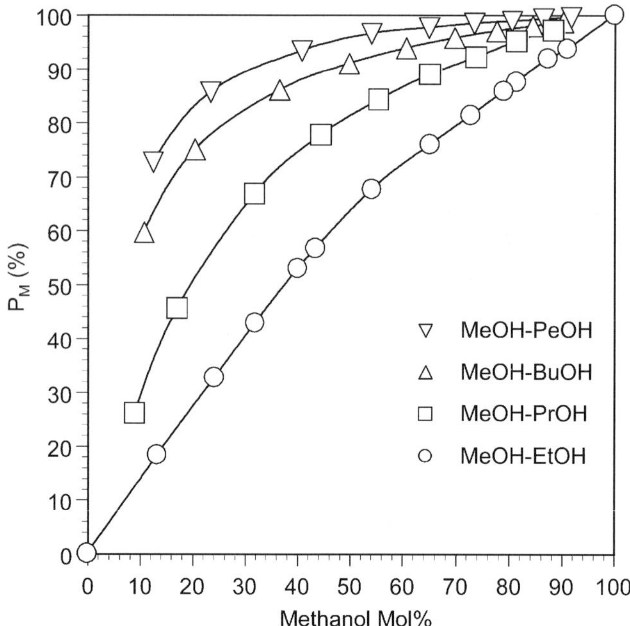

Fig. 10 Partial vapor pressures of methanol (P_M) in methanol–another alcohol binary mixtures. The values of P_M (%) in MeOH–EtOH, MeOH–PrOH, MeOH–BuOH and MeOH–PeOH under a constant total pressure of 760 Torr are plotted as functions of MeOH mol%. The values of P_M are referred from ref. 20.

methanol mole fractions are shown in Fig. 10.[20] The P_M of the MeOH–EtOH mixtures is the closest to a linear relationship with the methanol mole fraction. However, the deviation from Raoult's law is enhanced with the increasing molecular size of the alcohol in the mixture. The same trend was seen for the binary mixtures of ethanol with the larger alcohols (PrOH, BuOH or PeOH).

These results indicate that when the difference in the molecular size of two components in a mixture of alcohols is large, the deviation from an ideal solution increases. This is strongly related to the cluster structures observed in Fig. 9. In the binary mixtures that deviated furthest from ideal mixing, the molecular self-association was more prevalent. This non-ideal mixing is the result of the difference in the intermolecular interactions among the components. Owing to the difference in the intermolecular interactions in the binary mixtures, it is clear that the microheterogeneity in solutions in the form of molecular self-association is generated by the complementary relationship.

Conclusion

By using specially designed mass spectrometry for clusters isolated from liquid droplets, we have demonstrated that the macroscopic phase separation for a water–1-butanol binary mixture is related to the cluster structures. The 1-butanol self-association clusters were observed at a high frequency in both the water rich and the 1-butanol rich phases in the two-phase region. In both phases, the 1-butanol self-association clusters were the result of the presence of water. As a mechanism for this phase separation, we have pointed out that the 1-butanol rich phase is generated by the aggregation of 1-butanol self-association clusters. When the water rich phase, where the main structure is composed of hydrogen-bonded water, is saturated with the 1-butanol self-association clusters, a 1-butanol rich phase will be generated. In the cases of water–smaller alcohol (ethanol, or 1-propanol) binary mixtures, the formation of the alcohol self-association clusters was also promoted by the presence of water. However, the hydrogen-bonding network of water was broken to hydrate the alcohol self-association clusters of ethanol and 1-propanol. The macroscopic phase separation is thus controlled by the level of hydration for the alcohol clusters.

We have also discussed the role of water in the self-association of alcohol molecules. As observed in Figs. 3–5, the hydrogen-bonding network of water is not indispensable in the formation of the alcohol self-association clusters. The same kind of molecular self-association was observed for non-aqueous binary mixtures that deviated far from ideal mixing (Figs. 9 and 10). As an inherent property of solutions, we have demonstrated that molecular self-association is promoted by mixing with another liquid. The complementary relationship among the intermolecular interactions in binary mixtures determines the level of microheterogeneity in the solutions. The formation of 1-butanol clusters both in the water rich phase and in the 1-butanol rich phase can be explained by the complementary relationship of the intermolecular interactions.

Here we have also demonstrated that the physicochemical properties (viscosity and partial vapour pressure) of solutions can be explained by the cluster structures in the solutions. It should be noted that the microscopic interface at the cluster level is generally formed in macroscopically homogeneous solutions. The physicochemical properties of solutions can be explained more precisely when microheterogeneity is considered.

References

1   *Handbook of Chemistry (Fundamental)*, Maruzen, Tokyo, ed. The Chemical Society of Japan, 4th edn., 1993, ch. 8.
2   I. N. Levine, *Physical Chemistry*, McGraw–Hill Kogakusha Ltd. Tokyo, 1978, pp. 298–299.
3   A. Wakisaka, H. A. Carime, Y. Yamamoto and Y. Kiyozumi, *J. Chem. Soc., Faraday Trans.*, 1998, **94**, 369.
4   A. Wakisaka, S. Komatsu and Y. Usui, *J. Mol. Liq.*, 2001, **90**, 175.
5   T. Fukasawa, Y. Tominaga and A. Wakisaka, *J. Phys. Chem. A*, 2004, **108**, 59.
6   N. Nishi and K. Yamamoto, *J. Am. Chem. Soc.*, 1987, **109**, 7353.
7   N. Nishi, K. Koga, C. Ohshima, K. Yamamoto, U. Nagashima and K. Nagami, *J. Am. Chem. Soc.*, 1988, **110**, 5246.
8   M. Matsumoto, N. Nishi, T. Furusawa, M. Saita, T. Takamuku, M. Yamagami and T. Yamaguchi, *Bull. Chem. Soc. Jpn.*, 1995, **68**, 1775.
9   S. Dixit, J. Crain, W. C. K. Poon, J. L. Finney and A. K. Soper, *Nature*, 2002, **416**, 829.
10   K. Nishikawa, Y. Kodera and T. Iijima, *J. Phys. Chem.*, 1987, **91**, 3694.
11   K. Nishikawa, H. Hayashi and T. Iijima, *J. Phys. Chem.*, 1989, **93**, 6559.
12   P. G. Kusalik, A. P. Lynbartsev, D. L. Bergman and A. Laaksonen, *J. Phys. Chem. B*, 2000, **104**, 9526.
13   K. Nakanishi, K. Ikari, S. Okazaki and H. Touhara, *J. Chem. Phys.*, 1984, **80**, 1656.
14   H. Tanaka, K. Nakanishi and H. Touhara, *J. Chem. Phys.*, 1984, **81**, 4065.
15   W. Blokzijl and J. B. F. N. Engberts, *Angew. Chem., Int. Ed. Engl.*, 1993, **32**, 1545.
16   J. B. F. N. Engberts and M. J. Blandamer, *Chem. Commun.*, 2001, 1701.
17   S. Mochizuki and A. Wakisaka, *J. Phys. Chem. A*, 2002, **106**, 5095.
18   T. Megyes, T. Radnai and A. Wakisaka, *J. Phys. Chem. A*, 2002, **106**, 8059.
19   L. Benjamin and G. C. Benson, *J. Phys. Chem.*, 1963, **67**, 858.
20   V. B. Kogan, *Data Book of Vapour–Liquid Equilibrium*, Kodansha Scientific, Tokyo, 1974.
21   R. M. Salinas, F. G. Sánchez and G. E. Jiménez, *Fluid Phase Equilibria*, 2003, **210**, 319.
22   *Handbook of Chemistry (Fundamental)*, ed. The Chemical Society of Japan, Maruzen, Tokyo, 4th edn., 1993, pp. II–46.
23   P. Wang and A. Anderko, *Ind. Eng. Chem. Res.*, 2003, **42**, 3495.
24   A. Wakisaka, Y. Akiyama, Y. Yamamoto, T. Engst, H. Takeo, F. Mizukami, K. Sakaguchi and H. Jones, *J. Chem. Soc., Faraday Trans.*, 1996, **92**, 3539.
25   S. Wei, Z. Shi and A. W. Castleman, Jr., *J. Chem. Phys.*, 1991, **94**, 3268.
26   X. Yang and A. W. Castleman, Jr., *J. Am. Chem. Soc.*, 1989, **111**, 6845.
27   A. Wakisaka, Y. Yamamoto, Y. Akiyama, H. Takeo, F. Mizukami and K. Sakaguchi, *J. Chem. Soc., Faraday Trans.*, 1996, **92**, 3339.
28   S. Mochizuki, Y. Usui and A. Wakisaka, *J. Chem. Soc., Faraday Trans.*, 1998, **94**, 547.
29   F. Rastrelli, G. Saielli, A. Bagno and A. Wakisaka, *J. Phys. Chem. B*, 2004, **108**, 3479.
30   P. W. Atkins, *Physical Chemistry*, Oxford University Press, 6th edn., 1988, ch. 7.

Enzyme hyperactivity in AOT water-in-oil microemulsions is induced by 'lone' sodium counterions in the water-pool

Christopher Oldfield,*[a] Robert. B. Freedman[b] and Brian H. Robinson[c]

[a] Chemistry Research Group, MicroScience Technologies Ltd., Pentlands Science Park, Edinburgh, UK EH26 OPZ. E-mail: christopher.oldfield@microsciencetechnologies. co.uk; Tel: 0131 445 6211
[b] Department of Biological Sciences, University of Warwick, Coventry, UK CV4 7AL
[c] School of Chemical Sciences, University of East Anglia, Norwich, UK NR4 7RJ

Received 29th April 2004, Accepted 7th May 2004
First published as an Advance Article on the web 13th October 2004

Water-in-oil microemulsions are thermodynamically stable single-phase dispersions of water and surfactant within a continuous oil phase. The classical ternary system, based on the surfactant sodium bis(2-ethylhexyl)sulfosuccinate ('AOT'), water and an alkane such as n-heptane, is an optically transparent monodispersion of spherical water-droplets coated with a close-packed surfactant monolayer and the droplet radius is, to a good first approximation, directly proportional to the molar water:surfactant ratio, R. Enzymes dissolved in the water droplets retain activity and stability. These systems have attracted interest as media for biotransformations. Principally based upon studies in AOT-stabilized w/o microemulsions, a peculiar feature of the kinetics of enzyme-catalyzed reactions has long been apparent: the reaction rate characteristically increases from around zero at $R = 3$, through a maximum, in the range $R = 10$–20, and thereafter decreases again, so that plots of rate *vs.* R are characteristically 'bell-shaped'. Furthermore, at optimal R, enzymes seem to be 'hyperactive', *i.e.*, they are more active, by a modest but significant factor of 2–3-fold, than in aqueous solution. In this paper we propose the hypothesis that this kind of R-dependence arises because of the presence of freely mobile lone surfactant counterions (Na^+) within the water-pool. These ions have no charge partners within the water pool and consequently have a high electrochemical potential. According to our model, lone counterions facilitate the hydrolysis of ester or amide substrates, for example, by stabilizing the tetrahedral intermediate formed during the reaction through ion-pairing with the carbonyl oxygen of the substrate, thus facilitating transfer of negative charge from the carbonyl carbon as it is attacked by the incoming nucleophile. An expression for the relationship between the concentration of free counterions in the water-pool and the compositional parameter R leads directly, through Debye–Hückel theory, to an expression for the relationship between the reaction rate and R,

$$\log k(R) = \log k^{\circ} + C(1/R)^{\frac{1}{2}}$$

where $k(R)$ is the rate constant at some finite R, k° is the rate constant extrapolated to $R = \infty$ and C is an R-independent coefficient. For enzymes that display bell-shaped kinetics, such as bovine α-chymotrypsin and *Chromobacterium viscosum* lipase, the descending part of the plot (*i.e.* from optimal R to high R) obeys this equation very well. Inspection of the above equation shows that the rate constant, $k(R)$ is greater than k°. Furthermore it is reasonable to equate k° with k_{aq}, the aqueous solution value of k since the condition $R = \infty$ may be equated with the condition of infinite dilution with respect to counterions, so eliminating their specific effect on the kinetics. It follows from the inequality, $k(R) > k^{\circ} \cong k^{aq}$, that the enzyme is 'hyperactive' in the microemulsion compared with aqueous solution. We show that this is indeed the case for the chymotrypsin-catalyzed hydrolysis of N-*trans*-cinnamoylimidazole and the lipase catalyzed hydrolysis of 4-nitrophenyl acetate. The tailing off of enzyme activity at low R (<10) is most likely due to conformational immobilization, probably due to partial dehydration in these low-water preparations (water activity, a_{w}, drops off rapidly below $R = 15$). We show that the reaction of glycylglycine with 4-nitrophenyl acetate, a 'hyperactive' non-enzymic reaction, does not suffer from this effect and obeys the above equation across the whole range of R.

Introduction

Water-in-oil (w/o) microemulsions are thermodynamically stable dispersions of water and surfactant in a continuous oil medium. The oil–water interface is coated with a close-packed monolayer of surfactant molecules and as a result there is no direct oil–water contact. Herein lies the thermodynamic stability of microemulsions: since there is no direct oil–water contact, hydrophobic exclusion does not operate and the oil and water phases appear to be completely miscible.

Water-in-oil microemulsions based on the anionic surfactant sodium bis(2-ethylhexyl) sulfosuccinate (AOT) are very well-characterized, both physicochemically and as host media for enzyme-catalyzed reactions. AOT stabilizes microemulsions consisting of water (including solutions of enzymes and other chemicals at low or moderate ionic strengths) and an oil, which is typically a medium chain-length alkane such as *n*-heptane, at physiologically relevant temperatures without the need for a cosurfactant. Typically, the aqueous solution is added to a solution of the surfactant dissolved in the oil; the perfectly transparent microemulsion forms within seconds upon shaking.

Physical characterization of AOT w/o microemulsions using nmr spectroscopy,[1] low-angle X-ray scattering[2] and analytical ultracentrifugation[3] has shown that, for (water + AOT) volume-fractions <0.2 approx., the water is present as a monodispersion of approximately spherical droplets surrounded by a close-packed surfactant monolayer and that, to a good first-approximation, the droplet radius is directly proportional to the molar water : AOT ratio, R. In the physiologically relevant temperature range the maximum value for R is about 55 mol mol^{-1}. Furthermore, when R is fixed the droplet concentration is directly proportional to the (water + AOT) volume fraction; thus the droplet concentration can be varied independently of R (*i.e.* of the droplet size), so allowing these two parameters to be varied independently of each other. The well-behaved nature of this ternary system explains its popularity as a prototype microemulsion system for enzymological studies.

Enzymes solubilized in AOT w/o microemulsions retain activity and usually show good stability. The concept of the enzyme-containing water droplet as a microreactor stably dispersed within a continuous oil phase is an attractive one for biotechnologists interested in biotransformations of water-insoluble and amphipathic substrates. Consequently the development of these systems for such purposes is an area of great interest and active research.

One of the most interesting general aspects of enzyme behaviour in AOT microemulsions is the "bell-shaped" dependence of activity on R. This phenomenon was first described by Martinek and co-workers[4] for bovine α-chymotrypsin (α-CT) with N-*trans*-cinnamoyl imidazole (t-CNI) as the substrate, and since confirmed for both this enzyme[5] and many others (see ref. 6 for a review). For a typical enzyme, the activity increases from around zero at low R (<5), through a maximum at around $R = 10$ and thereafter decreases again, giving rise to the characteristic bell-shaped profile. Furthermore, the activity in the optimal range of R is typically higher than that in aqueous solution under comparable conditions (pH, temperature), by a modest but significant factor of 2- or 3-fold. This phenomenon is known as hyperactivity.

Despite the mass of work[6] describing the phenomenon, the origin of hyperactivity is poorly understood. Martinek and co-workers[5,7] proposed the interesting theory that in the microemulsion the enzyme becomes trapped in a unique highly active, conformation that it is not accessible to the enzyme (or at least not highly populated) in purely aqueous solutions.

From our point of view, a clue to an alternative explanation for this behaviour came from some of our earlier work.[8] We showed that in AOT microemulsions the pK_a of the highly water-soluble weak acid, 4-nitrophenol 2-sulfonate (NPS), $pK_a^o = 6.7$, varies linearly with $(1/R)^{\frac{1}{2}}$ according to the expression,

$$pK_a = pK_a^o - C_1 \, (1/R)^{\frac{1}{2}} \tag{1}$$

where C_1 is an R-independent coefficient and pK_a^o has the same value as that measured in aqueous solution at infinite dilution (zero ionic strength). The original plot is reproduced in Fig. 1.

We presented this result without comment, simply noting that it was intriguing (the focus of the paper was an analysis of the behaviour of 4-nitrophenol (4NP), which needed to be understood in order to correctly interpret the kinetics of enzyme-catalyzed reactions followed by monitoring the release of 4-nitrophenol (4-NP) from synthetic ester substrates).

Actually, in order to get to eqn. (1) we had noted, (i) that the pK_a of NPS was strongly dependent on R (the ΔpK_a is 1.5 between $R = 5$ and $R = 50$); (ii) that this neutral–acid equilibrium is anyway extremely sensitive to ionic strength in aqueous solution; (iii) in our measurements we had eliminated the possibility of an uncontrolled R-dependence of pH by using highest grade AOT (free of acidic impurities) and by paying careful attention to buffering, and (iv) we were convinced by UV-visible spectroscopic data that NPS is confined to the water pool and does not partition into the surfactant layer or oil-phase.

Therefore we concluded that the only reasonable explanation for the observed dependence of pK_a on R in microemulsions was an ionic strength effect where, specifically, the ions in question are freely mobile lone surfactant counterions (Na^+) within the water-pool. Eqn. (1) then proceeds directly from this conclusion, as explained in the Theory section below.

More importantly, we show here that this behaviour is not confined to acid–base equilibria, and that a variety of chemical and enzyme catalyzed reactions, including the αCT-catalyzed hydrolysis of tCNI, obey the empirical expression,

$$\log k(R) = \log k^o + C_2 \, (1/R)^{\frac{1}{2}} \tag{2}$$

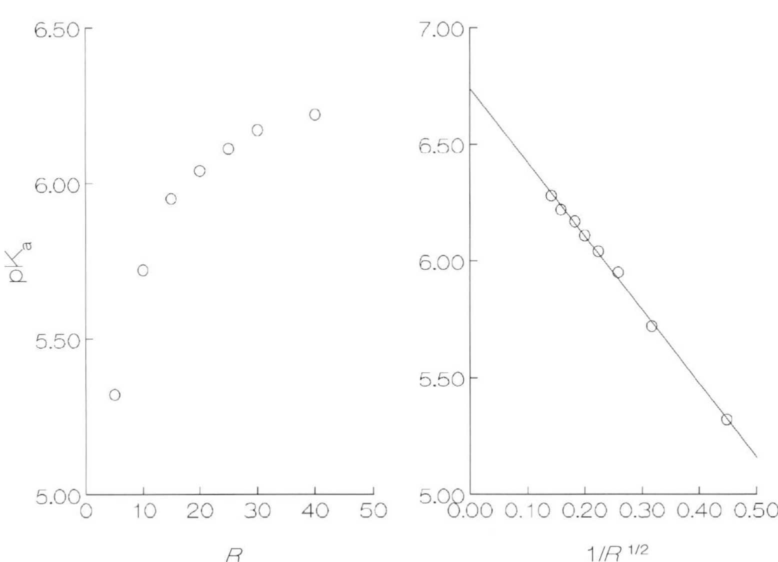

Fig. 1 Dependence of pK_a on R and on $(1/R)^{\frac{1}{2}}$ for NPS in microemulsions (0.1 mol dm^{-3} AOT) buffered with MES (1.8 mmol dm^{-3}) at 25 °C. Redrawn from ref. 3.

where $k(R)$ is a suitable rate constant measured at some finite R, k° is the ordinate intercept value, corresponding to the value of k extrapolated to infinite R, and C is an R-independent coefficient that can take a positive or negative value.

In contrast to the acid–base behaviour of NPS, the kinetics of enzyme catalyzed reactions would not *a priori* be expected to obey eqn. (2), since these reactions are usually not subject to a primary ionic strength effect in aqueous solution. In order to explain this behaviour we hypothesized that lone water-pool counterions are also able to modulate reaction rates directly, by acting as local high-potential positive charges that stabilize negatively charged transition states formed, for example, in the hydrolysis of esters and amide substrates by enzymes such as bovine α-chymotrypsin (αCT) and lipase, and that they are able to do this because they are unshielded by partner cations, these being confined at the interface. We reasoned that charge separation within the microemulsion environment both explains effects that are observed in microemulsions, but cannot be reproduced in aqueous solution, and also how enzymes in such an environment might be hyperactive.

In this paper we first of all show that another highly water-soluble weak acid, pyranine, also obeys eqn. (1), showing that the behaviour originally observed for NPS is a general phenomenon. We go on to show that the following chemical reactions, the aminolysis of 4-nitrophenyl acetate (NPC_2) by glycylglycine ('GLYGLY') and, the lipase catalyzed hydrolysis of NPC_2 and the αCT-catalyzed hydrolysis of tCNI all obey eqn. (2). We finally discuss how the microemulsion ionic environment is able to modulate the rates of actions that are insensitive to primary ionic strength effects in aqueous solution.

Nomenclature

Concentrations in microemulsions are given in mol dm^{-3ov}, *i.e.*, in overall (total microemulsion) concentration, or in mol dm^{-3aq}, *i.e.* as the concentration within the aqueous subphase, the latter used only for water-soluble (non-partitioning substrates). Rate constants in aqueous solution (water) are denoted by the suffix 'w', thus k_{+2w}; rate constants in microemulsions are likewise denoted by the suffix 'm'. Second-order rate constants for microemulsion reactions have units of 'dm^3 (overall microemulsion volume) $mol^{-1}\ s^{-1}$', thus the units of k_{+2m} are $dm^{3ov}\ mol^{-1}\ s^{-1}$.

It is not possible to measure the pH of a microemulsion directly, *e.g.* by means of a glass electrode, and we apply the convention that the pH of the microemulsion is equal to pH_{buff}, the pH of the buffered aqueous solution used to prepare it. Therefore apparent pK_a values, $pK_{a\ app}$, are reported.

Theory

The purpose of this section is firstly to recall the basic equations describing the relationship between acid–base behaviour, and the rates of chemical reactions, and ionic strength. We then show how these equations can be adapted to predict their behaviour in AOT microemulsions.

In aqueous solution, the pK_a of a neutral weak acid ($AH \rightleftharpoons A^- + H^+$) is dependent on the ionic strength, I, where the pK_a is defined simply as,

$$pK_a = pH - \log\left(\frac{[A^-]}{[HA]}\right) \tag{3}$$

According to the Debye–Hückel Limiting Law,[9] there is a linear relationship between the $pK_a(I)$ measured at finite ionic strength, I, and the square root of the ionic strength, as long as $I < 0.01$, approximately,

$$pK_a(I) = pK_a^{\circ} - A|z_1 z_2|I^{\frac{1}{2}} \tag{4}$$

where pK_a° is the value of $pK_a(I)$ extrapolated to zero ionic strength, coefficient $A = 0.509$ and $|z_1 z_2|$ is the absolute value of the product of the charges on the acid and base species (the proton and the conjugate base of the weak acid, in this case). For neutral weak acid equilibria, therefore, plots of pK_a *vs.* $I^{\frac{1}{2}}$ are linear with slope $-A|z_1 z_2|$ and ordinate intercept pK_a°, as long as the ionic strength is not too high. The ionic strength itself is defined in the usual way,

$$I = \frac{1}{2}\sum c_i z_i^2 \tag{5}$$

where c is the molar concentration and z is the charge of each of the i ionic species present in the solution.

The same theory provides a framework for understanding the influence of ionic strength on reaction rates. For a bimolecular reaction of the type $A + B \rightarrow$ products, where z_A and z_B are the charges on A and B in magnitude and sign,

$$\log k(I) = \log k^\circ + 2A z_A z_B I^{\frac{1}{2}} \tag{6}$$

so that a plot of $\log k(I)$ vs. $I^{\frac{1}{2}}$ is linear with ordinate intercept $\log k^\circ$ (corresponding to the rate constant at infinite dilution) and slope $2A z_A z_B$. Since this time the signs of the charges on the reactant are taken into account, the slope may take a positive or a negative value.

In w/o microemulsions stabilized by AOT, free sodium counterions are present within the water-pool[10] and these must, therefore, contribute to the ionic strength. In the absence of added electrolyte, the water-pool ionic strength must arise solely from the presence of these cations. Let $[\mathrm{Na}^+]^{\mathrm{ov}}$ be the overall concentration of free counterions, and let $[\mathrm{AOT}]^{\mathrm{ov}}$ be the overall surfactant concentration (units are mol dm^{-3} microemulsion, mol dm$^{-3\mathrm{ov}}$). Therefore, $[\mathrm{Na}^+]^{\mathrm{ov}} = \alpha_R [\mathrm{AOT}]^{\mathrm{ov}}$, where α_R is the fractional degree of dissociation for any given R [α_R must lie between 0 (no dissociation) and 1 (complete dissociation)]. Let ϕ^{aq} be the volume-fraction of the microemulsion aqueous sub-phase. Therefore, $[\mathrm{Na}^+]^{\mathrm{aq}} = [\mathrm{Na}^+]^{\mathrm{ov}}/\phi^{\mathrm{aq}}$. Since, by definition, $[\mathrm{H_2O}]^{\mathrm{ov}} = [\mathrm{AOT}]^{\mathrm{ov}}/R$, then $[\mathrm{Na}^+]^{\mathrm{aq}} = \alpha_R [\mathrm{H_2O}]^{\mathrm{ov}}/\phi_{\mathrm{aq}}\, R$. Since $[\mathrm{H_2O}]^{\mathrm{ov}}/\phi^{\mathrm{aq}} = [\mathrm{H_2O}]^{\mathrm{aq}} = 55.5$ mol dm^{-3},

$$[\mathrm{Na}^+]^{\mathrm{aq}}/\mathrm{mol\ dm}^{-3\mathrm{aq}} = 55.5 \alpha_R (1/R) \tag{7}$$

From eqn. (5), for a monovalent ion, the ionic strength of the water-pool due to this single ionic species is $[\mathrm{Na}]^{\mathrm{aq}}/2$, and therefore,

$$I_{\mathrm{Na}^+} = 27.8 \alpha_R (1/R) \tag{8}$$

Substitution of eqn. (8) in eqn. (4) gives the following expression,

$$pK_a = pK_a^\circ - A|z_1 z_2|(27.8 \alpha_R/R)^{\frac{1}{2}} \tag{9}$$

and if α_R is independent of R, all constants can be grouped together as a single constant, C_1:

$$pK_a = pK_a^\circ - C_1 (1/R)^{\frac{1}{2}} \tag{1}$$

Therefore, if free water-pool surfactant counterions determine microemulsion ionic strength, a plot of pK_a vs. $(1/R)^{\frac{1}{2}}$ should be linear with negative slope intersecting the ordinate at pK_a° (since $I_{\mathrm{Na}^+} \rightarrow 0$ as $R \rightarrow \infty$).

Similarly, substitution of eqn. (8) in eqn. (6) gives,

$$\log k(R) = \log k^\circ + 55.5 A z_A z_B \alpha_R (1/R)^{\frac{1}{2}} \tag{10}$$

which again simplifies, if α_R is independent of R, to

$$\log k(R) = \log k^\circ + C_2 (1/R)^{\frac{1}{2}} \tag{2}$$

Therefore, at least for reactions that are sensitive to ionic strength (*i.e.* they obey eqn. (6) in aqueous solution), if free water-pool surfactant counterions determine microemulsion ionic strength, a plot of $\log k$ vs. $(1/R)^{\frac{1}{2}}$ should be linear with ordinate intercept $\log k^\circ$ corresponding to the rate constant at infinite dilution with respect to counterions). As for eqn. (6) coefficient C_2 may take a positive or a negative value.

Experimental

Materials

The following compounds were obtained from Sigma: sodium bis(2-ethylhexyl) sulfosuccinic acid, sodium salt (AOT), 4-nitrophenol (NP), sodium cacodylate, sodium maleate, ACES, BES, CHES, EPPS, GLYGLY, HEPES, MES, MOPS, MOPSO, POPSO, TAPSO and tris(hydroxymethane) aminomethane 'tris'. Pyranine (trisodium 1-hydroxy 3,6,8-pyrene trisulfonate) was obtained from

Molecular Probes (Leiden, Netherlands). *Chromobacterium viscosum* lipase was obtained from Genzyme (Maidstone, Kent, UK). Doubly distilled deionised water was used throughout.

NPC$_2$ contained significant amounts of non-esterified 4NP which was removed by dissolving the crude material in a minimum volume of diethyl ether at room temperature and extracting with tris buffer pH 8.5 (10 mmol dm^{-3}) until no further yellow colour was extracted. The recovered organic phase was dried with MgSO$_4$ then poured into a glass dish. The ether was allowed to evaporate off at room temperature and the off-white crystals of NPC$_2$ were stored over silica gel in a dessicator at 4 °C.

The *C. viscosum* lipase preparation contained two lipase-active components with molecular weights approximately 150 and 30 kDa and no other protein components of any significance, as determined by chromatography on Sephacryl S-200 (Pharmacia, Uppsala, Sweden) in 10 mM ammonium hydrogen carbonate buffer. The 30 kDa protein fractions were pooled and freeze-dried. The protein, a pure white powder, $\varepsilon_{280} = 2.2 \times 10^4$ dm^3 mol^{-1} cm^{-1}, gave a single band at 34 kDa on SDS-PAGE stained with Coomasie dye, in good agreement with the value of 33.1 kDa obtained from amino acid sequencing.[11] The solid preparation is stable indefinitely at −80 °C.

Methods

Buffers were used at a single pH = pK_a prepared by half-neutralization, either of the sodium salt with HCl or of the free acid with NaOH, and diluted to 0.40 mol dm^{-3}. The mixed buffer CAPS + GLYGLY was prepared by neutralizing an equimolar solution of the free acids of CAPS and GLYGLY with 1.5 equivalents of NaOH (final pH = 10.9 = pK_a CAPS) and diluting to 0.40 mol dm^{-3}. Stock solutions of pyranine (6 mmol dm^{-3}) and NPS (2.5 mmol dm^{-3}) were prepared in water. All solutions were filtered through 0.45 μm Acrodisc® syringe filters (Pall Corporation), and stored at 4 °C when not in use. Stock solutions of NPC$_2$ were prepared in acetone (20 mmol dm^{-3}) for aqueous work and in *n*-heptane (15 mmol dm^{-3}) for microemulsion work and were stored at room temperature in foil-wrapped flasks. Stock solutions of tCNI were prepared in water at 15 mmol dm^{-3}). Stock solutions of AOT in *n*-heptane were prepared by dissolving AOT in *n*-heptane to a final concentration of 0.50 mol dm^{-3}, filtered through Whatman GF filter paper and stored at room temperature in foil-wrapped flasks.

Stock solutions of *C. viscosum* lipase (molecular weight 33.1 kDa) were prepared by dissolving about 5 mg of the freeze-dried protein in 5.0 ml of sterile distilled water and diluting this to a working concentration of about 30 μg ml^{-1} (0.9 μM) with distilled water, which was kept on ice. Freeze drying was found to result in some (*ca.* 10%) loss of activity which is reversible, full activity being restored following incubation of the solution on ice for 3 h. The main 1 mg ml^{-1} stock was stored at 4 °C and a fresh working solution was prepared each day.

Aqueous assay solutions were prepared by introducing stock buffer into a 5 ml volumetric flask using a glass pipette, followed by other components from microsyringes. Where NPC$_2$ was used, additional acetone was included to give a final acetone concentration of 2% v/v. The flasks were finally stoppered and inverted several times to mix. Microemulsions were prepared by introducing the required amount of stock AOT in *n*-heptane into a 10 ml volumetric flask, followed by approx. 3 ml *n*-heptane and then, from microsyringes, aqueous buffer and other components, including additional water to give the required *R*. The flask was stoppered, sonicated in a water-bath until completely transparent (< 1 min) and diluted to volume with *n*-heptane. Microemulsions containing enzymes were prepared in the same way except that the enzyme solution was added after sonication and the flask gently shaken to clarity (< 1 min) before being finally diluted to volume with *n*-heptane. This approach allowed rapid incorporation of the enzyme solution into the microemulsion, at the same time as avoiding the need for sonication, which causes enzyme inactivation (data not shown).

Spectroscopic measurements were made using either a Pye-Unicam Series PU8800 recording spectrophotometer, thermostatted at 25 °C (internal thermocouple) by circulating water from a remote bath (Haake series F3) or using a Beckmann series DU 7500 diode-array spectrophotometer, thermostatted at 25 °C using a Peltier cell-holder. Aliquots (*ca.* 2 ml) of samples were transferred to stoppered 3 ml cuvettes (Hellma) for the measurement. Microemulsion preparations were kept at 25 °C, for at least 24 h, after rate- or absorbance measurements, in order to check for stability. Measurements from samples showing evidence of phase-separation after this time (this may occur at too-high buffer concentrations) were excluded from the final analysis.

Results

1. Acid–base behaviour of pyranine

Pyranine is highly water-soluble and insoluble in n-heptane and solutions of AOT (0.1 M) in n-heptane. The microemulsion spectrum of pyranine is unchanged compared with that in aqueous solution. From these observations we conclude that pyranine in microemulsions is confined within the water-pool. The equilibrium under investigation is,

The ratio of base (pyr-O^{4-}) to acid (pyr-OH^{3-}) was determined using $[base]/[acid] = \varepsilon_{470}/(\varepsilon_{470\ max} - \varepsilon_{470})$, where ε_{470} is the extinction coefficient at 470 nm at a pH close to the pK_a and $\varepsilon_{470\ max}$ is the extinction coefficient at that wavelength at pH $\gg pK_a$. At 470 nm only pyr-O^{4-} absorbs and the Beer–Lambert law is obeyed, in aqueous solution and in microemulsions, for concentrations in the range 0–70 μmol dm^{-3} (overall concentration; data not shown). In aqueous solution $\varepsilon_{470\ max} = 14.5 \times 10^{3}$ dm^{3} mol^{-1} cm^{-1}. In microemulsions prepared with 20 mmol dm^{-3aq} NaOH (0.1 mol dm^{-3} AOT), $\varepsilon_{470\ max} = 16.5 \times 10^{3}$ dm^{3ov} mol^{-1} cm^{-1}, independent of R.

In buffered aqueous solutions of NaCl the pK_a of pyranine is dependent on the ionic strength according to the expression (data not shown),

$$pK_a = 7.6 - 1.55I^{\frac{1}{2}} \tag{11}$$

($I < 0.1$). This behaviour is independent of the buffer type. This equation has the form of eqn. (4), where $pK_a^{o} = 7.6$. The theoretical slope value can be calculated from eqn. (4), given $A = 0.509$, $z_+ = 1$ (H^+), $z_- = -4$ (the relevant ionic species in the equilibrium is the pyr-O^{4-} anion), giving a value of -2.04. We do not regard the measured value, -1.55, as being significantly different for this rather large molecule (and indeed a slope value of -1.55 would be obtained if the effective charge on pyr-O^{4-} anion is -3).

In AOT microemulsions buffered with different buffers, each at pH $= pK_a$ pyranine obeys the Henderson–Hasselbalch (H–H) equation, $i.e.$ for a given R data fits well to plots of log ([base]/[acid]) $vs.$ pH$_{buff}$ with slope $+1.0$ (Fig. 2). The obvious poor fit at $R = 5$ is attributed to buffer-specific changes in pK_a of both buffer (hence pH) and pyranine, in these low water-activity systems ($R < 10$). The water activity falls off strongly below $R = 10$,[12,13] almost certainly because in these preparations there is barely enough water to hydrate the surfactant and the Na$^+$ counterion, and the water-pool, as such, does not exist.[6,8]

The abscissa intercept of each H–H plot is the buffer-independent mean pK_a at that R. Plots of the derived pK_a plotted $vs.$ $(1/R)^{\frac{1}{2}}$, are linear according to the expression,

$$pK_a(R) = 7.6 - 4.0(1/R)^{\frac{1}{2}} \tag{12}$$

From theory (eqn. (1)), the ordinate intercept must be the pK_a^{o} and the value obtained, $pK_a^{o} = 7.6$ is identical to the value measured in aqueous solution.

This behaviour was also apparent when R was varied in microemulsions buffered with a single buffer. Referring specifically to data for microemulsions buffered with ACES (Table 1), this behaviour is independent of whether the overall- or aqueous (water-pool) buffer concentration is held constant as R is varied (Fig. 3).

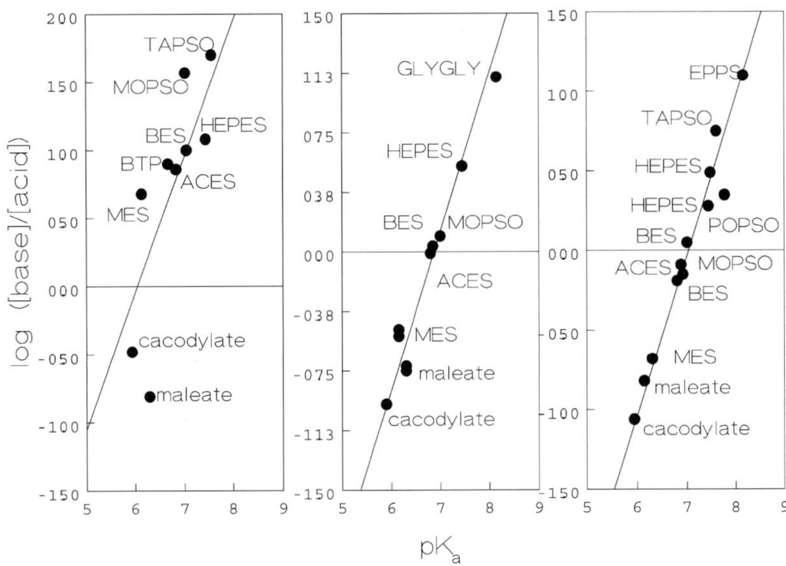

Fig. 2 Henderson–Hasselbalch plots for pyranine in microemulsions ar $R = 5$, 25 and 50 (left to right). Each buffer, as shown, was titrated to $pH_{buff} = pK_a$ by half neutralization. [buffer] = 1.8 mmol dm^{-3ov}; [pyranine] = 26 μmol dm^{-3ov}. The abscissa intercept is $pK_{a\ app}(R)$.

2.   The aminolysis of 4-nitrophenyl acetate by GLYGLY

The base, but not the conjugate acid of GLYGLY (pK_a^o 8.2) is a nucleophile which reacts with NPC$_2$ to give glyglyglycinyl acetate.[14] The first order rate constant for the reaction in aqueous solution k_{+1w} ($=$ rate/[NPC$_2$]) was obtained from integrated rate plots (log $\Delta A_{346\ nm}$ *vs.* time). In microemulsions rates were obtained by the method of initial rates. The spectrum of NP in microemulsions is subject to an R-dependent red-shift, and the pK_a is also strongly dependent on R.[8] Therefore measurements were made at a wavelength at which the extinction coefficient is independent of R for a given pH, as follows[8] (extinction coefficients, $\Delta\varepsilon/10^3$ dm^{3ov} mol^{-1} cm^{-1} in parentheses): 330 nm at pH_{buff} 8.2 (5.2); 336 nm at pH_{buff} 9.6 (4.5), and 373 nm at pH_{buff} 10.9 (13.8).

In aqueous solution at 25 °C the reaction between GLYGLY and NPC$_2$ is first order with respect both to GLYGLY and NPC$_2$ (data not shown). In aqueous 50 mmol dm^{-3} GLYGLY, at $pH = pK_a = 8.2$ (so that GLYGLY is also the buffer), the second order rate constant $k_{+2w} = 6.0 \times 10^{-2}$ dm^3 mol^{-1} s^{-1}. In the mixed buffer CAPS + GLYGLY (pH 10.9, so that CAPS is the buffer), $k_{+2w} = 12.0 \times 10^{-2}$ dm^3 mol^{-1} s^{-1} (data corrected for the background rate of alkaline

Table 1   Ordinate and slope values derived from plots of $pK_{a\ app}$ *vs.* $(1/R)^{\frac{1}{2}}$ for pyranine in w/o microemulsions stabilized by AOT (0.1 mol dm^{-3}) at 25 °C. Each pH corresponds to the aqueous solution pK_a of the named buffer, by half-neutralization. Multi-buffer H–H refers to buffer-independent $pK_{a\ app}$ data obtained from Henderson–Hasselbalch plots (*e.g.* Fig. 3)

	Buffer	Buffer concentration	pK_a^o	C_1
Pyranine	MULTIBUFFER H–H	1.8 mmol dm$^{-3\ ov}$	7.6	−4.0
	MES	1.8 mmol dm$^{-3\ ov}$	7.4	−4.5
	ACES	1.8 mmol dm$^{-3\ ov}$	7.5	−4.5
	ACES	20 mmol dm$^{-3\ aq}$	7.5	−4.2
	ACES	40 mmol dm$^{-3\ aq}$	7.6	−4.6
	MOPSO	1.8 mmol dm$^{-3\ ov}$	7.9	−5.5
	BES	1.8 mmol dm$^{-3\ ov}$	7.7	−3.8
	MOPS	1.8 mmol dm$^{-3\ ov}$	7.6	−4.6
	HEPES	1.8 mmol dm$^{-3\ ov}$	7.5	−4.6

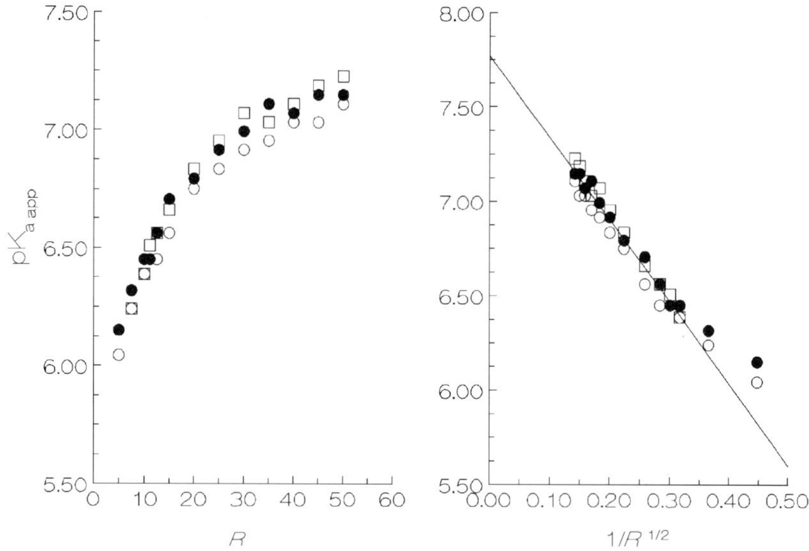

Fig. 3 Pyranine: Dependence of $pK_{a\,app}$ on R and on $(1/R)^{\frac{1}{2}}$ in microemulsions buffered with ACES at three different concentrations: 1.8 mmol dm^{-3m} ($\bigcirc$), 20 mmol dm^{-3aq} ($\square$) and 40 mmol dm^{-3aq} ($\bullet$). [pyranine] = 26 µmol dm^{-3ov}.

hydrolysis). That k_{+2w} doubles when the pH is increased from pH 8.2 to pH 10.9 ($\gg pK_a$) is simply consistent with the fact that only the base participates in this reaction.

In aqueous solution the reaction is only slightly dependent on the ionic strength. At pH 8.2, for example, increasing the ionic strength 11-fold, from 0.1 to 1.1 using sodium chloride resulted in only a 10% increase in k_{+2w} and no conditions could be found in which the Debye–Hückel limiting law was obeyed.

In microemulsions the reaction rate was directly proportional to [NPC$_2$] in the range 0.5–2 mmol dm^{-3ov} (data not shown) and to [GLYGLY], in the range 20–200 mmol dm^{-3aq}. The apparent second order rate constant is $k_{+2m} = (\text{rate})/([\text{NPC}_2]^{ov} \times [\text{GLYGLY}]^{ov})$. As shown in Fig. 4(a), k_{+2m} decreases with increasing R and plots of log k_{+2m} *vs.* $1/R^{\frac{1}{2}}$ are linear (Fig. 4b) according to the expression,

$$\log k_{+2m} = 0.61 + 2.3\,(1/R)^{\frac{1}{2}} \tag{13}$$

(pH$_{buff}$ 8.2), where the units of k_{+2m} are dm^{3ov} mol^{-1} s^{-1}/10^{-3}. At pH$_{buff}$ 8.2 eqn. (17) holds both when the overall GLYGLY concentration was held constant (at 1.8 mmol dm^{-3ov}; Fig. 4) and when the aqueous subphase concentration was held constant in the range 20–50 mmol dm^{-3aq} (data not shown). The same behaviour was observed at pH$_{buff}$ 10.9),

$$\log k_{+2m} = 0.97 + 2.1\,(1/R)^{\frac{1}{2}} \tag{14}$$

where the units of k_{+2m} are again dm^{3ov} mol^{-1} s^{-1}/10^{-3}.

The ordinate intercept is log k^{o}_{+2m}, the value of log k_{+2m} extrapolated to $R = \infty$. The corresponding values of k^{o}_{+2m} are 4.1×10^{-3} dm^{3ov} mol^{-1} s^{-1} (pH$_{buff}$ 8.2) and 9.3×10^{-3} dm^{3ov} mol^{-1} s^{-1} at pH$_{buff}$ 10.9. We expect these two values to be related through the microemulsion pK_a of GLYGLY. Since only the base participates in the reaction we may write,

$$pK_{a\,app} = pH_{buff} - \log_{10}\left(\frac{k^{o}_{+2m(\text{pH}\,8.2)}}{k^{o}_{+2m(\text{pH}\,10.9)} - k^{o}_{+2m(\text{pH}\,8.2)}}\right) \tag{15}$$

so that the $pK_{a\,app} = 8.3$, which we do not regard as being significantly different from the expected value 8.2. (N.B. This is not an estimate of the actual microemulsion pK_a of GLYGLY, since we make the assumption pH$_m$ = pH$_{buff}$, but it does show that the system behaves self-consistently).

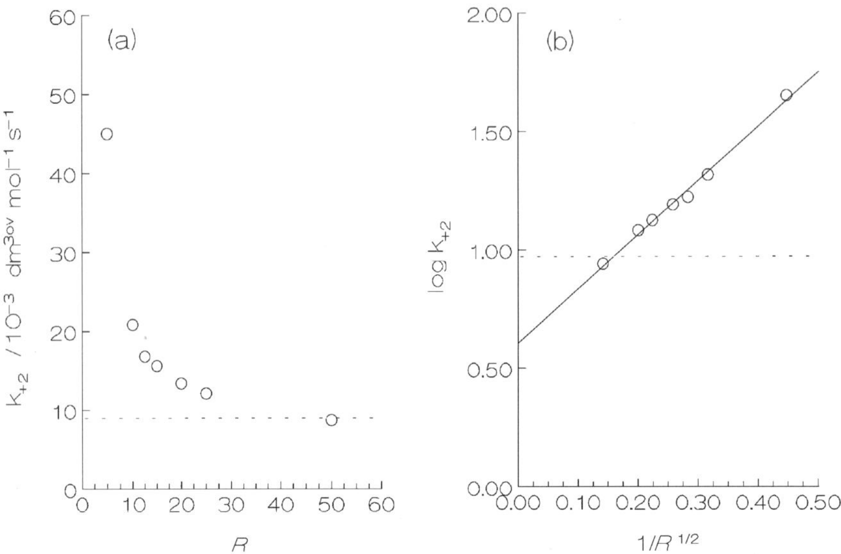

Fig. 4 Dependence of (a) k_{+2m} on R and (b) $\log k_{+2m}$ on $1/R^{\frac{1}{2}}$ for the reaction between GLYGLY and NPC$_2$. [GLYGLY] = 1.8 mmol dm$^{-3\text{ov}}$; pH$_{\text{buff}}$ = 8.2. [NPC$_2$] = 0.79 mmol dm$^{-3\text{ov}}$. The dotted line is the product $(k_{+2w} \times K_p)$ where k_{+2w} is the rate constant in aqueous solution and K_p = 0.16, the bulk water : n-heptane partition coefficient (see text).

It is reasonable to compare k^{o}_{+2m} with k_{+2w}, after correction for partitioning of NPC$_2$ between the aqueous and non-aqueous subphases of the microemulsion. We have no estimate for the microemulsion partition coefficient but we have measured the water : n-heptane partition coefficient precisely: at 25 °C, K_p = 0.16. The dashed lines in Figs. 4(a) and (b) correspond to $(k_{+2w} \times K_p)$ and $\log (k_{+2w} \times K_p)$, respectively. The equality $k^{\text{o}}_{+2m} \cong (k_{+2w} \times K_p)$ holds to a good first approximation (all experimental points lie above the line); and the values of the ratio $k^{\text{o}}_{+2m}/(k_{+2w} \times K_p)$ are, 0.41, at pH$_{\text{buff}}$ = 8.2, and 0.60 at pH$_{\text{buff}}$ 10.9, which we do not regard as being significantly different from the theoretical value of 1.0, and therefore we may accept the equality, $(k_{+2w} \times K_p) \cong k^{\text{o}}_{+2m}$.

In summary, the reaction between GLYGLY and NPC$_2$ in microemulsions obeys the general expression,

$$\log k_{+2m} = \log k^{\text{o}}_{+2m} + C_2(1/R)^{\frac{1}{2}} \tag{16}$$

where coefficient C_2 has a positive value and the inequality $k_{+2m} (R) > (k_{+2w} \times K_p) \cong k^{\text{o}}_{+2m}$ holds so that the rate constant in the microemulsion at any finite R is greater than the aqueous solution value at the same pH *i.e.* the reaction is "hyperactive" in the microemulsion.

3. The alkaline hydrolysis of NPC$_2$

The first order rate constants for the alkaline hydrolysis of NPC$_2$ in aqueous solution k_{+1} (= rate/[NPC$_2$]) were obtained from integrated rate plots (log $\Delta A_{346 \text{ nm}}$ *vs.* time). The measured first order rate constants, k_{+1} = (rate)/[NPC$_2$], were 2.0×10^{-4} s^{-1} in 50 mmol dm^{-3} CHES, pH 9.6 and 3.5×10^{-3} s^{-1} in 50 mmol dm^{-3} CAPS, pH 10.9. The apparent second order rate constant $k_{+2w} = k_{+1w}/[\text{OH}^-] = 4.3$ dm^3 mol^{-1} s^{-1}.

In microemulsions rates were obtained by the method of initial rates, using the same pH-dependent R-independent wavelengths and extinction coefficients used to follow the aminolysis of GLYGLY. Plots of k_{+1m}, the apparent first-order rate constant (= (rate)/[NPC$_2$]) *vs.* R, for the reaction at two pH values is shown in Fig. 5. The corresponding plots of $\log k_{+1}$ *vs.* $(1/R)^{\frac{1}{2}}$ are linear, and the expressions are,

$$\log k_{+1m} = 1.6 - 1.9 \, (1/R)^{\frac{1}{2}} \tag{17}$$

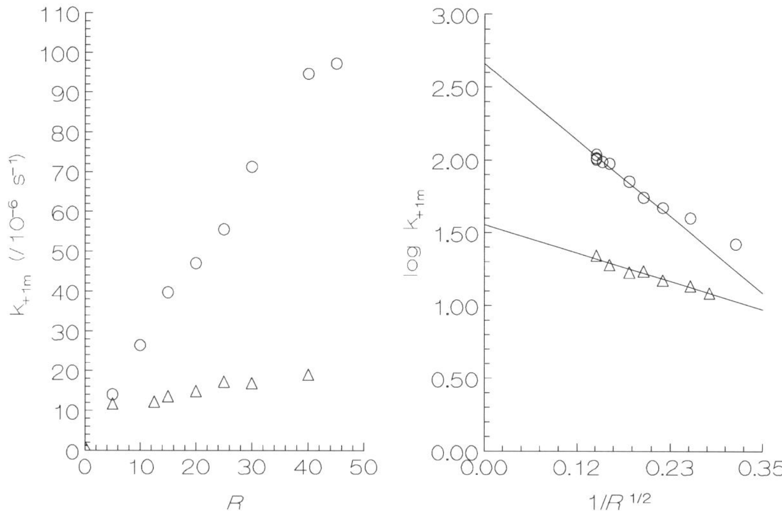

Fig. 5 Plots of k_{+1} *vs.* R and of $\log k_{+1}$ *vs.* $(1/R)^{\frac{1}{2}}$ for the alkaline hydrolysis of NPC_2 in microemulsions buffered with CHES, pH_{buff} 9.6 ($\triangle$) or CAPS pH_{buff} 10.9 ($\bigcirc$). [buffer] = 1.8 mmol dm^{-3ov}; [NPC_2] = 1.5 mmol dm^{-3ov}.

in CHES buffer (1.8 mmol dm^{-3ov}, pH 9.6), and

$$\log k_{+1m} = 2.7 - 4.54\,(1/R)^{\frac{1}{2}} \tag{18}$$

in CAPS buffer (1.8 mmol dm^{-3ov}, pH 10.9), where the units of k_{+1m} are s$^{-1}/10^{-6}$.

At pH_{buff} 10.9 there is some deviation at low R (Fig. 5) and eqn. (18) describes the limiting high R (>20) behaviour. By contrast with the aminolysis reaction (Fig. 4), the slopes of these plots are negative. The corresponding values of k^{o}_{+1m} are 4.0×10^{-5} s^{-1} and 5.0×10^{-4} s^{-1}, at pH_{buff} 9.6 and 10.9 respectively, and the second order rate constant k^{o}_{+2m} ($= k^{o}_{+1m}/[OH]^{-}$) $= 0.61$ dm^{3ov} mol^{-1} s^{-1}.

Again, it is reasonable to compare k^{o}_{+2m} with the aqueous solution value, k_{+2w} after correction for the microemulsion partitioning of NPC_2. The ratio, $k^{o}_{+2m}/(k_{+2w} \times K_{p}) = 0.88$ (theoretical value 1.0) and again we accept, as a first approximation, the equality $(k_{+2w} \times K_{p}) \cong k^{o}_{+2m}$.

In summary, therefore the alkaline hydrolysis of NPC_2 in microemulsions obeys the general equation,

$$\log k_{+2m} = \log k^{o}_{+2m} - C_2(1/R)^{\frac{1}{2}} \tag{19}$$

where coefficient C_2 has a negative value, and the inequality $(k_{+2w} \times K_{p}) \cong k^{o}_{+2m} > k_{+2m}(R)$ holds so that the rate constant in the microemulsion at $R < 50$ is smaller than that in aqueous solution.

4. The lipase-catalysed hydrolysis of NPC_2

In both aqueous solutions and microemulsions rate constants for the lipase-catalyzed hydrolysis of NPC_2 were determined using the method of initial rates. In aqueous solution the reaction was followed by monitoring the release of NP at 346 nm ($\Delta\varepsilon_{346} = 4.8 \times 10^3$ dm^{-3} mol^{-1} cm^{-1}), independent of pH. In microemulsions NP release was followed at 317 nm ($\Delta\varepsilon_{317} = 9.35 \times 10^3$ dm^{-3} mol^{-1} cm^{-1}), independent of R at pH_{buff} 6.0.

In both aqueous solutions and microemulsions the rate of the reaction is first-order with respect to the concentrations of both enzyme and NPC_2. Saturation (Michaelis–Menten) kinetics were not observed, presumably because the K_m for NPC_2 is far higher than the accessible concentration range. Therefore kinetic data were analysed in terms of the second order rate constant k_{+2} ($= k_{cat}/K_m$).

In aqueous MES buffer (50 mmol dm^{-3}), pH 6.0 $k_{+2w} = 5.4 \times 10^4$ dm^3 mol^{-1} s^{-1} increasing by about 10% on addition of NaCl to 1.0 mol dm^{-3}. No conditions could be found for which the reaction obeyed eqn. (6).

In microemulsions k_{+2m} increased steeply to a maximum at $R = 10$ and thereafter declined again (Fig. 6a). Plots of log k_{+2} vs. $(1/R)^{\frac{1}{2}}$ are linear in the range $R = 15$–50, according to the equations (Fig. 6b),

$$\log k_{+2m} = 3.6 + 2.5(1/R)^{\frac{1}{2}} \tag{20}$$

(in cacodylate buffer, pH$_{buff} = 6.0$), and

$$\log k_{+2m} = 3.5 + 2.4(1/R)^{\frac{1}{2}} \tag{21}$$

(in MES buffer, pH$_{buff} = 6.0$), where the units of k_{+2m} are dm^{3ov} mol^{-1}s^{-1} 10^{-3}, so that the ordinate intercepts, corresponding to k_{+2m}^{o} (k_{+2m} extrapolated to $R = \infty$) are 3.8×10^3 and 3.3×10^3 dm^{3ov} mol^{-1} s^{-1}, in the cacodylate and Mes-buffered systems, respectively. We conclude that buffer type does not significantly influence the activity of the enzyme.

The dashed line in Fig. 6(a) is the value for k_{+2m} that would be expected after correction of k_{+2w} (the rate constant in aqueous solution) for partitioning of NPC$_2$ within the microemulsion using $K_p = 0.16$. The ratio $k_{+2m}^{o}/(k_{+2w} \times K_p) = 3.3 \times 10^3/(5.4 \times 10^4 \times 0.16) = 0.38$ is rather low and we accept the equality $k_{+2m}^{o} \cong (k_{+2w} \times K_p)$ only to a crude first approximation. We conclude nevertheless (noting that all experimental points lie on or above the line) that the enzyme is hyperactive in the microemulsion compared with its aqueous solution behaviour. At optimal $R = 10$, the enzyme is around twice as active as in aqueous solution.

To summarize, the lipase-catalyzed hydrolysis of NPC$_2$ obeys the equation,

$$\log k_{+2m} = \log k_{+2m}^{o} + C_2(1/R)^{\frac{1}{2}} \tag{22}$$

where the inequality, $k_{+2m} > k_{+2m}^{o} \cong k_{+2w}$ holds (to a first approximation) and the enzyme in the microemulsion is hyperactive compared with aqueous solution in the range $10 < R < 50$.

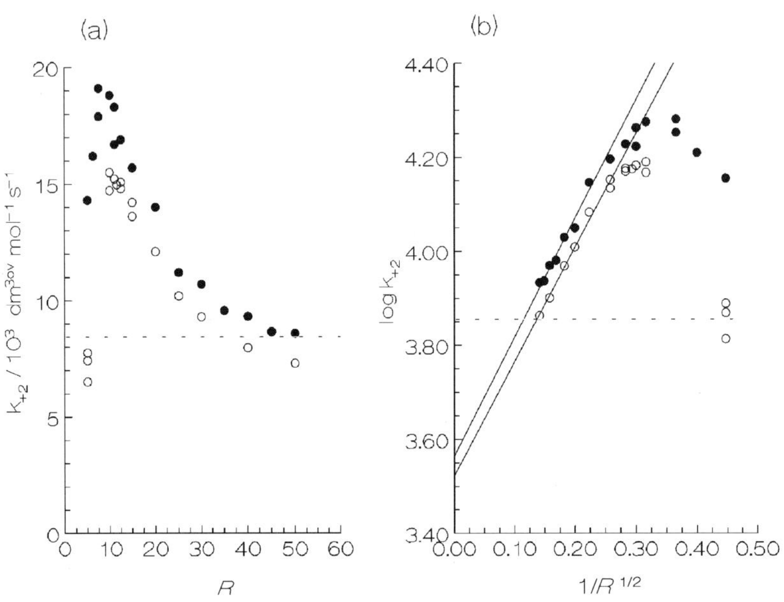

Fig. 6 The dependence of k_{+2} on R, and of log k_{+2} on $(1/R)^{\frac{1}{2}}$ for the hydrolysis of NPC$_2$ by *C. viscosum* lipase in 0.1 mol dm^{-3} AOT microemulsions buffered with cacodylate (●) and Mes (○), both at pH$_{buff} = pK_a = 6.0$. [lipase] = 22 nmol dm^{-3ov}; [NPC$_2$] = 603 μmol dm^{-3ov}. The dashed line corresponds to the value that would be expected in a microemulsion after correction for substrate partitioning (see text).

5. The hydrolysis of tCNI by α-chymotrypsin

In aqueous solution, in the neutral/alkaline pH range, αCT has an optimal pH of around 8, and its activity is insensitive to ionic strength and to the type of buffer used.[15] Hydrolysis of tCNI was followed by monitoring the decrease in absorbance at 335 nm ($\Delta\varepsilon = 7.1 \times 10^3$ dm^3 mol^{-1} cm^{-1} in both aqueous solution and microemulsions). The Michaelis constant, K_m, for tCNI is very low (*ca.* 1 μM), and therefore k_{cat} was determined directly by using working concentrations of tCNI = 60 μmol dm^{-3} in both aqueous solution and microemulsion (so [tCNI] $\gg K_m$). In aqueous solution $k_{cat} = 1 \times 10^{-2}$ s^{-1} (50 mmol dm^{-3} EPPS buffer, pH = pK_a = 8.0).

Plots of k_{cat} *vs.* R and of log k_{cat} *vs.* $(1/R)^{\frac{1}{2}}$ are shown in Fig. 7. The direct plot compares very well with the original study of Martinek and co-workers.[4] The activity rises very steeply from zero at approx. $R = 3.5$ to a maximum at $R = 6$ and thereafter declines again. The descending curve in the range $R = 6$–25 obeys the expression,

$$\log k_{cat\ m} = 0.74 + 1.52\ (1/R)^{\frac{1}{2}} \tag{23}$$

where the units of k_{cat} are s^{-1}/10^{-3}.

The ordinate intercept corresponds to a minimum value of $k_{cat\ m}^o = 5.5 \times 10^{-3}$ s^{-1}, at $R = \infty$. The microemulsion kinetics of tCNI hydrolysis are not complicated by partitioning (tCNI is insoluble in *n*-heptane), and the aqueous and microemulsion rates can be compared directly. The dotted line in Fig. 7(a) is the aqueous solution value and as with lipase we accept the equality $k_{cat\ m}^o \cong k_{cat\ w}$ only as a first approximation (the ratio $k_{cat\ m}^o/k_{cat\ w} = 0.55$). Nevertheless, we conclude the enzyme is hyperactive in microemulsions at $6 < R < 25$ and at optimal $R = 6$ is more than twice as active than in aqueous solution ($k_{cat\ m} = 2.2 \times 10^{-2}$ s^{-1}).

To summarize therefore, the αCT-catalyzed hydrolysis of tCNI in AOT microemulsions obeys the expression,

$$\log k_{cat\ m}\ (R) = \log k_{cat\ m}^o + C_5(1/R)^{\frac{1}{2}} \tag{24}$$

where the inequality, $k_{cat\ m} > k_{cat\ m}^0 \cong k_{cat\ w}$ holds (to a first approximation) and the enzyme in the microemulsion is hyperactive compared with aqueous solution in the range at $6 < R < 25$.

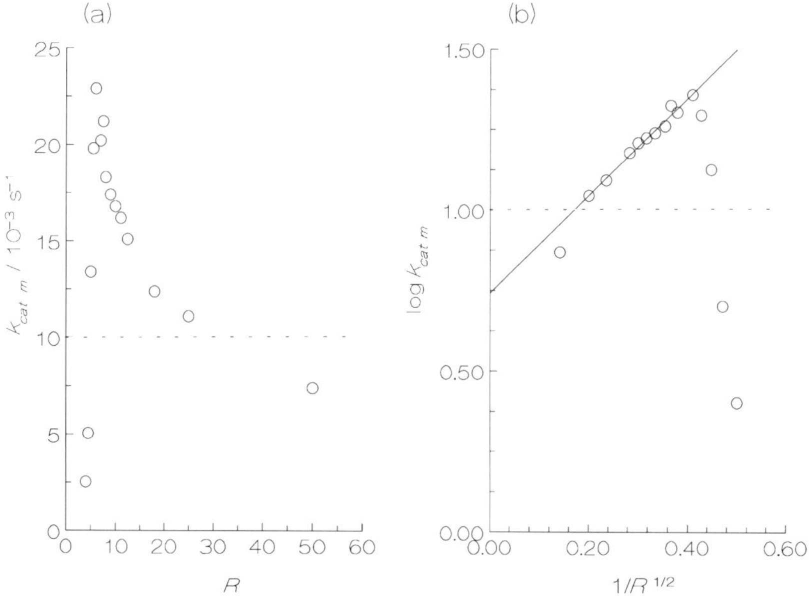

Fig. 7 Plots of (a) k_{cat} *vs.* R and (b) log k_{cat} *vs.* $(1/R)^{\frac{1}{2}}$ for the hydrolysis of tCNI by αCT in microemulsions buffered with EPPS (1.8 mmol dm^{-3ov} pH$_{buff}$ 8.0). [tCNI] = 60 μmol dm^{-3ov}; [αCT] = 1.6 μmol dm^{-3ov}. The dashed line corresponds to the aqueous solution value of k_{cat}.

Discussion

In AOT microemulsions at fixed R pyranine exhibits normal acid base behaviour. It obeys the Henderson–Hasselbalch equation (Fig. 2) and the $pK_{a\,app}$ is independent of the buffer-type. (Table 2). The buffer-set includes only highly water soluble and (with the exception of maleate) neutral base type buffers, so that changes in pK_a, and hence buffered pH, due to partitioning or primary ionic strength effects, respectively, are not expected.[16] Since there is no means of measuring the pH of the microemulsion directly, the actual pK_a cannot be calculated. However, this is less important than the observation that, in well-buffered microemulsions, the system behaves self consistently.

The $pK_{a\,app}$ of pyranine varies linearly with $(1/R)^{\frac{1}{2}}$ as predicted by eqn. (1), as previously observed for NPS (Fig. 1). This behavior is independent of whether overall or the aqueous phase concentration of buffer is held constant as R is varied (Fig. 3; Table 1) and we conclude that it is due principally to an R-dependent ionic strength effect, where the ions in question are lone surfactant counterions present in the water pool and not added electrolyte. ^{23}Na nmr spectrometry provides direct qualitative evidence for the presence of sodium counterions free in the aqueous subphase of AOT microemulsions.[10] We conclude that, since both NPS and pyranine obey eqn. (1) this is the general case for weak neutral acid equilibria in AOT microemulsions.

The alkaline hydrolysis of NPC$_2$ is likewise expected to be subject to a primary ionic strength effect, since the mechanism involves the rate-limiting attack of the negatively charged OH$^-$ on the δ +vely charged carbonyl carbon.[17] In aqueous solution, therefore, the rate decreases with increasing ionic strength due to progressively stronger shielding of the charged species from each other. In microemulsions the fact that $\log k_{+1m}$ decreases linearly with $(1/R)^{\frac{1}{2}}$ (eqn. (20)) leads us to conclude that shielding of the attacking OH$^-$ ion by lone counterions is the rate-determining factor (Fig. 8(a)).

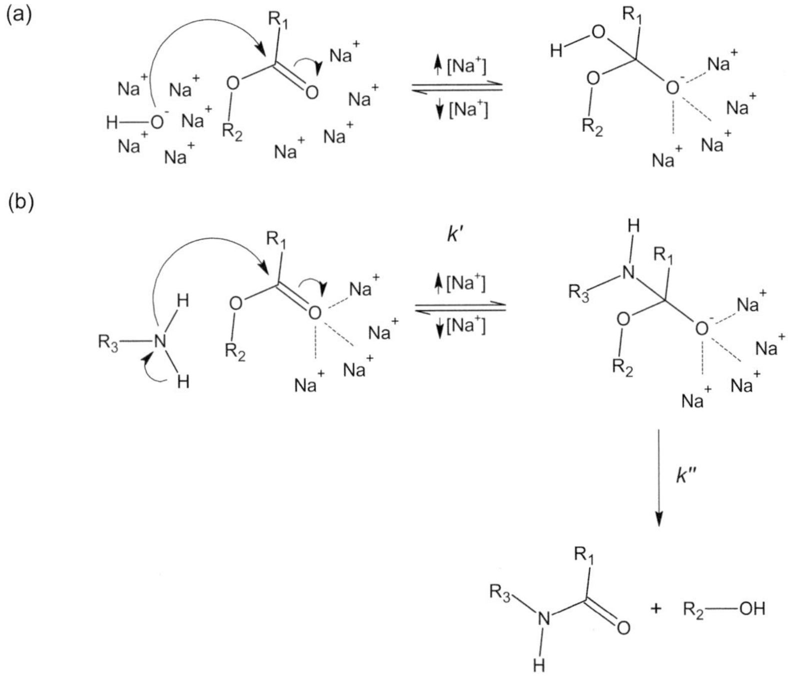

Fig. 8 The effect of lone counterions (Na$^+$) on the rates of alkaline hydrolysis (a) and the aminolysis (b) of an ester. In (a) attack by OH$^-$ is the rate-limiting step. Lone counterions reduce the rate by neutralizing the effective charge on OH$^-$, lowering its electrochemical potential. In (b) the attacking nucleophile is uncharged. The transition state is stabilized through ion-pair formation between lone counterions and the carbonyl oxygen as it accepts negative charge. Therefore as long as formation of the transition state is rate limiting in the overall reaction ($k' < k''$), the rate of the reaction is increased. A similar argument holds for the enzyme catalyzed hydrolysis of ester or amide substrates (see text).

The other reactions discussed here are insensitive to primary ionic strength effects in aqueous solution, yet they also obey eqn. (2). To explain this behaviour, we propose that water-pool counterions have an additional property which arises from the fact that these ions have no charge-partners (in terms of the Debye–Hückel model, the counterion ion, is 'bare'–not shielded by a cloud of anions). We postulate that these ions are able to act as positive charge centres with a high electrochemical potential and thus are capable of modulating reaction rates that are insensitive to ionic strength effects in ordinary 'charge neutral' aqueous solutions. The aminolysis of NPC_2, the lipase-catalyzed hydrolysis of NPC_2 and the αCT-catalyzed hydrolysis of tCNI all share a common mechanism in which nucleophilic attack on the carbonyl carbon of the ester or amide results in the formation of the tetrahedral intermediate (THI) as the carbonyl oxygen accepts negative charge. According to our theory, lone water-pool counterions are able to strongly ion-pair with the negatively charged carbonyl oxygen and thus stabilize the THI (Fig. 8b).

Actually, the consequences of stabilization of the THI for the rate of the reaction depends exactly on the reaction in question. For a reaction in which, referring to Fig. 8b, $k' < k''$, implying that formation of the THI is rate-limiting, factors that act to stabilize the THI are expected to increase k' and hence the overall rate. In this case the reaction is expected to obey eqn. (2) where the R-independent coefficient C has a positive value, so that $k(R) > k^\circ \cong k_{aq}$. On the other hand, if $k' \gg k''$, implying that the breakdown of the THI is the rate limiting step in the overall reaction, then stabilization of the THI is expected to decrease the rate of breakdown, and hence the overall rate. In this case the reaction is still expected to obey eqn. (2), but the R-independent coefficient C has a negative value, so that $k(R) < k^\circ \cong k_{aq}$.

The mechanism of the aminolysis of NPC_2 is known to be $B_{AC}2$, with attack of the nucleophile as the rate determining step[18–20] and the hydrolysis of tCNI by αCT likewise involves the rate limiting attack of the nucleophilic serine 195 on the carbonyl carbon.[21,22] Therefore the kinetic mechanisms of these reactions are both consistent with theory. The mechanism of *C. viscosum* lipase, on the other hand, has not been investigated. The enzyme, like αCT, is a serine hydrolase that relies on a Ser87-Asp263-His285 catalytic triad,[11] but this alone is not evidence that attack of Ser87 is rate limiting. Given the kinetic behaviour observed in AOT microemulsions, we make the prediction that the rate limiting step is the initial attack by Ser87 on the substrate.

We attribute the tailing off of the enzyme-catalyzed reactions at low R to deactivating conformational changes, arising perhaps from partial dehydration. We arrive at this deduction by noting that a_w is very low in $R < 15$ preparations[12,13] and the water-pool (*i.e.* free water) as such does not exist. Indeed, referring specifically to αCT, it is remarkable that the activity of the enzyme increases from zero at $R = 3.5$ to its maximum value at $R = 6$, and lipase likewise to a maximum at $R = 10$. Since neither the aminolysis nor the alkaline hydrolysis of NPC_2 displays this behaviour (compare Fig. (4) and (5) with Figs. (6) and (7)) the decline in a_w as $R \to 0$ clearly has no consequence for the reaction itself, even though water is involved in the latter reaction.

Martinek and co-workers attributed the hyperactive behaviour of enzymes in microemulsions to conformational freezing of the enzyme in a hyperactive form induced by a 'tight' fit of the enzyme into an optimally sized droplet.[23] Even accepting the physics of the argument (that there is sufficient surface energy in the surfactant-coated interface to exert a deforming pressure on the enzyme), it seems unlikely that this highly evolved piece of machinery could be distorted by these crude means into a more active form. The T–R analogy adopted by these workers[23] (borrowed from the Monod–Wyman–Changeux model for co-operative binding properties of allosteric enzymes) ignores the fact that conformational flexibility is a key factor determining why enzymes are such effective catalysts.

The counterion model has the benefit that it has predictive power and can therefore be tested further. We have already predicted, based on the microemulsion kinetics, that the *C. viscosum* lipase mechanism involves rate-limiting attack of the catalytic serine on the substrate, and we have noted that enzymes (and other reactions) utilizing a nucleophilic mechanism but for which breakdown of the THI, rather than its formation, is rate-limiting, will not show bell shaped kinetics or hyperactivity. Indeed (Fig. 8b) we expect such enzymes to be inhibited in AOT (and other anionic) microemulsions since the stabilized THI will break down more slowly. We may also predict the behaviour of a given reaction in other microemulsion types. For example, the activity of enzymes for which THI-formation is rate-limiting should be depressed in microemulsions prepared with cationic surfactants since lone counteranions are expected to have an inhibitory effect by making negative charge transfer to the carbonyl oxygen more difficult. Regarding *C. viscosum* lipase, for example, we have observed that, in w/o microemulsions stabilized by cetyltrimethylammonium bromide (CTAB), the enzyme is around 10-fold less active than in AOT microemulsions, and k_{+2m} is independent of R, *i.e.* the bell-shaped kinetics characteristic of its behaviour in AOT microemulsions is entirely absent.[24] Finally, we expect enzymes in non-ionic microemulsions (no counterions) to have roughly the same activity as in aqueous solution, after substrate partitioning is taken into account, and to display no R-dependent kinetics.

While the counterion model explains a widely observed general phenomenon, ionic strength-related effects may not be the only cause of apparent hyperactivity. We have shown that bilirubin oxidase is far more active in microemulsions stabilized by CTAB, with chloroform as a cosurfactant, than in aqueous solution, and that the activity increases by over 100-fold as $R \rightarrow 0$, but this is because the enzyme is virtually inactive in the absence of detergents and CTAB has an activating effect which can be duplicated in aqueous solution.[25] Therefore the key diagnostic test of the applicability of the counterion model is not merely the observation of an R-dependency in the kinetics, but that a plot of log k *vs.* $(1/R)^{\frac{1}{2}}$ is linear.

It remains to develop a quantitative theory for this behaviour, which should lead to a theoretical estimate of the value of the slopes, C, of these plots, for comparison with the experimental values. We do not expect the original Debye–Hückel limiting law to apply quantitatively to solutions containing an excess of a particular ion type, since the value of the constant $A = 0.509$ was calculated for strictly electroneutral solutions. While it seems reasonable that the model can be modified, and A can be recalculated without this condition in place, we note that comparison of theoretical and experimental values is complicated because we have no estimate of α_R, the fractional dissociation of NaAOT, and hence no estimate of $[Na^+]^{aq}$. A direct experimental route to the measurement of α_R is not obvious.

Conclusions

The data given here are consistent with the hyperactivation of *C. viscosum* lipase, and of α-CT due to the presence of freely mobile 'lone' surfactant counterions dissolved in the aqueous subphase (and hence accessible to the enzyme active site). It is suggested that these have the action of lowering the activation energy of the reaction by stabilizing the transition state, and the prediction is that any enzyme operating a nucleophilic mechanism in which formation of the THI is rate-limiting will exhibit hyperactivity in this w/o microemulsion system. This hyperactivation is not mediated indirectly, through conformational effects, since simple chemical reactions with a nucleophilic mechanism, such as the reaction between NPC_2 and GLYGLY, behave in the same way. The diagnostic test of this behaviour is that plots of log k *vs.* $(1/R)^{\frac{1}{2}}$ will be linear, where k is a suitable rate constant.

References

1 A. Maitra, *J. Phys. Chem.*, 1984, **88**, 5192.
2 M. P. Pileni, in *Structure and Reactivity in Reverse Micelles*, ed. M. P. Pileni, Elsevier, Amsterdam, 1989, p. 44.
3 C. Oldfield, R. B. Freedman and B. H. Robinson, *J. Chem. Soc., Faraday Trans.*, 1996, **92**, 73.
4 K. Martinek, A. V. Levashov, N. L. Klyachko, Y. L. Khelminsky and I. V. Berezin, *Dokl. Akad. Nauk SSSR*, 1977, **236**, 951.
5 K. Martinek, A. V. Levashov, N. L. Klyachko, V. I. Pantin and I. V. Berezin, *Biochem. Biophys. Acta*, 1981, **657**, 277.
6 C. Oldfield, in *Biotechnology and Genetic Engineering Reviews*, ed. M. P. Tombs, Intercept Ltd., Andover, UK, 1994, p. 12.
7 A. V. Levashov, N. L. Klyachko, N. G. Bogdanova and K. Martinek, *FEBS Lett.*, 1990, **268**, 238.
8 C. Oldfield, B. H. Robinson and R. B. Freedman, *J. Chem. Soc., Faraday Trans.*, 1990, **86**, 833.
9 P. W. Atkins, *Physical Chemistry*, 5th edn., Oxford, 1994, p. 320.
10 M. Wong, J. K. Thomas and T. Nowak, *J. Am. Chem. Soc*, 1977, **99**, 4730.
11 D. Lang, B. Hofmann, L. Haalck, H.-J. Hecht, F. Spener, R. D. Schmid and D. Schomburg, *J. Mol. Biol.*, 1996, **259**, 704–717.
12 R. Kubik, H.-F. Eicke and B. Jönsson, *Helv. Chim. Acta.*, 1982, **65**, 170.
13 E. Bardez, B.-T. Goguillon, E. Keh and B. Valeur, *J. Phys. Chem.*, 1984, **88**, 1909.
14 W. L. Koltun and F. R. N. Gurd, *J. Am. Chem. Soc.*, 1959, **81**, 301.

15 M. L. Bender, G. E. Clement, F. J. Kézdy and H. D'A. Heck, *J. Am. Chem. Soc.*, 1964, **86**, 3680.
16 N. E. Good, G. D. Winget, W. Winter, T. N. Conolly, S. Izawa and R. M. M. Singh, *Biochemistry*, 1966 **5**, 467.
17 P. Sykes, in *Mechanism in Organic Chemistry*, 6th edn., Longman, 1986, p. 384.
18 M. B. Smith and J. March, in *Advanced Organic Chemistry*, 5th edn., Wiley, 2001, p. 511.
19 T. C. Bruice, A. Donzel, R. W. Huffman and A. R. Butler, *J. Am. Chem. Soc.*, 1967, **89**, 2106.
20 M. J. Gregory and T. C. Bruice, *J. Am. Chem. Soc.*, 1967, **89**, 2121.
21 M. L. Bender, F. J. Kezdy and C. R. Gunter, *J. Am. Chem. Soc.*, 1964, **86**, 3714.
22 D. M. Blow, *Acc. Chem. Res.*, 1976, **9**, 145.
23 A. V. Levashov, N. L. Klyachko, N. G. Bogdanova and K. Martinek, *FEBS Lett.*, 1990, **268**, 238.
24 P. D. I. Fletcher, B. H. Robinson, R. B. Freedman and C. Oldfield, *J. Chem. Soc., Faraday Trans. 1*, 1985, **81**, 2667.
25 C. Oldfield and R. B. Freedman, *Eur. J. Biochem.*, 1989, **183**, 347–355.

Interfacial and mass transport enhancement effects on rates of styrene epoxidation catalyzed by myoglobin films in microemulsions

Abhay Vaze[a] and James F. Rusling*[b]

[a] Department of Chemistry, University of Connecticut, Storrs CT 06269-3060, USA
[b] Department of Pharmacology, University of Connecticut Health Center, Farmington CT 06032, USA. E-mail: James.Rusling@uconn.edu.; Fax: +860 486 2981; Tel: +860 486 4909

Received 14th April 2004, Accepted 5th May 2004
First published as an Advance Article on the web 24th September 2004

Biocatalytic surfaces hold promise for future clean methods of chemical synthesis. Biocatalyst films utilizing inexpensive redox proteins can operate in low-toxicity microemulsions with high capacities to dissolve nonpolar reactants. Crosslinked films of myoglobin (Mb) and poly(L-lysine) (PLL) attached to oxidized carbon cathodes gave up to 40-fold larger turnover rates in bicontinuous microemulsions compared to o/w microemulsions and micelles. Larger synthetic turnover rates are correlated with up to 10-fold faster diffusion of solutes in oil phases of the bicontinuous fluids, as measured by voltammetry. Visible and circular dichroism spectra suggest that myoglobin resides in the film in a largely aqueous environment. The reactant sites in films monitored by a fluorescent probe were more polar when films were in CTAB than in SDS microemulsions. These combined results suggest that larger mass transport rates in the bicontinuous fluids is a major factor for enhanced turnover rates. Electrostatic interactions between the charged films and oppositely charged surfactants may significantly influence turnover rates when reactant mass transport is not fast enough.

Introduction

Biocatalysis is predicted to be a major methodology for future clean synthetic routes.[1] Enzyme-catalyzed synthetic reactions[2–5] can be very efficient and are often regio- and stereo-selective. Additional advantages include mild reaction conditions and high turnover rates. Newly discovered or genetically engineered enzymes for specific stereoselective synthesis are reported frequently,[6] suggesting a versatile future for this approach.

Previous work in our laboratory showed how microemulsions can enhance rates and control pathways of electrochemical synthetic reactions using surface-bound metallopolymers as electro-chemical catalysts.[7,8] Microemulsions are clear, stable fluids made from oil, water and surfactant.[9] They are less toxic and less expensive than solvents in which syntheses of organic chemicals are often done. The technological impact of efficient catalytic electrodes designed for stereo- and regioselective syntheses in microemulsions could be large. They would enable clean reusable synthetic technology for synthesis of fine chemicals, such as pharmaceuticals, employing only electricity, consumable feed reactants, and recyclable low-toxicity fluids.

DOI: 10.1039/b405486e

Faraday Discuss., 2005, **129**, 265–274

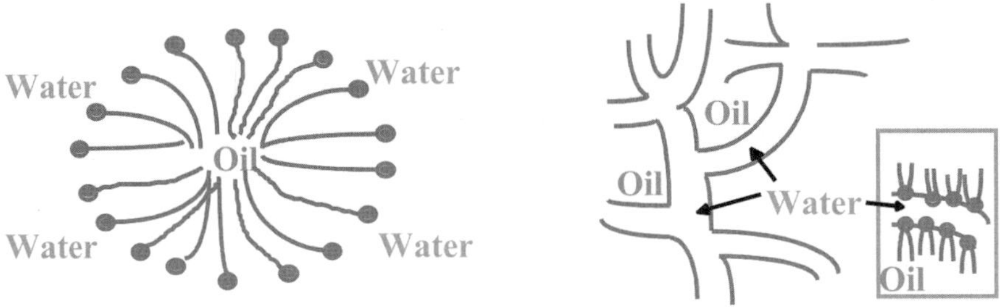

Scheme 1 Dynamic microemulsion structures.

Microemulsions are nano-heterogeneous in nature with dynamic internal structures stabilized by interfacial surfactant leading to ultra-low interfacial free energy. There are two types of micro-emulsions that are conductive by virtue of ionic surfactants and/or added salt that are consequently useful for electrochemical syntheses. *Oil-in-water (o/w) microemulsions* feature surfactant-coated oil droplets of diameter 10–100 nm dispersed in a continuous aqueous phase. Second, *bicontinuous microemulsion* have dynamic nano-scale aqueous and organic micro phases that are continuous with surfactant layer at the interfaces. These structures are represented in Scheme 1.

We sought to extend our work in microemulsions to stereoselective syntheses. For this reason, we began to evaluate catalytic films of inexpensive redox proteins. Recently, we have explored crosslinked films of native myoglobin and polyions that have shown a high degree of stability.[10] We developed prototype crosslinked films of the iron heme protein myoglobin (Mb) and poly (L-lysine) (PLL) that are covalently linked to carbon cathodes and specifically designed to be stable in microemulsions.[11]

In order to obtain films stable enough for synthetic applications in microemulsions, we covalently linked poly-(L-lysine) (PLL) with the iron heme protein myoglobin (Mb) and to carbon electrodes. These covalently linked films were physically stable and biochemically active in various micro-emulsions and were used to catalyze the electrochemical epoxidation of styrene. The pathway of this reaction has been studied with Mb dissolved in homogeneous solutions and microemulsions,[12] and

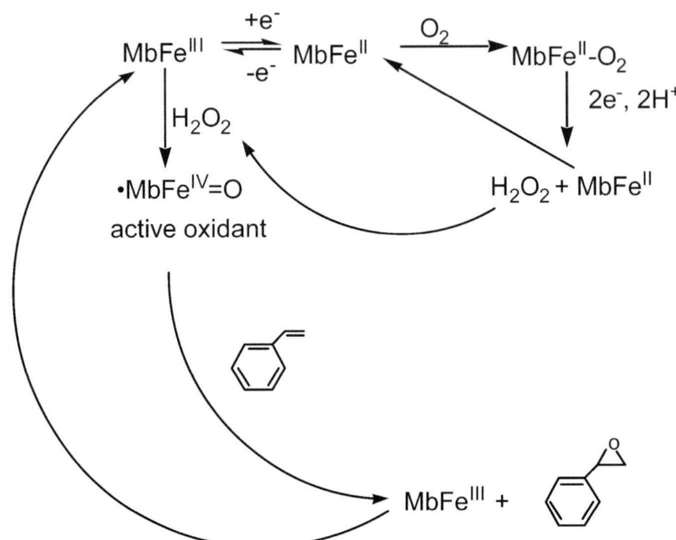

Scheme 2 Pathway for electrochemical and hydrogen peroxide mediated oxidation of styrene catalyzed by myoglobin (Mb).

Table 1 Micellar and microemulsion media used in this work[a]

Fluid type	Materials	Composition (μE by weight)	$\kappa/M\Omega^{-1}$	η/cP
Bicontinuous μE	CTAB/pentanol/tetradecane/H$_2$O	17.5/35/12.5/35	1.6	14.8
Bicontinuous μE	SDS/pentanol/tetradecane/H$_2$O	13.3/26.7/8/52	6.0	12.5
O/w μE	CTAB/butanol/hexadecane/H$_2$O	5/5.4/1/88.6	4.2	5.4
O/w μE	SDS/pentanol/dodecane/0.1 M NaCl	3.35/6.65/1.0/89	10.5	4.1
Micellar	CTAB/buffer + 50 mM NaCl	27 mM CTAB		
Micellar	SDS/buffer + 50 mM NaCl	34 mM SDS		

[a] CTAB = cetyltrimethylammonium bromide; SDS = sodium dodecyl sulfate.

in films in aqueous buffers (Scheme 2).[13,14] Ferrous Mb is generated by reduction as the electrode binds oxygen, and this complex is reduced to give hydrogen peroxide which activates the ferric Mb to a key ferryloxy intermediate. Ferryloxy Mb, which can also be produced by addition of a peroxide, transfers an oxygen atom to styrene to produce the epoxide.

We investigated the electrochemical epoxidation of styrene to styrene oxide with these crosslinked Mb/PLL films in bicontinuous and oil-in-water microemulsions and micellar solutions (Table 1). Results showed a large dependence on fluid type and composition, with 40-fold larger turnover rates for styrene oxide formation when bicontinuous microemulsions *vs.* o/w microemulsions or micellar solutions. The data were consistent with two possible explanations. First, a larger rate of styrene mass transport in the bicontinuous fluids, and second, a higher rate of styrene exchange from the oil phase to the water phase at the Mb reaction sites.

The present paper further addresses how microemulsion composition controls the rate of the catalytic epoxidation. Our approach includes measurement of the transport characteristics of the fluid used, catalytic current estimations of Mb oxidation by an organic peroxide, and fluorescence probe studies designed to characterize the reactant microenvironment in catalytic films in each fluid used.

Experimental

Materials and solutions

Horse heart myoglobin (Mb) from Sigma was dissolved in acetate buffer, pH 5.45 containing 10 mM NaCl, and filtered through a YM30 filter (Amicon, 30 000 MW cutoff)[15] giving 2.9 mg mL^{-1} Mb. 1-Butanol, hexadecane, dodecane, pentanol, 1-[3-(dimethylamino)propyl]-3-ethyl carbodiimide (EDC), ferrocene, *t*-butyl peroxide and cetyltrimethylammonium bromide (CTAB) were from Aldrich. Sodium dodecyl sulfate (SDS) was from Kodak. Tetradecane was from Acros, and poly (L-lysine) (MW 150 000–300 000) was from Sigma. Water was purified by a Hydro Nanopure system to specific resistance >15 MΩ cm^{-2}. The microemulsions used (Table 1) were characterized previously as having bicontinuous or oil-in-water structures.[16] Microemulsions were made by

Table 2 Turnover rates for 1 h styrene epoxidations with Mb/PLL films at 22 °C[a]

Fluid	Turnover/h^{-1} (±15%)[b]
CTAB bicontinuous μE	375
SDS bicontinuous μE	380
CTAB o/w μE	97
SDS o/w μE	8
CTAB micelles	73
SDS micelles	10

[a] Data from ref. 11, with permission. All reactions run by electrolysis with oxygen present for 1 h at −0.6 V *vs.* SCE except entry *CTAB bicon. μE + 10 mM H$_2$O$_2$. [b] Turnover rate as (nmol styrene oxide nmol Mb^{-1}/h^{-1}).

mixing appropriate weight ratios of components and stirring. Micellar solutions of CTAB and SDS were made in 50 mM tris buffer pH 7.4 containing 50 mM NaCl.

Electrode preparation

For voltammetry, films were made on pyrolytic graphite (PG) or small carbon cloth electrodes. Carbon electrodes were first oxidized electrochemically as reported previously[17] to form surface carboxylate groups. The oxidized electrode was immersed into freshly prepared 10 mL 24 mM EDC for 20 min, which was subsequently replaced with aqueous PLL 4 mM in lysine. This step forms amide linkages between carboxylates on carbon to amino groups on PLL.[18] After 4 h, the electrode was rinsed with water, and then covered with 10 mL Mb solution and 10 mL 48 mM EDC and left to stand overnight. Additional layers of PLL and Mb were added in a similar manner to make two bilayers of PLL/Mb, followed by a final layer of PLL.

Voltammetry

A three-electrode thermosatted cell was used at 22 °C with a BAS-100B/W electrochemical analyzer. The working electrode was protein-film coated carbon cloth or PG, the counter electrode was a Pt wire, and the reference electrode was saturated calomel (SCE). Surface coverage of active catalyst was measured by integrating CVs at low scan rates. Polished glassy carbon electrodes were used with a previously reported method[19] to obtain apparent diffusion coefficients of ferrocene (2 or 3 mM) in the various fluids. Rotating disk voltammetry was done with a Pine Electrode rotator.

Fluorescence studies

Fluorescence spectroscopy was done with a SLM AMINCO 8100 spectrometer. Pyrene dissolved (2 μM) in the microemulsions or various solvents were purged with pure nitrogen to remove oxygen before each spectral scan. The excitation wavelength was 332 nm and fluorescence emission spectra of pyrene were used to obtain peak ratios characteristic of polarity.[20] Films prepared on fused silica slides using the methods described above were also treated with pyrene in microemulsions. Alternatively, films were prepared as above but with the PLL and Mb solutions used to make the films containing 30 μM pyrene. The latter films were exposed to various fluids.

Results

Catalytic activity

We have already established that crosslinked Mb/polyion films have peroxidase activity character-istic of Mb by measuring characteristic catalytic voltammetry in the presence of hydrogen peroxide.[10,11] This catalytic reduction of peroxide has common features with the catalytic epoxidation pathway in Scheme 2. If such films are to be used as electrochemical epoxidation catalysts, a necessary property is that Mb must catalytically reduce oxygen to hydrogen peroxide (Scheme 2). Electrochemical signatures of catalytic reduction of oxygen by the Mb/PLL films are shown in Fig. 1. The reversible cyclic voltammograms of the Fe^{III}/Fe^{II} redox couple of Mb are shown in the curves marked "1". Curves marked "2" and "3" show the increase in current at the Fe^{III}/Fe^{II} reduction potential characteristic of catalytic reduction of oxygen to hydrogen peroxide, which has been identified in electrolyses using these films. The catalytic peak current from RDV is larger than that from CV consistent with dependence of the catalytic reaction on mass transport of oxygen. In RDV, mass transport is more efficient by virtue of forced convection that is not present in CV. Curves "4" show that nearly the same CV is obtained in fresh anaerobic fluid after the electrode is used in multiple scans for this catalytic reaction.

Mass transport within the fluids

Mass transport in the oil or micellar phase may be important since the reactant styrene will reside mainly in these phases. We measured an oil/water partition coefficient of 6.1. Ferrocene, which also

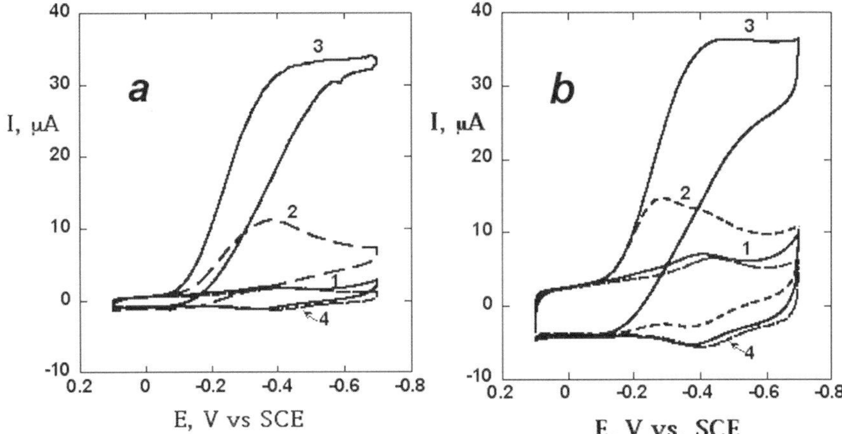

Fig. 1 Cyclic (CV) and rotating disk voltammetry (RDV) of Mb/PLL films at 20 mV s^{-1} in (a) in CTAB o/w microemulsion and (b) SDS bicontinuous microemulsion. Labels are for (1) CV in anaerobic fluid; (2) CV after injecting 25 mL oxygen; (3) RDV at 700 rpm after injecting 25 mL oxygen; and (4) CV in anaerobic fluid for films used for > 10 catalytic scans.

resides predominantly in the oil or micellar phases, was used as an electrochemical probe. Its electrode reaction is:

$$Fc = Fc^+ + e^- \tag{1}$$

Peak current (I_p) in micellar solution or microemulsion will depend on the apparent diffusion coefficient D' of Fc as given by the Randles–Sevcik equation[21]

$$I_p = 0.4463 nFAC^*(nF/RT)^{1/2}\nu^{1/2}D'^{1/2} \tag{2}$$

where n is the in the number of electrons transferred per molecule, F is Faraday's constant, A is electrode area, C^* is solute concentration, R is the gas constant, T is temperature in Kelvins, and ν is scan rate. Eqn. (2) predicts a linear plot of I_p *vs.* ν for the reversible oxidation, and slopes of these plots were used to estimate D' for ferrocene in micellar solutions and microemulsions.[19]

For aggregates with a single size distribution where equilibrium between bound and free solute is fast with respect to the time scale of the electrochemical experiment, D' is given by a two-state model:

$$D' = f_a D_o + f_b D_1 \tag{3}$$

where f_a is the mole fraction of Fc in the water microphase, f_b is the fraction of Fc in the oil microphase, D_o is the diffusion coefficient of the solute in water and D_1 is that of the micelle, droplet in the fluid, or Fc in a continuous oil phase.

Since ferrocene is strongly bound to micelles and almost exclusively present in oil phases of microemulsions, the ferrocene D' approaches D_1 as [Fc] increases.[8,19] At [Fc] > 2 mM, D' measured by voltammetry is a good approximation to D_1, and reflects the mass transport of the micelle or oil droplet, or within oil phase.

Fig. 2 shows representative cyclic voltammograms of ferrocene in a bicontinuous CTAB microemulsion and the o/w SDS bicontinuous microemulsions. Peak current in the bicontinuous microemulsion is much larger than in the o/w microemulsion. This reflects the much faster rate of mass transport within the continuous oil phase of the bicontinuous microemulsions. The discrete oil droplets in which Fc is dissolved in the o/w fluid decrease the mass transport rate of Fc. The oxidation peak has more positive potentials in the SDS microemulsions because of interactions of the product Fc$^+$ with dodecylsulfate ion which facilitates the Fc oxidation.[8]

Peak currents of reversible CVs of 2 or 3 mM Fc over a range of scan rates were used with eqn. (2) to obtain D' in the fluids used in this work. The D' values indicate up to 10-fold faster mass transport in the oil phases of bicontinuous microemulsions compared to oil droplets in o/w

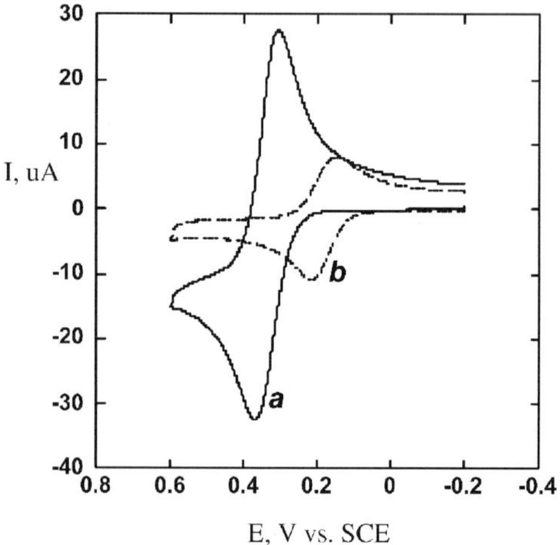

Fig. 2 Cyclic voltammograms at glassy carbon electrode (0.071 cm^2) at 0.2 V s^{-1} for 2 mM ferrocene dissolved in (a) CTAB bicontinuous microemulsion and (b) SDS o/w microemulsion.

microemulsions or micelles (Table 3). The more rapid mass transport in the bicontinuous fluids correlates with turnover rate of the epoxidation reaction (Table 1).

Catalytic reduction of a peroxide

Unfortunately, voltammetry cannot provide much information about the rate of epoxidation of styrene because the addition of styrene to microemulsions containing oxygen and Mb/PLL electrode gave only a small increase in the RDV limiting current, as previously reported in styrene/water emulsions.[13] From Scheme 2 we see that hydrogen peroxide activates ferric Mb to the active oxidant which is a ferryloxy species, which then transfers an oxygen atom to styrene. Addition of hydrogen peroxide to solutions containing Mb film electrodes also gives a catalytic current, and this process has been studied by voltammetry. However, the reaction is quite complicated.[22] Catalytic currents result from reduction of ferryloxy Mb, but hydrogen peroxide also reacts with this species to given dioxygen and ferric Mb, both of which can also be reduced. Thus, we chose to study the catalytic reduction of *t*-butyl peroxide, which also converts iron heme proteins to ferryloxy forms, but which does not react rapidly with these forms.[23,24] With *t*-butyl peroxide ROOH, we expect the predominant electrochemical reaction pathway to be:

$$MbFe^{III} + ROOH \rightarrow *MbFe{=}O + ROH \qquad (4)$$

$$*MbFe{=}O + 2\ e^- + 2\ H^+ \rightarrow MbFe^{III} + H_2O \qquad (5)$$

While it is possible that some MbFe^{III} reduction may also occur at the electrode, the predominant process is the activation of MbFe^{III} to the ferryloxy radical, which is analogous to the activation step in the electrochemical epoxidation Scheme 2.

Table 3 Apparent diffusion coefficient of ferrocene in different fluids

Solvent	$10^6 \times D'(Fc)/cm^2\ s^{-1}$
CTAB bicontinuous μmE	3.4
SDS bicontinuous μmE	2.9
CTAB o/w μmE	0.37
SDS o/w μmE	0.40
CTAB micelles	0.76
SDS micelles	0.95

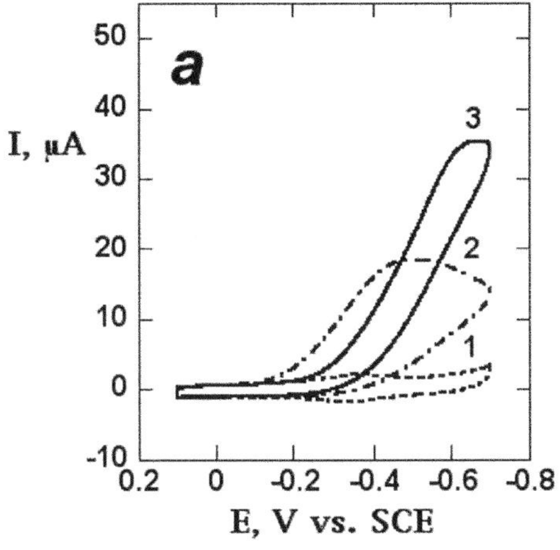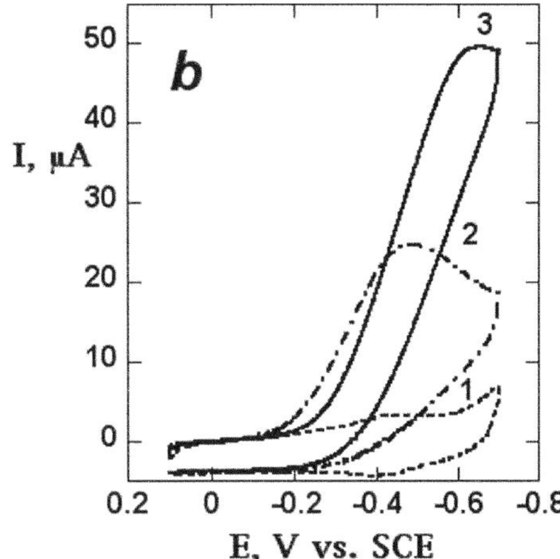

Fig. 3 CV and RVD of catalytic reduction of 0.5 mM *t*-butyl peroxide at Mb/PLL electrodes at 0.02 V s^{-1}: (a) CTAB bicontinuous microemulsion, and (b) SDS bicontinuous microemulsion. Labels on curves are (1) CV in anaerobic fluid; (2) CV in presence of 0.5 mM *t*-butyl peroxide; (3) RDV at 700 rpm in presence of 0.5 mM *t*-butyl peroxide.

Fig. 3 shows voltammograms of Mb/PLL films in microemulsions with and without *t*-butyl peroxide. In the presence of the peroxide, the large current increases at the MbFe^{III} peak and the disappearance of the reverse CV peak are consistent with eqn. (4) and (5). Catalytic efficiencies given by the ratio of the RDV limiting current to the peak current of the MbFe^{III} are listed in Table 4. The largest values are found for the bicontinuous microemulsions, but the difference from the o/w fluids is not as large as the differences in epoxidation turnover rate. The values are compared in Table 4 to catalytic efficiencies of reduction of oxygen, for which there are no large differences in different microemulsions. These catalytic efficiencies reflect the rate of the catalytic reaction, but also contain contributions from mass transport. Further studies are planned to obtain the pure catalytic current at a series of peroxide concentrations and rotation rates to enable estimation of the catalytic rate constant for eqn. (4) and (5) using the appropriate mathematical models.[25]

Spectroscopic studies

UV-visible Soret bands and UV circular dichroism spectra of the crosslinked Mb/PLL films were consistent with Mb residing in films in a near native state in a water-rich environment when the films were wet with buffer solutions or with microemulsions.[10,11] This suggests that the reaction site for epoxidation in the film is largely aqueous, since the ferryloxy heme is involved.

We also wish to know the residence site of the reactant styrene in the film, which it presumably enters from the microemulsion, where it resides mainly in the oil phase. For this, we have begun studies of the fluorescence of another hydrophobic molecule, pyrene, that we hope will report back on the polarity of its environment in the films. Fluorescence of pyrene has been used extensively as a probe for solubilization sites in aggregates such as surfactant micelles and vesicles.[20] The

Table 4 Catalytic efficiency of *t*-butyl peroxide (TBP) reduction from CV and RDV (700 rpm) at PLL/Mb electrodes in different fluids

Microemulsion	I_{Lim}/I_p TBP^a	I_{Lim}/I_p O$_2$^{b}
CTAB bicontinuous	46	16
SDS bicontinuous	51	9
CTAB O/W	19	12
SDS O/W	38	13

^{a} 0.5 mM TBP, 20 mV s^{-1}. ^{b} 25 mL O$_2$ bubbled into anaerobic solution.

Table 5 Pyrene fluorescence peak ratios specific for microenvironmental polarity

Solvent	Dissolved Py, no film	Py in film[a]	Py in μE[b]
Water	1.59		
Methanol	1.31		
CTAB bicontinuous	0.96	0.99	1.13
SDS bicontinuous	0.91	0.71	0.84
CTAB o/w	1.07	1.18	1.32
SDS o/w	0.89	1.03	1.00
Pentanol	0.95		
Hexadecane	0.60		
SDS micelles	1.3		1.09
CTAB micelles	1.3		1.12

[a] Films containing Py were subsequently treated with the microemulsion. [b] Films treated with 2 μM Py in the fluid in question.

fluorescence spectra of the pyrene monomer has five vibronic bands, the relative intensities of which show a strong dependence on solvent microenvironment. Ratios of intensities of the first and third vibronic peaks appearing at 373 nm and 384 nm (I_I/I_{III}) represent the environment experienced by pyrene.

Fig. 4 shows emission spectra of pyrene dissolved in a bicontinuous SDS microemulsion and in a Mb/PLL film that had been exposed to an SDS microemulsion. Table 5 shows that the more polar solvents give larger values of I_I/I_{III}. For example, water gives a value of 1.59 while a less polar solvent such as pentanol gives a value of 0.95. A totally non-polar hydrocarbon gives a value of 0.6. The I_I/I_{III} of pyrene dissolved in the microemulsions ranges from 0.89–1.07 suggesting microenvironment polarity a little less than of pentanol (Table 5), but not as large as methanol. Exposure of films to microemulsions and pyrene using 2 different methods gave qualitatively similar results. In the films exposed to bicontinuous SDS microemulsions, the ratio decreases relative to dissolved pyrene, suggesting that pyrene's microenvironment is less polar in the film. However, in w/o SDS microemulsions and both CTAB microemulsions, the data are consistent with small increases in solute polarity with respect to dissolved pyrene in the respective fluids.

With respect to the working film composition, Mb/PLL films built on quartz crystal electrodes of quartz crystal microbalances showed increase in weight of 16 μg cm^{-2} when soaked in SDS

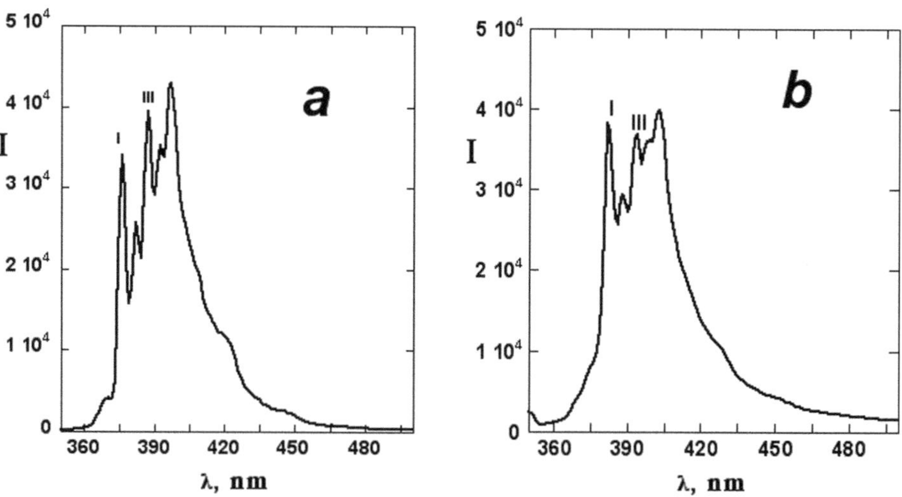

Fig. 4 Fluorescence spectra with excitation wavelength of 332 nm of pyrene (a) 2 μM dissolved in SDS bicontinuous microemulsion (2) pyrene in a Mb/PLL film exposed to SDS o/w microemulsion.

microemulsions as opposed to 0.07 µg cm^{-2} when exposed to CTAB microemulsions. These data are consistent with SDS micelle formation within the Mb/PLL films in SDS microemulsions as reported previously for other PLL-based films.[26]

In future work, we will covalently bond amine reactive pyrene derivatives to PLL, and also attach 1-aminopyrene to Mb, and make films of these materials that will have pyrene moieties bound on either the protein or PLL. Fluorescence micropolarity estimates of pyrenes bound to these specific sites will be made.

Discussion

Results presented above show that the mass transport rates of non-polar molecules in microemulsions as estimated by D' of ferrocene (Table 3) show a reasonably good correlation with reported epoxidation turnover rates (Table 2). The much more efficient turnover rates in bicontinuous fluids correlate with 10-fold better mass transport in the continuous oil phase compared to the o/w microemulsions. However, improved mass transport does not explain the 7- to 12-fold larger turnover rates for CTAB o/w microemulsions and micelles compared to the SDS fluids of comparable structure (Table 2). If we take the t-butyl peroxide catalytic efficiencies (Table 4) as relative measures of the rates of ferric Mb activation to the ferryloxy radical (*cf.* Scheme 2 and eqn. (4) and (5)), we see that although activity in the bicontinuous microemulsions is a bit larger, the differences are not of the scale of the differences in turnover rates. Moreover, part of the difference may be due to mass transport effects. Furthermore, the catalytic efficiency of t-butyl peroxide reduction in the o/w microemulsions is twice as large in the SDS system as in the CTAB system (Table 4). This is the opposite of the trend in the turnover rates. These data suggest that activation of ferric Mb to the key ferryloxy radical oxidant is not the limiting factor. In addition, the catalytic efficiency of the reduction of oxygen by Mb in the films does not depend strongly on microemulsion composition, and is unlikely to be rate limiting.

Based on the above discussion, the oxygen transfer (Scheme 2) from ferryloxy radical to the olefinic double bond is the most likely rate limiting step. All the other reactions in Scheme 2 are known to be rapid,[12–14] at least where microemulsions are not involved. Furthermore, the catalytic efficiencies of ferryloxy Mb formation and reduction of oxygen do not depend as strongly on microemulsion composition as epoxidation rate.

Perhaps a partial answer to reactivity control lies in the physical nature of the molecular environment of the reactive site. As mentioned above, spectroscopy studies showed that Mb resides in these films in a near native state in a water-rich environment.[10,11] What occurs on the molecular scale when a microemulsion comes into contact with the Mb/PLL film? We showed by quartz crystal microbalance studies and voltammetry that in-film micellization is likely to occur when vitamin B$_{12}$–PLL films are in contact with SDS microemulsions, but that CTAB probably does not enter these films from its microemulsions.[26] This is because PLL is protonated at acidic/neutral pH, and its positive charge attracts the negatively charged SDS, but not CTAB. As PLL is the major component of both types of films, it is likely that SDS enters the Mb/PLL films as indicated by the QCM data on these systems as well. We note that pyrene has a much more polar microenvironment in the Mb/PLL films exposed to CTAB microemulsions compared to those exposed to SDS microemulsions (Table 5). This enhanced polarity should be similar for styrene in films in the CTAB microemulsions, and correlates well with the faster epoxidation turnover rates in CTAB o/w microemulsions. Could the decreased polarity and reactivity be because SDS enters the films, forms in-film micelles, and inhibits styrene for reaching the polar reactive sites of Mb? We could envision a certain fraction of the styrene reaching the films, then becoming entrapped in the hydrophobic micelles which do not contain Mb reactive sites.

Why then are the turnover rates for the bicontinuous CTAB and SDS microemulsions nearly the same? One rationalization is that there is a subtle interplay between mass transport rate and non-productive reactant trapping in SDS micelles in the film. When the mass transport rate is large enough, the supply of styrene to Mb reactive sights is large, and reactant trapping in SDS micelles does not influence the rate of reaction. However, when the mass transport rate decreases as in the o/w and micellar systems, the amount of reactant trapped in the micelles may become a limiting factor and decreases the reaction rate.

Clearly our discussion above involves considerable speculation. There is more to be done to support or disprove our working hypothesis. Additional experiments on localized fluorescent probes and *t*-butyl peroxide reduction kinetics are planned as mentioned above. In addition, more detailed quartz crystal microbalance studies on the Mb/PLL films in the fluids in Table 1 should provide some insight into in-film micellization. We hope to achieve these results before the final paper is presented.

Conclusions

This paper explores the control of turnover rates for electrochemical styrene epoxidation in microemulsions using stable myoglobin films. High rates of mass transport of the reactant styrene are of paramount importance. Efficiency of the key transformation of ferric to ferryloxy radical appears not to be greatly influenced by microemulsion type.

Acknowledgements

This work was supported by Grants No. CTS-9982854 and CTS-0335345 from NSF.

References

1 H. E. Schoemaker, D. Mink and M. G. Wubbolts, *Science*, 2003, **299**, 1694–1697.
2 J. P. Lehman, L. Ferrin, C. Fenselau and G. S. Yost, *Drug. Metab. Dispos.*, 1981, **9**, 15–18.
3 H. Joo, Z. Lin and F. H. Arnold, *Nature*, 1999, **399**, 670–673.
4 M. Hara, S. Iazvovskaia, H. Ohkawa, A. Hideo and J. Miyake, *J. Biosci. Bioeng.*, 1999, **87**, 793–797.
5 P. Fernandez-Salquero, A. W. Bunch and C. Gutierrez-Merina, *Prog. Membr. Biotechnol.*, 1991, 291–305.
6 A. M. Rouhi, *Chem. Eng. News*, Feb. 18, 2001, 86–87.
7 J. F. Rusling, in *Reactions and Synthesis in Surfactant Systems*, ed. J. Texter, Marcel Dekker, N.Y., 2000, pp. 323–335.
8 J. F. Rusling and C. J. Campbell, in *Encyclopedia of Surface and Colloid Science*, ed. A. Hubbard, Marcel Dekker, New York, 2002, pp. 1754–1770.
9 M. Bourrel and R. S. Schechter, *Microemulsions and Related Systems*, Marcel Dekker, New York, 1988.
10 V. Panchagnula, C. V. Kumar and J. F. Rusling, *J. Am. Chem. Soc.*, 2002, **124**, 12 515–12 525.
11 A. Vaze, M. Parizo and J. F. Rusling, *Langmuir*, in press.
12 A. C. Onuoha, X. Zu and J. F. Rusling, *J. Am. Chem. Soc.*, 1997, **119**, 3979–3986.
13 X. Zu, Z. Lu, Z. Zhang, J. B. Schenkman and J. F. Rusling, *Langmuir*, 1999, **15**, 7372–7377.
14 B. Munge, C. Estavillo, J. B. Schenkman and J. F. Rusling, *Chem. Biochem.*, 2003, **4**, 82–89.
15 Alaa-Eldin F. Nassar, William S. Willis and James F. Rusling, *Anal. Chem.*, 1995, **67**, 2386–2392.
16 (*a*) J. Georges and J. W. Chen, *Colloid Polym. Sci.*, 1986, **264**, 896–902; (*b*) R. A. Mackay, S. A. Myers and A. Brajter-Toth, *Electroanalysis (N. Y.)*, 1996, **8**, 759–764.
17 A. Vaze and J. F. Rusling, *J. Electrochem. Soc.*, 2002, **149**, D193–D197.
18 D.-L. Zhou, C. K. Njue and J. F. Rusling, *J. Am. Chem. Soc.*, 1999, **121**, 2909–2914.
19 J. F. Rusling, C.-N. Shi and T. F. Kumosinski, *Anal. Chem.*, 1988, **60**, 1260–1267.
20 K. Kalyanasandaram, *Photochemistry in Microheterogeneous Systems*, Academic Press, Orlando, 1987.
21 A. J. Bard and L. R. Faulkner, *Electrochemical Methods*, Wiley, New York, 1980.
22 Z. Zhang, S. Chouchane, R. S. Magliozzo and J. F. Rusling, *Anal. Chem.*, 2002, **74**, 163–170.
23 *Cytochrome P450*, ed. J. B. Schenkman and H. Greim, Springer–Verlag, Berlin, 1993.
24 *Cytochrome P450*, ed. P. R. Ortiz de Montellano, Plenum, New York, 1995, pp. 159–206.
25 H. A. Heering, J. Hirst and F. A. Armstrong, *J. Phys. Chem. B.*, 1998, **102**, 6889–6902.
26 C. K. Njue and J. F. Rusling, *J. Am. Chem. Soc.*, 2000, **122**, 6459–6463.

General discussion

Dr Dagastine opened the discussion of Prof. Rathman's paper: Have the authors considered using systems other than lipid mixtures for the initial phase separated structures at the liquid interfaces? Perhaps a protein–surfactant system that separates into islands or a polymer–surfactant mixture would work. Would any of these systems provide a means to control the "patter" of the surface? The authors point out that the resulting collagen patterned surface has some surface roughness, which is suitable for cell adhesion as demonstrated in the last figure, but the cells don't seem to reflect the pattern of bio-film they are on. Are they're any means to better control the patterning method that would affect cell position on the film?

Prof. Rathman responded: To date we have only investigated lipid mixtures containing membrane lipids—lipids that are found in actual cell membranes—since this provides the best chance for biocompatibility. Unfortunately, many conventional surfactants are detrimental to living cells because they are easily able to penetrate the cell membrane. With regard to achieving better control of how cell growth proceeds on a collagen-patterned surface, one option we are pursuing is to vary the ratio of the dissimilar lipids such that collagen growth occurs in the segregated domains rather than in the continuous "phase" as in our current system. This should provide a more discrete and controllable patterning of the collagen.

Prof. Deutsch asked: The fact that the collagen could be spread on top of the DOPE/DPPE monolayer only within a very narrow surface pressure window is intriguing. The fact that this occurs at a very low surface pressure, *i.e.* at a rather rare state of the monolayer rather than at a dense state, is even more so, and to some extent counterintuitive. Clearly the effect is not dominated by the mechanical rigidity of the DOPE/DPPE monolayer. Could Prof. Rathman speculate on what might be the reason for this effect?

I would also like to point out that there is a liquid subphase that will allow easy spreading of the collagen, *with* or *without* the DOPE/DPPE monolayer, for any coverage and surface pressure. This subphase is mercury (the use of which will obviously make the safety officers of any laboratory rather unhappy). We have recently shown that sharply defined, reproducible isotherms can be measured for Langmuir films of a range of chain- and aromatic molecules on mercury.[1–3] Moreover, mercury is an ideal substrate for measuring the structure of the Langmuir films *in situ*, with sub-Ångström resolution, as a function of the area per molecule, using surface specific X-ray techniques like X-ray reflectivity and grazing-incidence diffraction, which we have also done for a range of molecules. Of course, the interaction of the film with the water and the mercury subphases are completely different: on mercury, the hydrophilicity and water-solubility of the collagen plays no role, and the same holds for the hydrophobicity of the alkyl tails of DOPE and DPPE. For example, the pure collagen, which could not be spread on water due to its solubility, would certainly spread on mercury. At low surface coverages it is most likely that the Langmuir films' structure will be different on water and mercury. The DOPE and DPPE molecules will probably lie flat on the surface of mercury, while on water the hydrophobic water–alkyl repulsion will align the molecules roughly perpendicular to the surface at all coverages. At high coverage, these molecules should stand up also on mercury, and phases similar to those observed on water can be expected, based on our results in the publications cited above.

1   H. Kraack, B. M. Ocko, P. S. Pershan, E. Sloutskin, and M. Deutsch, *Science*, 2002, **298**, 1404.
2   H. Kraack, B. M. Ocko, P. S. Pershan, E. Sloutskin, L. Tamam and M. Deutsch, *Langmuir*, 2004, **20**, 5375.
3   H. Kraack, B. M. Ocko, P. S. Pershan, E. Sloutskin, L. Tamam and M. Deutsch, *Langmuir*, 2004, **20**, 5386.

DOI: 10.1039/b416302h   *Faraday Discuss.*, 2005, **129**, 275–289   **275**

Prof. Rathman responded: It's important to note that the collagen monomer we used is soluble in aqueous subphases while the polymerized collagen is not. When deposited on pure water, the monomer dissolves into the subphase and no polymerization is observed. When deposited on a compressed lipid monolayer (a relatively hydrophobic surface), the monomer does not pass through the monolayer into the subphase and also does not diffuse laterally within the monolayer, resulting in localized "clumps" of collagen on the surface. A lipid monolayer at low surface pressure appears to provide a sufficient barrier to prevent partitioning of monomer into the subphase while at the same time the loosely-packed lipid monolayer allows for lateral diffusion of monomer on the surface, so that uniformly-distributed monomer concentrations are obtained.

The idea of using subphases other than water-based solutions is very good. Like nearly all researchers, we have focused on aqueous systems but there are indeed many other subphases that could provide interesting and novel results.

Prof. Schlossman asked: You stated that there is no surface pressure change when the collagen is first added on top of the lipid film at a low pressure of 1 mN m^{-1}. Does this imply that the addition of collagen does not influence the organization of the lipid monolayer, even at these low pressures?

Prof. Rathman replied: A more accurate answer on my part would be to say that we are not able to observe any significant change in the surface pressure when the collagen solution is deposited. I suspect this is primarily a consequence of our inability to measure small changes in Π with a Wilhelmy plate on the Langmuir trough. Random fluctuations of 0.5 mN m^{-1} are typical, so even though the collagen most likely does change the interfacial free energy, the magnitude of this effect is so small that we can't measure it reproducibly.

Dr Bain asked: You have stressed the sensitivity of the collagen film formation to surface pressure. It seems that the integrity of the lipid monolayer is critical to the success of the experiment, yet a lipid monolayer is unstable at an aqueous/aqueous interface—it will preferentially reconstruct to a bilayer or vesicle. Are the kinetics of reorganisation of the lipid monolayer a key feature of your experiment?

Prof. Rathman responded: Yes, the kinetics are important in many ways in this process, including rates of transport of molecules through the monolayer into the subphase, lateral diffusion of species within the monolayer, chemical reaction of monomeric tropocollagen to form polymerized collagen, and possible reorganization of lipids to form other structures such as bilayers. This could in part explain the fact that we cannot successfully prepare collagen films on compressed lipid monolayers; perhaps when the aqueous collagen is spread on a packed monolayer the water in the collagen spreading solution is unable to penetrate the monolayer, resulting in local "pools" of water on the surface that induce lipid reorganization, effectively destroying the monolayer.

Prof. Fermín asked: Fig. 5 illustrates some fascinating 2D patterns of collagen in mixed DOPE/DPPE monolayers. These patterns appear to be affected by the ratio of DOPE : DPPE : collagen. Can the authors propose a physical reason why interwoven sheets of polymerised collagen are only obtained for a ratio of 70 : 70 : 1, while other ratio show branching bundles?

Prof. Rathman answered: The total amount of lipid plus collagen was higher in Fig. 5(c) than 5(d). The purpose of this slide was to illustrate the variety of patterns that can be obtained by varying lipid : lipid : collagen ratio and/or total concentration. The concentration effect was not properly noted in the figure caption.

Prof. Rusling asked: This may be a sacrilegious question for this meeting, but why if you want to make patterned collagen films did you choose liquid interfaces? Most people making patterned films of biomolecules use solid substrates with pre-patterned chemistry for biomolecule adsorption. Ultra thin films can be made by adsorption on solids under the right conditions.

Prof. Rathman replied: Structural control of films formed in the presence of solid substrates is often limited because interaction between the film-forming molecules and the solid surface generally dominate. That's fine if you are able to prepare a solid surface that will pattern the film in the desired manner; however, if you wish to exploit the interactions between self-assembling molecules in a film without "interference" from a solid, a liquid/vapour or liquid/liquid interface is preferred. We've previously had success with this method for other systems. For example, mesoporous silica films of any desired structure (hexagonal, cubic, or lamellar pore geometries) can be selectively synthesized at a liquid/liquid interface by varying composition of the film-forming molecules. When synthesized directly at a solid interface, the pore geometry of the final product is most strongly dependent on the solid itself; one obtains only one type of material regardless of composition.

Dr Leermakers asked: When the DOPE layer with surface pressure 1 mN m^{-1} is on the surface there is already a significant number of lipids on the surface. So there is a nice layer of PE on the surface. I can imagine that the biopolymer adsorbs on the s PE layer; *i.e.* that it sits underneath the adsorbed lipid layer. In this case one would expect that the collagen could also spontaneously adsorb onto the DOPE monolayer if it is injected in the subphase. This adsorption can be sensitive to the density of PE (*i.e.* surface pressure). Is there experimental data for spontaneous adsorption of Collagen from the subphase to the DOPE lipid layer at surface pressures around 1 mN m^{-1}?

Prof. Rathman responded: Our first approach was to try exactly what is described in this question: to spread a lipid monolayer over a subphase containing dissolved collagen monomer, or to inject collagen monomer into a subphase underneath a pre-formed lipid monomer. No evidence of collagen penetration into the monolayer was found in any of these experiments. The collagen monomer is water-soluble and, if injected into the subphase, is so dilute that the rate of polymerization is negligible. The dynamics are key to our process: the lipid monolayer acts to slow the partitioning of collagen monomer into the subphase so that polymerization occurs in the monolayer. Once polymerized, the collagen is no longer water-soluble.

Prof. Fermín opened the discussion of Dr Sastry's paper: In Table 1, it can be observed that the fractal dimension of the $CaCO_3$ crystals increases with increasing viscosity of the organic phase. What is the physical principle behind this correlation? Can the viscosity dependence of the fractal dimension be modelled quantitatively under the current experimental conditions?

Dr Sastry replied: The correlation between the finger width in the Hele–Shaw cell and parameters such as difference in viscosity of the two liquids, flow rate, interfacial tension *etc.* is well understood and parametrized in terms of a capillary number (see for *e.g.* ref. 1). Change in the finger width due to variation in any of these parameters would then change the nature of the viscous fingering pattern formation in a radial Hele–Shaw cell experiment and consequently, the fractal dimension of the liquid–liquid interface.

1 D. Bonn, H. Kellay, M. B. Amar and J. Meunier, *Phys. Rev. Lett.*, 1995, **75**, 2132

Prof Lynden-Bell asked: I was struck by the change in morphology between the low viscosity and high viscosity examples of $CaCO_3$ crystals grown in the Hele–Shaw cell in Fig. 2. Is the change from an approximately hexagonal morphology to one with angles of 90° due to a change in the preferred axis of growth or is there some other explanation?

Dr Sastry answered: We do not have a satisfactory explanation for the change in morphology/ assembly of the calcite crystals formed under different reaction conditions in the radial Hele–Shaw cell. A number of factors could be affecting the morphology of the calcite crystals such as diffusivity of the reacting species (calcium and carbonate ions) at the liquid–liquid interface, local hydrodynamic flow variation *etc.* Clearly, what we have at this stage is correlations between fractal dimensions of the liquid–liquid interface and structure of the calcite crystals—further work on understanding the factors mentioned above would be required to make firm statements on crystal growth at dynamic liquid–liquid interfaces.

Dr Whitby asked: What was the size of the gold nanoparticles used in the experiments investigating the assembly of the nanoparticles into quasi-linear superstructures? Could you speculate about the origins of the different cluster sizes of nanoparticles formed by the two different assembly protocols?

Dr Sastry replied: The gold nanoparticles used in this study were standard, borohydride reduced nanoparticles of 5 nm average size. The aggregation of gold nanoparticles at the liquid–liquid interface was initiated by the bifunctional cross-linking molecule, ethylenediamine. The extent of cross-linking is observed to be greater for chloroform than toluene suggesting that the dielectric properties of the organic solvent plays an important role in modulating the strength of interaction of the amine groups (the amine groups bind to the gold surface) with the nanoparticles. This is currently being investigated by isothermal titration calorimetry measurements of the diamine binding with gold nanoparticles (for an illustration of this technique to ligand binding with nanogold, see ref.1).

1   H. Joshi, P. S. Shirude, V. Bansal, K. N. Ganesh and M. Sastry, *J. Phys. Chem. B*, 2004, **108**, 11535–11540.

Prof. Fermín asked: In Fig. 7, the absorption spectra of Au aggregates exhibit two well defined plasmon bands. As mentioned in the paper, this behaviour can be taken as evidence of a preferential 1D aggregation of the colloids. Have you attempted to estimate the distribution of lengths of the aggregates? Can the extent of aggregation be affected by the viscosity of the organic phase?

Dr Sastry answered: It is difficult to estimate the extent of aggregation from the position of the longitudinal plasmon absorption band in the UV-Vis spectrum of the aggregated gold nanoparticles; this is better done by TEM which clearly establishes that the aggregation is greater in the chloroform experiment. While we haven't done the measurements with varying viscosity of the two liquid phases, we believe that the extent of aggregation and indeed, the nature of aggregation (open *vs.* compact structures) can be modulated using viscosity (among others) as a parameter.

Dr Walker asked: Can you comment on the opportunities for using a Hele–Shaw cell to create new materials from building blocks that are individually soluble in the two immiscible liquids?

Dr Sastry replied: That is a very good idea and would significantly enhance the power of the approach. We have some very preliminary results on the formation of gold nanoparticles by reduction of organically modified gold ions with reducing agents present in water and the approach appears promising.

Prof. Fermín asked: The results shown in the paper presented by Dr Sastry illustrate interesting possibilities for the synthesis of materials with a controlled morphology employing the Hele–Shaw cell. In principle, this strategy can be extended to two-phase reaction processes, confining the nucleation of the new phase to the liquid–liquid boundary. An example of this kind of process can be found in the poster presented by R. Knake *et al.* entitled "*Electrochemical Synthesis of Polymer covered Gold Nanoparticles at a Liquid–Liquid Interface*", in which the simultaneous nucleation of Au particles and polymerisation of tyramine are induced at the water/1,2-dichloroethane interface.[1] The poster presented by Platt and co-workers[2] also provides an alternative way for generating metallic phases at the liquid–liquid boundary. Would you expect that the interfacial reduction of metal cations in the aqueous phase by reducing agent in the organic phase can generate elaborate metallic structures in these cells? An interesting outcome of this approach could be the synthesis of metallic wires at the two-phase boundary. This phenomenon can be compared with metallic structured generated at the boundary between two fluids in parallel laminar flow.[3]

1   R. Knake, A. W. Fahmi and V. J. Cunnane, Electrochemical Synthesis of Polymer covered Gold Nanoparticles at a liquid–liquid Interface, *Faraday Discuss.*, **129**, poster 24.
2   M. Platt, E. P. L. Roberts and R. A. W. Dryfe, Controlled Deposition at the Liquid|Liquid Interface, Growth of Metal and Binary Alloy Nanoparticles, *Faraday Discuss.*, **129**, poster 9.
3   P. J. A. Kenis, R. F. Ismagilov and G. M. Whitesides, *Science*, 1999, **285**, 83.

Dr Sastry replied: The posters referred to did present some interesting results on the formation of gold nanoparticles by reduction of gold ions at a liquid–liquid interface that is accompanied by oxidative polymerisation of the reducing agent. In fact, we have published a nice account of reduction of gold ions at the interface between water and chloroform bearing a molecule with aniline functionality at both ends—this process leads to the one-step formation of a polyaniline network with gold nanoparticles embedded in them[1] —unfortunately, not referred to by the authors in the posters! Yes, carrying out such reactions at a *dynamic, expanding liquid–liquid interface* could lead to interesting structures and are currently being investigated.

1   P. R. Selvakannan, P. S. Kumar, A. S. More, R. D. Shingte, P. P. Wadgaonkar and M. Sastry, *Adv. Mater.*, 2004, **16**, 966.

Prof. Unwin asked: Is it possible to operate the Hele–Shaw cell under conditions where the interface expands at a constant rate (by decreasing the volume flow rate with time)? What would be the likely effects on the crystal growth experiment?

Dr Sastry replied: It would be possible to modulate the flow rate such that the interface expands at a constant rate. This would certainly affect the nature of crystals formed both in terms of the morphology and crystallography. In an earlier study,[1] we found that the flow rate could be modulated to enable/prevent dissolution of aragonite and recrystallization into calcite. Such processes would clearly be affected by interfacial growth rates and could permit polymorph selectivity in crystal growth experiments in the Hele–Shaw cell.

1   D. Rautaray, A. Banpurkar, A. V. Limaye, S. R. Sainkar, N. R. Pavaskar, S. B. Ogale and M. Sastry, *Adv. Mater.*, 2003, **15**, 1273.

Prof. Unwin asked: I presume that the $CaCl_2$ and Na_2CO_3 solutions are mixed just before the entrance to the cell. Nonetheless, is there a chance that the interfacial crystal growth process is influenced by the formation of nuclei homogeneously in solution, particularly as the time of the experiment increases?

Dr Sastry replied: The concentrations of the $CaCl_2$ and Na_2CO_3 solutions are adjusted so that the time scale associated with induction of crystal growth in the bulk of the solution is greater than the time of the Hele–Shaw cell experiment. The crystals are collected for analysis soon after the injection process. We are thus confident that the crystals obtained in the cell are due to nucleation and growth at the interface.

Prof. Unwin asked: Does the induction time not depend on how it is measured? Indeed, induction time measurements *via* turbidity measurements will be insensitive to nanoscopic nuclei in solution that might influence crystal growth.

Dr Sastry answered: True, the induction time depends on the method of measurement. We have relied on SEM analysis of the reaction solution to check the time required for formation of calcite crystals. However, it is possible that nanoscopic nuclei could have been overlooked by this approach.

Dr Bain asked: The interpretation of Dr Sastry's beautiful experiments is complicated by the fact that the hydrodynamics are evolving continuously during the experiment. It would be interesting to look at an experimental system in which the interfacial dilation rate remained constant. In the 1990s, Dr A. Prins at Wageningen University in The Netherlands developed an overflowing cylinder as a means of generating a continuously expanding oil–water interface at a constant rate of strain. I do not know if this work ever resulted in a journal publication, but it is an interesting idea that could be useful in the study of nanoparticle synthesis at L/L interfaces.

Dr Sastry replied: I would like to thank Dr Bain for this important information and I will certainly try to obtain this publication.

Prof. Kornyshev asked: When you have measured fractal dimension of your viscous fingers, what was the lower self-similarity cut-off? In which length scale do you see this fractality?

Dr Sastry answered: The fractal dimensions were obtained on the images of the viscous fingering patterns in the Hele–Shaw cell and therefore, at macroscopic length scales.

Prof. Kornyshev suggested: making the Hele–Shaw cell with two immiscible electrolytes, not just immiscible solvents. Make electrodes at the sides of the cells, and try to see what will be the electric field effect on viscous fingering? This suggestion/discussion goes back to our discussion yesterday of field-induced corrugation and instability at ITIES.

Dr Sastry responded: It would be an interesting challenge to design a Hele–Shaw cell with electrodes on either side of the interface. There could be interesting effects on materials synthesis due to the presence of a field at the interface.

Prof. Karyakin opened the discussion of Dr Marken's paper: Have you tried different organic solvents in order to change a polarity?

Dr Marken replied: It would be very interesting to explore the effect of polarity or solvent dielectric constant in more detail in particular exploring specific ion–solvent interaction effects. However, at the moment data for other types of solvents are not available.

Dr Wakisaka commented: PPP (4-(3-phenylpropyl)-pyridine) will have an electron donating property. This can solvate Li^+, Na^+, K^+ *etc.*, preferentially. If so, these alkali cations will be easy to transfer from the aqueous phase to the organic PPP phase. I would like to suggest this preferential solvation effect as a reason for the Li^+, Na^+ and K^+ being deviated from the linear relationship in Fig. 7.

Dr Marken answered: This is a very good comment and we fully agree with the proposed explanation of the midpoint potential data for Li^+, Na^+ and K^+ (see Fig. 7).

Prof. Fermín addressed Dr Marken: It appears that one of the motivations of this work is to increase the electrochemical responses originating from charge transfer across the liquid/liquid interface by supporting the molecular boundary on a highly porous substrate. However, the results shown in Fig. 10 reveal that the magnitude of the current of the "porous" $ITO/TiO_2/Au$ electrode is smaller than the one obtained with the basal plane pyrolitic graphite electrode. These results effectively confirm that current enhancement requires that the thickness of the immobilised layer have to be comparable to the corrugation of the Au sputtered TiO_2 mesoporous film (*i.e.* nanometer dimensions). Has any strategy been envisaged to achieve this goal?

Dr Marken replied: The performance of basal plane pyrolytic graphite electrodes as so-called biphasic electrodes (in simultaneous contact with two liquid phases) is excellent in particular with the right amount of deposit per electrode area. The results obtained with the $ITO/TiO_2/Au$ electrode are preliminary and the interpretation given in terms of porous structure depth *versus* deposition volume is correct. Improved versions of this type of electrode will be necessary and the effect of the electrode coverage on the extent of the triple phase boundary zone will be explored in more detail.

Prof. Fermín added: In connection to the results in Fig. 10, no clear explanation is given for the absence of current responses with the ITO/TiO_2 electrode. I presume that electrons cannot be injected from the CoTPP into the conduction band of the insulating TiO_2 particles. In the presence of Au, oxidation of CoTPP will increase in the Fermi level of the metal allowing the injection into the conduction band of the TiO_2. Do the authors share this interpretation? An interesting aspect that could be investigated is whether photo-excitation of CoTPP could also promote electron injection into the TiO_2, which could manifest itself by anodic photocurrents.

Dr Marken responded: This is a very good comment and probably correct in terms of the physical picture. For the interpretation of data shown in Fig. 10c, it has to be assumed that the PPP phase is blocking access to the ITO electrode surface and conductivity through the TiO_2 layer is insufficient.

The gold layer introduces sufficient conductivity for the electron transfer/ion transfer process to proceed.

Photochemical studies at electrodes of this type (both with and without the gold layer) would be very interesting and should be conducted.

Prof. Scholz asked: Did you try to calculate the standard free energies of transfer of ions from water to PPP?

Dr Marken replied: We have chosen to represent the experimental data in terms of midpoint potentials in order to simplify the comparison of anion and cation transfer data (see Fig. 7). However, the calculation of (approximate) standard free energies would be possible and interesting. With additional measurements for the tetraphenylarsonium cation and the tetraphenylborate anion, scales for anion and cation data could be unified and compared to literature data.

Prof. Scholz asked: Do you think that ion pair formation in PPP may be responsible for the deviation of the dependence of formal potentials on ion activity?

Dr Marken replied: This is a good suggestion and very likely particularly in the case of highly hydrophilic anions and cations in non-polar liquids. The processes which are likely to cause deviation from Nernstian behaviour are (i) ion interaction/pairing, (ii) the formation of a separate phase, (iii) partitioning of ions not directly involved in the dominating redox/ion transfer process, (iv) co-insertion of water and (v) surface energy effects for very small droplets or microphases. The $CoTPP(L)_2^{+/0/-}$ redox system is highly soluble in PPP and unlikely to exhibit inner sphere ion pairing, but currently outer sphere ion interaction cannot be ruled out.

Prof. Girault asked: If the current at a 3-phase electrode occurs along a line, what is the typical current density in $A\ m^{-1}$? And then what is the maximum you have measured with the mesoporous 3 phase system?

Dr Marken replied: This is a very interesting question. Treatments of an experimental[1] and theoretical[2] nature describing currents at line interfaces have been published. In my view, the notion of a 'mathematical line' and a current density with dimensions $A\ m^{-1}$ is misleading and the physical reality of the process is closer to that of nanoband electrodes or heterogeneous surfaces encountered for example at graphite surfaces.[3] Our aim is to increase the extent of an 'active' triple phase boundary zone at the electrode surface to achieve conditions typical for heterogeneous electrode surfaces and consistent with conventional $A\ m^{-2}$ dimensions.

1 (a) F. Scholz, S. Komorsky-Lovric and M. Lovric, *Electrochem. Commun.*, 2000, **2**(2), 112–118; (b) J. C. Ball, F. Dr Marken, F. L. Qiu, J. D. Wadhawan, A. N. Blythe, U. Schroder, R. G. Compton, S. D. Bull and S. G. Davies, *Electroanalysis*, 2000, **12**(13), 1017–1025.
2 (a) M. Lovric and F. Scholz, *J. Electroanal. Chem.*, 2003, **540**, 89–96; (b) F. L. Qiu, J. C. Ball, F. Marken, R. G. Compton and A. C. Fisher, *Electroanalysis*, 2000, **12**(13), 1012–1016.
3 See for example: C. E. Banks, R. R. Moore, T. J. Davies and R. G. Compton, *Chem. Commun.*, 2004, (16), 1804–1805.

Prof. Scholz addressed Dr Marken: Your paper and other contributions to this conference show very well that there is arising a new field of research, *i.e.* the structure of triple-phase boundaries (liquid–liquid–solid, in most cases). This research will be even more difficult than that of liquid–liquid, *i.e.* two-phase interfaces, mainly because of the dimensionality. The triple-phase boundary is a line.

Dr Marken responded: The significance of this type of line interface in technical developments can be seen not only for solid | liquid | liquid interfacial processes but also for solid | solid | liquid and for solid | solid | gas type processes. In our work, we are aiming at maximising the extent of the triple phase boundary zone to allow conventional methods to be applied for the analysis of these processes. However, new approaches are required.

Dr Grayeff addressed Dr Marken: Does the speaker, consider the use of other electrodes apart from gold might be useful in the determination of ion transfer processes?

Dr Grayeff commented:

(1) A Li/polyethylene/vermiculite solder electrode can be utilised in micro emulsion ion transfer reactions.[1]

(2) For example an Al/C/alkyl ammonium vermiculite derivative film, in particular butyl ammonium and other similar matrices can be utilised in (pH) measurements using a calomel electrode to show varying types of ion transfer reactions in aqueous solutions. In this case the main reaction is thought to be the action of the (H^+) ion.[2]

(3) For example a silver/silver chloride electrode can be utilised in (pH) measurements of a bentonite polystyrene polymer aqueous derivative.[3]

1 S. G. Grayeff, *Vermiculite Modifications*, M. Phil. Thesis, 1992.
2 Compiled by S. G. Grayeff, *Research Report*, The Civil Engineering Dept., The Technion, 2002.
3 M. Brewer, J. Kiner and S. G. Grayeff, *The Physical Properties of a Polystyrene Matrix*, Internal Research Report, Unilever Research Labs, 1971.

Dr Marken replied: This is an interesting question and the answer is almost certainly, yes. The conductor, gold, seems unsuitable for more positive potentials where gold oxidation occurs and for more negative potentials where hydrogen evolution is expected. There are many possible designs for efficient triple phase boundary electrode systems and improved electrodes with optimised porosity, conductor coating, and surface wetting properties will become available for specific applications in the future.

Prof. Janata opened the discussion of Dr Wakisaka's paper: Would it be possible to use your approach to determine the size of the water cluster surrounding a solute (*e.g.* Li^+) and/or the composition of the solvatomers?

Could this approach be used to determine the Gibbs free energy of transfer by counting the number of respective molecules in the solvatomer?

Dr Wakisaka replied: If a solute–solvent interaction is relatively strong such as electrostatic, hydrogen-bonding, dipole–dipole interaction, *etc.*, the solute–solvent interaction can be observed as cluster formation. When the solvent is a binary mixture, the composition of the solvated molecules is determined by our mass spectrometry.[1] In the case that the solute is an ionic species, an electrospray-type nozzle is used for the formation of liquid droplets instead of a thermospray-type nozzle used here.[2]

The temperature effect on the observed solvated clusters should be useful to determine the Gibbs free energy for the solvation. We have already reported that the solvation for phenol in water–acetonitrile binary mixtures was changed by the increase of temperature.[3] The enthalpically stable solvated clusters were disintegrated with an increase of the temperature.

1 A. Wakisaka, Y. Shimizu, N. Nishi, K. Tokumaru and H. Sakuragi, *J. Chem. Soc., Faraday Trans.*, 1992, **88**, 1129.
2 H. Kobara, A. Wakisaka, K. Takeuchi and T. Ibusuki, *J. Phys. Chem. B*, 2003, **107**, 11827.
3 A. Wakisaka, S. Takahashi and N. Nishi, *J. Chem. Soc., Faraday Trans.*, 1995, **91**, 4063.

Prof. Girault asked: How can we distinguish structures stemming from the bulk properties and those stemming from the fragmentation process or from the different rates of vaporisation?

Dr Wakisaka replied: In our mass spectrometry for clusters, the clusters are generated through the fragmentation of liquid droplets *via* adiabatic expansion. Accordingly, the weakly interacting molecules are vaporized without forming clusters. In this sense, the observed clusters do not reflect the real structure in the bulk solution. However, the resulting clusters indicate that the molecules forming clusters have relatively strong intermolecular interactions in the bulk solution. From this mass spectrometry, we have information on the microheterogeneous structures caused by the difference in the intermolecular interactions in a solution.

The cluster-formation process is not directly related to the rate of vaporization. This is clearly demonstrated by the example shown in our paper presented here. From the water–alcohol (ethanol or propanol) mixtures, the alcohol which is more volatile than water forms self-association clusters predominantly. On the contrary, for the methanol—larger alcohol (1-butanol or 1-pentanol) mixtures, the larger alcohol which is less volatile than methanol forms self-association clusters. Therefore, the cluster formation is not dependent on the rate of vaporisation. It should be noted that the cluster formation is mainly controlled by the intermolecular interactions in the liquid state.

Dr Bain asked: In the two-phase co-existence regime of butanol–water mixtures, what do you think the interface between the butanol-rich and water-rich phase will look like? Will the butanol clusters in the water-rich phase be adsorbed or depleted at the interface?

Dr Wakisaka replied: This indicates a fundamental point of our paper. Although we observed cluster structures in the butanol-rich phase and the water-rich phase separately, it is difficult to speculate the clustering structure of the interface on the basis of the observed cluster structures. However, we will be able to discuss the phase separation mechanism in consideration with the behaviour of butanol clusters as you indicated.

When the water was added into the pure butanol, the self-association of butanol was promoted by the 'complementary relationship' as shown in our paper. The added water molecules existed as monomeric molecules. When the concentration of water attained *ca.* 80 wt%, the monomeric H_2O would be saturated in the mixture. The monomeric H_2O molecules will exist in small spaces generated by the self-association of butanol molecules. With further increasing the water concentration, the water molecules were localized to generate the water-rich phase. On the other hand, when the butanol was added into the pure water, butanol molecules without having hydrogen-bonding interactions with water molecules formed self-association clusters. At the butanol concentrations higher than 6 wt%, these butanol self-association clusters would form the butanol-rich phase accompanied with monomeric water molecules.

In the two-phase region, the interface will be formed between the 1-butanol self-association clusters in the butanol-rich phase and the hydrogen-bonded water clusters in the water-rich phase. The butanol self-association clusters also exist in the bulk of the water-rich phase; however, these butanol self-association clusters would not be adsorbed at the interface, because they exist in the water-rich phase as a result of the fact that the butanol is dissolved in water as much as possible.

Dr Jedlovszky asked: Have you also investigated the possible presence of self-association clusters in other aqueous binary mixtures, such as water–acetone or water–acetonitrile?

Dr Wakisaka replied: We have already reported on the cluster structures in several aqueous binary mixtures including water–acetone and water–acetonitrile binary mixtures.[1,2] In water–acetone and water–acetonitrile binary mixtures, the hydrogen-bonded water clusters were stabilized rather than the self-association clusters of acetone and acetonitrile. There is a striking contrast between the clustering in these binary mixtures and that in the water–alcohol binary mixtures.

1   A. Wakisaka, H.-A. Carime, Y. Yamamoto and Y. Kiyozumi, *J. Chem. Soc., Faraday Trans.*, 1998, **94**, 369.
2   D. N. Shin, J. W. Wijnen, J. B. F. N. Engberts and A. Wakisaka, *J. Phys. Chem. B*, 2002, **106**, 6014.

Prof. Fermín asked: The various mass spectra reported in this paper are expected to be extremely sensitive to the conditions of vaporisation/fragmentation. What are the key experimental conditions to be controlled in order to obtain reproducible mass spectra of the water–alcohol binary mixture?

Dr Wakisaka answered: The observed clusters are generated through fragmentation of liquid droplets *via* adiabatic expansion in a vacuum chamber. Therefore, the temperature of the liquid droplets and the pressure difference in the four-stage differentially pumped vacuum system worked as controlling factors. When these factors are constant, we can obtain the reproducible mass spectra. Furthermore, the cluster structures in the water–alcohol mixtures are sensitive to the mixing ratio between water and alcohol.

Dr Walker addressed Dr Wakisaka: Are you surprised at the absence of dimer, trimer, *etc.* formation in butanol (100% by weight)? Do the distributions in relative cluster intensities (and changes in distribution as a function of water content) contain any information about the thermodynamics/equilibrium properties of the cluster sizes?

Dr Wakisaka responded: As we explained here, the distribution of cluster is determined by the relative interaction force working on the molecules at the same time in a solution. Accordingly, the intermolecular interactions at an equilibrium condition should be reflected in the resulting clusters. The reason for the absence of dimer, trimer, *etc.* in butanol will be explained by the absence of hydrogen-bonding interaction in pure butanol. A randomly mixed structure having inter-alkyl interaction will prefer a hydrogen-bonded clustering structure in pure butanol. Hydrogen-bonded butanol cluster formation is promoted by mixing with water.

Dr Walker asked: If steric interactions play such an important role in inhibiting dimer formation (in butanol and higher C_n), do you expect to observe a similar behaviour for branched alcohols (*i.e.* 2-propanol, isobutanol, *tert*-butanol, *etc.*)?

Dr Wakisaka replied: The inhibition of dimer formation in pure butanol and higher alcohols is based on the increase of inter-alkyl group interaction as mentioned above. The dimer formation should be controlled by the balance between the inter-alkyl interaction and the hydrogen-bonding interaction. Since the secondary and the tertiary alcohols have high steric hindrance for the hydrogen-bonding interaction, for these alcohols it will be difficult to form their dimers in pure alcohol. On the other hand, the formation of alcohol self-association clusters through hydrogen-bonding interaction, promoted by mixing with water or other solvent of smaller size, is clearly influenced by the steric effect. The self-association of *iso-* and *tert*-butyl alcohol is less favourable than that of 1-butanol.

Prof. Karyakin opened the discussion of Dr Oldfield's paper: Hyperactivity is not the right term for the observed phenomenon, because whereas k_{cat} is increased, the K_m is also increased, and their ratio, which is always referred as enzyme activity is nearly unchanged.

Dr Oldfield responded: We did not coin the term 'hyperactivity', but it seems reasonable enough to use it, where it can be shown that the enzyme genuinely is more active in the microemulsion (or any other) environment than in aqueous solution. Most workers in the field by now accept that at least under some circumstances, enzymes are 'hyperactive', though there is an argument about the degree of that hyperactivity. In the cases we have investigated, including those reported here, hyperactivity (at optimal R), is modest, at the level of 2–3-fold relative to an aqueous solution of the same nominal composition as the microemulsion's aqueous subphase.

I don't agree that changes in k_{cat} and K_m somehow cancel each other out, even as a generalization, and its certainly not a defining feature in the studies we have made. Referring to the work on the *C. viscosum* lipase catalysed hydrolysis of 4-nitrophenyl acetate given in the paper, the activity reported as $k_2 = k_{cat}/K_m$ is clearly strongly dependent on R (so there are factors that do not cancel out)—and the enzyme is hyperactive relative to water by a factor of around 2–3, after partitioning of the substrate has been taken into account.

Prof. Karyakin commented: The mechanism for k_{cat} increase is now generally accepted: When micelle dimensions coincide with the enzyme dimensions the maximum for k_{cat} is observed. Some enzymes able to form aggregates display several extremes of k_{cat}, each corresponds to the stable aggregate dimensions.

Dr Oldfield answered: Well, I don't think it is generally accepted. As we state in the paper, we believe the 'tight-fit' model and its underlying assumption, that coincidence of enzyme- and droplet size somehow forces the enzyme into a more active conformation, to be much too simplistic. I think that the apparent coincidence of droplet size and enzyme activity maxima—reported for a small number of enzymes spanning a very small size range is exactly that—a coincidence. We did analytical ultracentrifugation studies of α-chymotrypsin in AOT microemulsions[1] which shows that below $R = 15$ there probably is not a complete droplet around the enzyme—the sedimentation coefficients are just too small for any reasonable model based on a complete shell of water then surfactant, and that it should be more rightly regarded as a 'hydrated enzyme–surfactant complex'. Thus, at $R = 7$ (the activity maximum for α-chymotrypsin), we have a picture of a hydrated enzyme–surfactant complex where the amount of water present is just sufficient to fully activate the enzyme, and the enzyme is then hyperactivated because of the field effect of lone counter-ions which are present at a very high concentration, in the $R = 7$ system.

1   C. Oldfield, R. B. Freedman and B. H. Robinson, *J. Chem. Soc., Faraday Trans.*, 1996, **92**, 73.

Dr Caruana commented: Regarding the proposed stabilisation of the transition state by the freely mobile 'lone' counterion to change the activity of the enzyme catalysed reaction. The transition state for the reaction is stabilised very precisely by the active site of the enzyme. It would be very difficult to expect an external ion to enter the active site and do something useful to stabilise the transition state and increase the apparent rate of reaction.

Dr Oldfield replied: We do not mean to suggest that these ions actually enter the active site, and we avoided speaking in such terms in the paper. If sodium ions were to stably penetrate the active site of the enzyme substrate complex, then a roughly spherical volume of 0.09 nm^3, equivalent to a diameter of about 0.56 nm is required to accommodate the ion. This is the value of the hydrated sodium ion plus a little to allow for stable (attractive) electrostatic interactions with neighbouring groups. Although, as X-ray crystallography shows very clearly, there are voids, or holes, in the active sites of enzyme substrate complexes, they are in general smaller than this. Therefore we reject a model based on hydrated ions entering the active site. Furthermore we would not accept a model based on penetration of the much smaller dehydrated sodium ion, since there is no case to suggest that sodium ions are dehydrated, or can be become so, in this rather water-rich system (water activity ~ 1, over the range of R for which the linear relationship between $\log k$ and $1/R^{1/2}$ applies).

Our preferred model is a field effect, that is, we believe that 'lone'—that is, unshielded, counterions in the water-pool generate an undamped, and therefore rather extended electric field gradient, and the enzyme sits in, and its activity is strongly influenced by, this gradient. This model has the dual attraction that it allows the phenomenon to be non-specific with respect to the enzyme, since it requires no assumptions about topology of the active site of the enzyme–substrate complex, and it explains why the phenomenon cannot be duplicated in bulk aqueous sodium chloride solutions (where chloride counter-ions severely dampen the field strength of the sodium ions).

To say much else at this stage would be arm-waving and more work needs to be done—some rather difficult modelling, for example—any takers?

Dr Caruana added: The conditions within the micelle can be different to that in the bulk solution or even from the solution used to fill the micelle. It is well known from partitioning of acids and bases in oil and water mixtures that apparent pK_a shifts are observed. These effects are attributed to nonequilibrium conditions within the diffusion layer (or the unstirred water layer). This is explained by protonation which occurs prior to transfer from water to oil as in the case of a base A^-, $H^+ + A^- \rightarrow HA$. This protonation reaction alters the equilibrium due to the depletion of the protonated base (HA) as it crosses the interface into the oil, as a consequence the pH decreases in the diffusion layer.

It is possible that a similar effect is occurring in these enzyme filled micelles. The actual pH inside the micelles can be very different to the pH of the solution that was used to fill the micelles even if the solution is strongly buffered. A shift of one pH unit may give a substantial change in reaction rate. This effect cannot be ruled out to describe the change in reaction rate observed.

Dr Oldfield answered: Well, I think it can. Firstly, there is no unstirred layer in a microemulsion system. The unique feature of microemulsions (as opposed to macroemulsions and other biphasic systems) is that the dispersed aqueous subphase exists as monodisperse droplets with radii of the order of nanometers. Correspondingly, the interfacial surface area for a typical microemulsion (0.1 M AOT, $R = 25$), is very large (of the order of 30 m^2 per millilitre of microemulsion). Given that the surface area occupied by a single AOT molecule is about 0.5 nm^2 then it is easy to calculate that an equivalent two-phase preparation would consist of an oil (n-heptane) layer about 31 nm thick and an aqueous phase about 1.5 nm thick. In such a system simple convection is more than sufficient to ensure the absence of stagnant layers.

The pK_a of a buffer may change due to partitioning, as you say, but such effects can be controlled using sufficient concentrations of highly water-soluble neutral-base buffers, as we have done in this work. The pH of a microemulsion cannot be measured directly using a glass electrode, but good evidence that we have buffering under control comes from our study of the R-dependence of 4-nitrophenyl sulfonate and pyranine, where the extrapolated infinite-R value of the pK_a for both of these dyes (ordinate intercepts in Figs. 1 and 3) is equal to their pK_a° measured in aqueous solution. We do not regard this as a coincidence: The result tells us that our microemulsions are well buffered, and that the R-dependence of the pK_a is due to the ionic strength contribution of water-pool lone counter-ions, which influences this neutral–acid equilibrium according to a simple Debye–Hückel mechanism—this being diluted out at infinite-R. The neutral–base buffers used in these experiments are of course insensitive to primary ionic strength effects.

In unbuffered microemulsions the principle mechanism determining pK_a for NPS and pyranine is ion-exchange of protons for sodium ions, so that the anionic AOT layer, acting as a proton sink, causes a lowering of the pK_a of dyes such as 4-nitrophenyl sulfonate or pyranine by around 1–2 pH units depending on R (unpublished data).

Dr Dryfe addressed Dr Oldfield: Your ionic strength model would suggest that the surfactant counter-ion's charge should have a strong effect on reaction rate. Have you attempted to use, for example, alkaline earths as the counter-ion?

Dr Oldfield replied: Absolutely, I agree. We would indeed expect to see some interesting effects according to the field strength of the ion, which is a function of its charge : radius ratio and degree of hydration. Although unfortunately we have not carried out any experiments with the other salts of AOT, we hope to get round to looking at these at some point. The difficulty in doing this work, apart from preparing the appropriate AOT-salt, is identifying the water + organic solvent + AOT salt composition ranges that correspond to microemulsion-phases for each of the AOT salts in question, and this basically requires the construction and interpretation of equilibrium temperature-phase diagrams, before any useful enzymology can be done. This is quite an undertaking.

Prof. Janata asked: Is the activity of "free" (*i.e.* uncoordinated) water inside the droplet dependent on its size?

Dr Oldfield responded: It is. The presence of free water can be equated with high water activity, measured for example, using vapour pressure osmometry.[1] Water activity is low below about $R =$ 10 (*i.e.* below 10 mol water per mol AOT) because that water hydrates the AOT sulfonate cation and the sodium counter-ion. As R is increased above 10, water activity increases as free water accumulates until it reaches 1, that is it has the same value as bulk water, at around $R = 20$.

1   R. Kubik, H.-F. Eicke and B. Jönsson, *Helv. Chim. Acta*, 1982, **65**, 170.

Prof. Janata added: Activity of "free" water can have a more profound effect on the activity of a solute than the electrostatic contribution resulting from ionic strength.

Dr Oldfield replied: I think that's exactly what we are saying. At high R ($R > 10$), the water activity is high—close to that of bulk water, as I indicated in my answer to your previous question, and that's where we see the linear dependence of log k on $1/R^{1/2}$ that we attribute to the ionic strength effect of lone counter-ions. But, this behaviour holds only up to a point and below a critical value of R, in the range 7–12, depending on the particular enzyme, k decreases dramatically, and linearly, with decreasing R, falling to zero at around $R = 3$. This low-R deviation from the linear dependence of log k on $1/R^{1/2}$ we attribute to the impaired operation of the enzyme in a low water-activity environment, that is, we regard the enzyme in this system as being partially dehydrated.

Dr Bain asked: Branka Ladanyi has simulated water/alkane/AOT microemulsions and finds that almost all the Na^+ counterions are associated with sulfonate head groups and are dehydrated. Can you comment on the results of these simulations in the context of your model?

Dr Oldfield answered: That study[1] covers the range $R = 1$–10 only, so it is not surprising that the Na^+ in these systems is largely dehydrated. The hydrated sodium ion is typically co-ordinated by 4–6 water molecules, and the sulfonate head-group by about 4 molecules of water, so full hydration cannot be expected until at least $R = 8$. ^{23}Na NMR spectroscopy by Thomas and co-workers[2] indicates that the sodium ions are dehydrated below $R = 6$, but above that R, they are fully hydrated and distributed between the interface and the water pool, with a maximum value for α_R of 0.2 (*i.e.* at least 80% of Na^+ bound at the interface, the remainder free in the water-pool).

The key issue seems to me though, to be not whether or not the ions are dehydrated, but the (negative) electrical field strength within the core of the droplet, where we perceive the reaction to take place. Therefore, extending the modelling to including the potential distribution, and at higher R would be most welcome.

1   J. Faeder and B. M. Ladanyi, *J. Phys. Chem. B*, 2000, **104**, 1033.
2   M. Wong, J. K. Thomas, and T. Nowak, *J. Am. Chem. Soc.*, 1977, **99**, 4730.

Dr Bain asked: If there is only around one free Na^+ ion per micelle, how useful is it to use a mean field picture?

Dr Oldfield replied: I do believe that a mean field theory can be applied to this system. Consider that a typical microemulsion preparation with 0.1 mol dm^{-3} AOT at $R = [H_2O]/[AOT] = 25$, contains 450 µl aqueous solution per ml (so volume-fraction $\phi_{aq} = 0.045$)—which is substantial. 1 ml of this preparation contains around 10^{17} droplets. However, the picture of a microemulsion as a collection of very tiny (nanometer-diameter) water droplets stably suspended in a continuous oil phase is not a very accurate one because it ignores the dynamics of the system. Actually, the droplets collide with each other and exchange their contents on the millisecond timescale, which is comparable with the lifetime of a typical enzyme–substrate complex. Therefore the enzyme, during the catalytic cycle, is not confined within a single droplet, but is free to explore the volume of the aqueous subphase, where 'subphase' is a term which we have chosen with care to emphasize the fact that this water really does behave as a separate phase, although it is contained within another phase (for example, the water activity, a_w, may be measured by vapour pressure osmometry—as one might do for a *bona fide* 2-phase system—and is not significantly different from bulk water when $R > 15$, approx.). Therefore, from the enzyme's point of view, it is the free ion concentration within the significant volume of the entire aqueous subphase, and not that within a single droplet, that is relevant to a simulation of the microemulsion relevant to its effects on enzyme activity, and this seems to me accessible to mean field theory.

I think, in addition, that I can go a little further and try to give an insight into the concentration of free sodium ions in the aqueous subphase. This value is actually buried within the slope values of the linear plots of pK_a (of pyranine or 4-nitrophenyl sulfonate) *vs.* $(1/R)^{1/2}$ shown in Figs. 1 and 3 of the paper. From eqn. (9) the slope C_1 is given by,

$$C_1 = -A \; |z_A z_B| \; (27.8 \; \alpha_R)^{1/2}$$

where α_R is the fractional dissociation of Na^+ from the surfactant, $|z_A z_B|$ is the absolute value of the product of the charges on the acid z_A, and its conjugate base, z_B and A is the coefficient of the Debye–Hückel limiting law (in our simple derivation). For pyranine, $|z_A z_B| = 4$. Now, we don't know what the numerical value of A is right now, because the well-known value of $A = -0.509$ was derived strictly for dilute electroneutral solutions, and the microemulsion aqueous subphase, if you accept our 'lone-counter-ion' model, cannot be also electroneutral. But let us assume for the moment that the actual value is not very different from -0.5. Then, for a typical value of $C_1 = -4.0$ (Table 1 in the paper), we get a value for α_R of 0.14, independent of R.

Returning to our exemplar $R = 25$ system then, since $[AOT] = 0.1$ mol dm^{-3}, $0.14 \times 0.1 = 0.014$ mol Na^+ per litre microemulsion is dissociated from the interfacial AOT layer. Since for this system $\phi_{aq} = 0.045$, the corresponding aqueous subphase concentration is $(0.014/0.045) = 0.3$ M.

In other words, then, taken at face value, the enzyme in the microemulsion is immersed in a solution of 300 mM lone Na^+ at $R = 25$, and by similar calculation, about 800 mM at $R = 10$ and 150 mM at $R = 50$. Now, of course the Debye–Hückel limiting law only holds, for electroneutral solutions at rather low ionic strengths ($I < 0.01$) and we do not expect it to hold very well for none-dilute microemulsion systems, either. Indeed, if you study Fig. 3 in particular you can see that there is deviation from the limiting slope at low R (high $[Na^+]$) which we attribute to the breakdown of the limiting law. However, we do not know what the threshold would be for a non-electroneutral system, anymore than we can assign a reasonable value of A, at the present time. Therefore please accept these numbers as an approximation, and that there is the need for considerable refinement. I merely point out at this point that the numbers are not 'fantastic'.

Finally, it is instructive perhaps to convert these values into Na^+ per droplet. The values are 16 ($R = 10$), 60 ($R = 25$) and 200 ($R = 50$). (At fixed [AOT] the droplet concentration decreases, and the droplet size increases, with increasing R. The numbers used for this calculation are in Oldfield *et al.*[1]

1   C. Oldfield, R. B. Freedman and B. H. Robinson, *J. Chem. Soc., Faraday Trans.*, 1996, **92**, 73.

Prof. Rusling added: With respect to Colin Bain's question, I don't think you can compare simulations of w/o micelles without protein to the experimental situation with protein (enzyme) in the water pools. There would be coupled equilibria in the latter case between Na^+ ions in the pool, bound to AOT head groups, and bound to the enzyme. (There may also be AOT bound to the enzyme.)

Dr Oldfield responded: That's absolutely right. I agree entirely and it's no simple task. But remember that non-enzyme reactions can behave in the same way, that is show a linear dependence of log k on $(1/R)^{1/2}$, as we have shown in this paper for the aminolysis of 4-nitrophenyl acetate by glycylglycine. On this basis simulation of a no-protein system has value since it is directly relevant to this simple bimolecular reaction.

Prof. Fermín opened the discussion of Prof. Rusling's paper: The low solubility of ferrocene in water will confine most of the reactant to the oil droplets or the bicontinuous film. However, the ferricenium ion has non-negligible solubility in water and it is expected to transfer to the aqueous phase.[1] How stable are the voltammetric responses (*e.g.* Fig. 3) as a function of the number of cycles?

1   D. J. Fermín and R. Lahtinen, Dynamic Aspects of Heterogeneous Electron Transfer at Liquid|Liquid Interfaces, in *Liquid Interfaces in Chemical, Biological and Pharmaceutical Applications*, ed. A. G. Volkov, Marcel Dekker Inc., New York, 2001, ch. 8, pp. 179–227.

Prof. Rusling replied: This is correct. Ferricenium is likely to be ejected into the water phase rapidly after it forms. However, we measured the voltammetry of *ferrocene* to get the diffusion coefficients, and the peak size and diffusion coefficient estimated from CV will not depend on the occurrence of this event subsequent to the electron transfer. It's similar to a so called EC reaction. The CVs are quite stable and reproducible, especially since we polish the electrode after each scan.

Dr Dryfe asked: You attribute the differences in voltammetric response between the micelle and bicontinuous emulsion to differences in mass transport. How do these large differences arise?

Prof. Rusling answered: Mass transport of non-polar reactant in the bicontinuous microemulsions occurs within the tubules of the continuous oil phase, and is relatively less restricted that in the other fluids. In micelles and o/w microemulsions, the reactant is bound to a micelle or within an oil droplet, and these larger particles are necessarily transported more slowly than a solute in a continuous phase.

Dr Dryfe asked: Do the changes in mass transport correlate with other properties of the fluids, such as—for example—their viscosities?

Prof. Rusling replied: Mass transport rates in microemulsions generally do not correlate with bulk viscosities but with so called "microviscosities" representative of the residence site of the solute. Mass transport rates of solutes depend on where they are located in the microheterogeneous fluid. Solutes in continuous phases generally have relatively faster mass transport, but a bit slower than in the related pure solvent because of a percolation effect, such as in a porous solid material. Solutes in the dispersed phase, *e.g.* oil droplets, have much slower mass transport because the entire droplet must be transported.

Antibiotic assisted molecular ion transport across a membrane in real time

Jian Liu, Xiaoming Shang, Rebecca Pompano and Kenneth B. Eisenthal*

Department of Chemistry, Columbia University, New York, NY 10027 USA. E-mail: eisenth@chem.columbia.edu

Received 14th April 2004, Accepted 5th May 2004
First published as an Advance Article on the web 9th November 2004

The transport of an organic cation across a 4–5 nm liposome bilayer is observed in real time using second harmonic generation. It is proposed that an electrostatic barrier between the inside and outside of the liposome develops as the cation crosses the bilayer. This would explain why the SHG signal does not approach zero at long times. To test this mechanism, the antibiotic valinomycin, which can transport alkali ions across a phospholipid bilayer, is introduced into the system. It is found that the transport time is reduced by a factor of three from 90 ± 2 s to 30 ± 1 s with 1.25×10^{-8} M valinomycin concentration, and a factor of fifteen to 6.2 ± 0.2 s with 1.25×10^{-8} M valinomycin concentration. In addition, the SHG signal approaches zero, which further supports the presence of an electrostatic barrier that can be eliminated by the alkali ion transporter valinomycin.

1. Introduction

Phospholipid bilayers are the basic structural feature of biological membranes, which most importantly serve to control the passage of chemical species between the inner enclosed aqueous region of the cell and the outer aqueous solution.[1] Structures similar to membranes, such as liposomes, can be prepared in the laboratory. These membrane mimetic structures, consisting of a bilayer, which are made up of amphiphiles, such as phospholipids, have been extensively used as simplified models of biological membranes.[1] Basic research studies of various chemical and physical phenomena as well as applications such as solar energy conversion, catalysis, cosmetics and drug delivery are areas of considerable interest.[2,3] An issue of special importance is the transport of chemical species, ranging from inorganic and organic ions to neutral species, across the bilayer of a biomembrane.[4–13] With liposomes one can investigate the transport of chemical species across bilayers of differing phospholipid composition and net charge. This latter point is of interest because many biological membranes are negatively charged due to negatively charged phospholipids in the bilayer.

In earlier studies we have shown that the nonlinear spectroscopic method of second harmonic generation (SHG) can be used to selectively probe the surfaces of both centrosymmetric and asymmetric microscopic particles and nanoparticles.[14–22] Of relevance to this paper is the real time measurement of the transport kinetics of an organic cation, malachite green (MG), across a unilamellar (only one bilayer) liposome.[14–17] Unlike other optical methods the second harmonic method does not require the addition of quenching or other marker molecules in order to

DOI: 10.1039/b405410e

differentiate the molecules transported across the bilayer from the much larger population of the same molecules in the external bulk solution. It should be noted that SHG, being a second order optical effect is electric dipole forbidden in centrosymmetric systems. Despite this apparent prohibition we have shown that SHG is electric dipole allowed and can be generated at the interface of centrosymmetric structures when the surface is not locally centrosymmetric. In a series of experiments on centrosymmetric microparticles,[14–22] including polymer beads, emulsions and liposomes, we have demonstrated that second harmonic light can be generated selectively from their surfaces provided the size of the microparticle is not much smaller than the coherence length of SHG, roughly the wavelength of the incident light. For SHG measurements the molecules in the bulk solution do not generate a coherent second harmonic signal because they are randomly oriented. It is only the adsorbed molecules, which being polar, align themselves in a preferred orientation at the interface and thereby can generate an SHG signal. The underlying idea of the SHG method to study the transport kinetics of molecules across a liposome bilayer is as follows. On rapidly adding (≤ 1 s) the molecules of interest to a solution containing liposomes, a very rapid rise in the SHG signal is achieved as the molecules rapidly adsorb onto the outer surface of the liposome. If the liposome is permeable to the adsorbed molecules they will migrate across the bilayer and adsorb to the inner surface of the bilayer. By symmetry the adsorbed molecules at the inner and outer surfaces of the liposome are oppositely oriented. Because the thickness of the bilayer is 4–5 nm, which is much less than the SHG coherence length, the second order polarizations induced by the incident light have opposite phases for the oppositely oriented molecules and therefore cancel. Consequently the second harmonic field $E_{2\omega}$ generated by the light field E_ω, incident on a liposome is proportional to the second order susceptibility $\chi^{(2)}$, which contains information of the absorbate population. For liposomes the second harmonic field $E_{2\omega}$ is proportional to the difference in the populations of the molecules located on the outer surface, $N_0(t)$, and the inner surface, $N_i(t)$, at a time t after mixing the liposome solution with the MG solution, containing the solute molecules of interest. The generated second harmonic field at 2ω is given by,

$$E_{2\omega} \propto [N_o(t) - N_i(t)]E_\omega E_\omega \qquad (1)$$

Thus by monitoring the SHG signal, which is proportional to the square of the second harmonic field, $E_{2\omega}$, makes it possible to observe the transport of molecules across the bilayer in real time.

In this report we demonstrate that the transport of the positively charged cation malachite green (MG) across the bilayer of the negatively charged dioleoylphosphatidylglycerol (DOPG) liposome

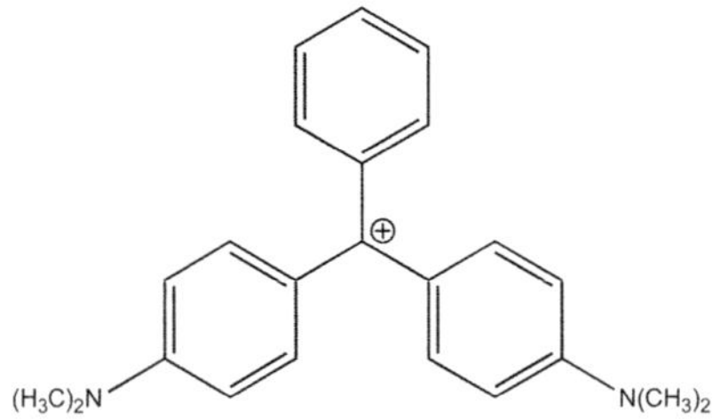

Malachite green cation (MG)

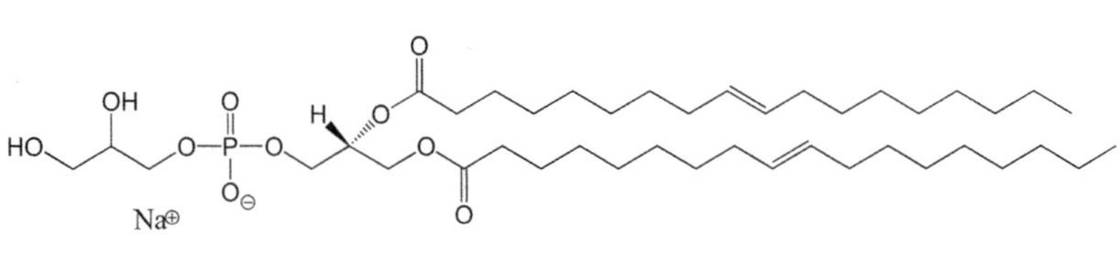

Scheme 1 The structures of MG and DOPG.

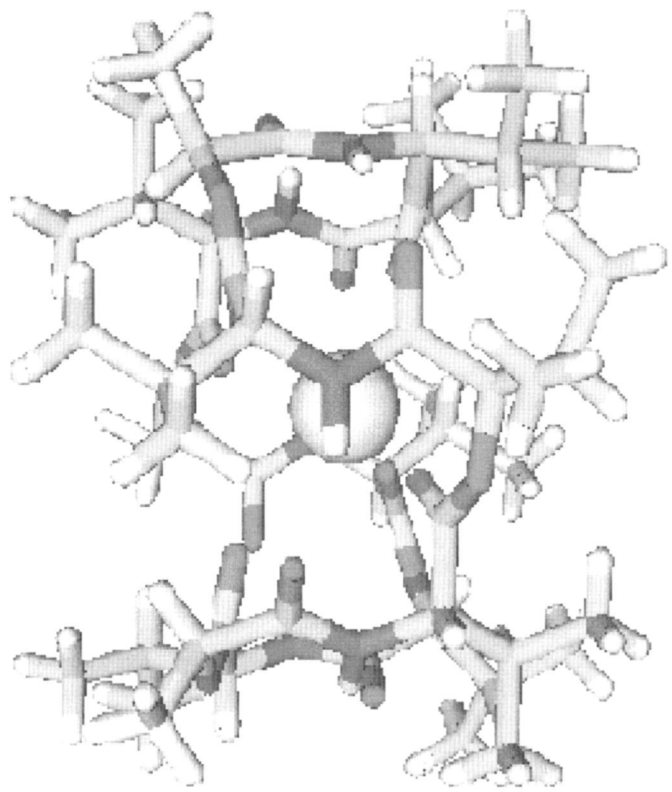

Scheme 2 The structure of valinomycin.

is incomplete because of electrostatic forces. The molecular structures of MG and DOPG are shown in Scheme 1. To support this electrostatic mechanism, we use the antibiotic valinomycin (Scheme 2), which can carry alkali ions across a liposome bilayer. As MG crosses to the liposome interior, a relatively negative potential external to the liposome is generated, which can serve as the driving force for valinomycin to transport a net population of Na^+ to the exterior. In this way, the positive potential that would otherwise have been generated by MG crossing to the inner compartment of the liposome can be reduced or eliminated. As a result, more MG molecules would then cross the bilayer, which would reduce the SHG signal to a value approaching zero.

2. Experimental

2.1 Sample preparation

DOPG phospholipid (Avanti Polar Lipids, Inc.) was used to make a negatively charged liposome. The procedures to prepare the liposome are described in detail elsewhere.[14] The sizes of liposomes were determined by UV-vis turbidity or dynamic light scattering measurements. All the liposome samples used in this study were very close in size, ranging from 100 to 104 nm in diameter.

The probe molecule, malachite green chloride (Aldrich) was used as received after the purity was checked by the HPLC method. The malachite green chloride solutions were adjusted to a pH of 4 and the NaCl concentration was 10 mM. The concentration of malachite green chloride in all experiments was 20 μM. The liposome density was calculated to be 1.8×10^{11} cm^{-3}. The valinomycin purchased from Molecular Probes, Inc. was added to the liposome solution before mixing with the MG solution. The transport of MG across the liposome bilayer was initiated by mixing the liposome solution containing valinomycin with an equal volume of MG solution. After mixing, the concentrations of both lipid and MG, diluted by a factor of 2, are 12.5 μM and

10 μM, respectively. For all experiments, we kept the same concentration of MG, NaCl and lipid, whereas the concentration of valinomycin in the liposome solution was varied.

2.2 SHG setup

The experimental configuration of SHG measurements has been described in detail previously.[15] Briefly, a Nd:YVO$_4$ solid state laser (Spectral-physics, Millennia Vs) pumped Ti:sapphire laser (KMLab) provided 30 fs pulses with an energy of 4 nJ pulse^{-1} at 840 nm at a repetition rate of 82 MHz. The femtosecond laser pulse train passes through a polarizer, and a color filter used to block all the stray light at twice the frequency of the laser light. It was then focused into the sample in a 1 cm rectangular cuvette by a lens with a focal length of 5 cm. A pair of extra-cavity prisms was used to compensate for dispersion due to the sample measurement optics. The insertion of the glass prisms were optimized by maximizing the SHG signal. The incident light was S polarized. The generated SHG signal was collected at 90° relative to the incident light direction and sent into a monochromator. Before the monochromator, an analyzer was used to allow for selectively detecting the signal with S polarization. The data acquisition system includes a PC connected to a PMT and single photon counter. The injection of the liposome solution into the MG solution was completed within one second.

3.   Results and discussion

A.   Kinetics of MG crossing the bilayer in the absence of valinomycin

Fig. 1(a) shows a typical time profile of the SH intensity before and after a liposome solution was injected into a MG solution in the absence of valinomycin. The initial constant signal level is from the MG solution alone and is dominated by the two-photon excited fluorescence of MG. This conclusion is supported by the experimental evidence that the two-photon fluorescence tail of MG (500 counts s^{-1}) at the second harmonic frequency 2ω, is much larger than the intensity of the hyper-Rayleigh scattering of water (~ 20 counts s^{-1}).

Immediately after the liposome solution was injected into the MG solution, a sharp increase (~ 1 s) in the SHG signal was observed. This indicates that adsorption of MG molecules onto the outer surface of the liposome bilayer is faster than the mixing process, which is limited by the injection time. Following the initial increase of the SHG signal, there is a decrease until equilibrium is reached. The decrease of the SHG signal is the result of MG molecules migrating across the liposome bilayers and adsorbing at the inner surface with an orientation opposite to that at the outer surface.

The analysis of the experimental data involves the background correction and calculation of the SH field, $E_{2\omega}$. It can be described by the following equation

$$E_{2\omega}(t) = \sqrt{I_{\text{MG+lipsome}}(t) - I_{\text{background}}} \tag{2}$$

where $I_{\text{MG+lipsome}}(t)$ is the total signal detected at twice the frequency of the fundamental incident laser, 2ω, at time t after mixing of the MG solution and the liposome solution. $I_{\text{background}}$ represents the contributions from the factors other than the SH field generated by the MG adsorbed on the DOPG liposome bilayer.

The data shown in Fig 1(b) is fitted to a single exponential function,

$$E_{2\omega}(t) = a_0 + a_1 \exp(-t/\tau_1) \tag{3}$$

A time constant of 90 ± 2 s is obtained, which is in agreement with our previous result of 107 ± 11 s.[14]

A notable feature of the SHG decay kinetics is the observation that the decay levels off at a SHG value that is well above zero. This indicates that there is a significantly greater population of MG at the outer surface than the inner surface at the conclusion of the MG transport kinetics. This result can be expressed quantitatively using the fact that the SHG kinetics contains information about the MG population at the bilayer interfaces as well as the transport rate constant. Combining eqns. (1) and (3) and the initial condition that there is no MG at the inner surface at $t = 0$, i.e. $N_{\text{i}}(t = 0) = 0$, one obtains

$$\frac{a_0}{a_0 + a_1} = \frac{N_{\text{O}} - N_{\text{i}}}{N_{\text{O}}} \tag{4}$$

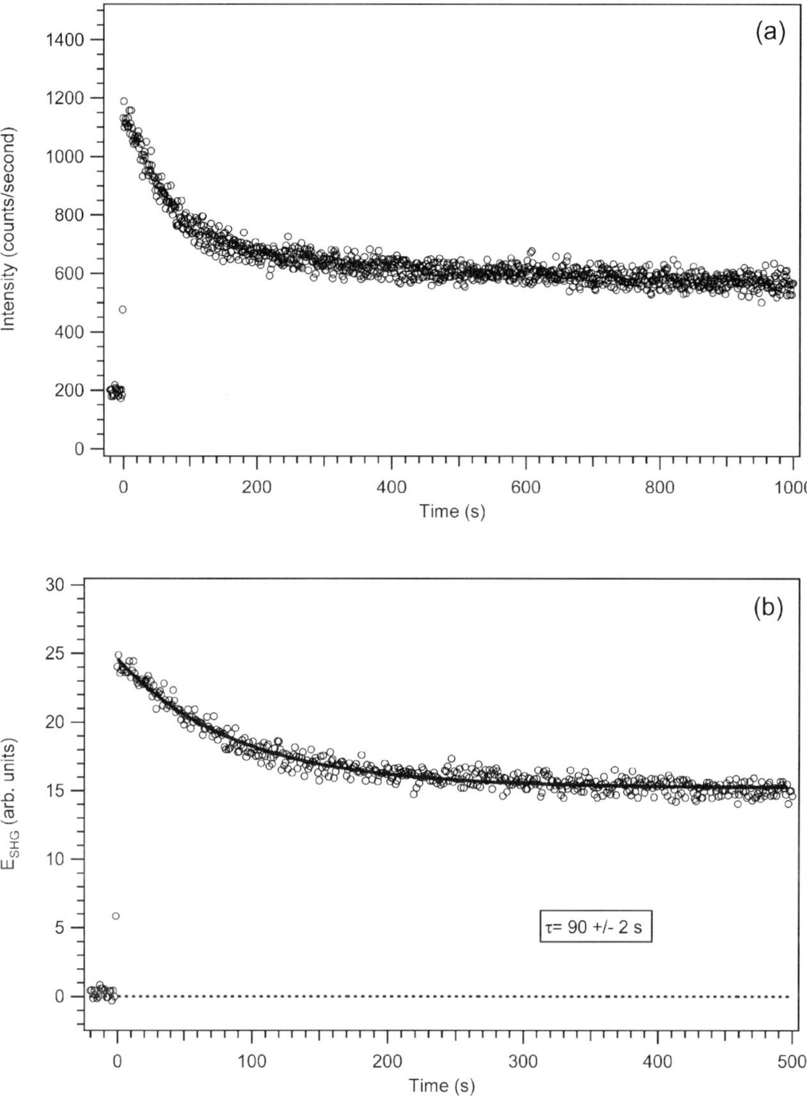

Fig. 1 Temporal profile of SHG intensity (a) and E field (b) before and after injection of liposome into MG solution. The data shown is for an aqueous 10 mM solution of NaCl at pH 4.0. The solid line represents the biexponential fit to the data.

where N_o is the number of MG molecules at the outer surface and N_i at the inner surface after the transport has been completed. It should be noted that at the bulk concentration of 10 μM the number of MG at the outer surface is time independent and equal to its maximum value. This is inferred from our observations that the adsorption of MG at the outer surface is saturated at the MG concentration of 10 μM.[23] The fit of the data shown in Fig. 1(b) gives $a_0/(a_0 + a_1) = 0.62$, which leads to $N_i/N_o = 0.38$. Thus, it is clear that there is a difference in the number of adsorption sites at the outer and inner interfaces. This disparity cannot be due to the difference in the areas of the surfaces, which differ by roughly 10%. Why then does MG not continue to cross the bilayer to yield an SHG signal that is within 10% of zero?

A key element to understand these results is the progressive buildup of a positive potential at the inner surface relative to the outer surface, as the positively charged MG crosses the bilayer. This increasing positive potential opposes the transport of additional MG molecules to the inner surface of the bilayer. The transport of MG across the bilayer would then cease when the concentration gradient driving MG to cross to the inner liposome region comes into balance with the opposing inner positive potential generated by the MG that have already crossed the bilayer. We can calculate this potential using Gauss's law. We find that the potential developed by the MG that have crossed the bilayer in the elapsed time t after the addition of MG to the liposome solution is given by

$$V_{\text{in}} - V_{\text{out}} = \Delta V(t) = \frac{N_i(t)e}{4\pi\varepsilon_0\varepsilon_r}\left(\frac{1}{r_{\text{in}}} - \frac{1}{r_{\text{out}}}\right) \qquad (5)$$

where e is the magnitude of the electron charge, ε_0 is the permittivity of the vacuum, ε_r is the dielectric constant inside the bilayer, r_{in} and r_{out} are the radius of the inner and outer surfaces, respectively, and $N_i(t)$ is the number of MG inside the liposome at time t. Eqn. (5) shows that ΔV is proportional to the number of MG at the inner surface. Furthermore we see that as the population of $N_i(t)$ increases with time the positive potential inside the liposome increases, which in turn more strongly opposes the transport of MG across the bilayer. From this description it follows that the SHG signal would not approach zero, as we have experimentally observed.

One way to determine if this explanation is correct would be to effect the transfer of a positive ion, *e.g.* Na$^+$, across the bilayer from the inner compartment of the liposome as the positively charged MG crosses the bilayer to the inner region of the liposome. The driving force for a positive ion such as Na$^+$ to transfer from inside the liposome is the relative negative potential outside the liposome that is produced as MG crosses the bilayer to the inner compartment of the liposome. If Na$^+$ could cross the bilayer from the inner compartment there would not be a buildup of a relative positive potential as MG crosses to the inner compartment. Thus additional MG molecules could cross to the inner region and the SHG signal would approach zero as the inner and outer surface populations of MG approach equality. However, because Na$^+$ is strongly hydrophilic it cannot cross the bilayer on the time scale of these experiments. To circumvent this problem we can use a Na$^+$ transporter to ferry Na$^+$ across the bilayer. Such alkali ion transporters are well known. Macrocyclic antibiotics, such as valinomycin and monactin, have been shown to increase alkali ion permeability by several orders of magnitude.[24,25] These macrocyclic transporters share common structural features, which are that the interior is hydrophilic and can accommodate an alkali ion whereas the exterior is hydrophobic, which permits the antibiotic to cross the hydrophobic alkyl chain region of the bilayer.

In the next section, we report on how the presence of valinomycin affects the transport of MG across a liposome bilayer and indeed find that our results support the idea that an increasing relative positive potential that develops as MG crosses the bilayer inhibits the crossing of additional MG molecules.

B. Effect of valinomycin on transport kinetics

The SH electric field $E_{2\omega}$ as a function of time t in the presence of valinomycin at two concentrations are shown in Figs. 2 and 3, respectively. Similar to the experiment without valinomycin, the SHG kinetics exhibits a sharp spike immediately after adding the solution containing liposomes and valinomycin (*i.e.*, $t = 0$) followed by a decay. A control experiment shows that valinomycin itself in liposome solution does not generate an observable SHG signal. However, the SHG kinetics of MG crossing the bilayer are seen to be strongly dependent on the presence of valinomycin. Figs. 1–3.

The MG transport kinetics with and without valinomycin were fitted to single exponential decays Figs. 1–3. The fit results are summarized in Table 1. For comparison, Table 1 also lists the fitting parameters for MG transport kinetics in the absence of valinomycin. Our results show that the presence of the alkali ion carrier, valinomycin, has significant effects on the MG transport across the DOPG liposome bilayer. At a valinomycin concentration of 1.25×10^{-8} M, the MG transport time ($\tau = 30$ s $\pm$ 1 s) is three times faster than in the absence of valinomycin, *i.e.* $\tau = 90$ s ± 2 s. Increasing the valinomycin concentration by a factor of ten decreases the transit time ($\tau = 6.2 \pm 0.2$ s) by an additional factor of five. The number of valinomycin carriers per liposome that are present in the bilayer or near the liposome surfaces is roughly estimated to be 20 and 200 at the two concentrations.

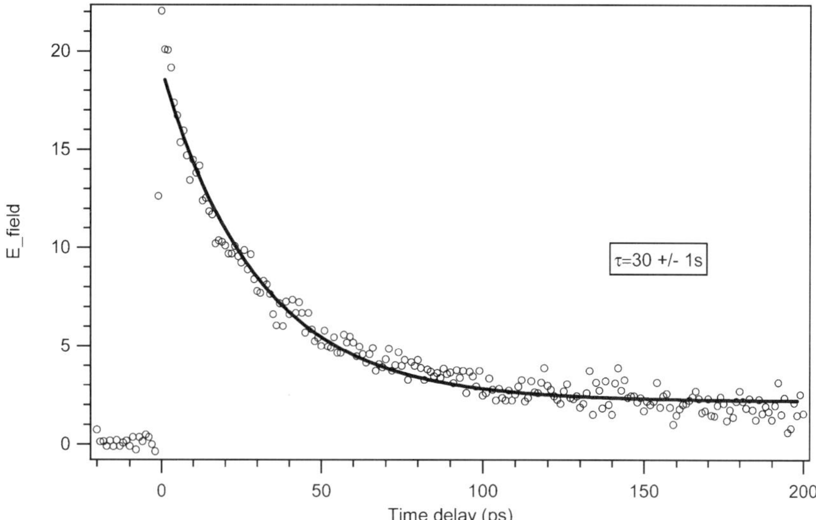

Fig. 2 Measured and fitted time evolution (solid line) of SH electric field upon injection of liposomes (with low concentration of valinomycin 1.25×10^{-8} M) into a MG solution at 10 mM of NaCl at pH 4.0; (a) and (b) are from the same data but drawn in different time scales.

In addition Figs. 2 and 3 show that the SHG electric field $E_{2\omega}$ decays to values much closer to the background in the presence of valinomycin which is contrary to the decay in the absence of valinomycin. These findings suggest that in the presence of valinomycin, not only is the transport more rapid, but also that more MG molecules cross the bilayer. With regard to the magnitude of the $E_{2\omega}$ decay we can calculate the ratio of the MG population at the inner surface to that at the outer surface with the fitting parameters, a_0 and a_1, as described earlier. The ratios of 0.88 for $C_{\mathrm{Val}} = 1.25 \times 10^{-8}$ M and 0.89 for $C_{\mathrm{Val}} = 1.25 \times 10^{-7}$ M are much larger than that in the absence of valinomycin (0.38). They are comparable to the ratio of the inner surface area to outer surface

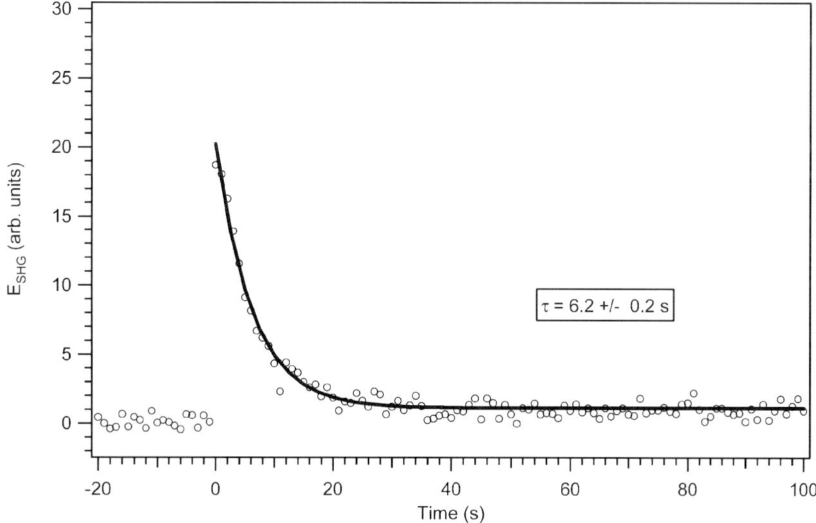

Fig. 3 Measured and fitted time evolution (solid line) of SH electric field upon injection of liposomes (with low concentration of valinomycin 1.25×10^{-7} M) into a MG solution at 10 mM of NaCl at pH 4.0; (a) and (b) are from the same data but drawn in different time scales.

Table 1 Summary of the fit parameters to the SHG kinetics

$C_{\text{valinomycin}}$	τ/s	N_i/N_o
0	90 ± 2	0.3836
1.25×10^{-8} M	30 ± 1	0.88
1.25×10^{-7} M	6.2 ± 0.2	0.89

area (0.85). These numbers are consistent with the description that valinomycin assists the MG to cross the bilayers and occupy the adsorption sites at the inner liposome surface.

Although the valinomycin–Na^+ complex may be formed at both surfaces of the bilayer, the direction of net complex transport depends on the potential between the two surfaces. Before the MG solution is added to the liposome valinomycin solution, there is no potential across the bilayer. The transport of Na^+ per unit time by the valinomycin carrier is equal in the opposing directions. Following the addition of the MG solution, a positive inner potential relative to the outer surface develops as MG crosses the bilayer. This initial development of a negative potential at the outer surface relative to the inner surface acts as the driving force for a net transfer of the Na^+– valinomycin complex from the inner surface to the outer surface. As a result, valinomycin can move a net population of Na^+ from the inner compartment to the outer aqueous solution. This net transport of Na^+ outward reduces or eliminates the development of an inner positive potential, which is the key factor in retarding and limiting the population of MG that crosses the bilayer. At time t, the potential at the inner surface relative to the outer surface can be written as,

$$\Delta V(t) = \frac{(N_i(t) - N_{\text{Na}^+}(t))e}{4\pi\varepsilon_0\varepsilon_r}\left(\frac{1}{r_{\text{in}}} - \frac{1}{r_{\text{out}}}\right). \tag{6}$$

where $N_{\text{Na}^+}(t)$ is the net number of sodium ions that have crossed from the inner aqueous compartment to the outer aqueous solution up to time t, and $N_i(t)$ is the net number of MG that crossed the bilayer to the inner liposome compartment up to time t. From eqn. (6) we see that the decrease in the net number of cations in the inner compartment lowers the positive potential developed by the transport of the MG molecules. Therefore the MG transport rate across the bilayer would be expected to be faster in the presence of valinomycin as we have observed.

We are currently developing a quantitative model based on the mechanism proposed in this paper to describe the transport of MG across a phospholipid bilayer. With no valinomycin, the model includes both the concentration gradient that drives MG to cross the bilayer and the development of an electrostatic barrier that retards and limits the further transport of MG. With valinomycin the model is modified to include the valinomycin mediated Na^+ transport.

Conclusion

The transport of the organic cation, malachite green (MG), from the bulk aqueous solution across a dioleoylphosphatidylglycerol liposome bilayer (4–5 nm) to the inner aqueous compartment is observed in real time using second harmonic spectroscopy. It is proposed that both the transport kinetics and the total number of MG located in the inner liposome compartment are limited by the positive potential that develops as MG crosses the bilayer. The inner positive potential opposes additional MG crossing the bilayer. To investigate this proposal the alkali ion carrier valinomycin, which is known to carry alkali ions across phospholipid bilayers, is introduced into the system. As MG crosses the bilayer, valinomycin can transport Na^+ ions in the opposite direction, driven by the outer negative potential due to MG that have crossed to the interior. In this way any positive potential due to MG inside the liposome is reduced or eliminated by the outward passage of Na^+ ions. At a valinomycin concentration of 1.25×10^{-8} M, the MG transit time (30 ± 1 s) is reduced by a factor of three compared with the transit time (90 ± 2 s) in the absence of valinomycin. At a higher valinomycin concentration, 1.25×10^{-7} M, the MG transit time (6.2 ± 0.2 s) is reduced by a factor of 15 with respect to the transit time when valinomycin is absent. At long times the second harmonic signal approaches zero ($\sim 10\%$ above) due to the cancellation of the second order polarization of MG at the outer surface by a slightly smaller number of MG oriented in the opposite direction at the inner interface. In summary, the observation of more rapid transport kinetics of the organic cation MG and reduction of the SHG signal in the presence of valinomycin, supports the mechanism of an electrostatic "barrier" limiting MG transport. This barrier can be eliminated by a small alkali ion carrier, such as valinomycin.

Acknowledgements

This work was supported by the National Science Foundation and by Chemical Sciences, Geosciences and Biosciences Division, Office of Basic Energy Sciences, Office of Science of the US Department of Energy.

References

1 R. B. Gennis, *Biomembranes: Molecular Structure and Function*, ed. C. R. Cantor, Springer–Verlag, 1989.
2 D. D. Lasic, *Liposomes: From Physics To Applications*, Elsevier, Amsterdam, 1993.
3 J. R. Philiport and F. Schuber, in *Liposomes as Tools in Basic Research and Industry*, CRC Press, Boca Raton, FL, 1995.
4 A. Parsegian, *Nature*, 1969, **221**, 844.
5 R. C. Macdonald, *Biochim. Biophys. Acta*, 1976, **448**, 193.
6 D. W. Deamer and J. Bramhall, *Chem. Phys. Lipids*, 1986, **40**, 167.
7 R. T. Hamilton and E. W. Kaler, *J. Phys. Chem.*, 1990, **94**, 2560.
8 S. Paula, A. G. Volkov, A. N. VanHoek, T. H. Haines and D. W. Deamer, *Biophys. J.*, 1996, **70**, 339.
9 S. Kaiser and H. Hoffmann, *J. Colloid Interface Sci.*, 1996, **184**, 1.
10 S. Paula, A. G. Volkov and D. W. Deamer, *Biophys. J.*, 1998, **74**, 319.
11 F. Bordi, C. Cametti and A. Naglieri, *Biophys. J.*, 1998, **74**, 1358.
12 F. Bordi, C. Cametti and A. Naglieri, *Colloids Surf., A*, 1999, **159**, 231.
13 F. Bordi, C. Cametti and A. Motta, *J. Phys. Chem. B*, 2000, **104**, 5318.
14 X. M. Shang, Y. Liu, E. Yan and K. B. Eisenthal, *J. Phys. Chem. B*, 2001, **105**, 12816.
15 Y. Liu, E. C. Y. Yan and K. B. Eisenthal, *Biophys. J.*, 2001, **80**, 1004.
16 E. C. Y. Yan and K. B. Eisenthal, *Biophys. J.*, 2000, **79**, 898.
17 A. Srivastava and K. B. Eisenthal, *Chem. Phys. Lett.*, 1998, **292**, 345.
18 E. C. Y. Yan and K. B. Eisenthal, *J. Phys. Chem. B*, 2000, **104**, 6686.
19 H. Wang, E. C. Y. Yan, E. Borguet and K. B. Eisenthal, *Chem. Phys. Lett.*, 1996, **259**, 15.
20 H. F. Wang, E. C. Y. Yan, Y. Liu and K. B. Eisenthal, *J. Phys. Chem. B*, 1998, **102**, 4446.
21 E. C. Y. Yan, A. Srivastava and K. B. Eisenthal, *Chem. Abstr.*, 1999, **218**, U403.
22 E. C. Y. Yan, Y. Liu and K. B. Eisenthal, *J. Phys. Chem. B*, 1998, **102**, 6331.
23 E. Yan, PhD Thesis, Columbia University, 2000.
24 P. Lauger and G. Stark, *Biochim. Biophys. Acta*, 1970, **211**, 458.
25 P. Lauger, R. Benz, G. Stark, E. Bamberg, P. C. Jordan, A. Fahr and W. Brock, *Q. Rev. Biophys.*, 1981 **14**, 513.

Dynamics of phospholipid monolayers on polarised liquid–liquid interfaces

Dynamics of phospholipid monolayers on polarised liquid–liquid interfaces

Zdeněk Samec,* Antonín Trojánek and Petr Krtil

J. Heyrovský Institute of Physical Chemistry, Academy of Sciences of the Czech Republic, Dolejškova 3, 182 23 Prague 8, Czech Republic. E-mail: zdenek.samec@ jh-inst.cas.cz; Fax: +420 286 582 307; Tel: +420 266 052 017

Received 14th April 2004, Accepted 28th May 2004
First published as an Advance Article on the web 22nd September 2004

Quasi-elastic laser light scattering (QELS) is used to investigate dynamics of the polarised water|1,2-dichloroethane (DCE) interface in the presence of adsorbed DL-α-dipalmitoyl-phosphatidylcholine (DPPC) over a range of the interfacial potential differences (±0.3 V) and DPPC concentrations (0–20 μM). An analysis of the frequency of thermally excited capillary waves reveals some novel features in the adsorption of DPPC. The effect of the capillary wavenumber on the capillary wave frequency and the damping factor suggest that the dynamic behaviour of ITIES is consistent with the theoretical predictions for a sharp liquid–liquid interface.

1. Introduction

Phospholipid monolayer has been regarded as a suitable model system for one half of the lipid backbone of biomembranes.[1–6] Monolayers at the water–air[7–9] and water–non polar liquid[3,10,11] have been studied mainly by the Langmuir balance technique,[12] which allows measuring the surface pressure *versus* the area per adsorbed molecule. These studies provided information about phase transitions from a two-dimensional gas (G), through liquid-expanded (LE) and liquid-condensed (LC) phases, to a tightly packed, two-dimensional solid as a function of the phospholipid structure, nature of the non-polar liquid phase and temperature, see *e.g.* refs. 2,3 and 6–9 for a review. More recently, fluorescence microscopy of labelled adsorbed phospholipids has been used to investigate the changes in monolayer morphology during the phase transition at both water–air[7–9] and water–non polar liquid[11] interfaces. Structural changes that accompany the transition have been followed using infrared and Raman spectroscopy,[13–15] while the vibrational sum frequency spectroscopy has provided direct information about the orientation and the degree of order among the acyl chains of the adsorbed phospholipid,[16,17] as well as about induced changes in water structure at a water–organic solvent interface.[18]

An increasing interest in the properties of phospholipid monolayers at the polarised interfaces between two immiscible electrolyte solutions (ITIES)[19–32] has been invoked by the possibility to investigate the stability and ion permeability of the monolayer as a function of the well-defined interfacial potential difference (electric field), see *e.g.* refs. 5 and 6 for a review. Unlike a Langmuir monolayer, the phospholipid monolayer at ITIES is in a thermodynamic equilibrium with the bulk phospholipid dissolved in the liquid (usually organic) phase. Interfacial tension[19,22,23,25–27] and capacitance[20,21,24] measurements have shown that the stability of the phospholipid monolayer on ITIES strongly depends on the Galvani potential difference $\Delta_o^w\phi = \phi(w) - \phi(o)$ between the

DOI: 10.1039/b405377j

aqueous (w) and the organic (o) phase, and also on the pH of the aqueous phase, as well as on the nature of the aqueous cation, both affecting the ionic state of the phospholipid. While a relatively stable monolayer can be formed from the zwitterionic phospholipid, the monolayer readily breaks down when the phospholipid is transformed to its ionic form either through the acid–base reaction,[19] or through the ion association with the aqueous cation.[23] The role of the aqueous cation was corroborated by voltammetric measurements pointing to the cation transfer facilitated by the adsorbed phospholipid.[24,25,28,30] An interesting result is an observation that the temperature of the phase transition from the LE to the LC state at ITIES can be considerably lower compared to that measured for the water-interface.[5] The disparity apparently stems from the reduced attraction between the adsorbed phospholipid associated with the solvation of the acyl chains by the organic solvent,[5] in an agreement with conclusions drawn from the vibrational spectroscopic study of the phospholipid monolayer at a water–carbon tetrachloride interface.[16]

Although the experimental data about phospholipid monolayers at ITIES are detailed and comprehensive, the direct information about the monolayer structure is lacking and the structural considerations have to rely on the relevant spectroscopic data for other systems *e.g.* the neat water–organic solvent interface.[16,18] On the other hand, first attempts to examine the nano-scale dynamics of ITIES have been made. Thus, confocal fluorescence correlation spectroscopy was used to measure and to compare the lateral diffusion coefficients of dioleoylphosphatidylcholine (DOPC) in the supported bilayer and in a DOPC monolayer at a polarised ITIES with the results pointing to a lower friction and, hence, to a lower ordering in the latter system.[32] Light scattering experiments[33–36] indicated that the adsorption of the ionic surfactants on ITIES has a remarkable effect on the interfacial dynamics, essentially on the frequency of thermally excited surface fluctuations (capillary waves) with 0.1–10 nm in amplitude and with the wavelength ranging from 0.1 nm to a value of the linear dimension of the interface.[37–39] Their existence on the liquid–liquid interfaces was demonstrated by molecular dynamics[40] and Monte Carlo[41] computer simulations. Quasi-elastic light scattering (QELS) measurements can provide information on capillary waves with a wavelength of 0.1 mm in order of magnitude.[33,42,43] The theory of capillary waves at liquid surfaces and interfaces has been reviewed by several authors.[38,44,45]

The aim of this work has been to clarify the effects of dipalmitoylphosphatidylcholine (DPPC) on the dynamics of the polarised water|1,2-dichloroethane interfaces. Power spectra of the capillary waves of a defined wavelength have been obtained by using the optical heterodyne technique.[42] In a previous work,[27] we have already used the QELS method to infer the interfacial tension data for a thermodynamic analysis of the binding of the aqueous cation to the DPPC monolayer at ITIES. In the present study we have focused on a detailed analysis of the capillary wave frequency and damping in relation to the interfacial potential difference and DPPC concentration. This analysis has allowed us to investigate thermodynamics of the DPPC adsorption, as well as to verify the theory of capillary waves that has been developed for the model of the sharp liquid–liquid interface.

2. Theoretical

The interpretation of the QELS measurements on liquid–liquid interfaces requires theoretical expressions for the power spectra of the thermally excited fluctuations. The light-scattering cross-section is usually proportional to the mean-square perpendicular displacement of the surface from its flat, equilibrium position.[38,44] The geometrical arrangement of two liquids considered for the theoretical analysis is shown schematically in Fig. 1. Here, it is assumed that the two liquids are separated by a thin film *e.g.* the monolayer of adsorbed molecules, which can be characterised by the surface dilatational modulus ε,[38]

$$\varepsilon = -\partial\gamma/\partial \ln \Gamma \tag{1}$$

where γ is the interfacial tension and Γ is the relative surface excess concentration of adsorbed molecules. Both γ and ε can be expanded as response functions to include the interfacial viscosities: $\varepsilon = \varepsilon_0 - i\omega\varepsilon'$ and $\gamma = \gamma_0 - i\omega\gamma'$, where ω is the angular frequency of the capillary wave.[38,44] A dispersion equation that would be applicable to a diffuse interfacial region has been derived,[46] but has not been able to provide better agreement with experimental observations than the relation for a sharp interface.[44]

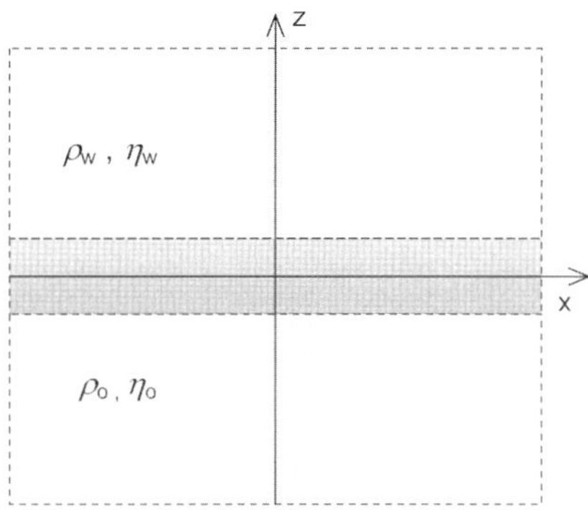

Fig. 1 Model of two liquids separated by a thin film: ρ and η denote respectively the density and kinematic viscosity of the liquid phase.

The general dispersion equation [38] accounts also for the surface acoustic waves, the frequencies of which are much greater than those of the thermally excited capillary waves. Neglecting the acoustic-wave part, the fluctuation spectrum takes the form[38]

$$\left\langle |u_z(0)|^2 \right\rangle_{k,\omega} = \frac{k_B Tk}{\pi A \omega} \text{Im} \left(\begin{array}{c} \gamma k^3 + (\rho_o - \rho_w)gk - (\rho_w + \rho_o)\omega^2 + \\[4pt] + k\omega \dfrac{4\omega(k\eta_o - i\eta_w T_w)(k\eta_w - i\eta_o T_o)}{i\omega[\eta_o(k - iT_o) + \eta_w(k - iT_w)] - \varepsilon k^2} \\[4pt] + k\omega \dfrac{i\varepsilon k^2[\eta_o(k - iT_o) + \eta_w(k - iT_w)]}{i\omega[\eta_o(k - iT_o) + \eta_w(k - iT_w)] - \varepsilon k^2} \end{array} \right)^{-1} \tag{2}$$

where $u_z(0)$ is the vertical displacement at $z = 0$, k_B is the Boltzmann's constant, T is absolute temperature, $k = 2\pi/\Lambda$ and Λ represent the wavenumber and wavelength of the capillary wave, respectively, g is gravitational acceleration, A is the interfacial area, and

$$T_{w,o} = [(i\rho_{w,o}\omega/\eta_{w,o}) - k^2]^{1/2} \tag{3}$$

Fig. 2 shows the fluctuation spectrum for the water|1,2-dichloroethane interface calculated using eqn. (2) for the wavenumber $k = 55608$ m^{-1} and two different values of the interfacial tension $\gamma = 30$ mN m^{-1} (a) and 1 mN m^{-1} (b). The spectrum has a Lorentzian lineshape with the peak frequency $\omega_0 = 2\pi f_0$ and the linewidth (full width at half maximum height) $\Delta\omega = 2\pi\Delta f$, which both apparently depend on the interfacial tension. Since the amplitude of the capillary wave is proportional to $\exp(-\Delta\omega t/2)$,[47] the linewidth has the physical meaning of a damping factor, which characterizes the decay of the capillary wave with time t due to the viscosity effect.[38,44] The spectrum peak frequency is slightly shifted with respect to the frequency of the capillary wave ω_c,[38]

$$\omega_0 \approx \omega_c - (1/2)\Delta\omega \tag{4}$$

$$\omega_c^2 = [\gamma k^3 + (\rho_o - \rho_w)gk]/(\rho_o + \rho_w) \approx \gamma k^3/(\rho_o + \rho_w) \tag{5}$$

The approximate form of eqn. (5) is known as the Lamb's equation.[47,48] Limiting expressions for the linewidth $\Delta\omega$ have been derived for the case of absence of any film ($\varepsilon = 0$), *i.e.*

$$\Delta\omega = \frac{1}{\rho_w + \rho_o} \frac{2^{3/2}\omega_c^{1/2}k(\rho_w\rho_o\eta_w\eta_o)^{1/2}}{(\rho_w\eta_w)^{1/2} + (\rho_o\eta_o)^{1/2}} \tag{6}$$

and for the case of a very stiff film ($\varepsilon \gg \gamma$), *i.e.*[38]

$$\Delta\omega = \frac{\omega_c^{1/2}k[(\rho_w\eta_w)^{1/2} + (\rho_o\eta_o)^{1/2}]}{2^{1/2}(\rho_w + \rho_o)} \tag{7}$$

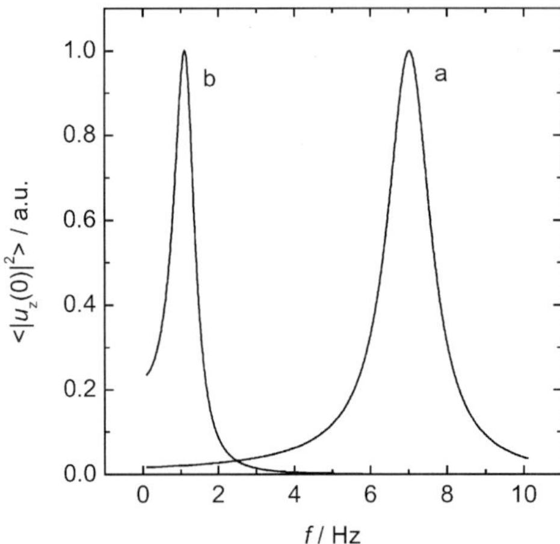

Fig. 2 Calculated fluctuation spectrum ($f = \omega/2\pi$) for the water|1,2-dichloroethane interface for $k = 55608$ m^{-1} and the interfacial tension $\gamma = 30$ mN m^{-1} (a) and 1 mN m^{-1} (b). Both liquid phases are assumed to have the density and kinematic viscosity of the pure solvent, *i.e.* $\rho_\text{w} = 0.997 \times 10^3$ kg m^{-3}, $\rho_\text{o} = 1.246 \times 10^3$ kg m^{-3}, $\eta_\text{w} = 0.89 \times 10^{-3}$ kg m^{-1}s^{-1} and $\eta_\text{o} = 0.78 \times 10^{-3}$ kg m^{-1} s^{-1}.

Neither eqn. (6) nor eqn. (7) reduces to the single-liquid linewidth when the parameters of one liquid are set equal to zero.[38] Actually, the damping of fluctuations at the interface of two liquid phases of comparable viscosity exceeds that of either isolated liquid surface.[38] It is noteworthy that for the particular case of the interface between water and 1,2-DCE, eqn. (6) and eqn. (7) give practically the same value for $\Delta\omega$. A comparison of the values of $\Delta\omega$ calculated from either equation with those obtained by fitting of the theoretical spectra such as those shown in Fig. 2 to Lorentz function reveals that both equations underestimate the Lorentzian linewidth by *ca.* 20%. The reason for this discrepancy is probably that the representation of the fluctuation spectrum by the Lorentz function is rather an approximation.[44] Therefore, we have based the comparison of the experimental linewidth with that predicted by the theory on the fitting of either the theoretical or experimental spectrum to the Lorentz function.

In the previous studies[34,35] we have used a different expression for $\Delta\omega$,

$$\Delta\omega = 4k^2 \frac{\eta_\text{w} + \eta_\text{o}}{\rho_\text{w} + \rho_\text{o}} \tag{8}$$

which presumably is valid in the low-viscosity limit.[39] Unlike eqns. (6) or (7), eqn. (8) can reduce to the single-fluid linewidth when the parameters of one fluid are set equal to zero. Apparently, eqn. (8) can be derived from eqn. (2) as the low frequency limit, for which the imaginary part $\text{Im}\,T \approx k$ is much larger than the real part.[38]

The total mean-square surface displacement can be obtained by summation over all wavevectors of the mean-square interface displacement including the contribution of the fluctuation of all frequencies that have a given surface vector. This gives[37,39,41]

$$\left\langle |u_z(0)|^2 \right\rangle = \xi^2 \approx \frac{k_\text{B}T}{2\pi\gamma} \ln \frac{L}{l_\text{b}} \tag{9}$$

where L is the linear dimension of the interface and $l_\text{b} \approx 1$ nm is the bulk correlation length. Square root of the total mean-square displacement has been considered as a measure of the mean (diffuse) thickness of the interface ξ.[37–39] An estimate based on eqn. (9) shows the mean thickness of the water|1,2-dichloroethane interface ($L = 1$ cm) can increase from 0.6 nm to 3.2 nm when the interfacial tension decreases from 30 mN m^{-1} to 1 mN m^{-1}.

3. Experimental

LiCl, tetraethylammonium chloride (TEACl), bis(triphenylphosphoranylidene) ammonium chloride (BTPPACl), potassium tetrakis(4-chlorophenyl)borate (KTPBCl), tris(hydroxymethyl)aminomethane

(TRIS buffer) were purchased from Fluka as reagent grade chemicals; DL-α-dipalmitoylphosphatidylcholine (99%) was purchased from Sigma. Bis(triphenylphosphoranylidene)ammonium tetrakis(4-chlorophenyl)borate (BTPPATPBCl) was prepared by metathesis of BTPPACl and KTPBCl. The aqueous and the organic solutions were prepared with purified water from a Milli-Q system (Milli-Q Gradient, Millipore) and 1,2-dichloroethane (DCE, puriss. p.a., Fluka).

Voltammetric and QELS measurements were carried out in a four-electrode quartz glass cell, which was similar to one used previously,[36] *cf.* Fig. 3. The cell can be represented by the scheme

$$Ag|AgCl|0.1 \text{ M LiCl, } 0.05 \text{ M TRIS (pH 8.9), } H_2O|5 \text{ mM BTPPA TPBCl, } x \text{ μM DPPC,}$$

$$DCE| 5 \text{ mM BTPPACl, } H_2O \text{ }|AgCl|Ag'$$

where $x = 0$–20. The cell potential E was controlled by a four-electrode potentiostat (1287 Electrochemical Interface Solartron, Solartron, UK). We have been aware of the possibility of the co-adsorption of DPPC on the wall of the electrolytic cell, which can lead to a depletion of DPCC from the organic phase and to invalidation of experimental data obtained at low DPPC concentrations. Therefore all measurements were carried out following a pretreatment procedure. The cell was first cleaned by using chrom-sulfuric acid and ultrapure water and then filled with the organic phase containing DPPC of the required concentration. After 30 min under stirring, the organic solution was poured out, and the cell was filled with the fresh organic phase containing DPPC of the same concentration to form an interface with the aqueous phase.

Light scattering was measured using the optical heterodyne technique[42] after a time delay of 2 min elapsed from adjusting the potential E stepwise (40 mV steps). The experimental set-up for the QELS measurements has been a modification of that described previously.[35] The beam from a 25 mW diode pumped laser (wavelength $\lambda = 532$ nm, beam diameter 0.36 mm, CrystaLaser, USA) formed the incident beam spot with the diameter $D = 1.35$ mm at the water|DCE interface with the geometric area of 15.6 cm^2. The scattered light was optically mixed with the light diffracted on a diffraction grating and focused at the photodiode detector (S1133, Hamamatsu), which was connected through a wide-band amplifier (13AMP005, Melles-Griot) to a FFT signal analyser (SR780, Stanford Research). The grating consisted of the dark lines 40 μm wide and 305 μm spaced on a photographic plate, *i.e.* the nominal grating constant $d = 345$ μm. The intensity maximum of the diffracted light (the diffraction spots) is found at the angle α, which fulfils the equation

$$n\lambda = d \sin \alpha \tag{10}$$

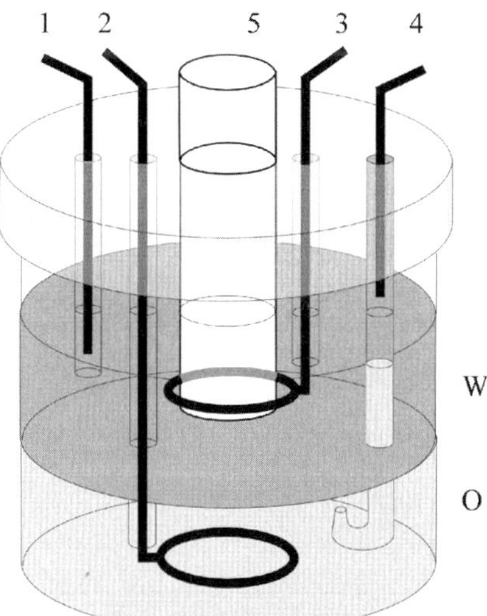

Fig. 3 Scheme of the four-electrode cell for the QELS measurements: (1) reference electrode for the aqueous phase (w), (2,3) counter electrodes, (4) reference electrode for the organic phase (o), (5) open glass cylinder with optical window located close to the interface. Geometric area of the interface: 15.6 cm^2.

where the order of the diffraction spot $n = 1,2,3$, and 4. In order to check the grating constant we measured the angle α at various diffraction orders. These measurements indicated that the grating constant is actually slightly smaller, $d = 339$ μm. The particular diffraction order for the corresponding capillary wavenumber k,[33]

$$k = \frac{2\pi n}{d} \tag{11}$$

was selected by a pinhole positioned in front of the detector. The spectra were fitted to the Lorentz function yielding the peak frequency f_0 and the experimental linewidth Δf_{exp}. Linewidth Δf_{exp} was corrected for the instrumental bandwidth Δf_i by using the relationship[42]

$$\Delta f = \Delta f_{\text{exp}} \left(1 - \left(\frac{\Delta f_i}{\Delta f_{\text{exp}}} \right)^2 \right) \tag{12}$$

where $\Delta f_i \approx (6.66/D)(\mathrm{d}f/\mathrm{d}k) = (6.66/D)(3f_0/2k)$.

Both electrochemical and interfacial tension measurements were carried out at the ambient temperature of 23 ± 2 °C. The aqueous and the organic solvent phases were equilibrated prior to measurements.

4.   Results and discussion

4.1.   Cyclic voltammetry

Fig. 4 shows the cyclic voltammogram of the water|DCE interface in the absence and the presence of DPPC in the organic phase. The potential scale can be converted to the scale of the Galvani potential differences using the measured value of the half-wave potential $E_{1/2}^{\text{rev}} = 0.43$ V for the TEA$^+$ ion transfer and the estimated value of the half-wave potential difference, $\Delta_o^w \phi_{1/2}^{\text{rev}} \approx 0.02$ V.[49] Hence, the zero potential difference is found at $E = 0.41$ V. In the presence of DPPC, the reversible current enhancement is observed at positive potentials, which is probably associated with the adsorption of phospholipid and/or the transfer of the aqueous metal cation facilitated by the adsorbed DPPC.[24,25,28,30] A tentative mechanism has been proposed involving the adsorption of the zwitterionic form $L^\pm$ of the phospholipid prevailing at more negative potentials,[28]

$$L^\pm \text{ (o)} \leftrightarrows L^\pm \text{ (}\sigma\text{)} \tag{13}$$

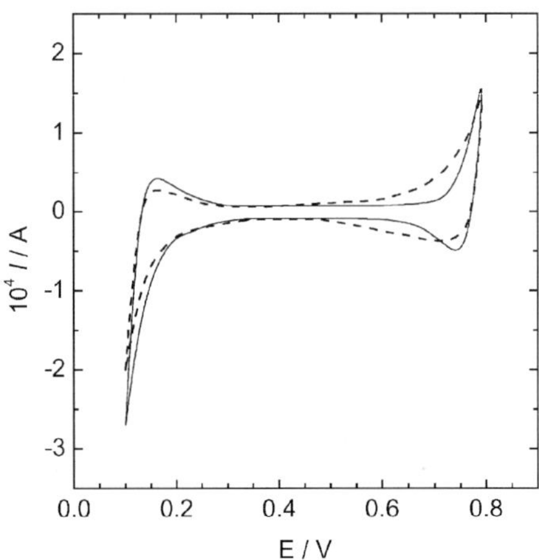

Fig. 4   Cyclic voltammogram of 0.1 M LiCl in water (pH 8.9) and 5 mM BTPPA TPBCl in DCE at a sweep rate of 25 mV s^{-1} in the absence (solid line) and presence (dashed line) of 10 μM DPPC in DCE.

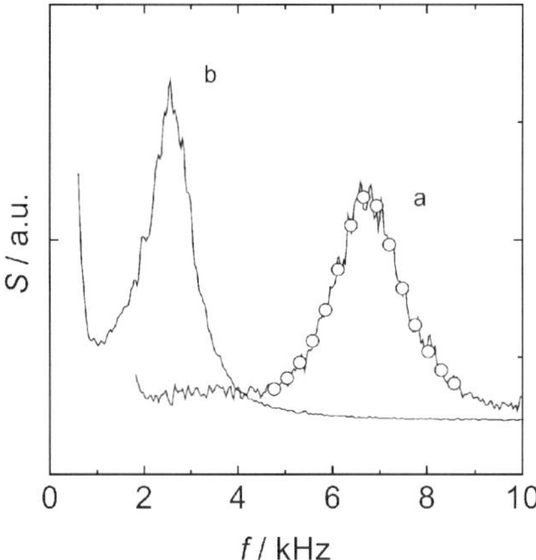

Fig. 5 Measured power spectrum for the interface between 0.1 M LiCl in water (pH 8.9) and 5 mM BTPPA TPBCl in DCE in the absence (a) and presence (b) of 10 µM DPPC in DCE for $k = 55608$ m^{-1}. Circles represent a fit of the spectrum (a) to the Lorentz function.

and the facilitated transfer of the aqueous cation M$^+$ at positive potentials

$$L^\pm (\sigma) + M^+(w) \leftrightarrows ML^+(\sigma) \leftrightarrows M^+(o) + L^\pm (o) \tag{14}$$

with the intermediate formation of the adsorbed cationic form ML$^+$ in the interphase σ. The effect of the aqueous cation on both the electric current and the interfacial tension follows the sequence sequence H$^+$ > alkaline earth metal cations $\gg$ Na$^+$, Li$^+$ > K$^+$, Cs$^+$, which reflects the change in the ion association (binding) constant.[25]

4.2. Verification of the theory

Fig. 5 shows the typical power spectra of the water|DCE interface in the absence and the presence of DPPC in the organic phase for $k = 55608$ m^{-1} ($n = 3$). These spectra have the Lorentzian shape, *cf.* the fit points (empty circles) for spectrum (a) in Fig. 5. As can be seen from Fig. 5, the presence of DPPC leads to a decrease in both the peak frequency (*i.e.* frequency of the capillary wave) and the linewidth. Such behaviour apparently corresponds to a decrease in the interfacial tension, *cf.* Fig. 2.

In order to check whether eqns. (5) and (6), or eqn. (7) can be reliably used for the interpretation of the peak frequency and linewidth data, we have carried out light scattering experiments for various wavenumbers. Table 1 summarizes the results of the evaluation of the interfacial tension at a constant potential in the absence of DPPC and for two DPPC concentrations. A very good agreement between the interfacial tension data at different values of k confirms that the Lamb's equation is applicable.

Insertion of the capillary wave frequency as given by eqn. (5) into eqn. (6) or eqn. (7) yields an expression indicating that Δf should be proportional to $k^{7/4}$. Fig. 6 shows that indeed the experimental Δf *vs.* $k^{7/4}$ plots at various DPPC concentrations are reasonably linear. As it follows from eqns. (6) or (7), the slope of these plots should decrease when the interfacial tension decreases,

Table 1 Interfacial tension γ evaluated from the QELS measurements for various wavenumbers k and DPPC concentrations in DCE at $E = 0.43$ V

n	k/m^{-1}	$\gamma/\text{mN m}^{-1}$ 0 µM DPPC	2.5 µM DPPC	5.0 µM DPPC
1	18530	27.1	9.81	2.67
2	37060	27.2	9.79	2.87
3	55608	29.1	9.96	2.82
4	74122	28.7	9.49	2.80
mean ± SD		28.0±1.0	9.8±0.2	2.8±0.1

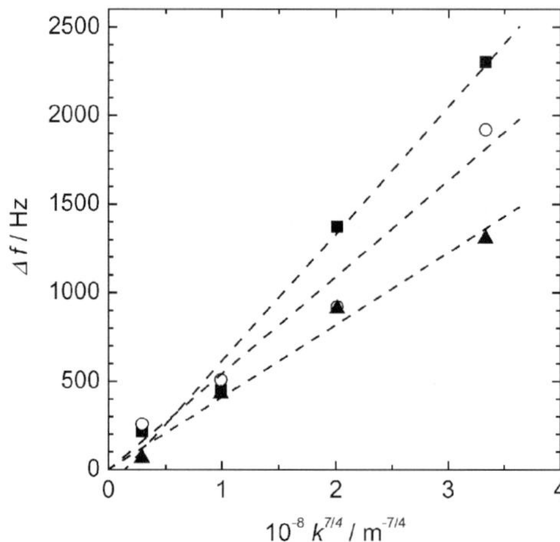

Fig. 6 Corrected power spectrum linewidth Δf vs. $k^{7/4}$ for the interface between 0.1 M LiCl in water (pH 8.9) and 5 mM BTPPA TPBCl in DCE in the presence of DPPC in DCE at concentrations (in μM): 0 (■), 2.5 (○) and 5 (▲). Potential $E = 0.43$ V.

which obviously is the case. Experimental slopes are compared with those calculated using eqn. (6) in Table 2. We have already noted in the theoretical part that this equation somewhat under-estimates the Lorentzian linewidth. Much better agreement is reached when a comparison is made with the theoretical slope, which is inferred by fitting of the theoretical spectra to the Lorentz function, *cf.* the slopes given in Table 2 in parentheses.

4.3. Effects of DPPC concentration and applied potential on interfacial tension

Fig. 7 shows the electrocapillary curves of the water|DCE interface in the absence and presence of DPPC. The interfacial tension γ was evaluated from the peak frequency f_0 of the measured power spectrum by using the Lamb's equation, eqns. (5) and (4). In the absence of DPPC, the electrocapillary maximum is found at $E = 0.43$ V, *i.e.* the zero-charge potential difference $\Delta\phi_{pzc} = 0.02$ V.

Effects of the DPPC concentration and the applied potential on the interfacial tension are qualitatively similar to those reported previously for the phosphatidylcholine adsorption at the water|DCE[19] and water|nitrobenzene[22,23] interfaces. However, the interfacial tension decays to a value near zero at the much lower phospholipid concentration, as compared with literature data, which can be a consequence of the depletion effect, *vide infra*. The effect of the potential can be reasonably interpreted in terms of two stability domains for the adsorbed phosphatidylcholine (PC):[19,22] a stable adsorbed layer of the zwitterionic PC is formed at potentials negative to the critical value (here *ca.* 0.6 V), while the binding of the aqueous cation to the adsorbed PC leads to its desorption at more positive potentials following probably the mechanism described by eqn. (14). Voltammetric behaviour shown in Fig. 4 and reported in the literature[28] suggests that the adsorption/desorption process is rather fast and reversible. Indeed, when the potential is changed

Table 2 A comparison of the experimental and theoretical slopes of the Δf vs. $k^{7/4}$ dependences at various DPPC concentrations c in DCE and $E = 0.43$ V. Theoretical values were calculated using eqn. (6) or by fitting the theoretical spectra to the Lorentz function (values in parentheses)

		Slope/10^{-8} Hz m$^{7/4}$	
$c/\mu M$	$\gamma/$mN m^{-1}	Experiment	Theory
0	28.0±1.0	717	578 (652)
2.5	9.8±0.2	545	442 (518)
5	2.8±0.1	408	323 (394)

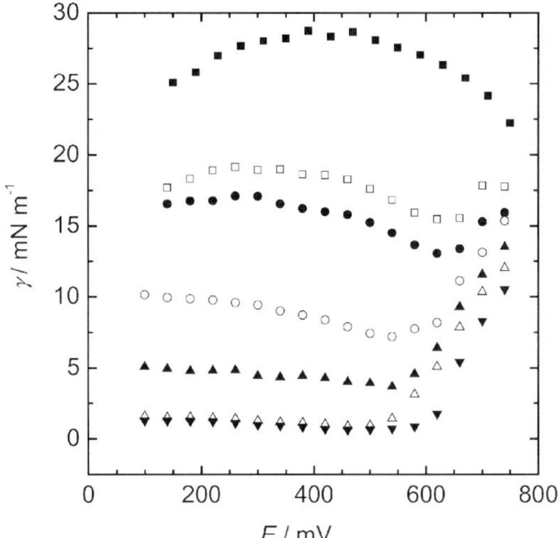

Fig. 7   Electrocapillary curves evaluated from the QELS measurements at the interface between 0.1 M LiCl in water (pH 8.9) and 5 mM BTPPA TPBCl in DCE in the presence of DPPC in DCE at concentrations (in μM): 0 (■), 0.5 (□), 1.0 (●), 2.5 (○), 5 (▲), 10 (△) and 20 (▼).

stepwise from the former region to the latter and reverse, a fast increase or decrease of the interfacial tension to a stationary value is observed within *ca.* 2 min from the step change in the potential, *cf.* Fig. 8. These results are consistent with dynamic measurements of electrocapillary curves indicating that the interfacial tension measured on the forward and reverse potential sweeps superimpose when the sweep rate is lower than 1 mV s^{-1}.[26]

Thermodynamic analysis of the ITIES was outlined, which accounts for the adsorption of phosphatidylcholine (PC) both as a zwitterion and a cation formed by the association of PC with the aqueous cation.[27] The electrocapillary equation has the form

$$-\mathrm{d}\gamma = Q\mathrm{d}E_{o+}^{w-} + (\Gamma_{L^\pm} + \Gamma_{ML^+})\mathrm{d}\mu_{L^\pm} + (\Gamma_{M^+} + \Gamma_{ML^+})\mathrm{d}\mu_{MX} + \Gamma_{Y^-}\mathrm{d}\mu_{SY} \tag{15}$$

where MX and SY represent the aqueous and organic electrolyte, respectively, $\mathrm{d}E_{o+}^{w-}$ is the differential of the applied potential difference relative to the aqueous reference electrode reversible

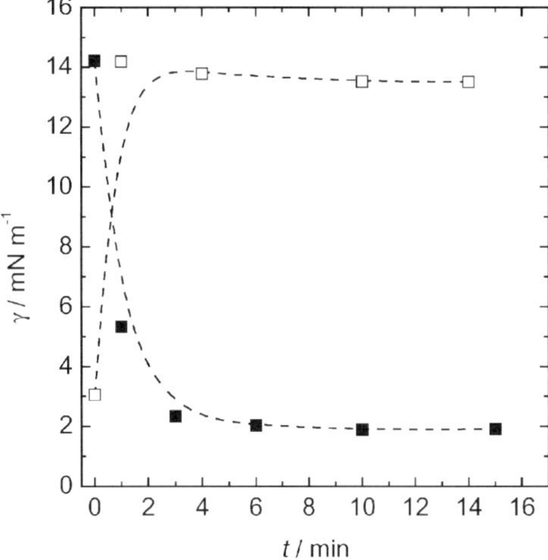

Fig. 8   Variation of the interfacial tension γ with time t for the interface between 0.1 M LiCl in water (pH 8.9) and 5 mM BTPPA TPBCl in DCE in the presence of 10 μM DPPC in DCE after stepping the potential E from 0.75 V to 0.15 (■) and the reverse (□).

to the anion X^- and the organic reference electrode reversible to the cation S^+, and the total surface charge density Q sums up the contributions from both the free charge and the charge associated with the adsorbed PC,[27]

$$Q = -\left(\frac{\partial \gamma}{\partial E_{o+}^{w-}}\right)_{\mu_i} = F(\Gamma_{M^+} - \Gamma_{X^-} + \Gamma_{ML^+}) = -F(\Gamma_{S^+} - \Gamma_{Y^-}) \tag{16}$$

It has been clear that the first derivative of the interfacial tension can have the meaning of the charge density if only the composition of both phases does not vary with the potential.[27,30] This condition is apparently not fulfilled at far positive potentials, where the double layer is essentially in the non-equilibrium state.[22] However, because the PC adsorption/desorption process is reversible, the changes in composition can be related to the potential using the Nernst equation, and the shape of the electrocapillary curve in both potential regions can be simulated using the Damaskin's model of a compound adsorbed in two different forms.[27] The critical value of the potential separating the two potential regions, is related to the binding (association) constant for the aqueous cation.[27]

4.4. Surface pressure and adsorption isotherm

In order to check whether the DPPC depletion effect due to DPPC co-adsorption on the wall of the cell plays any role, the light scattering experiments were also carried out with the cell that was carefully cleaned by using only chrom-sulfuric acid and ultrapure water. Results obtained after both cell pretreatments are compared in Fig. 9, which shows the effect of the apparent DPPC concentration on the surface pressure $\pi = \gamma_0 - \gamma$, γ_0 is the interfacial tension measured in the absence of DPPC, at a constant potential. The depletion effect is obviously rather significant, because the surface pressure measured using the clean electrolytic cell is significantly lower.

   The surface pressure in Fig. 9 reaches a limiting value of *ca.* 27.5 mN m^{-1} when the DPPC concentration exceeds 10 µmol dm^{-3}. Adsorption of dilauorylphosphatidylcholine (DLPC) at the neat water|carbon tetrachloride interface exhibits a similar behaviour:[16] the surface pressure levels out at 42 mN m^{-1}, *i.e.* close to the maximum value of 45 mN m^{-1}, when the bulk DLPC concentration is higher than *ca.* 2 µmol dm^{-3}. It has been proposed that DLPC is irreversibly adsorbed, and that once a tightly packed DLPC monolayer is formed, the surface pressure becomes independent of the bulk DLPC concentration.[16] However, this explanation is not applicable to the DPPC adsorption at the water|DCE interface, which is a reversible process. It is more likely that close to the limiting surface pressure the adsorbed layer reaches the collapse point, beyond which the multi-layer adsorption of DPPC or spontaneous emulsification takes place. This conclusion seems to be in accord with the results of the combined voltammetric and Langmuir study of

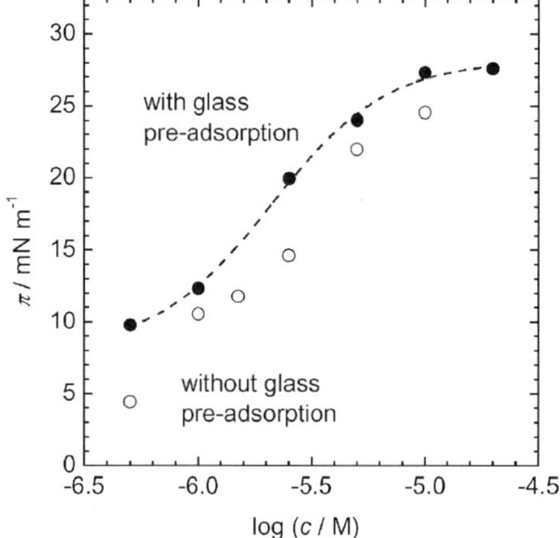

Fig. 9   Surface pressure π *vs.* the logarithm of concentration c of DPPC in DCE at $E = 0.43$ V measured in the cleaned glass cell ($\bigcirc$), and in the cell with DPPC pre-adsorbed on the cell wall ($\bullet$).

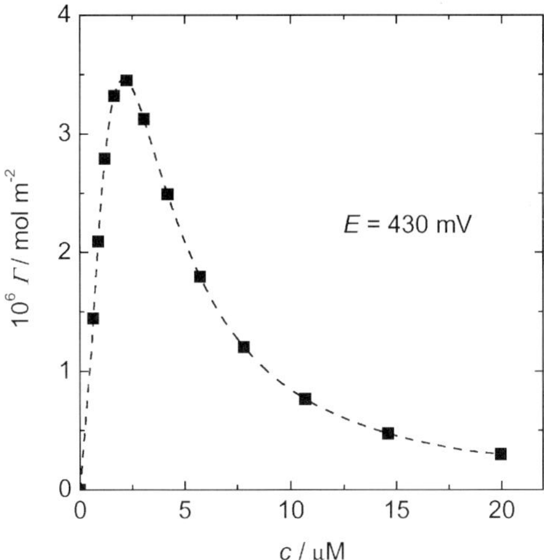

Fig. 10 Adsorption isotherm of DPPC at the interface between 0.1 M LiCl in water (pH 8.9) and 5 mM BTPPA TPBCl in DCE at $E = 0.43$ V.

phospholipid adsorption at the water|DCE interface indicating that the collapse point is near the molecular area of *ca.* 50 Å^2,[30] a value that can also be derived from the present data, *vide infra.*

Eqn. (15) indicates that the first derivative of the interfacial tension with respect to the logarithm of the phospholipid concentration gives the total relative surface excess concentration of phospholipid,[27]

$$\left(\frac{\partial \gamma}{\partial \ln c_{L^\pm}}\right)_{E_{o+}^{w-},\mu_{MX},\mu_{SY}} = -RT(\Gamma_{L^\pm} + \Gamma_{ML^+}) = -RT\Gamma \tag{17}$$

Fig. 10 shows the plot of Γ *vs.* the DPPC concentration (adsorption isotherm) evaluated from the data shown in Fig. 9. Note that at the applied potential $E = 0.43$ V, the DPPC adsorption in the zwitterionic form should prevail. The relative surface excess concentration of DPPC reaches the maximum value $\Gamma^m \approx 3.5 \times 10^{-6}$ mol m^{-2}, which corresponds to the molecular area of 47.4 Å^2 molecule^{-1}. The latter value can be compared with the cross-sectional area of 47 Å^2 molecule^{-1} for the choline headgroup based on a molecular model,[7] or with the value of 47.4 Å^2 molecule^{-1} measured for the DPPC in the gel phase of a bilayer lipid membrane.[50] Comparison points to the formation of a tightly packed DPPC monolayer at the water|DCE interface.

The rising part of the adsorption isotherm shown in Fig. 10 can be fitted to the Frumkin isotherm,[51]

$$Bc = \frac{\theta}{1-\theta}\exp(-2a\theta) \tag{18}$$

where $B = \exp(-\Delta G_{ads}^0/RT)$ is the adsorption coefficient, ΔG_{ads}^0 is the standard adsorption Gibbs energy, a is the interaction parameter and θ is the relative surface coverage. The fitting yields $\Delta G_{ads}^0 = 31.9$ kJ mol^{-1} and $a = 1.3$ indicating a weak repulsion between the adsorbed DPPC molecules. While both values are rather close to those obtained for the DPPC adsorption at the water|nitrobenzene interface,[23] the maximum surface excess concentration Γ^m is about twice as high.

4.5. Effects of DPPC concentration and applied potential on the damping factor

A test of the theory based on the analysis of the effect of the wavenumber k has already indicated that the linewidth (damping factor) decreases with increasing DPPC concentration. In an agreement with the theoretical prediction, this decrease appears to be a consequence of the decreasing interfacial tension, *cf.* the discussion in the chapter 4.2 above. Here, we extend the comparison with the theory by including also the effect of the potential. Fig. 11 shows the variation of the experimental linewidth Δf with the applied potential in the absence and the presence of DPPC. The

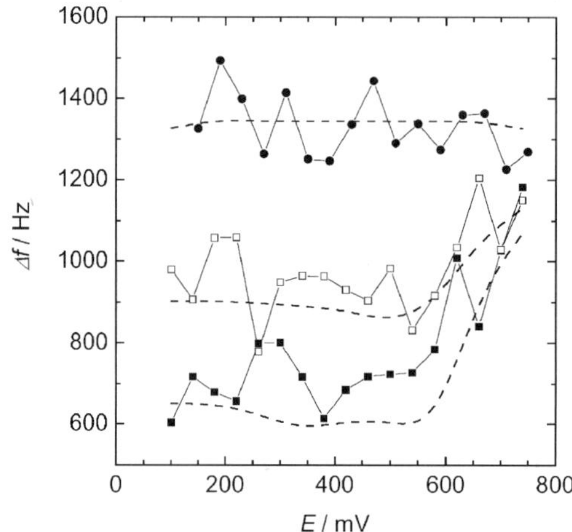

Fig. 11 Corrected power spectrum linewidth $\Delta f (k = 55608 \text{ m}^{-1})$ *vs.* the potential E for the interface between 0.1 M LiCl in water (pH 8.9) and 5 mM BTPPA TPBCl in DCE in the presence of DPPC in DCE at concentrations (in μM): 0 ($\bullet$), 5 ($\square$) and 20 ($\blacksquare$). Dashed lines show the theoretical dependences inferred from fitting the theoretical spectra to the Lorentz function for the corresponding values of the interfacial tension.

theoretical dependences have been inferred from fitting the theoretical spectra to the Lorentz function for the corresponding values of the interfacial tension. The agreement between the measured and theoretically predicted data is very good.

A correlation of the surface pressure and adsorption data shown in Figs. 9 and 10, respectively, provides an estimate of the dilatational modulus ε, which makes the value approximately equal to that of the surface pressure at a given DPPC concentration. A simulation of fluctuation spectra based on eqn. (2) for these values of ε indicates that the effect of the dilatational modulus on the damping factor is negligible. Hence, also the effects of interfacial viscosities representing the imaginary parts of the interfacial tension and dilatational modulus are probably too weak to be revealed safely by the light scattering measurements.

5. Conclusions

Light scattering measurements provide a valuable insight into the dynamics of the polarised water|DCE interface in the presence of adsorbed DPPC. Upon accounting for the effect of the DPPC co-adsorption on the wall of the electrolytic cell, an analysis of the measured frequency of capillary waves reveals some novel features in the adsorption of DPPC. First, the interfacial tension of the water|DCE interface decreases to a value near zero at much lower DPPC concentration than ever reported. Second, the surface excess concentration of DPPC reaches the value of 3.5×10^{-6} mol m^{-2}, which corresponds to the molecular area of 47.4 Å^2 molecule^{-1} not far from that estimated for a tightly packed phospholipid monolayer from molecular models (47 Å^2 molecule^{-1}). Third, the surface pressure approaches a limiting value, which is likely to correspond to the collapse of the phospholipid monolayer and multi-layer DPPC adsorption or emulsification near the zero interfacial tension. A detailed analysis of the effect of the capillary wavenumber on the capillary wave frequency and the damping factor suggests that the dynamic behaviour of ITIES is consistent with the theoretical predictions for a sharp liquid–liquid interface over the whole range of the DPPC concentrations studied. Therefore, the spontaneous formation of emulsion droplets at higher surface pressure appears to be less likely than the multi-layer DPPC adsorption.

Acknowledgements

This work was supported partly by the Grant Agency of the Czech Republic (Grant no. 203/01/0946) and partly by the Grant Agency of the Ministry of Education, Youth and Sports (Grant no. KONTAKT/ME 502).

References

1   M. C. Phillips and D. Chapman, *Biochem. Biophys. Acta*, 1968, **163**, 301.
2   M. N. Jones andD. C. Chapman, *Micelles, Monolayers and Biomembranes*, Wiley–Liss, New York, 1995.
3   K. S. Birdi, *Lipid and Biopolymer Monolayers at Liquid Interfaces*, Plenum Press, New York and London, 1989.
4   J. Koryta, L. Q. Hung and A. Hofmanová, *Stud. Biophys.*, 1982, **90**, 25.
5   T. Kakiuchi, in *Liquid–Liquid Interfaces. Theory and Methods*, ed. A. G. Volkov, D. W. Deamer, CRC Press, Boca Raton,1996, pp. 317–331.
6   L. Murtomäki, J. A. Manzanares, S. Mafé and K. Kontturi, in *Liquid Interfaces in Chemical, Biological and Pharmaceutical Applications*, ed. A. G. Volkov, M. Dekker, New York, 2001, pp. 533–551.
7   H. Möhwald, *Annu. Rev. Phys. Chem.*, 1990, **41**, 441.
8   H. M. McConnell, *Annu. Rev. Phys. Chem.*, 1991, **42**, 171.
9   C. M. Knobler and R. C. Desai, *Annu. Rev. Phys. Chem.*, 1992, **43**, 207.
10 B. Y. Yue, C. M. Jackson, J. A. G. Taylor, J. Mingins and B. A. Pethica, *J. Chem. Soc., Faraday Trans.*, 1976, **72**, 2685.
11 M. Thoma and H. Möhwald, *J. Colloid Interface Sci.*, 1994, **162**, 340.
12 I. Langmuir, *J. Am. Chem. Soc.*, 1917, **39**, 1848.
13 A. Gericke, D. J. Moore, R. K. Erukulla, R. Bittman and R. Mendelsohn, *J. Mol. Struct.*, 1996, **379**, 227.
14 R. D. Hunt, M. L. Mitchell and R. A. Dluhy, *J. Mol. Struct.*, 1989, **214**, 93.
15 R. A. Dluhy, N. A. Wright and P. R. Griffiths, *Appl. Spectrosc.*, 1988, **42**, 138.
16 R. A. Walker, J. A. Gruetzmacher and G. L. Richmond, *J. Am. Chem. Soc.*, 1998, **120**, 6991.
17 M. R. Watry, T. L. Tarbuck and G. L. Richmond, *J. Phys. Chem. B*, 2003, **107**, 512.
18 R. A. Walker, D. E. Gragson and G. L. Richmond, *Colloids Surf., A*, 1999, **154**, 175.
19 H. H. J. Girault and D. J. Schiffrin, *J. Electroanal. Chem.*, 1984, **179**, 277.
20 T. Wandlowski, S. Račinský, V. Mareček and Z. Samec, *J. Electroanal. Chem.*, 1987, **227**, 281.
21 T. Kakiuchi, M. Yamane, T. Osakai and M. Senda, *Bull. Chem. Soc. Jpn.*, 1987, **60**, 4223.
22 T. Kakiuchi, M. Nakanishi and M. Senda, *Bull. Chem. Soc. Jpn.*, 1988, **61**, 1845.
23 T. Kakiuchi, M. Nakanishi and M. Senda, *Bull. Chem. Soc. Jpn.*, 1989, **62**, 403.
24 S. G. Chesniuk, S. A. Dassie, L. M. Yudi and A. M. Baruzzi, *Electrochim. Acta*, 1998, **43**, 2175.
25 Y. Yoshida, H. Yoshinaga, N. Ichieda, A. Uehara, M. Kasuno, K. Banu, K. Maeda and S. Kihara, *Anal. Sci.*, 2001, **17**, (Suppl.), i1037.
26 R.-J. Roozeman, P. Liljeroth, C. Johans, D. E. Williams and K. Kontturi, *Langmuir*, 2002, **14**, 8318.
27 Z. Samec, A. Trojánek and H. H. Girault, *Electrochem. Commun.*, 2003, **5**, 98.
28 V. Mareček, A. Lhotský and H. Jänchenová, *J. Phys. Chem. B*, 2003, **107**, 4573.
29 D. Grandell and L. Murtomäki, *Langmuir*, 1998, **14**, 556.
30 D. Grandell, L. Murtomäki, K. Kontturi and G. Sundholm, *J. Electroanal. Chem.*, 1999, **463**, 242.
31 R. M. Allen, K. Kontturi, L. Murtomäki and D. E. Williams, *J. Electroanal. Chem.*, 2000, **483**, 57.
32 A. Benda, M. Beneš, V. Mareček, A. Lhotský, W. Th. Hermens and M. Hof, *Langmuir*, 2003, **19**, 4120.
33 Z. Zhang, I. Tsuyumoto, S. Takahashi, T. Kitamori and T. Sawada, *J. Phys. Chem. A*, 1997, **101**, 4163.
34 A. Trojánek, P. Krtil and Z. Samec, *Electrochem. Commun.*, 2001, **3**, 613.
35 A. Trojánek, P. Krtil and Z. Samec, *J. Electroanal. Chem.*, 2001, **517**, 77.
36 H. Nagatani, Z. Samec, P.-F. Brevet, D. J. Fermin and H. H. Girault, *J. Phys. Chem. B.*, 2003, **107**, 786.
37 J. C. Rowlinson and B. Widom, *Molecular Theory of Capillarity*, Clarendon, Oxford, 1982.
38 R. Laudon, in *Surface Excitations*, ed. V. M. Agranovich, R. Laudon, North Holland, Amsterdam, 1984, pp. 591–638.
39 L. I. Daikin, A. A. Kornyshev and M. Urbakh, *J. Electroanal. Chem.*, 2000, **483**, 68.
40 I. Benjamin, *Annu. Rev. Phys. Chem.*, 1997, **48**, 407.
41 S. Frank and W. Schmickler, *J. Electroanal. Chem.*, 2004, **564**, 239.
42 S. Hard, Y. Hamnerius and O. Nilsson, *J. Appl. Phys.*, 1976, **47**, 2433.
43 J. C. Earnshaw and R. C. McGivern, *J. Phys. D: Appl. Phys.*, 1987, **20**, 82.
44 J. C. Earnshaw, in *Fluid Interfacial Phenomena*, ed. C. A. Croxton, John Wiley & Sons Ltd, New York, 1986, pp. 437–468.
45 D. Langevin and J. Meunier, in *Photon Correlation Spectroscopy and Velocimetry*, ed. H. Z. Cummins, E. R. Pike, Plenum, New York, 1977.
46 M. S. Jhon, R. C. Deasai and J. S. Dahler, *J. Chem. Phys.*, 1978, **68**, 5615.
47 V. G. Levich, *Physicochemical Hydrodynamics*, Englewood Cliffs, Prentice Hall, NJ, 1962, p. 603.
48 H. Lamb, *Hydrodynamics*, Dover, New York, 1945, p. 348.
49 A. Sabela, V. Mareček, Z. Samec and R. Fuoco, *Electrochim. Acta*, 1992, **37**, 231.
50 J. F. Nagle and S. Tristram-Nagle, *Biochim. Biophys. Acta*, 2000, **1469**, 159.
51 A. N. Frumkin, *Z. Phys. Chem.*, 1925, **116**, 466.

Steady-state analysis of polymer adsorption at and transport across an interface between two polymer phases

S. M. Scheinhardt-Engels, F. A. M. Leermakers* and G. J. Fleer

Laboratory of Physical Chemistry and Colloid Science, Wageningen University, Dreijenplein 6, 6703 HB Wageningen, The Netherlands

Received 7th April 2004, Accepted 28th June 2004
First published as an Advance Article on the web 23rd September 2004

We consider a mutually incompatible polymer mixture composed of two major components A_N and B_N and a third minority component C_N. The interactions, parameterized by short-range Flory–Huggins interaction parameters, are chosen such that C wets the A/B interface completely at three-phase coexistence. At sub-saturated conditions the adsorption of C remains necessarily microscopic. We study such a system in a stationary off-equilibrium state: due to imposed chemical–potential gradients, polymer A_N diffuses from the A-rich bulk phase through the interface to the A-poor bulk phase. Polymer B_N travels in the opposite direction. Symmetric conditions are selected for which the polymer C_N that accumulates at the A/B-interface has no net flux in the stationary state. This system is described by the Mean Field Stationary Diffusion (MFSD) model, an approach that solves the Scheutjens–Fleer self-consistent-field (SF-SCF) equations with the boundary condition that the chemical potentials in the two bulk phases are different (but constant) so that a stationary state can be described. When the chemical potentials in the two bulk phases are the same, MFSD reduces to the equilibrium SF-SCF results. From the MFSD method we obtain the stationary volume fraction profiles and segmental fluxes. By forcing the system further from three-phase coexistence, i.e. by imposing larger concentration gradients, the adsorption of C goes unexpectedly from a thin adsorption layer to a thick adsorption film. The susceptibility $\partial J_A/\partial \Delta \varphi_A$ of the flux of A (equal to minus the flux of B) with respect to the imposed concentration gradient changes abruptly at the transition in adsorption behaviour. Interestingly, upon variation of the concentration gradients, the fluxes of A and B are enhanced by the accumulation of C at the interface. This means that the adsorbed C-film does not behave as an inert barrier.

Introduction

Most theoretical studies on adsorption at interfaces focus on equilibrium. This is not surprising since the equilibrium is a well-defined state to which all ergodic systems tend to evolve. However, it may take a very long time before true equilibrium is achieved. Moreover, during the evolution from off-equilibrium to equilibrium, systems may encounter local minima or extended local valleys in the free energy landscape, which may cause a system to be in an off-equilibrium state for a long time. To analyse these systems systematically it is useful to study systems in their simplest nonequilibrium form, i.e. in their stationary off-equilibrium state. Understanding the behaviour of stationary systems is a necessary step towards describing dynamic processes since the properties of a stationary

DOI: 10.1039/b405292g

state are governed by the dynamics and, even more importantly, these dynamic states are well-defined (do not depend on the history of the system).

We investigate the steady state of adsorption at, and diffusion through an off-equilibrium interface between two immiscible polymer liquids A_N and B_N. The interface is off-equilibrium by imposing and maintaining driving forces to these polymers. The interface develops due to the mutual immiscibility of A_N and B_N, but there is transport across this interface due to imposed gradients in chemical potentials. A similar system, but in equilibrium state, was previously studied.[1] There we were interested in the transitions from thin to thick adsorption layers induced by variation of the intermolecular interactions. Below we will review some essential results that set the stage for the present steady state analysis.

The equilibrium adsorption was studied by means of the Scheutjens–Fleer Self-Consistent Field (SF-SCF) method.[2] This mean-field method has proved to be a powerful tool to investigate a large variety of inhomogeneous polymer systems at equilibrium. For off-equilibrium interfaces we use an extension of the SF-SCF method, namely the Mean-Field Stationary Diffusion (MFSD) method. The details of this method were outlined in ref. 3. The SF-SCF and MFSD methods use the same language in the description of the system, the polymers, and the interactions. The methods differ in the input requirements (boundary conditions). In the equilibrium SF-SCF calculations the composition of a bulk system that is in equilibrium with the interface is uniquely defined. For the MFSD calculations one can select arbitrary bulk compositions, from which the driving forces for each component in the system follow. When these driving forces happen to be zero the solutions of the MFSD and SF-SCF calculations coincide. The MFSD method selects a solution in which the material fluxes are constant throughout the system, whereas the SF-SCF method focuses on the homogeneity of chemical potentials, pressure and temperature in the system. The fluxes in MFSD calculations are related to the free energy functional of SF-SCF calculations through the definition of the segment chemical potentials. Gradients in the segment chemical potential are the driving force for transport.

In the next section we first introduce our system. A connection is made to the equilibrium study of adsorption and wetting (see ref. 1) since the results of that study were used to select the appropriate boundary conditions for the stationary-state analysis. The stationary diffusion occurs between two bulk mixtures. The driving forces are defined by choosing the compositions of these bulk mixtures. The compositions were chosen such that the mixtures are stable, *i.e.*, within the bulk phases no phase separation will occur. After this we present the essential features of the MFSD method. The relation between the composition of the bulk mixtures and the driving forces (the chemical potential differences) is given. In the results section we focus on the consequences of varying concentration gradients for the adsorption layer and the stationary fluxes. In a short outlook we point at the possibility of using the MFSD method for biologically interesting systems. The conclusions are summarised at the end of the paper.

Equilibrium self-consistent field theory (SCF)

Self-consistent field (SCF) methods describe inhomogeneous polymer systems in equilibrium on a mean-field level. To evaluate the density distributions they usually incorporate the Edwards diffusion eqn. (1) for polymer trajectories in a self-consistent potential field.[4] This field contains contributions from short-range nearest-neighbour contact energies (parameterized by Flory–Huggins interaction parameters) and a Lagrange contribution coupled to an incompressibility constraint.

To obtain exact self-consistent-field solutions one has to resort to numerical methods. Then some discretisation scheme must be implemented. Here we apply the Scheutjens–Fleer (SF-SCF) method in which all types of segments have the same length b, which defines the characteristic size of a lattice site onto which the segments are placed. The lattice sites are arranged in layers over which the mean-field approximation is applied.

Here we use a flat lattice with layer numbers z running from $z = 1, \ldots, M$. At the system boundaries we employ mirror-like reflecting boundary conditions. Only concentration gradients perpendicular to the layers are considered. In the SF-SCF method the interaction energies are non-local, meaning that segments in layer z not only interact with neighbouring segments within the same layer, but also with segments in adjoining layers. This feature is essential to describe the interface between two immiscible polymer solutions.

The equilibrium SF-SCF method[2] provides an easy way to calculate volume fraction profiles for inhomogeneous (multicomponent) systems at equilibrium. Here we will consider only homodisperse homopolymers and the number of components is equal to the number of segment types, so we can refer to a component just by referring to its constituent segment type A, B and C. The chain length of homopolymer A is N_A; its segments have ranking numbers $s = 1, \ldots, N_A$. The conformation c of a chain is defined by the z-coordinate of each of its segments. The SF-SCF method optimises the partition function Q for a lattice in which each layer is fully occupied by segments or solvent (incompressibility constraint); in our present system there is no solvent. Then the optimisation of the partition function must be performed under M constraints:

$$\sum_X \varphi_X(z) = 1 \quad \forall z \in [1, M] \tag{1}$$

where the sum over X denotes summation over all components (or over all segment types) $X = $ A, B, C. Therefore, M Lagrange-parameters $\alpha(z)$ are introduced in the equilibrium SF-SCF method, which may be interpreted as the space-filling potentials. In the equilibrium method, these potentials do not depend on the segment type. The requirements for equilibrium then become

$$\frac{\partial}{\partial n_X^c} \left[\ln Q + \sum_z \alpha(z) \left\{ \sum_X \varphi_X - 1 \right\} \right] = 0 \quad \forall n_X^c \tag{2a}$$

$$\frac{\partial}{\partial \alpha(z)} \left[\ln Q + \sum_z \alpha(z) \left\{ \sum_X \varphi_X - 1 \right\} \right] = 0 \quad \forall z \in [1, M]. \tag{2b}$$

The parameter n_X^c denotes the number of molecules X in a specified conformation c. Obviously, eqn. (2b) ensures the constraint of incompressibility to be fulfilled. Eqn. (2a) dictates the way in which the volume fractions $\varphi_A(z)$ must be calculated from given segment potentials $u_A(z)$ to obtain the conformation distribution with minimal free energy. The volume fractions depend on the potentials $\varphi[u(z)]$, but the potentials are also dependent on the volume fractions $u[\varphi(z)]$, for example due to unfavourable segment–segment contacts. The equilibrium SF-SCF algorithm is an iterative procedure which leads to a fixed point for which the potentials are consistent with the volume fractions that obey the constraints.

In Fig. 1 we present some selected results for a system that was discussed in much more detail elsewhere.[1] It comprises three types of polymers all of equal length A_N, B_N and C_N that are mutually incompatible. Here and below we limit ourselves to $N = 10$. We choose the interaction parameter χ_{AB} between the two bulk polymers A and B such that a relatively wide interface is formed. The minority component C_N, which is assumed to have symmetric interactions with respect to A and B ($\chi_C = \chi_{AC} = \chi_{BC}$) adsorbs onto this interface. The adsorption isotherm depends on the two control parameters in the system χ_{AB} and χ_C.

In Fig. 1a we present a set of adsorption isotherms for which the width of the A/B interface was approximately kept fixed by setting $\chi_{AB} = 0.34$ and where the driving force for adsorption was varied by changing χ_C. Here the adsorbed amount (in equivalent monolayers) is computed as the excess $\theta_C^{ex} = \sum_z (\varphi_C(z) - \varphi_C^b)$, where φ_C^b is the bulk concentration of C_N in both bulk phases (phase A and phase B). The adsorption isotherms are presented as a function of a normalised bulk concentration Φ_C, which is the ratio between the fraction of C in (both) bulk phases and the bulk volume fraction of C at three-phase coexistence φ^*_C: $\Phi_C = \varphi_C^b / \varphi^*_C$. When a monomer C adsorbs at the interface, the unfavourable A–B contact is replaced by more favorable A–C and C–B interactions. The lower χ_C, the higher the driving force for adsorption. In line with this it is found that for small values of χ_C, i.e. for $\chi_C \leq 0.25$, the adsorption increases monotonically and diverges at the three-phase coexistence value (the interface is wetted by C). For values $\chi_C > 0.25$ all the isotherms cross the coexistence value $\Phi_C = 1$: the interface is partially wet. The adsorbed amount at coexistence increases monotonically as a function of χ_C, which proves that the wetting transition at $\chi_C \approx 0.25$ is continuous (second order). As shown in ref. 1 the wetting transition becomes first order when the A/B interface is significantly sharper ($\chi_{AB} > 0.9$).

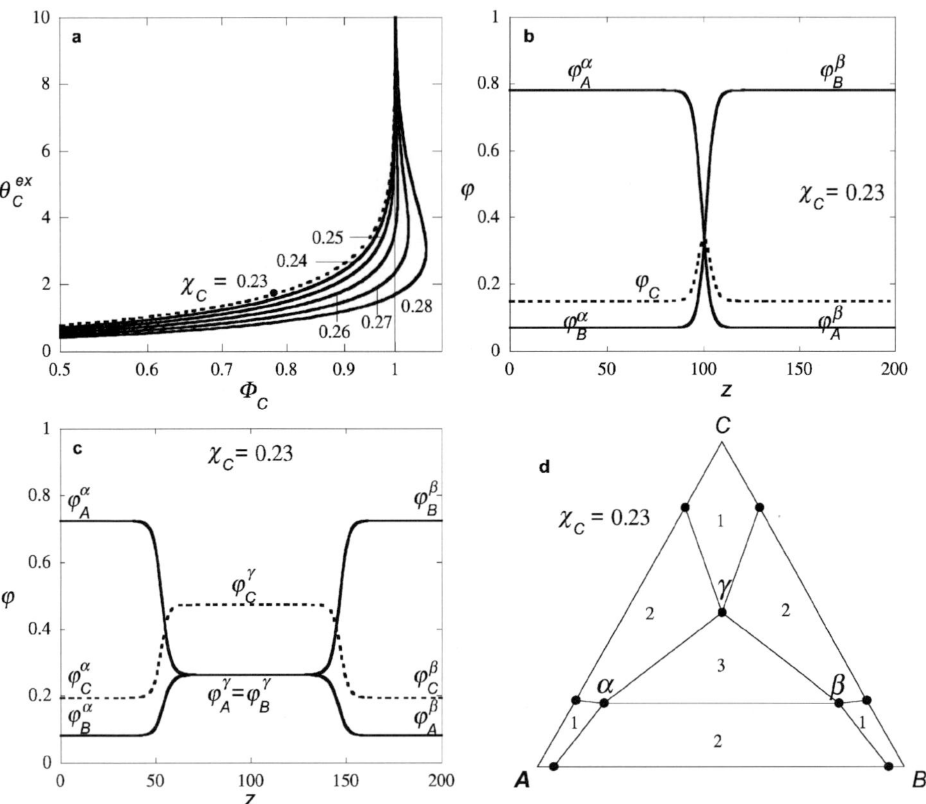

Fig. 1 Equilibrium behavior for $N_A = N_B = N_C = 10$ and $\chi_{AB} = 0.34$. (a) Excess adsorbed amount of the C polymer on an A/B interface as a function of the normalized bulk volume fraction $\Phi_C = \varphi_C^b/\varphi_C^*$ where φ_C^* the volume fraction of C in the A- or B phase at three-phase coexistence, for six values of χ_C as indicated. The curve for $\chi_C = 0.23$ (the χ_C value in panels b, c, and d) is dotted, the other adsorption isotherms are solid. The solid dot on this isotherm represents the bulk concentration $\varphi_C^b = 0.15$, for which the density profiles are shown in panel b. (b) The volume fraction profiles across the A/B interface for a sub-saturated bulk volume fraction $\varphi_C^b = 0.15$ and $\Delta\varphi^{eq} \equiv \varphi_A^\alpha - \varphi_A^\beta = 0.7107$. (c) A typical density profile across the interface at three-phase coexistence. The indices α, β and γ indicate the three coexisting phases; the profiles for the A and B components are solid curves, that for C is dotted. (d) A schematic phase triangle extracted (in part) from the volume fractions of the three coexisting phases given in c and three sets of data found from binary coexisting systems A/B, B/C, A/C. There is one three-phase region, three two-phase regions, and three one-phase regions, labeled 3, 2 and 1, respectively.

For first-order wetting transitions (not shown) there are more features in the adsorption isotherms (*e.g.* there may be a pre-wetting step). Such features typically complicate the system and we choose to avoid these problems in the present study. For this reason we selected the conditions $\chi_{AB} = 0.34$ and $\chi_C = 0.23$ which represent a system that, in equilibrium, leads to a wetting layer of C at three-phase coexistence and the adsorption isotherm has no special features (see the dotted curve in Fig. 1a).

In Fig. 1b, c we give two sets of volume fraction profiles which will be used as a reference for the steady state systems discussed below. When the bulk volume fraction of C is sub-saturated, *i.e.* $\varphi_C^b = 0.15$ the adsorption has the relatively low value of $\theta^{ex} \approx 1.7$ (Fig. 1a and 1b). However, at three-phase coexistence ($\Phi_C = 1$) there is a macroscopic thick film of C in between the phases A and B, as shown in Fig. 1c. From the latter graph it is possible to identify the compositions of the three phases (α, β, γ) that are in mutual equilibrium. We use the Flory–Huggins theory to obtain the binodals for the three binary systems A/B, A/C, B/C. In this way we obtain nine characteristic points in the phase triangle. These points are tentatively connected by straight lines in the phase diagram presented in Fig. 1d. There are three single phase regions, three two-phase coexistence regions and one three-phase coexistence region. We note that it is possible to obtain an accurate phase diagram, but the schematic one given in Fig. 1d suffices for our purpose.

Mean-field stationary diffusion (MFSD)

The starting point is a continuity relation $\partial\varphi_A/\partial t = -\nabla J_A(z)$. In words such an equation says that the change in local concentration in time is given by the disparity between the inward flux and outward flux of material. Below we consider steady states only and disregard the process towards such steady states. This means that we will directly focus on the condition that $\partial\varphi_A/\partial t = 0 = -\nabla J_A(z)$. This means that at each coordinate the inward and outward flux has the same value. Below we will give expressions for the flux.

To study adsorption at off-equilibrium interfaces, we put the selected systems of Fig. 1b,c into such a (off-equilibrium) steady state condition. This is done by defining two infinitely large bulk mixtures each with a composition corresponding to a one-phase region of the phase composition diagram (Fig. 1d). An example of such a system is shown in Fig. 2. Mixture I is rich in A_N with concentrations $\varphi_A{}^I > \varphi_A{}^\alpha$, $\varphi_B{}^I < \varphi_B{}^\alpha$, and $\varphi_C{}^I < \varphi_C{}^\alpha$. Mixture II is rich in B_N with $\varphi_A{}^{II} < \varphi_A{}^\beta$, $\varphi_B{}^{II} > \varphi_B{}^\beta$, and $\varphi_C{}^{II} < \varphi_C{}^\beta$. The composition differences between mixtures I and II for A_N and B_N are denoted by $\Delta\varphi = |\varphi^{II} - \varphi^I|$. We restrict ourselves to the situation that $\varphi_C{}^I = \varphi_C{}^{II}$, thus $\Delta\varphi_C = 0$. The bulk mixtures are brought into contact. Since their compositions do not correspond to coexisting phases, the polymer components will start to diffuse in order to make mixtures I and II identical or coexisting. However, we keep the compositions of mixtures I and II constant (ideal sink–source system). Eventually, a steady state will occur, where all components diffuse with a flux that is constant in time and space.

The steady state is studied by means of the Mean-Field Stationary Diffusion (MFSD)-method. This method yields the stationary volume fraction profiles, chemical potential profiles, material fluxes and (average) chain conformations. That information is obtained for the region between bulk mixtures I and II as depicted in Fig. 2.

As in the SF-SCF method, the region is described by a one-dimensional lattice, whose layers lie perpendicular to the diffusion direction and are numbered $z = 1, \ldots, M$. The flux of segments A through layer z is denoted by $J_A(z)$. Due to the imposed volume fraction gradients and the constraints $\nabla J_A = \nabla J_B = \nabla J_C = 0$, the C-rich wetting layer (*cf.* Fig. 1c) is replaced by a C-rich adsorption layer that has a different width and composition. For example, the mesoscopically thick wetting layer may shrink to a small accumulation of C_N at the interface as in Fig. 1b. The presence of an adsorption layer is expected to influence the material fluxes. Intuitively it reduces the available space for the diffusing components and may thereby act as a barrier. The effect of a similar, but static, barrier on the chain conformations was studied elsewhere.[5] Below we show that in the case of a 'dynamic' barrier it is highly nontrivial to relate the fluxes of A and B to the amount of C at the A/B-interface.

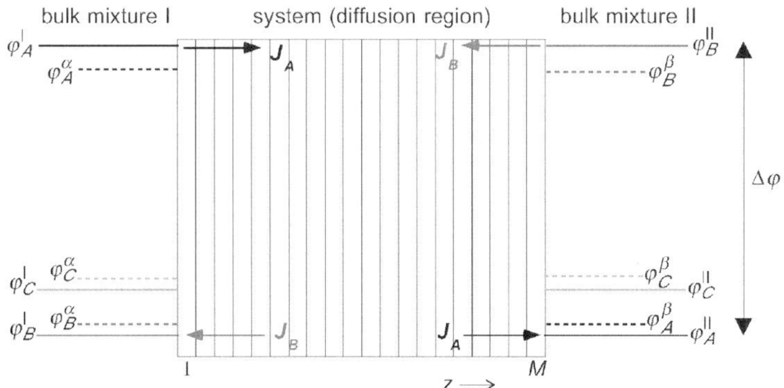

Fig. 2 The system used to study adsorption of C at off-equilibrium A/B-interfaces and the associated stationary material fluxes, with imposed concentrations in I and II corresponding to the levels indicated by the solid horizontal lines. A symmetric system, typical for this study, is shown with $\Delta\mu_A = -\Delta\mu_B$ and $\Delta\mu_C = 0$. The dashed levels are the coexistence concentrations of Fig. 1d.

Note that we have chosen a system with a high degree of symmetry: $N_A = N_B = N_C = N$, $\varphi_C^I = \varphi_C^{II}$, and $\chi_{AC} = \chi_{BC}$. In the results that we present we have chosen the compositions of mixtures I and II such that $\Delta\varphi_A = -\Delta\varphi_B$. Then, due to the other symmetric properties, $\Delta\mu_A = -\Delta\mu_B$, $\Delta\mu_C = 0$, the adsorption layer will be situated exactly in the center of the system. Even for such symmetric systems we obtain a rich adsorption behaviour.

In the MFSD method, the volume fractions are calculated in a way similar to that in the SF-SCF method. Thus, the volume fractions in the stationary state correspond to that conformation distribution of all molecules for which the free energy is minimal. We apply the SF-SCF free energy functional which is valid for equilibrium systems. It is common to use equilibrium functionals for off-equilibria, since usually the true free energy functionals are unknown.[6] We do not consider this as a serious approximation, since we are only interested in the steady state and not in the evolution towards the steady state. There is a small difference between the calculation of φ in SF-SCF and in MFSD. This is due to the extended set of constraints for the stationary state. For the stationary state, we have the constraints

$$\varphi_X(0) = \varphi_X^I \; \forall X \tag{3a}$$

$$\varphi_X(M + 1) = \varphi_X^{II} \; \forall X \tag{3b}$$

$$\sum_X \varphi_X(z) = 1 \quad \forall z \in [1, M] \tag{3c}$$

$$\frac{\partial \varphi_X(z)}{\partial t} = 0 \quad \forall X, z \in [1, M]. \tag{3d}$$

The first two constraints (given by (eqns. (3a) and (3b)) are treated separately by the boundary conditions. The next M constraints eqn. (3c) are obeyed by additional stop-criteria for the iterations which must lead to the consistency between the potentials and the volume fractions. The number of constraints remaining is M times the number of segment types (see (eqn. (3d)). As noted at the beginning of this section, the variable 'time' does not enter explicitly, as the continuity equation gives the translation to fluxes which, in turn, are calculated from the profiles. We assume that there exists only one volume fraction profile, which obeys all constraints and has minimal free energy. If this profile is given by $\varphi_A^{stat}(z)$ the constraints in eqn. (3d) may be summarized by:

$$\varphi_X(z) = \varphi_X^{stat}(z) \; \forall X, z \in [1, M]. \tag{4}$$

The requirements for the stationary state become:

$$\frac{\partial}{\partial n_X^c}\left[\ln Q + \sum_{X,z} \alpha_X(z)\left\{\varphi_X^{stat}(z) - \varphi_X(z)\right\}\right] = 0 \quad \forall n_X^c \tag{5a}$$

$$\frac{\partial}{\partial \alpha_X(z)}\left[\ln Q + \sum_{X,z} \alpha_X(z)\left\{\varphi_X^{stat}(z) - \varphi_X(z)\right\}\right] = 0 \; \forall X, z \in [1, M]. \tag{5b}$$

This equation may be compared to eqn. (2) where the multipliers $\alpha(z)$ are independent of the segment type. However in the stationary state case we only have the correct number of Lagrange parameters if we take $\alpha(z)$ to be dependent on the segment type as in eqn. (5b).

The MFSD-method starts, as the SF-SCF method,[2] with a guess for the segment potentials $u_A(z)$.[3] These potentials can be expressed in terms of Boltzmann-weighting factors $G_A(z)$ through the relation $u_A(z) = -\ln G_A(z)$. Here and below we express the segment potentials and chemical potentials in units of $k_B T$. The segment potentials depend on the volume fractions φ of all other segments since energetic interactions, quantified by the Flory–Huggins parameters χ_{ij}, chain connectivities, and space filling determine the probability to find segment A in layer z. Thus $u_A = u_A[\varphi]$. On the other hand, the volume fraction $\varphi_A(z)$ of segments A depend on the Boltzmann-weighting factors and therefore on the segment potentials: $\varphi_A = \varphi_A[u]$. The MFSD-calculation comes to a solution when self-consistency is obtained between the potentials and the volume fractions. When the weighting factors or segment potentials are known, the volume fractions φ_A and segment chemical potentials μ_A of all segments can be calculated as a function of position z.

From these, we may calculate other system properties such as adsorbed amounts and stationary segment fluxes.

In the MFSD-method the segment weighting factors are calculated under the constraints of incompressibility and stationary volume fractions. Moreover, the solution of the MFSD-calculation must obey the imposed compositions for bulk mixtures I and II at $z = 0$ and $z = M$. By defining the compositions of bulk mixtures I and II we fix the chemical potentials for the segments at both sides of the system. Unlike in equilibrium, these chemical potentials are not the same. We thus impose a driving force $\Delta\mu_A = \mu_A^{II} - \mu_A^{I}$ for each segment type to diffuse. The segment chemical potential for segment type A in bulk mixture I is given by:

$$\mu_A^{I} = \frac{1}{N_A}\ln\varphi_A^{I} + \chi_{AB}\varphi_B^{I} + \chi_{AC}\varphi_C^{I}. \tag{6}$$

In the following, we vary the chemical potentials in bulk mixtures I and II by variation of the compositions of mixtures I and II at constant χ-parameters, such that $\Delta\varphi_A = -\Delta\varphi_B$ and $\Delta\varphi_C = 0$. Of interest to the diffusion are the consequences of these variations for the imposed driving forces $|\Delta\mu_A|$ and $|\Delta\mu_B|$. We note that $\Delta\mu_C = 0$ due to the symmetry of our system. With the varying compositions of mixtures I and II, $|\Delta\mu_A|$ and $|\Delta\mu_B|$ are rather complex functions of these compositions. However, as long as both bulk mixtures have a stable composition, *i.e.* if they fall into the one-phase region of a phase composition diagram, the driving forces increase monotonically with increasing $|\Delta\varphi_A| = |\Delta\varphi_B| = |\Delta\varphi|$.

We study the effect of varying driving forces on the adsorption and on the stationary segment fluxes. These fluxes were derived for multicomponent systems in ref. 3. Two different diffusion mechanisms were put forward, the so-called slow-mode and fast-mode diffusion. According to the slow-mode model, the diffusion occurs through swapping the positions of two different segments.[7] According to the fast-mode model the slow-mode diffusion mechanism must be accompanied by a drift flux.[8,9] In terms of Onsager coefficients Λ_i, the slow-mode flux and fast-mode flux are given by, respectively,

$$J_A^s = -\frac{\Lambda_A}{\sum_i \Lambda_i} k_B T \sum_j \Lambda_j \nabla(\mu_A - \mu_j) \tag{7a}$$

$$J_A^f = -k_B T \sum_j \left(\varphi_j \Lambda_A \nabla\mu_A - \varphi_A \Lambda_j \nabla\mu_j\right). \tag{7b}$$

The Onsager coefficient Λ_i equals $\tilde{B}_i\varphi_i$, where $\tilde{B}_i$ is the mobility of segment type i. In this paper we assume equal mobilities for all types of segments ($\tilde{B}_i = \tilde{B}\ \forall_i$), so that the slow-mode and fast-mode fluxes become identical:

$$J_A^s = J_A^f = -\tilde{B}k_B T \sum_j \varphi_A \varphi_j \nabla(\mu_A - \mu_j). \tag{8}$$

We do not specify the conversion factor between our (dimensionless) fluxes and real fluxes and thus focus on the trends of the flux behaviour.

Results and discussion

Fig. 3 presents the stationary volume fraction profiles for two distinct situations in very similar systems. Fig. 3a and 3b both refer to a system with $N = 10$, $\chi_{AB} = 0.34$, and $\chi_C = 0.23$, the same parameters as used in Fig. 2. Moreover, we have chosen for both systems $\varphi_C^{I} = \varphi_C^{II} = 0.15$ and $\Delta\varphi = -\Delta\varphi = \Delta\varphi_B$. There is no net diffusion of C-segments ($J_C = 0$). The only difference between the two cases is the value of $\Delta\varphi$. In Fig. 3a the imposed driving force ($\Delta\varphi = 0.73$) is substantially smaller than in Fig. 3b ($\Delta\varphi = 0.84998$). For comparison, in equilibrium $\Delta\varphi^{eq} = 0.7107$ (see Fig. 1b); at this concentration difference the driving force for diffusion is zero. In fact, the true measure for the driving force is $\Delta\mu_A$ instead of $\Delta\varphi_A$, but in the present system $\Delta\mu_A$ grows monotonically with $\Delta\varphi_A$.

For the smaller driving force, the volume fraction profiles of all components are nearly flat up to a small distance from the interface. Consistent with the fact that the system is nearly at equilibrium

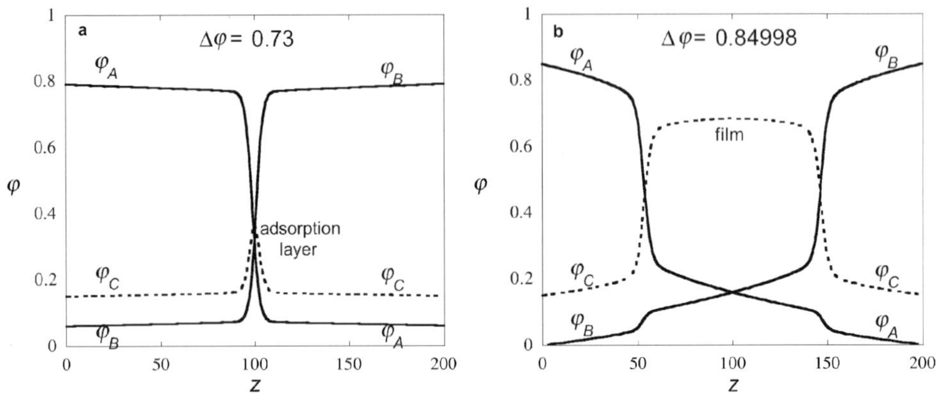

Fig. 3 Stationary volume fraction profiles for adsorption at off-equilibrium interfaces in two similar systems; $N = 10$, $\chi_{AB} = 0.34$, $\chi_C = 0.23$, $\varphi_C^I = \varphi_C^{II} = 0.15$, $\varphi_A^I = \varphi_B^{II}$, $\varphi_A^{II} = \varphi_B^I$. (a): $\Delta\varphi = -\Delta\varphi_A = \Delta\varphi_B = 0.73$. (b): $\Delta\varphi = -0.84998$.

and (with respect to the χ-values) identical to the system in Fig. 1d, the adsorption layer of C_N remains microscopically small. This is expected because the volume fraction of C is sub-saturated. The profile is very similar to that given in Fig. 1b. Indeed, the system is sufficiently close to a two-phase coexistence with an A- and B-rich phase and can not form a thick C film. In other words, the system remains everywhere subsaturated with C.

The larger driving force in Fig. 3b significantly changes the adsorption layer. It grows to mesoscopic scales (a liquid film develops) and the volume fraction of C within this film even exceeds the value of φ_C^{γ} in the equilibrium γ-phase.

The origin of this quite different behaviour for seemingly similar systems is clarified by Fig. 4. In this figure the volume fraction profiles of Fig. 3a and 3b are plotted as composition profiles in the equilibrium phase composition diagram of Fig. 1d. Each point in Fig. 4 corresponds to the composition of one lattice layer in Fig. 3. The starting point to the left of a composition profile in Fig. 4 is the composition of mixture I in the A-rich phase. The end point to the right corresponds to the composition of mixture II in the B-rich phase. The composition profile for small driving forces (Fig. 3a and gray dots in Fig. 4) runs through the A/B -binodals and the three-phase coexistence region. The profile does not cross the C-rich one-phase region; therefore there is only a (micro-scopic) adsorption layer. From the small number of points within the three-phase region it is clear that only a few layers have a composition which would be unstable in homogeneous macroscopic systems. These layers form the interface between the two stable portions of the system that are rich

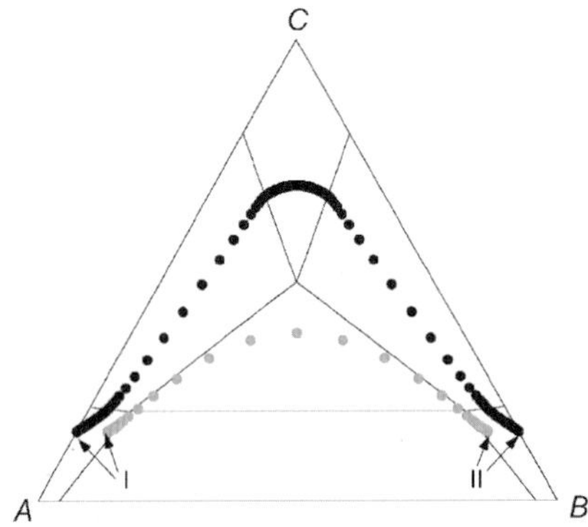

Fig. 4 The composition profiles of Fig. 3 plotted in the equilibrium phase composition diagram. The gray curve corresponds to Fig. 3a, the black curve to Fig. 3b. Each dot is the composition of one lattice layer.

in A and B, respectively. The gradients of φ within the interface stabilize the instability, as in van der Waals' theory of liquid/liquid interfaces.

For a larger driving force (Fig. 3b and black dots in Fig. 4) the composition profile first crosses the A/C-binodals so that a stable C-rich phase develops, which is the liquid film at the A/B interface. Then the B/C-binodals are crossed to arrive in the stable B-rich phase. Within the one-phase regions the composition profiles 'travel' more slowly through the composition diagram than within two- or three-phase regions.

We conclude that whether a mesoscopic film will develop at the A/B interface or not depends in general on the location of one-phase regions in the phase composition diagram and on the driving forces between two points in this diagram. When the C-rich one-phase region or the driving forces are small, or when mixtures I and II are far from the A/C and B/C binodals, the adsorption layer will remain of microscopic size. We call the adsorption layer a 'film' when the composition profile enters the stable C-rich phase region of the phase composition diagram.

From the composition profiles in Fig. 4 it can be anticipated that when these profiles cross a phase line, that is, when they reach the stable C-rich phase, the adsorbed amount of C may increase more readily. Fig. 5a shows how the film develops with increasing driving force. The lower curve in Fig. 5a (right-hand scale) gives the excess adsorbed amount θ_C^{ex} of component C at the interface. This quantity is calculated similarly as in the SCF model by

$$\theta_C^{ex} = \sum_z \left[\varphi_C(z) - \varphi_C^I \right] = \sum_z \left[\varphi_C(z) - \varphi_C^{II} \right]. \tag{9}$$

Note that the adsorbed amount of C also has contributions relatively far from the interface, since the φ_C-gradients in the A- and B-rich phases also contribute (to a minor extent) to θ_C^{ex} (cf. Fig. 3). We define the susceptibility of θ_C^{ex} with respect to a change in $\Delta\varphi$ as

$$X_{\Delta\varphi}^{\theta} = \frac{\partial \theta_C^{ex}}{\partial \Delta\varphi}, \tag{10}$$

so that $X_{\Delta\varphi}^{\theta}$ follows immediately from the slopes of the lower curve in Fig. 5a. In this figure it is seen that the excess amount of C grows continuously with increasing $\Delta\varphi$, but the susceptibility $X_{\Delta\varphi}^{\theta}$ changes suddenly for $\Delta\varphi \approx 0.77$. We can distinguish three contributions to the increase of θ_C^{ex}: (i) the volume fraction of C within the adsorption layer is higher, (ii) the adsorption layer widens, (iii) the gradients of φ_C near the system boundaries increase. The first contribution to θ_C^{ex} is related to the concentration φ_C^{center} in the middle of the adsorption layer; this quantity is plotted as upper curve (left-hand scale) in Fig. 5a. It is seen that as soon as φ_C^{center} exceeds the value of φ_C^{γ}, in other

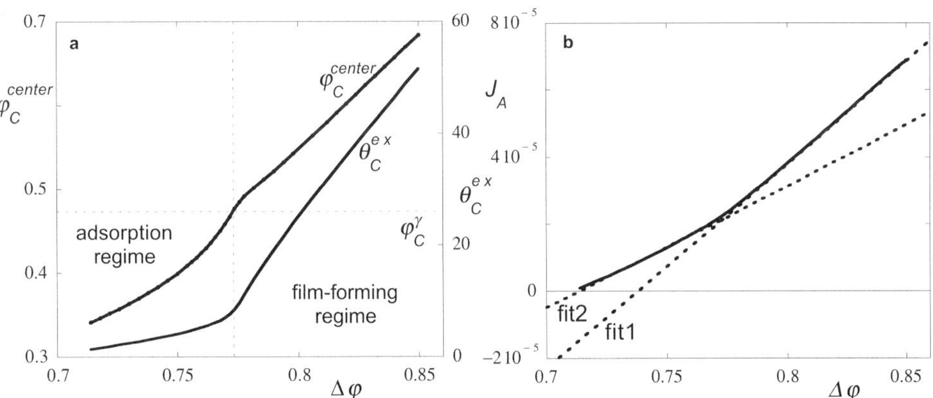

Fig. 5 (a) The volume fraction of C in the center of the system (left axis; line with labels) and the excess amount of C (right axis) as a function of the imposed volume fraction gradient $\Delta\varphi$. In all cases $|\Delta\varphi_A| = |\Delta\varphi_B|$ and $\varphi_C^I = \varphi_C^{II} = 0.15$. The system has the same interaction parameters as in Figs. 3 and 4. The horizontal dotted line gives the volume fraction of C at three phase coexistence in phase γ. The vertical dotted line gives the value for $\Delta\varphi$ where $\varphi_C^{center} = \varphi_C^{\gamma}$. (b) Corresponding stationary flux of segments A (in units $\tilde{B}k_BT$) as a function of the imposed $\Delta\varphi$ with two linear fits (dotted lines).

words, as soon as the composition profiles enter the stable C-rich phase region, the adsorbed amount starts to grow faster. This supports the observations from Fig. 4.

In passing we note that when $\Delta\varphi = \Delta\varphi^{eq} = 0.7107$ the excess adsorbed amount of C nicely converges to the equilibrium adsorbed amount found by the SCF analysis which is given by $\theta_C^{ex} = 1.7$ (black dot in Fig. 1a). This shows that the MFSD method indeed reproduces the SCF results in the limit that the driving force vanishes.

At the fixed bulk concentration $\varphi_C^I = \varphi_C^{II}$ the adsorption at the A/B interface will be higher when the (dynamic) surface tension is higher. Indeed, the A/B interfacial tension will be a function of how much the interface is driven away from equilibrium: the higher the value of $\Delta\varphi$, the higher the (dynamic) surface tension. This is another way to rationalise the increase in adsorption with increasing driving force, *i.e.* increasing $\Delta\varphi$.

In Fig. 5b the stationary flux J_A of A segments is plotted as a function of the imposed volume fraction differences. As expected, the flux increases with increasing driving force. The susceptibility of the flux with respect to $\Delta\varphi$, $X_{\Delta\varphi}{}^J = \partial J_A/\partial\Delta\varphi_A$, suddenly increases at $\Delta\varphi = 0.77$ (the value at which φ_C^{center} reaches φ_C^{γ}). This is shown by the slopes of the two dotted lines that are linear fits to the upper and lower part of the J_A-curve. The slopes differ by a factor of 1.7.

Again, when the driving force for diffusion vanishes, *i.e.* when $\Delta\varphi = \Delta\varphi^{eq} = 0.7107$ we find that the flux accurately vanishes, *i.e.* $J_A = 0$.

Combination of Fig. 5a and 5b indicates that the flux increases with the growing adsorption layer that accompanies the increasing driving force. Apparently, the adsorption layer does not act as a barrier to the segmental fluxes. This result shows that the C-segments participate in the swap–diffusion mechanism. The flux of segments A can be decomposed through eqn. (8) into two partial fluxes J_{AB} and J_{AC} which denote the contributions of A/B and A/C position swappings to the overall diffusion of A:

$$J_A = \sum_i J_{Ai} \tag{11a}$$

$$J_{Ai} = -\tilde{B}k_B T\varphi_A\varphi_i\nabla(\mu_A - \mu_i). \tag{11b}$$

Although the overall flux J_A is independent of z, the partial fluxes J_{AB} and J_{AC} may strongly vary. Fig. 6a and 6b show the separate contributions to J_A as a function of z for $\Delta\varphi = 0.73$ and $\Delta\varphi = 0.84998$, respectively. It is clearly seen that segments C have an important contribution to the overall flux of A. This contribution even exceeds the contribution of A/B-swappings in the film in Fig. 6b. Fig. 6a and 6b also show that $J_{AB} = -J_{BA}$, as it should according to the definition in eqn. (11b). Moreover, $J_{AC} = -J_{BC}$ due to equal interactions ($\chi_{AC} = \chi_{BC}$) and the symmetry of mixtures I and II. As a consequence, $J_C = 0$ and $J_A = -J_B$. Fig. 5b would change significantly if we would only allow position swappings between segments A and B so that $J_{AC} = J_{BC} = 0$. Polymer C_N would statically occupy a part of the diffusion space. The adsorption layer or film would then act as a barrier and reduce the fluxes of A and B. In the static-C_N case $\Delta\varphi$ must be larger than 0.77 (as in Fig. 5a) in order to have a thick adsorption layer. The static-C_N case may be approximated by choosing the segment mobilities $\tilde{B}_A$ and $\tilde{B}_B$ much larger than $\tilde{B}_C$.

Our three-component system could be used to study the diffusion over a barrier that consists of polymers. The adsorption layer acts as a barrier to the diffusion of A and B if the polymer segments

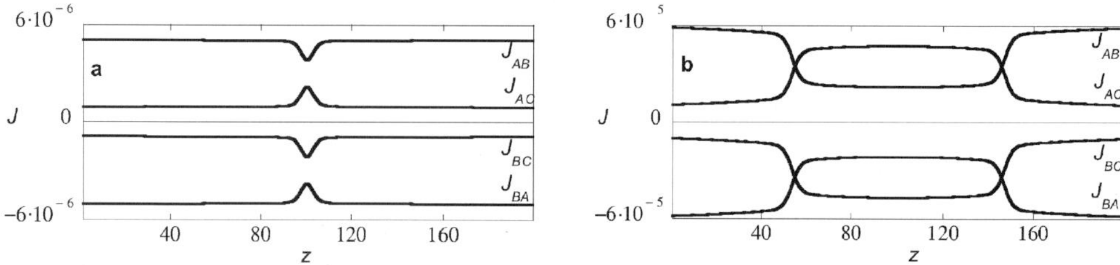

Fig. 6 (a): The contributions J_{AB} and J_{AC} to the flux of A and the contributions J_{BA} and J_{BC} to the flux of B as a function of z for $\Delta\varphi = 0.73$, $J_A = -J_B = 5.953 \ 10^{-6}$ (*cf.* Figs. 3a, 5a). (b): the same for $\Delta\varphi = 0.84998$ $J_A = -J_B = 6.902 \ 10^{-5}$ (*cf.* Figs. 3b and 5b).

that constitute the adsorption layer do not contribute to the fluxes of A and B. These fluxes should then be calculated as $J_A = -\tilde{B}k_B T \varphi_A \varphi_B \nabla(\mu_A - \mu_B) = -J_B$. The polymer chain conformations of the barrier material may still respond to the diffusing segments.

Above we have focused on the case that the two bulk systems are an ideal source and sink, respectively. This means that the two bulk phases are assumed to be large and well mixed, so that addition or depletion of material does not influence its properties. When these conditions are not met, true steady states will not occur. However, when the changes in the two bulk phases proceed more slowly than the time needed to come to a steady state in the diffusion zone, we may construct, from the steady state results as a function of the driving force, the entire dynamic path towards equilibrium. If during such a process the bulk concentration of C remains fixed and the two bulk phases A and B go gradually to their equilibrium composition, we should first expect a thick film of C at the off-equilibrium A/B interface, which decreases in time until its size is microscopic at the true equilibrium. In other words, the thick C-film can only be found transiently. The initially seen mesoscopically thick films of C should not be confused by wetting films of C in between the A/B interface that can only occur at three-phase coexistence.

The dynamic phase transition discussed above was rationalised using the equilibrium composition diagram. We saw that the transition was triggered because the local composition became unstable. The transition is therefore analogous to a liquid–liquid demixing. For macroscopic systems this would be a strongly first-order phase transition. However in the present system, the unstable composition occurs only locally in the system. As a consequence the transition is not first-order and becomes continuous (we were unable to find van der Waals' loops). A similar effect is known for equilibrium first-order transitions that become continuous when the system size is small. A more detailed investigation of this phenomenon will be published elsewhere.

Outlook

The MFSD method is a new powerful tool to study off-equilibrium systems. The geometry of two bulk phases separated by a diffusion region is ideally suited to test the performance of the method. For highly simplified systems, *i.e.* athermal monomeric systems, we checked that the numerical solutions follow known analytical predictions. Already for athermal mixtures of monomers and polymers analytical solutions become more difficult. When there are contact interactions (non-zero χ values) between the components in the system, analytical solutions are most likely out of reach and we thus have to resort to numerical calculations. Fortunately, the limit where the driving forces vanish are well defined and accurate predictions are given by the SCF solution. Indeed the results of the MFSD method reduce to the SCF results when the driving forces vanish.

The equilibrium SCF theory has been used to model molecularly complex inhomogeneous systems. In this paper we considered only short oligomers composed of segments of the same type (homopolymers). When however short copolymers are used and the solvent is selective for one of the moieties, one may study self-assembly of these surfactants into association colloids. Depending on the ratio polar to apolar in the surfactant the preferred curvature of the micelles may be tuned from spherical, cylindrical to flat. The latter objects are called bilayer membranes. The equilibrium structural, thermodynamical and mechanical properties of lipid bilayer membranes have extensively been studied using the equilibrium SCF theory because of its biological relevance.[10–12] The results are almost quantitatively similar to all-atom MD simulations as was proved (and understood) recently.[13,14]

The working hypothesis of most (if not all) modeling efforts of lipid bilayer membranes is that the relevant membrane is in full equilibrium. In biological systems we know that membranes are not in equilibrium. Biological membranes can at best be considered to be in some sort of steady state. How much the off-equilibrium biological bilayer differs from an equilibrated one is currently unknown. The extension of the MFSD method towards copolymers in selective solvents will open up the possibility to study steady state properties of lipid bilayer membranes. For example, we may consider the transport of ions across a membrane composed of charged lipid molecules by imposing different ionic strength conditions on the two sides of the bilayer. In this case the head groups will be screened differently on the two sides of the bilayer and the membrane will inherently be asymmetric (and curved). In turn these membrane adaptations will influence the transport properties of the bilayer membranes. These and related problems will be the topic of future applications of the MFSD method.

Conclusions

We have shown that off-equilibrium interfaces in simple symmetric systems exhibit a very rich adsorption behaviour. The adsorption and associated material fluxes have been studied in stationary three-component homopolymer systems. Two of these components A_N and B_N form an interface at which the third component C_N adsorbs. Due to imposed concentration gradients, the systems are off-equilibrium. Polymer A_N diffuses from the A-rich phase through the interface to the B-rich phase. Polymer B_N diffuses in the opposite direction. Polymer C_N does not have any net diffusion since we imposed symmetric driving forces to A_N and B_N such that the driving force on C_N remained zero. The driving force is expressed in terms of $\Delta\mu = |\Delta\mu_A| = |\mu_A{}^{II} - \mu_A{}^{I}| = |\Delta\mu_B|$. The symmetry applies only when the compositions at both sides of the interface are symmetric ($\Delta\varphi = |\Delta\varphi_A| = |\varphi_A{}^{II} - \varphi_A{}^{I}| = |\Delta\varphi_B|$, $\Delta\varphi_C = 0$), when symmetric Flory–Huggins interaction parameters are used ($\chi_{AC} = \chi_{BC} = \chi_C$), and when the chain lengths and segment mobilities are equal. The interaction parameters were chosen such that all systems studied would display complete wetting at three-phase coexistence at equilibrium. We quantify the adsorbed amount of C by $\theta_C{}^{ex}$, *i.e.* the excess amount of C compared to the bulk mixtures.

The excess amount of C increases with increasing $\Delta\varphi$. As the concentration of C in the bulk phases was kept constant, the increase in adsorption indicates a larger affinity for C, implying a higher (dynamic) surface tension.

The susceptibility $X_{\Delta\varphi}{}^{\theta} = \partial\theta_C{}^{ex}/\partial\Delta\varphi$ increases if the composition in the center of the adsorption layer corresponds to the composition of a stable C-rich phase. From phase composition diagrams it can be verified that this should occur for sufficiently large values of $\Delta\varphi$. The changing susceptibilities of $\theta_C{}^{ex}$ indicate a transition from molecular adsorption to film formation (in other words, from thin to thick adsorption layers). The transition is continuous because the instability occurs only locally in space.

The stationary segment fluxes decrease, as expected, with decreasing driving forces. The fluxes and $\theta_C{}^{ex}$ decrease simultaneously for $\Delta\varphi$ closer to equilibrium. Although the accumulation of C at the interface can assume significant values, it does not hinder the transport of A or B over the interface as an inert barrier would have done; C participates in the swapping mechanism.

We conclude that the MFSD method is a promising new tool to investigate the properties of off-equilibrium interfacial systems in molecularly complex systems.

References

1   S. M. Engels and F. A. M. Leermakers, *J. Chem. Phys.*, 2001, **114**(9), 4267–4276.
2   G. J. Fleer, J. M. H. M. Scheutjens, M. A. Cohen Stuart, T. Cosgrove and B. Vincent, *Polymers at Interfaces*, Chapman & Hall, London, 1993.
3   S. M. Scheinhardt-Engels, F. A. M. Leermakers and G. J. Fleer, *Phys. Rev. E*, 2003, **68**, 11802.
4   S. F. Edwards, *Proc. Phys. Soc.*, 1965, **85**, 613.
5   S. M. Scheinhardt-Engels, F. A. M. Leermakers and G. J. Fleer, *Phys. Rev. E*, 2004, **69**, 21808.
6   H. Morita, T. Kawakatsu and M. Doi, *Macromolecules*, 2001, **34**, 8777.
7   F. Brochard, J. Jouffroy and P. Levinson, *Macromolecules*, 1983, **16**, 1638–1641.
8   E. J. Kramer, P. Green and C. J. Palmstrøm, *Polymer*, 1984, **25**, 473–480.
9   H. Sillescu, *Makromol. Chem., Rapid Commun.*, 1984, **5**, 519–523.
10   F. A. M. Leermakers and J. M. H. M. Scheutjens, *J. Chem. Phys.*, 1988, **89**, 3264.
11   F. A. M. Leermakers and J. M. H. M. Scheutjens, *J. Chem. Phys.*, 1988, **89**, 6912.
12   L. A. Meijer, F. A. M. Leermakers and J. Lyklema, *J. Chem. Phys*, 1999, **110**, 6560.
13   A. L. Rabinovich, P. O. Ripatti, N. K. Balabaev and F. A. M. Leermakers, *Phys. Rev. E*, 2003, **67**, 11909.
14   F. A. M. Leermakers, A. L. Rabinovich and N. K. Balabaev, *Phys. Rev. E*, 2003, **67**, 11910.

Fusion and fission of fluid amphiphilic bilayers

Martin Gotter,*[a] Reinhard Strey,[a] Ulf Olsson[b] and Håkan Wennerström[b]

[a] *Institut für Physikalische Chemie, Universität zu Köln, Luxemburgerstr. 116, Köln D-50939, Germany. E-mail: Martin.Gotter@uni-koeln.de; Fax: +49 221 470 5104; Tel: +49 221 470 4550*
[b] *Phys. Chem. 1, Center for Chemistry and Chemical Engineering, Lund University, P.O. Box 124, Lund SE-22100, Sweden. E-mail: Ulf.Olsson@fkem1.lu.se; Fax: +46 46 222 4413; Tel: +46 46 222 8159*

Received 14th April 2004, Accepted 1st June 2004
First published as an Advance Article on the web 11th November 2004

The system water–oil (n-decane)–nonionic surfactant ($C_{12}E_5$) forms bilayer phases in a large concentration region, but, for a given oil-to-surfactant ratio, only in a narrow temperature range. In addition to the anisotropic lamellar phase (L_α) there is also, at slightly higher temperature, a sponge or L_3-phase where the bilayers build up an isotropic structure extending macroscopically in three dimensions. In this phase the bilayer mid-surface has a mean curvature close to zero and a negative Euler characteristic. In this paper we study how the bilayers in the lamellar and the sponge phase respond dynamically to sudden temperature changes. The monolayer spontaneous curvature depends sensitively on temperature and a change of temperature thus provides a driving force for a change in bilayer topology. The equilibration therefore involves kinetic steps of fusion/fission of bilayers. Such dynamic processes have previously been monitored by temperature jump experiments using light scattering in the sponge phase. These experiments revealed an extraordinarily strong dependence of the relaxation time on the bilayer volume fraction ϕ. At $\phi < 0.1$ the relaxation times are so slow that experiments using deuterium nuclear magnetic resonance (^{2}H-NMR) appear feasible. We here report on the first experiments concerned with the dynamics of the macroscopic phase transition sponge–lamellae by ^{2}H-NMR. We find that the sponge-to-lamellae transition occurs through a nucleation process followed by domain growth involving bilayer fission at domain boundaries. In contrast, the lamellae-to-sponge transformation apparently occurs through a succession of uncorrelated bilayer fusion events.

Introduction

Aqueous systems containing nonionic surfactant of the alkyl oligoethylene type show strongly temperature dependent properties both for the binary systems[1,2] and the ternary water–oil–surfactant ones.[3,4] The Helfrich curvature free energy concept[5,6] is very useful in the description of the physical properties and it clarifies that the temperature sensitivity of these systems can be attributed to the temperature dependence of the spontaneous curvature of the surfactant film.[7,8] The ternary system water–n-decane–$C_{12}E_5$ has been studied particularly carefully with respect to both equilibrium and dynamic properties.[9–12] This has led to accurate knowledge of the

DOI: 10.1039/b405363j

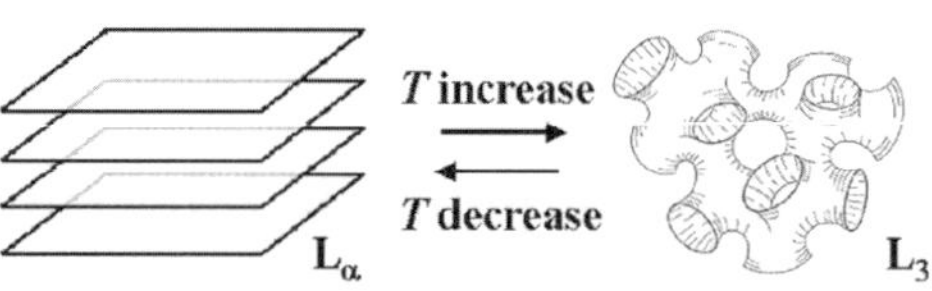

Fig. 1 Schematic illustration of the lamellar and sponge phase structures studied. The stack of sheets represents surfactant bilayers swollen with *n*-decane. The picture of the sponge phase is taken from ref. 14.

spontaneous curvature and its temperature dependence and additionally to reliable estimates of the values of the bending constants. For the understanding and quantitative description of dynamic effects, it has turned out to be very fruitful to have detailed knowledge of the equilibrium properties. A particularly useful property of the studied system is that within a readily accessible temperature range one can go from a state with oil-swollen spherical micelles to the sponge phase consisting of a complex bilayer topology, while keeping a unified theoretical description of the system.[10]

In the temperature range from 35 to 45 °C and at a surfactant-to-oil volume ratio of 0.82 the system water–*n*-decane–$C_{12}E_5$ forms bilayer structures that are either lamellar liquid crystalline (L$_\alpha$) or as in the sponge phase. Fig. 1 illustrates schematically sponge and lamellar phase structures. A balanced state with zero spontaneous curvature occurs at 38.3 °C (36.8 °C in D$_2$O) and the L$_\alpha$-phase has its maximum stability around this temperature. Upon increasing T, the spontaneous curvature becomes negative. In order to realize an actual curvature close to the spontaneous one, the system has to change its topology,[13] a process which involves the introduction of an unfavourable splay curvature free energy. However, for a given composition there exists a narrow (≈ 1 °C) temperature range where the isotropic sponge phase is the equilibrium phase. In previous studies we have shown by equilibrium light scattering experiments that even within the narrow stability range there are major changes in the equilibrium topology of the surfactant/oil bilayer.[10,11]

Within the harmonic approximation, the curvature free energy density g_c is often written as[5]

$$g_c = 2\kappa(H - H_0)^2 + \bar{\kappa}K \tag{1}$$

where H and K are the mean and Gaussian curvatures respectively. H_0 is the so called spontaneous curvature, κ is the bending modulus and $\bar{\kappa}$ is the saddle modulus. For the present system and temperature range, H_0 varies with the temperature as $H_0 = \beta(T - T_0)$, with $\beta = 5 \times 10^{-4}$ Å^{-1} °C^{-1} and $T_0 = 38.3$ °C ($T_0 = 36.8$ °C in D$_2$O), and κ and $\bar{\kappa}$ have been determined as 1×10^{-20} and -8×10^{-21} J, respectively.[10] Here we use the convention that curvature towards oil is defined as positive. The stability of these self-assembly structures depends crucially on how closely the mean curvature of the surfactant monolayer matches the spontaneous curvature H_0. The lamellar phase has its maximum stability when $H_0 = 0$, while the sponge phase forms when H_0 is weakly negative.[13]

Accordingly, one can go from the L$_\alpha$-phase to the sponge phase by a temperature change since changes in the spontaneous curvature of the monolayers induce topology changes. The equilibration process following the temperature change thus involves bilayer fusion/fission events necessary to generate the topological change. The fusion/fission of bilayers is a central dynamic event in complex liquids and fusion has been studied extensively for vesicular sytems.[15–17] Using the sponge phase as a basis, one can extend the studies of these processes to situations with more complex bilayer topologies also exhibiting bilayer fission. In addition, for the present system, there is a characterization of bilayers in terms of known parameters in the Helfrich curvature free energy.

In this paper, we first briefly review temperature jump experiments within a stable or metastable sponge phase, which were the subject of an earlier study of ours.[12] This is followed by a section describing the experimental details, which in turn is succeeded by a presentation and discussion of the experimental results concerning the sponge-to-L$_\alpha$-phase transition. We then focus on the reverse transition from sponge to L$_\alpha$ phase and present some preliminary results. In a final discussion, we compare the experimental observations of the dynamics of the different processes.

Temperature jumps inside the sponge phase

By recording the relaxation of the light scattering signal after a fast (sub-millisecond) temperature jump within the sponge phase, Le *et al.*[12] found that the relaxation time showed a remarkably strong dependence on the bilayer volume fraction ϕ.

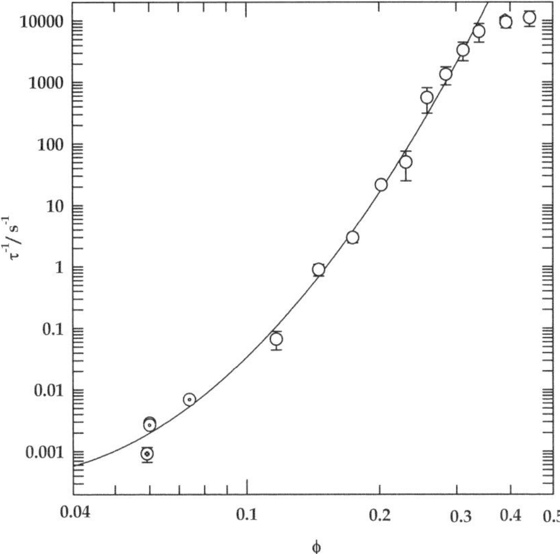

Fig. 2 Plot of the relaxation rates τ^{-1} obtained from small temperature jumps performed within the sponge phase. Data are taken from refs. 12 and 18.

This strong variation is shown in Fig. 2 where we have plotted the reciprocal relaxation time τ^{-1} as a function of ϕ. For $0.1 < \phi < 0.4$ fast temperature jumps were performed using the Joule-heating technique.[12] At lower bilayer concentrations ($\phi < 0.1$) the relaxation times become greater than the temperature reequilibration time of approximately 20 s of the T-jump apparatus. The data points at $\phi < 0.1$ were therefore measured by transferring samples from an external water bath to a spectrometer set at a slightly higher temperature.[18]

The relaxation time in these experiments was found to be independent of the probing length scale (*i.e.* the magnitude of the scattering vector). It was concluded that the relaxation time had to be associated with structural change. Based on the theoretical model of the sponge phase,[13] a temperature increase should result in the formation of a bilayer structure with more passages, *i.e.* a more negative Euler characteristic. In that case, the rate limiting step would then be a membrane fusion process resulting in the formation of additional passages. A remarkable feature of the system is the fact that at low concentrations ($\phi \approx 0.05$) the relaxation times reach the scale of several minutes. This allows using slower techniques to study the structural changes, which in turn could possibly be more informative. Additionally, it is reasonable to expect that global phase transformations sponge-to-lamellar, or *vice versa*, should not occur at a faster rate in this concentration regime. Taking this into consideration, we have used deuterium NMR spectroscopy to study the rate of the formation of the new phase, following a temperature change leading from a stable lamellar phase to a sponge phase, as well as the reversed process.

Experimental

Materials

The surfactant pentaethyleneoxide dodecyl ether ($C_{12}E_5$, purity $>99\%$) was purchased from Nikko Chemicals Co. Ltd. (Tokyo, Japan). The hydrophobic component *n*-decane (purity $>99\%$) came from Lancaster Synthesis (Eastgate, GB), while the D_2O (degree of deuteration $>99.8\%$) was obtained from Dr. Glaser AG (Basel, Switzerland). All components were used as purchased.

Sample preparation

The three components were mixed to form a stock solution with a bilayer volume fraction $\phi = 0.08$. The mass distribution within the bilayer corresponded to 52.1% $C_{12}E_5$ and therefore 47.9% *n*-decane, which in turn corresponds to a surfactant-to-oil volume ratio of 0.82. Three samples of approximately 2 ml with $\phi = 0.07$, 0.06, and 0.05 were created by appropriate dilution with D_2O. They were each immediately flame sealed in *ca.* 6 cm long test-tubes of 8 mm outer radius. Volume fractions were calculated using the following densities of the components: $C_{12}E_5$: 0.97 g cm^{-3}, D_2O: 1.11 g cm^{-3} and *n*-decane: 0.73 g cm^{-3}.

Phase behaviour

The phase boundaries of the L_α- and sponge phases were determined visually by keeping the samples in a thermostated water bath and varying the temperature. The anisotropic lamellar phase was identified through its optical birefringence by observing the samples between crossed polarizers.

^{2}H-NMR experiments

^{2}H-NMR experiments were performed on a Bruker DMX-200 spectrometer operating at a deuterium frequency of 30.8 MHz. The flame sealed samples were introduced into 10 mm precision NMR-tubes that were then inserted into the spectrometer. For better thermal conduction, the minute space between the sample glass and the NMR-tube was filled with water. The experiments were performed on stationary, *i.e.* non-spinning, samples. The samples were thermally equilibrated through an air flow. Temperature changes were performed while the sample remained in the spectrometer. The rate of temperature equilibration was checked in separate experiments, for both cooling and heating, using a thermocouple. Cooling and heating curves are shown in Fig. 3 (right) for a temperature change of 1.1 °C. Temperature equilibration times are on the order of 100–200 s. This represents the time resolution for the phase transition kinetics that it is possible to achieve with this simple temperature variation procedure.

The deuterium nucleus is a spin $I = 1$ particle and carries an electric quadrupole moment. The interaction between this quadrupole moment and the intramolecular electric field gradient results in an orientation dependent shift of the Zeeman energy levels. When D_2O molecules experience a macroscopically anisotropic environment, as is the case in the L_α-phase, the quadrupolar interaction has a non-zero average and the ^{2}H spectrum consists of a doublet with a frequency separation $\Delta \nu_Q$ given by[19]

$$\Delta \nu_Q = \Delta \nu_Q^0 |3 \cos^2 \theta - 1| \tag{2}$$

Here θ is the angle between the bilayer normal and the external magnetic field, and $\Delta \nu_Q^0$ is the (reference) frequency separation for the $\theta = 90°$ orientation, which often is referred to as the quadrupolar splitting. In the samples studied, the bilayers were so dilute, and the viscosity therefore so low, that the lamellar phase orients itself in the magnetic field with $\theta = 90°$ due to the anisotropy in the magnetic susceptibility. The peak separation in our experiments thus corresponds to $\Delta \nu_Q^0$.

Only a fraction P_b of the total number of water molecules are motionally perturbed by the bilayer interfaces. Assuming that the number of perturbed water molecules is proportional to the number of surfactant molecules, a two-site discrete exchange between perturbed and non-perturbed (bulk) sites implies $\Delta \nu_Q^0 \sim P_b \sim \phi_s/\phi_w \approx \phi.$[20]

Hence, the quadrupolar splitting can be used to estimate the surfactant concentration in the lamellar phase even when it coexists with other phases, *e.g.* the sponge phase. In the isotropic sponge phase the quadrupolar interaction is averaged to zero and only a single resonance is obtained. Schematic equilibrium NMR spectra are shown in Fig. 3 (left) to illustrate the change in the spectrum when passing from an isotropic sponge to an anisotropic L_α-phase. When the two phases coexist, the spectrum consists of a superposition of a central singlet originating from the isotropic phase and a doublet arising from the anisotropic one.

Equilibrium phase behaviour and ^{2}H-NMR properties

Fig. 3 (middle) shows the phase equilibria in the temperature range around the L_α- to sponge phase transition for the composition range investigated. The sponge phase is stable only within a narrow temperature interval of approximately 0.5 °C around 37 °C (using D_2O). At higher temperatures, the sponge phase is in equilibrium with excess, almost pure, water. At lower temperatures, it is separated from the L_α-phase by a narrow (approximately 0.7 °C wide) coexistence region. The

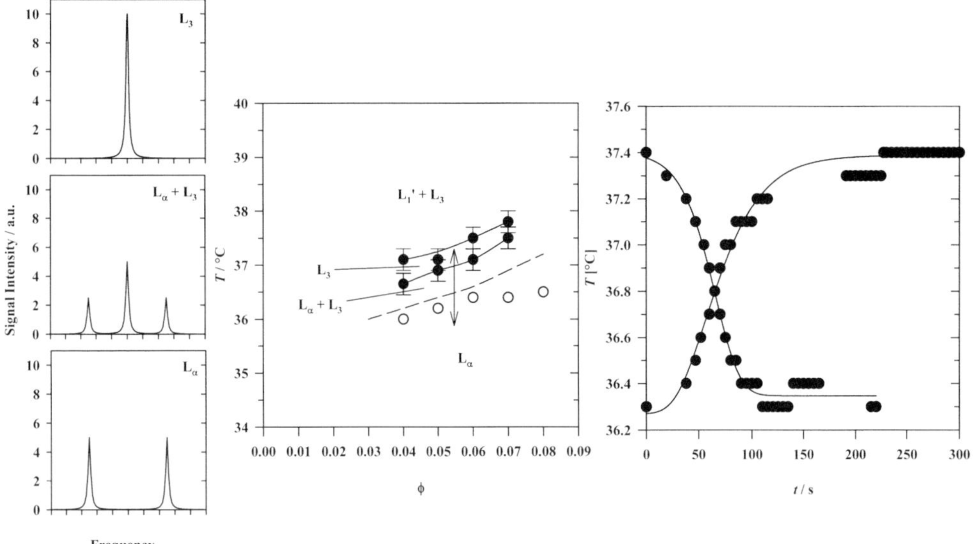

Fig. 3 *Centre*: phase diagram of the D_2O–n-decane–$C_{12}E_5$ system at a surfactant to bilayer-volume ratio $\phi_s/\phi_o = 0.82$ as a function of the membrane volume fraction ϕ and temperature T. The narrow L_3(sponge)-channel is represented by filled circles. A two phase region is located at higher temperature ($L_1' + L_3$) and at lower temperature ($L_\alpha + L_3$). The dashed line indicates our suspected position of the two phase region to the L_α-phase boundary. The empty circles mark the highest temperatures where the L_α-phase was determined. The double arrow denotes that temperature jump experiments were made from the L_3 to the lamellar phase as well as in the reverse direction. *Left*: schematic illustration of NMR-spectra for the different phase regions showing a singlet, a triplet and a doublet for the sponge phase, the two phase region, and the L_α-phase respectively. *Right*: Illustration of the heating and cooling kinetics for a sample. Temperature equilibrium is reached within 200 s.

estimated upper boundary of the L_α-phase is indicated by the broken line, while the boundaries of the L_3-phase are marked with filled symbols. The open symbols indicate where the pure lamellar phase is certain to exist in equilibrium. The phase boundary temperatures are in good agreement with those found in a previous study[21] and for a full phase diagram of the present system the reader is referred to ref. 22.

The ^{2}H-NMR spectrum of D_2O consists of a narrow singlet in the case of the isotropic sponge phase. In the L_α-phase, on the other hand, the signal is split into a doublet, a so called quadrupolar splitting $\Delta\nu_Q^0$, due to the anisotropy of the phase. The quadrupolar splitting, *i.e.* the frequency separation of the doublet, depends on the water-to-surfactant ratio. For the present samples the quadrupolar splitting was rather small, on the order of 10–20 Hz, due to the high water content.

In Fig. 4 we have plotted the equilibrium quadrupolar splittings in the lamellar phase as a function of the bilayer volume fraction ϕ. In accordance with the simple two-site discrete exchange model described above, the quadrupolar splitting is approximately proportional to the bilayer concentration.

The L_3-to-L_α phase transition kinetics

By following the evolution of the ^{2}H-NMR spectrum in time resolved experiments, we can take advantage of the fact that the NMR spectra of the sponge and lamellar phases are qualitatively different. This phenomenon contains information not only on the kinetics of the phase transition, but also on the underlying mechanism of the phase transformation process. To study the transition from the sponge to L_α-phases, the samples were first allowed to equilibrate inside the NMR magnet while being in the L_3-state. Then the temperature was reduced to a new value at which the L_α-phase was stable and the phase transformation process was observed by subsequently recording ^{2}H-NMR spectra.

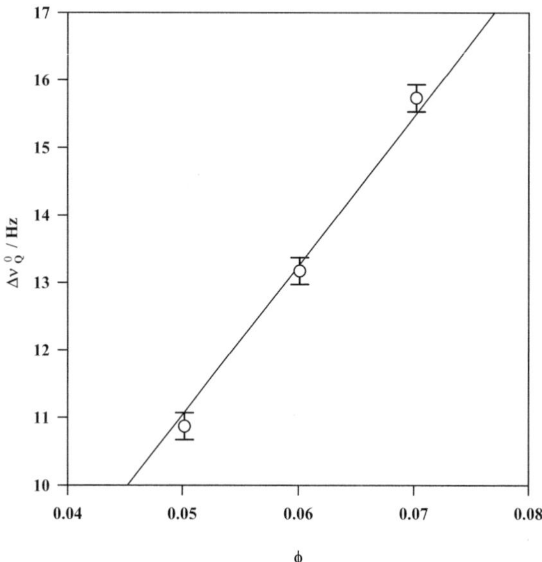

Fig. 4 Variation of the equilibrium ^{2}H quadrupolar splitting $\Delta\nu_Q^0$ of D_2O with the bilayer concentration ϕ.

A typical result is shown in Fig. 5 for a sample of $\phi = 0.060$, in which case the temperature was decreased by 1.5 °C from an initial temperature of 37.5 °C. In Fig. 5 (left) the temporally subsequent NMR spectra are stacked in a 3D-plot to visualize the evolution. The new equilibrium in the L_α-phase is established after approximately 1700 s, which is about an order of magnitude greater than the time required for the temperature equilibration. At time $t = 0$ the spectrum consists of a singlet characteristic for the initial isotropic phase. After approximately 600 s two additional peaks, resulting from anisotropic lamellar domains, begin to be clearly visible. As time progresses, the intensity of the doublet increases while the intensity of the central singlet decreases until it eventually vanishes and only the doublet, characteristic for the equilibrium lamellar phase, remains.

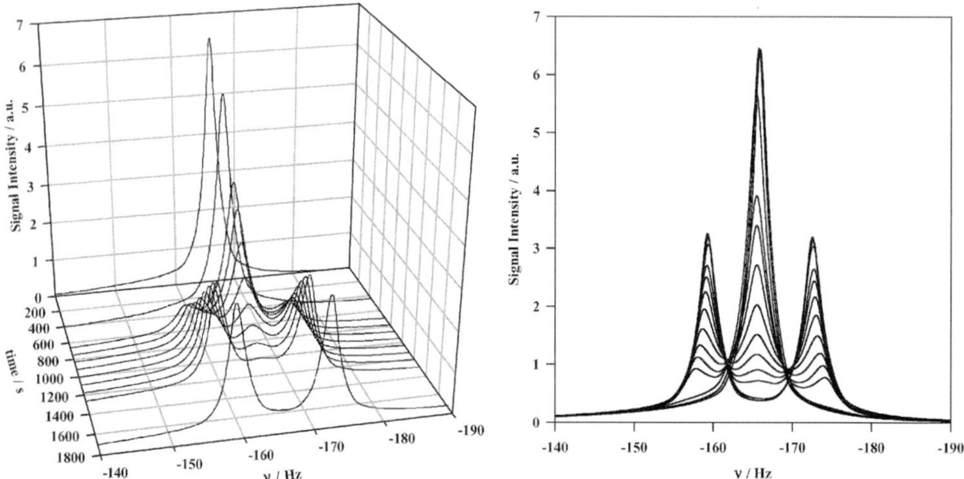

Fig. 5 *Left*: 3D-graph of the L_3 to L_α-phase transition experiment with $\Delta T = -1.5$ °C at a composition of $\phi = 0.060$ displaying the NMR-spectra and thus the peak development as a function of time t. *Right*: superposition of the NMR-spectra given on the left, eliminating the time axis. The position of all peaks remains roughly constant. Two isosbestic points are clearly visible.

The same general trend was also observed for the other compositions investigated in the present study ($\phi = 0.05$ and 0.07). It indicates that the lamellar phase forms from the sponge phase through a nucleation and growth mechanism. In Fig. 5 (right) we have superimposed all the spectra. Two frequencies are clearly visible, one on either side of the singlet maximum, where the intensity remains essentially constant throughout the transformation process. The existence of these isosbestic points indicates that the spectra recorded at intermediate times can be seen as weighted sums of two constant subspectra, in this case the L_3-singlet and the L_α-phase doublet.

The observation of a quadrupolar splitting allows us to estimate a minimum lateral extension of the bilayers in the emerging lamellar domains. The fact that we obtain separate signals from the sponge and lamellar domains implies that the lifetime τ_l of D_2O molecules within the lamellar domains is greater than the inverse of the quadrupolar splitting, which is approximately 15 Hz. While the bilayers represent a barrier for the diffusion of the water molecules, they are essentially free to diffuse parallel to the bilayers, with a two dimensional mean-square displacement given by

$$\langle r^2 \rangle = 4Dt \tag{3}$$

where D is the diffusion coefficient and t is the time. With $t = \tau_l \approx 1/(2\pi15)$s and assuming free two-dimensional diffusion of water with $D \approx 2 \times 10^{-9}$ m^2 s^{-1}, we obtain with eqn. (3) a root mean square displacement of approximately 10 µm. Thus, when we observe a quadrupolar splitting after 650 s, it comes from lamellar domains where the bilayers are larger than approx. 10 µm. This value is about 100 times the structural length scale of the sponge-like structure of the L_3-phase which is approximately given by $1.5\delta/\phi$, where $\delta \approx 50$ Å is the bilayer thickness.[10]

The area under an NMR peak is proportional to the concentration of molecules in the particular state. Hence, analyzing the time dependent variation of the areas of the lamellar doublet and the central singlet allows us to follow the growth of the lamellar domains at the expense of the sponge phase domains. The intrinsic linewidth of the peaks, due to magnetic field inhomogeneity and relaxation, is approximately 5 Hz, while the quadrupolar splitting is 15 Hz and thus there is significant overlap between the resonances. In order to evaluate the areas, we have fitted the NMR spectra with the sum of a number of Lorentzian lines corresponding to the number of peaks. Eqn. (4) is the general ^{2}H-NMR lineshape for a single peak in the isotropic phase or an oriented liquid crystal one.[20]

$$I_i(\nu) = \frac{I_i^{\mathrm{max}}}{1 + \left(2(\nu - \nu_i)/\Delta\nu_{1/2,i}\right)^2} \tag{4}$$

Here, ν is the frequency, ν_i is the peak centre, I_i^{max} the peak maximum, and $\Delta\nu_{1/2,i}$ is the peak width at half height. With I_i^{max} and $\Delta\nu_{1/2,i}$ obtained from the fit, the integrals were calculated from

$$\int_{-\infty}^{+\infty} I_i(\nu)d\nu = \frac{1}{2}\pi I_i^{\mathrm{max}}\Delta\nu_{1/2,i} \tag{5}$$

In Fig. 6 we show some selected spectra recorded at different times together with the corresponding fits.

In Fig. 7 we have plotted the peak integrals as a function of time for the corresponding data presented in Fig. 5. Here, we show both the decay of the central peak as well as the rise of the doublet intensity. Note that in the case of the doublet we report the sum of the two integrals. The total sum of the singlet and doublet integrals, which is supposed to be a constant value, is also shown. The small deviations from that constant value provide a measure of the inaccuracy of the analysis.

The relative integral of the lamellar doublet corresponds essentially to the volume fraction of the lamellar phase φ, although not exactly. The integral is proportional to the amount of D_2O, while the quadrupolar splitting is a measure of the D_2O-to-surfactant ratio. At closer inspection, one sees that the quadrupolar splitting is not constant during the transformation process. When a splitting is first detected it has a higher value ($\approx 20\%$) than the final equilibrium value, measured for the completely transformed sample. The splitting gradually decays to the equilibrium value during the process.

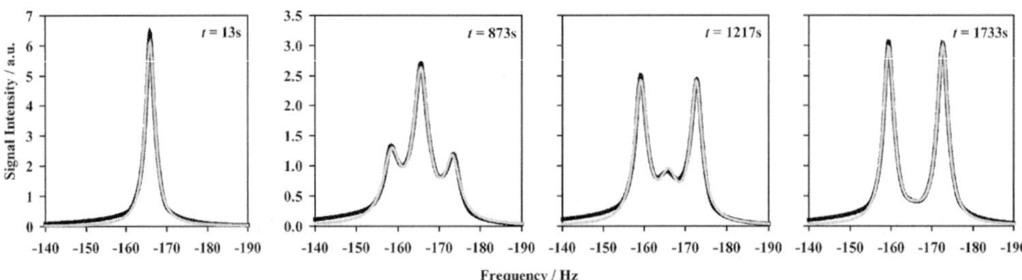

Fig. 6 Four of the NMR-spectra of Fig. 3 presented individually (black) along with their corresponding fit function (grey) obtained from the addition of one, two or three Lorentzian functions. Shown are the results for the spectra at $t = 13$, 873, 1217, and 1733 s. In all cases the fits are satisfactory.

This indicates that the lamellar domains that first form have a higher bilayer concentration compared to the overall average concentration. This is consistent with the equilibrium phase behaviour (Fig. 2). In the coexistence region the sponge phase coexists with a lamellar phase having a higher bilayer concentration. However, if we neglect this small variation in bilayer concentration we can identify the doublet integral with the lamellar phase volume fraction φ.

Kolmogorov, Johnson, Mehl and Avrami have derived a simple and general model for the formation rate of a new phase.[23,24] This model considers the growth of domains starting from infinitesimal size at time $t = 0$ and takes into account the termination of growth in regions where two domains touch and would overlap if growth would have continued. The model predicts a sigmoidal time dependence of the volume fraction of the new phase with an initial acceleration, due to the increase in interfacial area where the transformation occurs, and a deceleration at an advanced point in time, due to more and more domains having grown into contact. The model also considers the two cases with a fixed number of nuclei at $t = 0$ (short nucleation period) or a continuous formation of nuclei during the transformation with a constant rate (constant nucleation or long nucleation period). Furthermore, it considers domain growth in three, two or only one dimensions.

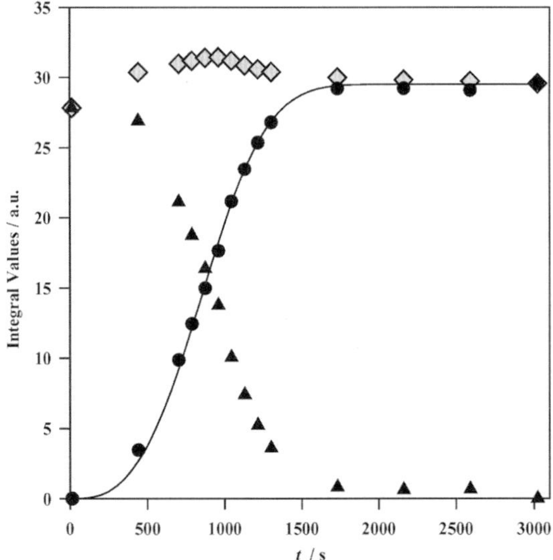

Fig. 7 Plot of the development of the Lorentzian integrals as a function of time for the phase transition experiment with $\Delta T = -1.5\ °C$ at a composition of $\phi = 0.060$. The circles depict the sum of the integrals of the two lamellar peaks, the triangles that over the L_3-peak, and the diamonds the sum over all integrals. The lamellar and sponge plots both are of sigmoidal shape. The former was fitted using "Avrami's Law" (eqn. (6)). The phase transition was completed after *ca.* 1700 s.

A general expression for the time evolution of φ is[23,24]

$$\varphi = 1 - \exp\{-(kt)^{\alpha}\} \tag{6}$$

where the exponent $\alpha = n + \psi$. Here, n is the dimensionality of the domain growth and $\psi = 0$ in the case of a short nucleation period, while $\psi = 1$ in the case of continuous nucleation. For $\psi = 0$, the rate constant k depends both on the growth rate of individual domains and on the concentration of those domains and additionally on the formation rate of domains in the case of continuous nucleation, $\psi = 1$.

At the present stage we have not elaborated on the different possible values of the exponent α, but rather fitted all the data keeping $\alpha = 3$. In principle, if the sponge phase domain is of constant composition throughout the phase transition process, one would expect $\psi = 1$, as the super-saturation then remains constant. On the other hand, we have evidence from the quadrupolar splittings that the lamellar domains initially formed have a higher bilayer concentration and the corresponding decrease of the bilayer concentration in the sponge domain may terminate the nucleation process.

In Fig. 8 we have plotted the time evolution of φ derived from five different experiments, a–e, together with the best fits using eqn. (6). The five experiments were performed at different bilayer concentrations ϕ and quench depths ΔT. The results from the fits are summarized in Table 1, where we report the determined values of k.

Although the data set is limited, we can see the following trends: k is sensitive to the quench depth and increases with increasing ΔT. It also appears to increase with increasing bilayer concentration. Increasing ϕ from 0.05 to 0.07 leads approximately to a doubling of k. However, the concentration dependence at constant ΔT appears to be more subtle. For $\Delta T = -1.0\ ^{\circ}\mathrm{C}$ we see a clear increase in k when ϕ increases from 0.05 to 0.07. On the other hand, we observe similar rate constants for $\phi = 0.05$ and 0.06 when $\Delta T = -1.5\ ^{\circ}\mathrm{C}$. This indicates that, as the driving force for the phase change increases, the concentration dependence of the rate becomes less pronounced. However, this needs to be confirmed by additional experiments.

The L_{α}-to-L_3 phase transition kinetics

The reverse process, *i.e.* the transformation from planar bilayers in the L_{α}-phase to the multiply connected bilayer structure of the sponge phase, involves the fusion of bilayer patches. In contrast to the sponge-to-L_{α} transition, the emerging new phase does not seem to form through a nucleation and growth process. Rather, it develops through a number of random fusion events, which create the "passages" in the multiply connected bilayer structure in such a way that the sample is homogeneous over distances probed by the diffusing water molecules on a time scale determined by the inverse quadrupolar splitting. In Fig. 9 we present a series of time resolved ^{2}H-NMR spectra recorded after a temperature jump from the L_{α} to the sponge phase. The bilayer concentration corresponds to $\phi = 0.06$ and $\Delta T = +1.5\ ^{\circ}\mathrm{C}$. It is clearly visible that this transition is characterized by a gradual coalescence of the doublet into a singlet, without the simultaneous existence of a singlet and a doublet. The overall time required to complete the transition in this experiment is approximately 600 s. The transition here is faster by a factor of two when compared with the reverse one of Fig. 4. This difference in rate can both be due to the difference in final temperature and to the fact that a homogeneous process is expected to proceed faster than a process involving nucleation and growth.

The quadrupolar splitting $\Delta\nu_Q^{\ 0}$ results from a residual quadrupolar interaction which is not averaged to zero by the molecular reorientation. In the experiments presented here, this is due to the hydration of the polar headgroups of the surfactant molecules with $\mathrm{D}_2\mathrm{O}$ and as a result the $\mathrm{D}_2\mathrm{O}$ has a preferred orientation with respect to the interface. When the interface is curved, the $\mathrm{D}_2\mathrm{O}$ molecules can reach new surface orientation by diffusion. When all orientations are accessible, as is the case in the sponge structure, the quadrupolar interaction can be averaged to zero and this results in a singlet. The averaging of the interaction also causes the interaction to become time dependent, and the motion of the molecules along the curved bilayer thus contributes to the relaxation of the magnetization. Hence, besides the initial gradual decrease of the quadrupolar splitting, there is also a gradual broadening of the doublet resonances.

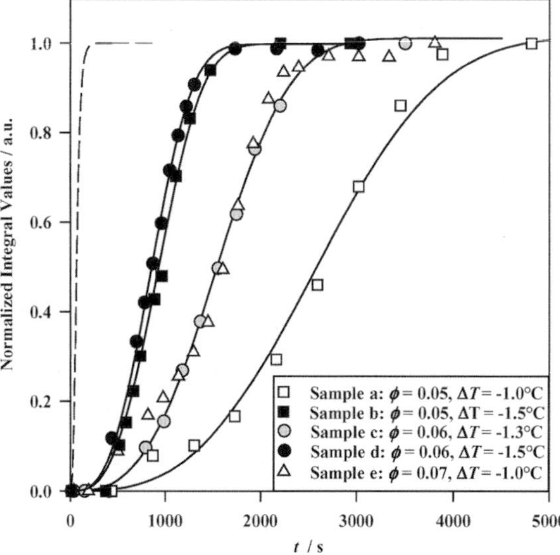

Fig. 8 Plot comparing the result of five different phase transition experiments. As in Fig. 7, the development of the sum of the integrals of the lamellar peaks is shown. The curves were each normalized with their corresponding maximum value. All are of sigmoidal shape and were fitted satisfactorily with "Avrami's Law" (eqn. (5)) using an exponent of $n = 3$. The fits for c and e are essentially the same. The broken line indicates the temperature equilibration of Fig. 3.

The evolution of the spectra can be seen as a result of a gradual decrease of a correlation time τ describing the change of the preferred orientation of the D_2O molecules as they diffuse along a curved surface. A rough estimate of τ is

$$\tau \approx \frac{\lambda^2}{D} \tag{7}$$

where λ is the average distance between passages, which more formally can be expressed in terms of an Euler characteristic density, and $D \approx 2 \times 10^{-9}$ m^2 s^{-1} the water diffusion coefficient.

As the passage density increases, λ and thereby τ decrease. The coalescence of the doublet occurs when $\tau \approx 1/\Delta\nu_Q^0 \approx 0.1$ s, which is observed experimentally after *ca.* 300 s. Using eqn. (7) at this point, we obtain the estimate that $\lambda \approx 10^{-5}$ m. This value can be compared to that in the equilibrium sponge phase, which is approximately equal to the average pore size, *i.e.* $\delta/\phi \approx 10^{-7}$ m. A more quantitative analysis requires the fitting of the NMR band shapes using slow motion theory[25] for the water dynamics.

The lamellar phase can also be induced from the sponge phase at constant temperature by the application of shear flow and the relaxation back to the equilibrium sponge phase can be monitored after cessation of flow by *e.g.*, birefringence measurements. Such experiments have recently been

Table 1 Kinetic results from the sponge-to-lamellar phase transition

Sample	Experiment	T_i^a/°C	T_f^b/°C	k^{-1}/s
a	$\phi = 0.05$, $\Delta T = -1.0$ °C	37.1	36.1	2900
b	$\phi = 0.05$, $\Delta T = -1.5$ °C	37.1	36.1	1050
c	$\phi = 0.06$, $\Delta T = -1.3$ °C	37.5	36.2	1750
d	$\phi = 0.06$, $\Delta T = -1.5$ °C	37.5	36.0	1000
e	$\phi = 0.07$, $\Delta T = -1.0$ °C	37.3	36.3	1800

[a] Initial temperature. [b] Final temperature.

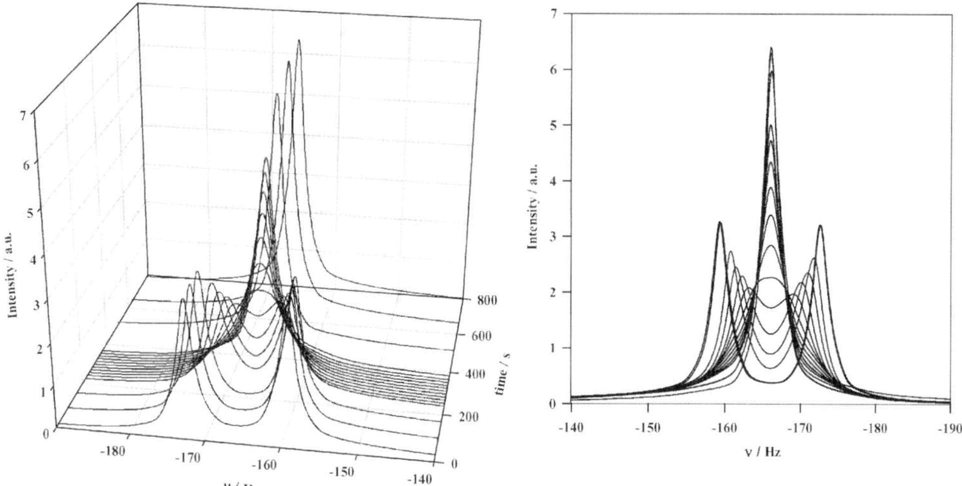

Fig. 9 *Left*: 3D-graph for the L$_\alpha$ to sponge phase transition experiment with $\Delta T = +1.5\,^\circ$C at a composition of $\phi = 0.060$ displaying the NMR-spectra and thus the peak development as a function of time t. *Right*: superposition of the NMR-spectra given on the left, eliminating the time axis. In contrast to the results of Fig. 5, the simultaneous existence of three peaks is not visible. Instead, the lamellar doublet converge and finally coalesce to form the singlet of the sponge phase. Furthermore, the relaxation process is quicker by more than a factor of two.

performed on the same system.[26] For $\phi = 0.06$, a relaxation time of ≈ 200 s was observed. This is in rough agreement with the data of Fig. 9.

Conclusion

The focus of this investigation is the kinetics of the phase transformation between the lamellar and the sponge phase. Both phases have extended bilayer structures, but of different topology. Changes in the topology involve creation or annihilation of passages which in turn involves fusion or the fission of bilayers. The equilibrium stability of the individual phases is rationalized by considering the curvature free energy of the surfactant monolayer film. In this study we have introduced a novel method for the investigation of the dynamics of a phase transformation by using ^{2}H-NMR. Since the lamellar phase spontaneously orients during the transformation, the NMR spectrum of the water deuterons is particularly informative.

In a previous study we have tentatively attributed the slow relaxation process within the sponge phase to bilayer fusion. For a change of sponge to lamellar, or *vice versa*, such fission/fusion has to occur. The fact that we observe phase transition dynamics occurring on the same time scale as for the relaxation within the sponge phase indicates that the basic rate limiting step is of a similar character. The most obvious alternative is that the rate is determined by fission/fusion.

An intriguing aspect of the measurements is that the transition sponge to lamellar goes through a nucleation and growth path, while in the reverse case of lamellar to sponge the sample remains homogeneous on the NMR length scale throughout the transition.

For lipid bilayers fusion is generally considered to occur through a so called Stalk intermediate structure. There are then two stages where first two opposing monolayers fuse, giving a structure with a locally high negative monolayer curvature. In the second step the high curvature energy is released through a second fusion event. This model should be equally applicable to the reverse event of membrane fission. For the present system we have reliable estimates of the parameters entering the Helfrich expression for the curvature energy. However, before we can quantitatively test the Stalk model more detailed experimental data has to be collected. Work along this direction is in progress.

Acknowledgements

M.G.'s stay in Lund was supported by an EU Marie Curie training site scholarship. This work was also supported by the Swedish Research Council (VR).

References

1 D. J. Mitchell, G. J. T. Tiddy, L. Waring, T. Bostock and M. P. McDonald, *J. Chem. Soc., Faraday Trans.*, 1983, **79**, 975.
2 R. Strey, R. Schomäcker, D. Roux, F. Nallet and U. Olsson, *J. Chem. Soc., Faraday Trans.*, 1990, **86**, 2253.
3 H. Kunieda and K. Shinoda, *J. Dispersion Sci. Technol.*, 1982, **3**, 233.
4 M. Kahlweit and R. Strey, *Angew. Chem.*, 1985, **24**, 654.
5 W. Helfrich, *Z. Naturforsch., C: Biochem., Biophys., Biol., Virol.*, 1973, **28**, 693.
6 S. A. Safran, *Statistical Thermodynamics of Surfaces, Interfaces and Membranes*, Addison–Wesley, Reading, MA, 1994.
7 U. Olsson and H. Wennerström, *Adv. Colloid Interface Sci.*, 1994, **49**, 113.
8 R. Strey, *Colloid Polym. Sci.*, 1994, **272**, 1005.
9 U. Olsson, H. Bagger-Jörgensen, M. Leaver, J. Morris, K. Mortensen, R. Strey, P. Schurtenberger and H. Wennerström, *Prog. Colloid Polym. Sci.*, 1997, **106**, 13.
10 T. D. Le, U. Olsson, H. Wennerström and P. Schurtenberger, *Phys. Rev. E.*, 1999, **60**, 4300.
11 T. D. Le, U. Olsson, H. Wennerström, P. Uhrmeister, B. Rathke and R. Strey, *Phys. Chem. Chem. Phys.*, 2001, **3**, 4346.
12 T. D. Le, U. Olsson, H. Wennerström, P. Uhrmeister, B. Rathke and R. Strey, *J. Phys. Chem. B*, 2002 **106**, 9410.
13 D. Anderson, U. Olsson and H. Wennerström, *J. Phys. Chem.*, 1989, **93**, 4243.
14 R. Strey, W. Jahn, G. Porte and P. Bassereau, *Langmuir*, 1990, **6**, 1635.
15 D. P. Siegel, *Biophys. J.*, 1993, **65**, 2124.
16 Y. Kozlovsky and M. M. Kozlov, *Biophys. J.*, 2002, **82**, 882.
17 L. Yang and H. W. Huang, *Science*, 2002, **297**, 1877.
18 P. Uhrmeister, PhD Thesis, University of Köln, 2002.
19 B. Halle and H. Wennerström, *J. Chem. Phys.*, 1981, **75**, 1928.
20 A. Abragam, *Principles of Nuclear Magnetism*, Clarendon, Oxford, 1961.
21 U. Olsson and P. Schurtenberger, *Langmuir*, 1993, **9**, 3389.
22 M. S. Leaver, U. Olsson, H. Wennerström, R. Strey and U. Würz, *J. Chem. Soc., Faraday Trans.*, 1995 **91**, 4269.
23 M. Avrami, *J. Chem. Phys.*, 1939, **7**, 1103.
24 D. Kashchiev, *Nucleation Basic Theory with Applications*, Butterworth–Heinemann, 2000.
25 E. Meirovitch and J. H. Freed, *Chem. Phys. Lett.*, 1979, **64**, 311.
26 F. Nettesheim, C. B. Müller, U. Olsson and W. Richtering, *Colloid Polym. Sci.*, 2004, **282**, 918.

Surface freezing of chain molecules at the liquid–liquid and liquid–air interfaces

Eli Sloutskin,[a] Colin D. Bain,[b] Benjamin M. Ocko[c] and Moshe Deutsch*[a]

[a] *Physics Department, Bar-Ilan University, Ramat-Gan 52900, Israel.*
E-mail: deutsch@mail.biu.ac.il; Fax: +972 (0)3 5353298; Tel: +972 (0)3 5318476
[b] *Department of Chemistry, Oxford University, Mansfield Rd., Oxford, UK OX1 3TA*
[c] *Physics Department, Brookhaven National Laboratory, Upton NY 11973, USA*

Received 21st April 2004, Accepted 9th July 2004
First published as an Advance Article on the web 27th September 2004

Surface freezing (SF) is the formation of a crystalline monolayer at the free surface of a melt at a temperature T_s, a few degrees *above* the bulk freezing temperature, T_b. This effect, *i.e.* $T_s > T_b$, common to many chain molecules, is in a marked contrast with the surface melting effect, *i.e.* $T_s \leq T_b$, shown by almost all other materials. Depending on chain length, n, the SF layer shows a variety of phases, in some cases tuneable by bulk additives. The SF behaviour of binary mixtures of different-length alkanes and alcohols is governed by the relative chain length mismatch, $|\Delta n/n|^2$, yielding a quasi-"universal" behaviour for the freezing of both bulk and surface. While SF at the liquid–air interface was studied rather extensively, Lei and Bain (*Phys. Rev. Lett.*, 2004, **94**, 176103) have shown only very recently that interfacial freezing (IF) can be induced also at the water:tetradecane interface by adding the ionic surfactant CTAB to the water phase. We present measurements of the interfacial tension of the water:hexadecane interface, as a function of temperature and the ionic surfactant STAB, revealing IF at a STAB-concentration-dependent temperature $T_i > T_b$. The measurements indicate that a single frozen monolayer is formed, with a temperature-existence range of up to 10 °C, much larger than the 1.2 °C found for SF at the free surface of the melt. We also find a new effect, where the IF allows tuning of the interfacial tension between the two bulk phases to zero for a range of temperatures, $\delta T = T_{mix} - T_b \leq T_i - T_b$ by *cooling* the system below T_i. We discuss qualitatively the factors stabilizing the frozen layer and their variation from the liquid–air to the liquid–liquid interfaces. The surfactant concentration dependence of T_i is also discussed and a tentative theoretical explanation is suggested.

Introduction

The phase behaviour of matter is, in general, a function of the dimensionality. Thus, the phase sequences of thin films and surfaces were expected,[1] and found experimentally,[2] to differ from those of the three-dimensional bulk. In particular, for melting of a solid both theory and experiment show that with very few exceptions it is the *less ordered surface* which coexists with the *more ordered bulk*, *i.e.* the quasi-2D surface melts at a temperature lower than, or, at most, equal to that of the 3D bulk. This phenomenon, called surface melting, has been found for almost all solids studied to date.[2] This is easy to understand, since a molecule at the surface is less restricted, and hence has a higher entropy, than a molecule residing deep in the bulk. A much less common, and less understood, phenomenon is surface freezing, where an *ordered surface layer* coexists with a *disordered bulk* liquid. No experimental observation of this effect at a free surface was reported, to the best of our knowledge, for any material prior to 1992, when the first measurements revealed the SF effect in melts of normal alkanes ($CH_3(CH_2)_{n-2}CH_3$, hereafter denoted Cn).[3] A somewhat related effect of a smectic surface layer coexisting with a less-ordered nematic or isotropic bulk, has been observed previously in liquid crystals.[4] However, the surface layer did not show any lateral *crystalline*-like order, as did the SF layer of alkane melts.

Since this paper focuses on the alkane–water interface, only the SF behaviour of alkanes will be discussed in some detail below.[5] However, it should be noted that numerous chain molecules (*e.g.* alcohols, semi-fluorinated alkanes, alkyl-thiols, α,β-diols, alkyl-oligoethyleneglycols, 1-alkenes, *etc.*), though not all, also exhibit SF.[6] Fatty acids and other normal-alkane derivatives with bulky headgroups, and chain molecules with non-linear tails, do not seem to undergo SF. The structure of the SF layer varies considerably from one molecule to another, and depends also on the existence of additional intermolecular interactions. For example, for 1-alcohols,[7] where hydrogen bonding exists between the hydroxyl headgroups, SF yields a crystalline bilayer, rather than the monolayer found in alkanes. This, in turn, allows for the intercalation of either water,[8] or diols[9] into the center of the bilayer, inducing new phases and structures.

Several theoretical explanations have been proposed for the occurrence of SF in chain molecules. The experimental results strongly point to the chain-like structure of the molecule as the major cause for surface freezing. One approach argues that the lower density of the CH_3 end groups imparts them a slightly higher surface activity. The surface enrichment of the end groups induces a preferential vertical alignment of the alkane chains even at $T > T_s$, leading eventually to a more ordered phase at the surface.[10] Density functional theories for molecules with weak anisotropy also yield a preferential surface-normal molecular alignment at the liquid surface, inducing, in turn, the SF effect.[11] Molecular simulations with various models also find SF for chain molecules, and investigate its length dependence.[12] Tkachenko and Rabin[13] suggested an entropic stabilization. Since the molecules are long, large vertical thermal fluctuations, and hence a large entropy, are allowed without violating the Lindemann criterion for crystal melting. This entropy stabilizes the SF layer. This approach explains both the role of the chain structure in the occurrence of the surface freezing effect and the lower chain length limit for the occurrence of SF. Recently, conformational entropy due to chain-end disorder, argued to be higher at the surface than at the bulk, was also suggested as the stabilizing agent of the SF layer.[14] While the question "what drives SF in chain molecules" is far from being settled, it should be noted that SF can be explained as a pure wetting effect, resulting from a particular balance of free energy excesses and deficits at the surface, without the need to resort to a molecular-level theory.[5,15]

SF is not restricted to the liquid–vapour interface. It has been detected also in the freestanding thin films of bubble walls, where SF increases significantly the bubble's lifetime and changes its coalescence behaviour.[16] SF has also been detected and studied in thin alkane films at the solid–vapour interface,[17] where it influences significantly the phase sequence measured, and the structures formed in the films with temperature variation.

SF at the liquid–liquid interface is much more elusive. For the *alcohol*–water interface, temperature-dependent surface tension measurements, $\gamma(T)$, reveal a sharp break in the curve a few degrees above the bulk freezing temperature of the alcohol. This break has been interpreted as either a transition from an expanded to a condensed interfacial phase,[18] or as an interfacial crystallization transition.[19] However, detailed X-ray and surface tension studies by Schlossman *et al.*[20] (see also contribution in this volume) on long-chain alkanol surfactants at the water : hexane interface indicate that the break observed in $\gamma(T)$ is in fact due to an adsorption transition from a low to a high density monolayer of alkanols at the interface. They also find that the alkanols' chains, residing in the hexane, are progressively disordered with distance from the interface even in the low-temperature, high-density phase. An intercalation of water molecules into the alkanols' headgroup layer, residing in the water phase, is also conjectured.

For the *alkane*–water interface, no SF was detected in spite of a rather extensive search at several laboratories. Considering that replacing the vapour phase by a water phase changes completely the energetics at the surface, and hence the balance allowing SF to occur at the liquid–air interface,[15] this is not surprising. However, Lei and Bain[21] have recently demonstrated that adding a small amount of the ionic surfactant CTAB to the water phase of a water : tetradecane system can re-shift the balance, causing IF to emerge. We report here on surface tension measurements on this effect for a different system, water : hexadecane, with a different surfactant, STAB. The interfacial layer, which undergoes IF, consists of a binary mixture of two types of chains: the C18 tails of the STAB surfactant, adsorbed at the surface due to its amphiphilic nature, and the C16 molecules of the alkane phase. We will therefore discuss the present results within the context of the SF observed at the alkane–air interface, with particular reference to our recently-published theoretical and experimental study of SF at the free surface of melts of binary alkane mixtures.[22]

Experimental

The details of the experiment, materials, and data analysis are available in refs. 5–9, 22–25 and will not be repeated here. The surface tension measurements employ the Wilhelmy plate method in a two-stage oven under computer control. The X-ray measurements described in the next section were carried out at beamline X22B, NSLS, Brookhaven National Laboratory, USA The X-ray reflectivity (XR) and grazing-incidence diffraction (GID) measurements allow an Ångström-resolution determination of the SF layer's structure in the surface-normal, and the surface-parallel directions, respectively.

The air–melt interface

Pure alkanes. In spite of its relative simplicity, surface tensiometry is a sensitive tool for observing surface structure variations in layers as thin as a single monolayer. Fig. 1 shows a $\gamma(T)$ scan for the air–melt interface of C26. γ is the excess free energy at the surface over the bulk, $\gamma = \varepsilon_s - \varepsilon_b - T(S_s - S_b)$, where ε and S denote, respectively, the molecular energy and entropy and the subscripts s and b – the surface and bulk.[23] For simple liquids $S_s > S_b$ as discussed above, and hence, $d\gamma/dT = -(S_s - S_b) < 0$. An ordering of the surface, as occurs upon surface freezing, switches the slope to $d\gamma/dT > 0$, since now $S_s < S_b$. This is indeed observed in Fig. 1, where the abrupt slope change at T_s marks the SF transition, which is of first order, as expected from a freezing transition. The absence of additional slope changes for $T_b < T < T_s$ indicates that no other transitions occur in the surface-frozen layer from T_s down to T_b. The surface entropy change upon SF, ΔS, is derived from the slope change as $\Delta S = [d\gamma(T < T_s)/dT - d\gamma(T > T_s)/dT]$. The coincidence of this value, $\Delta S = 1.51$ mN m^{-1} K^{-1}, with the 1.52 mN m^{-1} K^{-1} calculated from the known *bulk* freezing entropy change for a single monolayer,[5] indicates that the ordered surface layer is just one molecule thick.

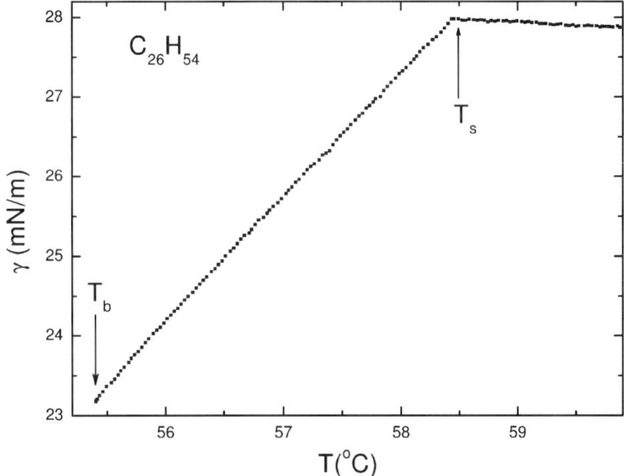

Fig. 1 Surface tension scan $\gamma(T)$ for C26 at the melt–air interface. T_b and T_s indicate the bulk and surface freezing temperatures. The change from a negative to positive slope upon decreasing T indicates an ordering transition at the surface.

This conclusion is indeed supported by the X-ray measurements[5] which probe directly the structure of the SF layer, and reveal a hexagonal packing with molecules aligned along the surface normal.

The phase diagram in the (n,T) plane is shown in Fig. 2, along with additional quantities derived from the surface tension and X-ray measurements. As the figure indicates the SF layer exists only over a limited range in temperature and chain length. However, within this range several different phases are found. For $15 < n < 44$, the SF phase is a rotator, with vertically-aligned molecules up to $n < 30$, and tilted towards nearest neighbours by an n-dependent angle ($5°$ for $n = 30$, $23°$ for $n = 44$) for $30 \leq n \leq 44$. For $n \geq 44$, the layer is a true herringbone-ordered crystal, with tilt in the next-nearest neighbour direction. The transitions with n between these phases are clearly observed, for example, in the changes they induce in ΔS in Fig. 2(b), where the slight slope decrease of the linear $\Delta S(n)$ marks the onset of the tilt at $n \approx 30$, and the large jump at $n = 44$ is due to a transition from a rotator to a crystalline surface phase. The measured layer thickness, D, and in-plane intermolecular distances, d, shown in Fig. 2(c,d), also reveal clearly the transition from untilted to tilted molecules, as well as the rotator-to-crystal transition. The known n-dependence of T_b, S_b, S_s and γ allows the derivation of a simple expression for the existence range, $\Delta T = T_s - T_b$, of SF: $\Delta T = a/n - b/n^3$. This functional form,[23] shown in a line in Fig. 2(a) accounts well for $\Delta T(n)$. Free-energy considerations,[5] based on the balance among the various interfacial energies (SF layer/liquid bulk, SF layer/vapour, liquid surface/vapour) also allow accounting for the occurrence of SF, and for its n- and T-ranges of existence. The tilt and tilt-direction transition depend on more minor changes in the chain–chain interaction with n, and are more difficult to account for. The rotator-to-crystal transition in the SF layer as n is increased follows the same transition in the bulk for the first phase appearing upon freezing below the melt.

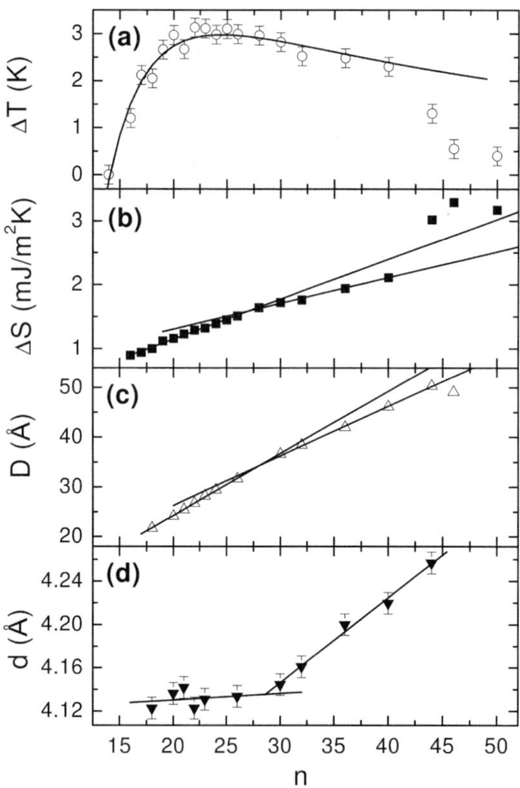

Fig. 2 (a) Length (n)—existence range ($\Delta T = T_s - T_b$) phase diagram (b) surface entropy change upon freezing (ΔS) (c) layer thickness (D) and in-plane nearest-neighbour spacing (d), for the surface-frozen layer of alkanes. Points denote measured values. Lines are guides to the eye, except in (a) where the theoretical line is discussed in the text.

Binary mixtures of alkanes. Binary mixtures of different-length alkanes also show SF. The variation of the fractional molar concentration ϕ of the longer component provides an additional "knob" allowing one to tune the SF effect to regions of phase space not reachable in pure materials. This results in the observations of new effects like a ϕ- or T-induced demixing transition in the solid surface layer manifested in an abrupt change in the thickness of the SF layer[24] and the only transition from a single to a double surface-frozen layer observed to date.[25]

Fig. 3(a) and (c) show the measured ϕ-dependence of T_b and T_s for two typical mixtures, one with a small relative length mismatch, $\delta = \Delta n / \bar{n} = (n_1 - n_2)/[(n_1 + n_2)/2] \approx 0.095$, and one with a large one: $\delta \approx 0.33$. The two mixtures show a markedly different T_s and T_b behaviour. For small δ, the ϕ-variation is almost linear and monotonic, while for large δ the variation is more complex. The ϕ-variation of ΔS, the entropy change upon SF, derived from the slope change in the measured $\gamma(T)$ above and below T_s, is shown in Fig. 3(b) and (d). While the ϕ-variation of ΔS is continuous for small δ, it is clearly bimodal for large δ.

For the large-δ case, shown in Fig. 3(c,d), the two constant ΔS values, observed in the small ($0 \leq \phi \leq 0.18$) and large ($0.6 \leq \phi \leq 1$) ϕ-regions, equal those measured for the two pure components, C26 and C36, respectively. This indicates that a phase separation occurs at the surface in this case, and the SF layer of the mixture consists entirely of a single component: C26 at low ϕ and C36 at high ϕ. Indeed, X-ray measurements[5] show a structure identical in each ϕ-region with that found for the SF layer of the melt of the corresponding pure component: an untilted rotator phase for C26, and a tilted rotator phase, with $18°$ tilt in the nearest-neighbour direction, for C36. As we show below, this phase separation is driven by the strong repulsion between the two species due to their large length mismatch, an effect well known for bulk polymer mixtures.[26] Note also the shaded region in Fig. 3(d), $0.18 \leq \phi \leq 0.4$, where no SF is observed, since it is preempted by bulk freezing, as observed in the coincidence of $T_s(\phi)$ and the $T_b(\phi)$ values in this range in Fig. 3(c). Also, in the range $0.4 \leq \phi \leq 0.5$, the exceptionally high ΔS values indicate the appearance of a different SF phase, having a smaller tilt, $13°$, and a different tilt direction: towards the next-nearest neighbours.[27]

For the small-δ case, shown in Fig. 3(a,b), the variation of ΔS with ϕ is continuous and monotonic. This indicates a mixed surface phase, where the mismatch interaction is too weak to induce phase separation. The X-ray results indeed show that the measured thickness of the SF layer coincides with that expected from ϕ assuming uniform mixing of the components.[22]

These results, and similar ones measured for binary mixtures of alkanes spanning a broad range of δ, can be accounted for within the theory of mixtures, using the properties of the pure components and taking into consideration the mixing entropy and the molecular interactions.[26] In the liquid (l) state, where the molecules are flexible and space filling, we treat the bulk (b) and surface (s) phases as ideal mixtures, neglecting the very small mixing enthalpy expected. The free energy F_l^j per mole of a $Cn : Cm$ mixture is:[26]

$$F_l^j = \phi_l^j f_l^j(n) + (1 - \phi_l^j) f_l^j(m) + k_B T [\phi_l^j \ln(\phi_l^j) + (1 - \phi_l^j) \ln(1 - \phi_l^j)] \tag{1}$$

where j = s or b, $f_l^j = \varepsilon_l^j(i) - T S_l^j(i)$, $\varepsilon_l^j(i)$ and $S_l^j(i)$ are the molar free energy, energy and entropy, respectively, of pure Ci, $i = n$ or m, ϕ_l^j and $(1 - \phi_l^j)$ are the mole fractions of Cn and Cm, respectively. The term in the square brackets is due to the mixing entropy.[26] For the crystalline (c) phase, the chains are extended and aligned in parallel. The mismatch repulsion now entails a free energy cost of ω^j for interchanging a long and a short molecule.[27] The free energy is then:[22,24,26]

$$F_c^j = \phi_c^j f_c^j(n) + (1 - \phi_c^j) f_c^j(m) + k_B T [\phi_c^j \ln(\phi_c^j) + (1 - \phi_c^j) \ln(1 - \phi_c^j)] + \omega^j \phi_c^j (1 - \phi_c^j) \tag{2}$$

where the notation of eqn. (1) is employed, with "crystalline" (c) replacing "liquid" (l). The last term in eqn. (2) is the repulsion term due to the interchange energy, ω^j, in the zeroth-order approximation (nearest neighbour interactions only) of the "strictly-regular" mixture theory.[26] The solid phase behaviour is now determined by the balance between the mixing entropy term and the interchange term. For $\omega^j \leq 2k_B T$ the mixing entropy dominates and a uniformly mixed crystalline phase is obtained at all ϕ_c^j. For $\omega^j \geq 2k_B T$, the repulsive term dominates and induces phase separation in the solid phase for compositions $\phi_c^j \approx 0.5$. Once phase separation occurs, eqn. (2), and the theory discussed here, cease to be valid, since for eqn. 2 a uniform mixing is required, with different-length molecules residing side by side. It is also important to note that the only quantity controllable experimentally is the *bulk* liquid composition, ϕ_l^b. All other quantities, *i.e.*, the

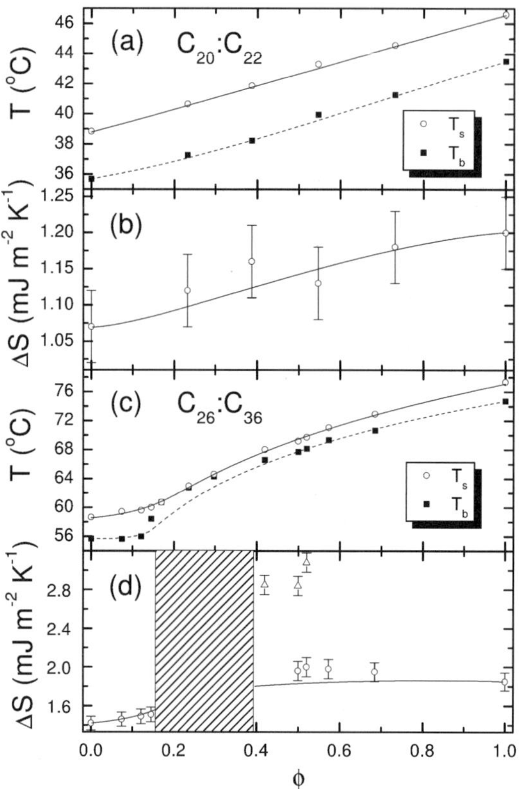

Fig. 3 (a,c) The measured (symbols) surface (T_s) and bulk (T_b) freezing temperatures for the mixtures indicated (b,d) the measured (symbols) surface entropy change upon freezing for the mixtures in (a) and (c). The lines in (a) and (c) are fits by the theory discussed in the text. This theory yields the lines shown in (b) and (d) without any adjustable parameters. The shadowed region in (d) indicates a fractional concentration (ϕ) range where surface freezing is not observed due to its pre-emption by bulk freezing.

liquid and crystalline surface compositions, ϕ_1^s and ϕ_c^s, and the crystalline bulk composition, ϕ_c^b, are determined by the thermodynamics of the system. In particular, ϕ_1^s can be calculated from the known ϕ_1^b using the Gibbs adsorption rule, which yields, in our case, an enrichment of the surface by the shorter of the two components.

The chemical potentials calculated from eqns. (1) and (2) for the liquid and solid phases must be equal for each component at the freezing temperatures T_b and T_s. This yields equations for T_b and T_s in terms of the properties of the pure components (transition temperatures and entropy changes) appearing in eqns. (1) and (2), with ω^j as the only unknown parameter in each equation. For Cn:

$$T_j(\phi_i^j) = [T_{j,n}\Delta S_{j,n} - \omega^j(1 - \phi_c^j)^2]/[\Delta S_{j,n} + k_B\ln(\phi_c^j/\phi_1^j)] \tag{3}$$

where $T_{j,n}$ and $\Delta S_{j,n}$ are the known freezing temperature and the entropy change upon freezing of the pure component n. ω^j can now be obtained by fitting eqn. (3) to the measured $T_b(\phi_1^b)$ and $T_s(\phi_1^b)$ values, using a single ω^j (different for j = b and j = s) for all concentrations ϕ_1^b of a given pair of alkane molecules. The equivalent expression for Cm, obtained by replacing $n \rightarrow m$ and $\phi_i^j \rightarrow (1 - \phi_i^j)$, is used to solve for ϕ_c^j numerically. The fits are shown in lines in Fig. 3(a,c), and clearly agree very well with the measured freezing temperatures in spite of the single fit parameter. Moreover, eqns. (1) and (2) allow also the calculation of $\Delta S(\phi_1^b)$, the concentration-dependent entropy change upon surface freezing of the mixture, without introducing additional parameters. This calculated $\Delta S(\phi_1^b)$ is shown as a solid line in Fig. 3(b,d). The good agreement, achieved without any adjustable parameters, further supports the validity of the analysis presented here.

An intriguing feature of the ω^j values derived for all the binary alkane mixtures studied is that they conform to a δ^2 dependence, $\omega^j = A\delta^2$, for both bulk and surface (albeit with a different prefactor A), as clearly demonstrated in Fig. 4. This is easy to rationalize as follows. The interchange energy ω^j should obviously depend on the chain length mismatch, Δn. However, a given mismatch Δn, which provides free volume for kinks, *gauche* conformations, *etc.*, should be relatively more important for a shorter molecule than for a longer one. It is therefore reasonable to assume that ω^j depends on $\delta = \Delta n/\bar{n}$, rather than on Δn only. Since $\delta < 1$ for all mixtures studied here, we can expand ω^j in a power series in δ, where the constant term vanishes, as identical molecules do not repel each other. Since only the absolute value of the mismatch counts, odd-power terms also vanish, leaving δ^2 as the lowest-order non-zero term. The series can be truncated after this term, since for $\delta \rightarrow 1$, where higher-order terms may be significant, phase separation of the components occurs anyway, and the theory above is not valid any more, as discussed above.[24]

Fig 4 demonstrates that the ω^j values obtained from the measured data indeed conform exceedingly well to the expected linear δ^2-dependence for ω^b up to $\delta^2 \approx 0.13$, which corresponds to $\omega^b \approx 2.5k_BT$, right at the limit of phase separation, which is also the limit of validity of our theory. For the surface, all mixtures studied, up to $\delta^2 \approx 0.23$, correspond to $\omega^s \leq 2.5k_BT$, and thus only the highest-δ^2 mixtures are close to phase separation. The higher prefactor A in the bulk (17.8 *vs.* 11.6) clearly indicates a higher inter-chain repulsion. This could possibly reflect larger strains in the crystal structure upon packing different chain lengths in a multilayered 3D solid, as compared to those in packing them in a single surface mono or bilayer.

A striking feature of the ω^j values of alkanes, and of those of binary mixtures of hydrated and dry alcohols, and deuterated-protonated alkanes (labeled D_n in Fig. 4),[22] is the fact that all points fall on the *same* line, regardless of the components of the mixtures. This "quasi-universal" behaviour is surprising considering the presence of additional interactions in some of the mixtures: hydrogen bonding in alcohols and isotope mismatch repulsion in the deuterated-protonated alkane mixtures. While the isotope effect is expected to be small, hydrogen bonding is strong enough to induce the

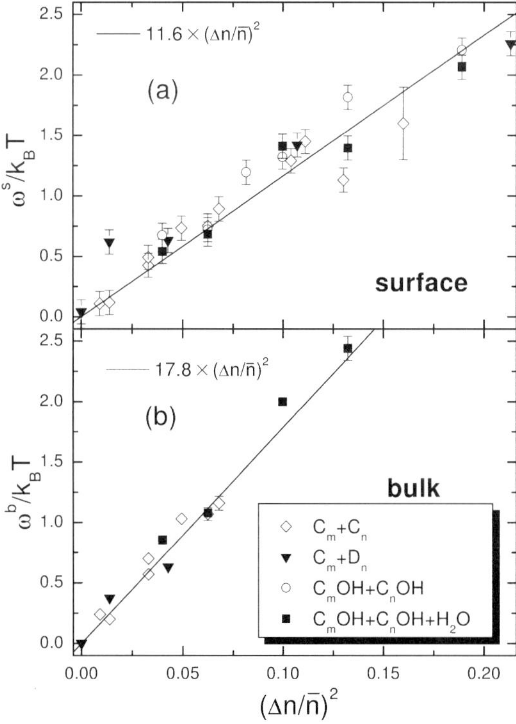

Fig. 4 The interchange parameter ω^j (symbols) derived from the theoretical fits shown in Fig. 3(a,c). Their quasi-"universal" linear dependence on the relative mismatch squared, δ^2, is discussed in the text.

formation of a SF *bilayer* in alcohols, rather than the monolayer observed in alkanes. Moreover, hydration increases the strength of this interaction, as indicated by the higher transition temperatures measured for hydrated alcohol mixtures as compared to dry ones.[8] Nevertheless, the "universal" behaviour found in Fig. 4 indicates that the length-mismatch repulsion energy of the chains dominate over any other interaction for the molecules studied.

Finally, we note that while the symmetry considerations above explain the observed δ^2 behaviour of the interchange energy ω^j, to the best of our knowledge, no theory predicting this behaviour from molecular-level considerations is available in the literature. The available molecular-level theories for polymer mixtures predict a $|\delta|$ dependence[28] or a $|\Delta n|$ dependence,[22] both disagreeing with our measurements. A more sophisticated molecular-level theoretical treatment, taking into consideration the presence of voids, *gauche* kinks, short-range clustering of equal-length molecules, *etc.* may achieve a better agreement with experiment.

The water–melt interface

The temperature dependence of the interfacial tension. Measurements at several laboratories, including ours, failed to detect interfacial freezing (IF) at the alkane melt : water interface. This, however, is not surprising. The large-amplitude surface normal molecular vibrations invoked to provide entropic stabilization of the SF layer at the free surface of the melt[13] may be greatly damped at an interface separating two dense liquid phases. A different proposed explanation for SF, a favourable balance between several surface tensions which drives the wetting of the free surface by a solid monolayer[5,15] at $T_s > T_b$, is certain to unbalance when several of the surface tensions involved change drastically upon replacing the vapour phase with water. However, *in principle*, the free energy balance against the formation of an IF layer at the alkane melt–water interface could be re-tuned to one favouring IF by adding a new, judiciously chosen interaction to the system. Lei and Bain[21] demonstrated recently that this can be achieved by adding a cationic surfactant, cetyl-trimethylammonium bromide (CTAB), at sub-mM concentrations, to the water phase of a water : C14 system. The resultant interfacial tension curves, $\gamma_i(T)$, are very similar to that in Fig. 1 where an abrupt slope change implies the occurrence of IF. They have also shown that the transition is not a CTAB adsorption/desorption transition of the type found by Schlossman *et al.*[20] for alcohols, since the surface excess of CTAB, derived from the variation of γ_i with concentration through the Gibbs rule, changes at the transition only insignificantly. These results, obtained at Oxford, have been subsequently verified at Bar-Ilan.

We have carried out interfacial tension measurements, using the Wilhelmy plate method in a two-stage computer controlled oven, on a different system, the water : C16 interface, using STAB (stearyl-trimethylammonium bromide) as a surfactant. The surfactant molecules adsorb at the interface, with their hydrophilic TAB headgroup residing in the water and their lipophilic stearyl (C18) tails protruding into the C16 phase. At the free surface of an alkane melt, sum frequency generation measurements indicate that the (fairly extended) chains align roughly along the surface-normal even in the liquid surface phase.[29] Similar measurements for CTAB at the water : C16 interface[30] show, however, that the interface-adsorbed CTAB chains are conformationally disordered. The chains adopt a more extended, upright conformation only when the saturation coverage is approached at the cmc. Fig. 5 shows measured interfacial $\gamma_i(T)$ curves in the absence and presence of STAB in the water phase. For pure water : C16 $\gamma_i(T)$ exhibits a constant negative slope down to T_b, as expected of a simple interface between two immiscible, mutually non-wetting, liquids. This indicates that no IF effect occurs at this interface, as found also for other water : Cn interfaces.[20] When STAB is present, a slope change occurs in $\gamma_i(T)$ at a $T = T_i$ which depends on the STAB concentration, c. For $c = 0.133$ mM, a somewhat rounded slope-change occurs at $T_i \approx$ 24.3 °C. For $c = 1.1$ mM a sharper slope change occurs at $T_i = 26$ °C. Both curves are very similar to that observed at the free surface of the melt, shown in Fig. 1, and to those reported for the water/ C14/CTAB system.[21] We assign, therefore, the break in $\gamma_i(T_i)$ here also to an IF transition. The slope changes yield an IF entropy change of $\Delta S^i = (0.86 \pm 0.05)$ and (0.78 ± 0.08) mJ m^{-2} K^{-1}, for $c = 1.1$ mM and 0.133 mM, respectively, both close to the $\Delta S = 0.896$ mJ m^{-2} K^{-1} measured for the SF monolayer of C16 at the free surface.[5] The coincidence between these ΔS values further supports the identification of the slope change as an IF effect, and suggests that the IF layer is a single monolayer thick. A point to note in Fig. 5 is that the Wilhelmy plate measuring the IF at the

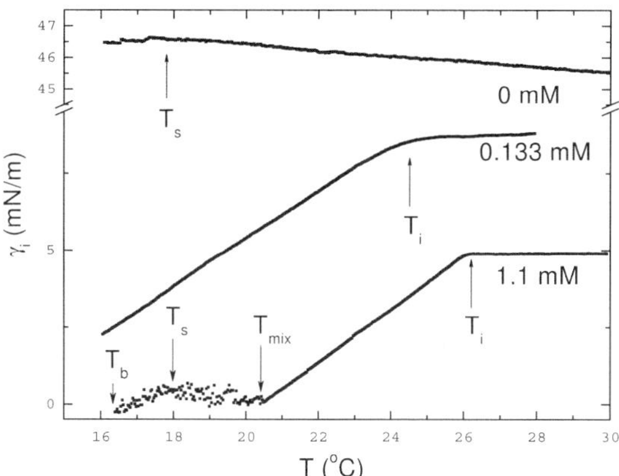

Fig. 5 The measured interfacial tension at the water:hexadecane interface without, and with the cationic surfactant STAB dissolved in the water phase at the indicated concentrations. Note, the abrupt slope change at T_i due to interfacial freezing, and the vanishing of the interfacial tension at T_{mix} for a STAB concentration of $c = 1.1$mM. For discussion, see text.

water:C16 interface is hanging from the film balance on a 0.6 mm diameter wire. This goes through the free surface of the C16, and thus senses the SF at the free surface, albeit as a very weak effect, due to the small diameter of the wire. This is observable in the $c = 0$ and $c = 1.1$ mM curves at $T_s \approx$ 17.8 °C as a very small slope change.

The appearance of IF at the water:Cn interface raises some interesting possibilities concerning crystal nucleation of alkanes from the melt. One of the most widely used methods for achieving homogeneous nucleation from a melt is the dispersion of the melt into droplets in a carrier medium.[31] If the melt can be divided finely enough, part of the droplets may be sufficiently pure and free from heterogeneous-nucleating agents to allow them to nucleate homogeneously. Alkanes-in-water emulsions, where the Cn droplets are stabilized by non-ionic surfactants, were shown to undergo homogeneous nucleation.[32] However, the interfacial energetics in these droplets is radically different from that of a free surface of a melt, and IF is not expected, nor observed. This is manifested by the fact that an alkane melt having a free surface does not supercool, since SF produces a crystalline template already a few degrees above T_b on which the bulk molecules can crystallize, and thus the barrier for crystal nucleation is eliminated. By contrast, a Cn emulsion's droplets, stabilized by non-ionic surfactants, e.g. Igepal, have been shown to undercool consider-ably, implying a nucleation barrier and no IF.[32] However, using alkyl-TAB surfactants, rather than non-ionic ones, to stabilize the droplets should restore IF at the droplets' surface. This should, in principle, eliminate the supercooling in such emulsions. First calorimetry measurements indicate a reduction in the undercooling range of CTAB-stabilized C14 and C15 in water as compared to those stabilized by the non-ionic surfactant Igepal. The incomplete elimination of the nucleation barrier, indicated by the fact that undercooling does not vanish completely, is probably due to an imperfect matching between the structures of the IF layer and the critical nucleus, particularly since the latter may have a transient, non-equilibrium structure not matching that of a macroscopic crystal.[33]

An even more intriguing possibility, provided by the formation of a IF layer, is the appearance of superheating. As pointed out by Frenkel[34] a solid with a free surface can not be superheated because surface melting invariably provides a liquid nucleus below the bulk melting point, which reduces to zero the nucleation barrier for the bulk liquid state. However, if surface freezing occurs, no liquid nucleus forms, and the non-zero nucleation barrier for the liquid state will allow one to superheat the bulk. A macroscopic-sized solid of alkane will always have heterogeneous nucleation sites like container walls, impurities, etc. Thus, even though SF will prevent formation of a liquid nucleus at the free surface upon heating past the equilibrium melting point, these sites will nucleate the melting. However, in Cn-in-water emulsions stabilized by alkyl-TAB surfactants, where IF occur over the full surface area of the (solid) emulsion particles, some particles may be free from liquid-nucleating agents. For those, IF will provide a kinetic barrier for melting, and thus allow superheating. The barrier should yield a superheating range of up to the IF existence range, $\leq 10\ ^\circ\mathrm{C}$ in our case. A search for this effect is in progress.

The $\gamma_i(T)$ curve for $c = 1.1$ mM in Fig. 5 shows, in addition to IF, another interesting effect. The reduction of the surface tension of a liquid to zero by *heating* is common. It occurs, *e.g.*, at the liquid's critical point. Here we observed the opposite effect, the reduction of the interfacial tension to zero by *cooling*. This reduction of $\gamma_i(T)$ to zero is a new mechanism for the vanishing of the interfacial tension. It occurs here since below T_i we have $\gamma_i(T) = \gamma_i(T_i) - (T_i - T)(\mathrm{d}\gamma_i/\mathrm{d}T)$. Thus, $\gamma_i \to 0$ at $T_{\mathrm{mix}} = T_i - \gamma_i(T_i)/(\mathrm{d}\gamma_i/\mathrm{d}T)$ *unless preempted by bulk freezing* of the alkane phase. At the free surface of a melt, and also at the solution : alkane interface at low surfactant concentrations, $\gamma_i(T_i)$ is usually large, and T_i is close to T_b, so that preemption indeed occurs, and vanishing of γ_i is not observed, as can be seen in the $c = 0.133$ mM curve in Fig. 5. The vanishing of γ_i can be induced in our system only because of the drastic reduction in γ_i from its high value at $c = 0$ to the much lower values obtained for large STAB concentrations. As observed in Fig. 5, for $c = 1.1$ mM γ_i vanishes at $T_{\mathrm{mix}} = 20.4\ ^\circ\mathrm{C}$, $4.2\ ^\circ\mathrm{C}$ above T_b. We find the same vanishing of γ_i for all curves measured at $c \geq 0.17$ mM. For $c \geq 0.2$ mM, we find a practically c-independent $T_{\mathrm{mix}} = 20.4\ ^\circ\mathrm{C}$, resulting from the c-independent $T_i = 26\ ^\circ\mathrm{C}$ and $\gamma_i(T_i) \approx 5$ mN m^{-1} for all $c \geq 0.2$ mM.

The ability to tune the interfacial tension γ_i to zero has several interesting implications. Usually, when γ_i between two separate phases vanishes, thermal diffusion will eventually lead to molecular-level mixing and form a uniform single phase. In our case, however, γ_i is zero only in the presence of the IF monolayer. Thus, the mixing can only occur by the formation of surfactant-coated Cn droplets. Thus, an emulsion is expected to form when the system is cooled below T_{mix}. However, in contrast with the usual emulsions, where the droplets are only metastable thermodynamically, and tend to coalesce with time, our "self-assembling" emulsion is stable under fluctuations, since $\gamma_i = 0$ and coalescence can not reduce the system's free energy. In our case, coalescence would reduce the entropy, without yielding any energetic compensation. We therefore expect that as $\gamma_i \to 0$ a micro-emulsion will form at the interface, where the smallest droplets which still exhibit IF will be the thermodynamically stable ones, since the larger the droplet, the larger is the entropy loss. Moreover, using surface energy cost consideration, tuning the temperature from above to below T_{mix} should change the shapes of the emulsion's droplets from a spherical shape, which has a minimal surface area, at $\gamma_i > 0$ to an elongated shape with a larger surface area, at $\gamma_i = 0$. We are currently exploring these emulsions by calorimetry and microscopy methods.

The $c = 1.1$ mM curve in Fig. 5 reveals a very large T-range of existence for the IF effect, reaching a maximal value of $\Delta T_i^{\mathrm{max}} = T_i^{\mathrm{max}} - T_b \approx 10\ ^\circ\mathrm{C}$, much larger than the $\sim 1.2\ ^\circ\mathrm{C}$ observed for SF of C16 at the free surface of the melt.[5] Of course, as $c \to 0$, T_i decreases, and so does the existence range, $\Delta T_i = T_i - T_b$, as we discuss below. Measurements on water/Cn/CTAB ($n = 13$–16)[21] and on water/Cn/STAB ($n = 14, 16$), show that as the chain length n of the alkane decreases below that of the alkyl-TAB's, m, both T_b and T_i^{max} decrease with increasing difference $\delta n = m - n$. However, T_b decreases faster, in agreement with measurements on mixed monolayers of CTAB/Cn ($n = 11$–16) at the air : water interface. This results in an increase in $\Delta T_i^{\mathrm{max}}$ with δn. This, in turn, allows the vanishing of γ_i to appear at increasingly lower surfactant concentrations c as the length difference between surfactant and alkane, δn, increases. Indeed, preliminary measurements on the water/C14/STAB show that the $\gamma_i(T) = 0$ region appears already at $c = 0.12$ mM, well below, and thus most probably unrelated to, the cmc or the solubility limit, which will be discussed below.

The STAB concentration dependence. The measured variation of T_i with STAB concentration is plotted in Fig. 6 (points). The cmc of STAB in water is also marked. We observed that at $c > $ cmc, a constant value of T_i is found. This point is further discussed below in the context of the phase diagram. Here we just note that this indicates that the areal coverage, Γ, of the solution : C16 interface by the surfactant is saturated at a constant value independent of c. Unfortunately, our measurements, still in progress, do not allow at this stage a confident determination of the absolute coverage Γ, *e.g.* through the Gibbs adsorption rule $\Gamma = -[\mathrm{d}\gamma_i/\mathrm{d}(\ln(c))]/RT$. However, in the water/ C14/CTAB system, the IF effect is induced even by a ten-fold lower Γ_{CTAB} than the ~ 5 molecules nm^{-2} found in SF monolayers at the free surface of the melt, which is practically the same as that in the solid bulk alkane. It is therefore reasonable to assume that the incorporation of C16 molecules into the monolayer of the STAB's stearyl chains, which protrude into the alkane phase, reduces the

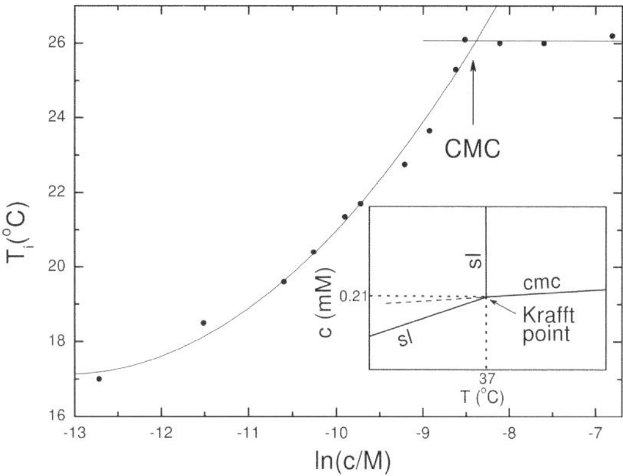

Fig. 6 The measured (points) variation of the IF onset temperature T_i with the molar concentration of STAB. Note the sharp change at the CMC to a constant T_i, due to the saturated, constant interface coverage with STAB. The solid lines are phenomenological linear (above CMC) and parabolic (below CMC) fits to the data. The inset is a scheme of the phase diagram topology near the Krafft temperature. The solubility limit line is marked by sl and the dash line is the metastable continuation of the cmc line below the Krafft temperature.

system's free energy. This creates an interesting new situation, where the interaction parameter between the two types of chains is attractive (negative), driving towards mixing, rather than the repulsive (positive) interaction parameter, driving towards phase separation, for different-length chains at the free surface of alkane mixtures, as discussed above. Since the surface coverage by the surfactant does not change significantly at the transition[21] the mixing entropy change upon IF is expected to be very small.

As observed in Fig. 6, T_i does not follow the linear dependence on $\ln(c)$, found for the water/C14/CTAB system.[21] Rather, a phenomenological fit by a polynomial reveals a $T_i \propto [\ln(c)]^2$ dependence, shown by the solid line in the figure. Within the regular solution model, this would indicate a non-zero enthalpy for our system, while that in the water/C14/CTAB should be negligible. We can therefore invoke the theory discussed above for the free surface of mixtures, neglecting the mixing entropy, which changes only marginally in the transition. Equating the chemical potential of the hexadecane in the crystalline (c) and liquid (l) interfacial phases yields for the dependence of T_i on the STAB concentration, an equation equivalent to eqn. (3):

$$T_i = T_{i,0}[1 - \lambda k_B(\phi_1^s)^2/\Delta S] \qquad (4)$$

where $\lambda = \omega^i/k_B T$, ω^i is the interaction parameter between the STAB's tail and the C16 molecule in the IF monolayer, and ϕ_1^s is the STAB concentration in the liquid surface phase. ϕ_1^s is obtainable in principle from Gibbs rule, if the c-variation of the absolute γ_i is known accurately. ΔS is the entropy loss upon freezing of C16, and $T_{i,0}$ is the IF temperature of the pure water : C16 interface, even if unobservable due to preemption by bulk freezing. By analogy with the water/C14/CTAB system, we expect that $\phi_1^s \sim \Gamma$ is approximately linear in $\ln(c)$[21]. Eqn. (4) then yields a parabolic dependence of T_i on $\ln(c)$, as indeed observed in Fig. 6 for $c <$ cmc. Unfortunately, a fit of eqn. (4) to the measured points is not possible since the absolute ϕ_1^s are not known, as mentioned above.

Equilibrium or metastable? The question of thermodynamic equilibrium in these measurements deserves some consideration. An accurately measured (c,T) phase diagram is not available for STAB in water, to the best of our knowledge. However, a qualitative scheme near the Krafft point is shown in the inset of Fig. 6. The dash line is the (almost T-independent) metastable cmc line, which is the continuation of the equilibrium cmc line below the Krafft point. For STAB in water, this line is always close to the cmc(T_K) ≈ 0.21 mM,[35] where, $T_K = 37$ °C is the Krafft temperature.[36] Above T_K, increasing c past the cmc drives the excess molecules to aggregate in micelles, leaving the bulk concentration, and thus the interface coverage by the surfactant, Γ, constant at the cmc value.[37] Below T_K this description still holds, except that now the excess bulk molecules aggregate as a precipitating solid. However, the constancy of the excess surface coverage Γ, and, therefore, the invariance of $\gamma_i(T)$, upon increasing c above the sl-line are still maintained. It is important to note that all $\gamma_i(T)$ scans in this study are carried out at $T < T_K$. Consequently, we expect the saturation of Γ, and the concomitant onset of the invariance of the $\gamma_i(T)$ curves with increasing c, to be at the sl-line. Even though the position of the sl-line in our system is not known, it is expected to lie considerably lower in c than the $\mathrm{cmc}(T_K) \approx 0.21$ mM since the sl-line is temperature dependent, as shown in the inset to Fig. 6, and our T_i's are at least 10 °C below the Krafft point. Nevertheless, as Fig. 6 clearly shows, the onset of the $\gamma_i(T)$ invariance occurs at the cmc. It seems therefore that for concentrations below 0.21 mM down to the (unknown) sl-line, the $\gamma_i(T)$ scans are carried out, at least partly, on a metastable STAB solution. This occurs since the high kinetic barrier against crystal nucleation from the solution of alkyl-TAB surfactants prevents precipitation of the excess molecules upon crossing the sl-line, yielding a metastable, supersaturated solution, the behaviour of which mimics that of a solution above T_K. In particular, it exhibits a crossing of the metastable cmc line upon increasing c, with a consequent onset at this c of the invariance of the $\gamma_i(T)$ scans. This is indeed observed in Fig. 6.

Alkane length dependence. First results on the dependence of the effects discussed above on the alkane length, n, are shown in Fig. 7, for a STAB concentration of 0.15 mM, close to the cmc. The temperature at which the interfacial tension vanishes, T_{mix}, is also shown. As can be observed, both T_i and T_b decrease with n. However, the faster decrease of T_b causes $\Delta T(n) = T_i - T_b$ to grow with decreasing n. Since data for $n < 13$ is unavailable, it is not clear whether T_i remains constant with decreasing n, as the values for $n = 13$ and 14 seem to indicate. This, however is not likely. At the other end of the n-range shown in the figure, T_i and T_b approach each other, so that ΔT is

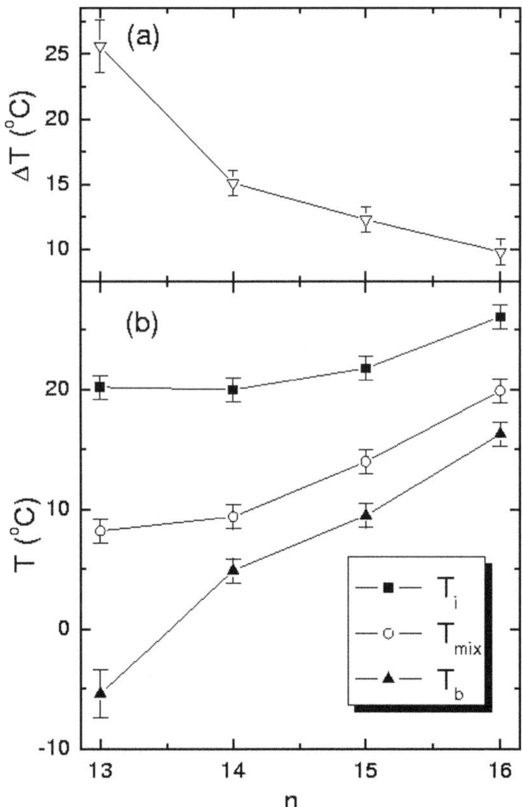

Fig. 7 (a) The temperature difference between the bulk and interface feezing temperatures as a function of the alkane length. (b) The measured (points) variation of the bulk and interface freezing temperatures for various alkane lengths n, for a STAB concentration of 0.15 mM. T_{mix} is also shown.

expected to vanish at some n, yet to be determined. This general trend is in agreement with recent measurements by one of us (C.D.B) on CTAB-induced surface freezing of alkane monolayers at the free surface of water.[38] For CTAB-induced *interfacial* freezing at the water–alkane interface, C16 (which has a chain length equal to that of CTAB) does not show IF. However, at a 0.6 mM CTAB concentration, C13, C14 and C15 do show IF, with existence ranges of $\Delta T = 13, 5, 3$ °C, respectively.[21] The trend in $\Delta T(\Delta n = n - 16)$ for CTAB is similar to $\Delta T(\Delta n = n - 18)$ for STAB, albeit the latter is of a larger absolute magnitude. Clearly, measurements over a more extended n range are called for to establish the n-dependence of T_i and T_{mix} (T_b is available in the literature)[39] before the construction of a theory, even at the thermodynamic level, can be attempted.

Concluding remarks

The results presented here, and, in particular, the rather tentative and approximate calculation of T_i in the previous section, while somewhat preliminary and available for a few systems only at this stage, do provide a glimpse into an intriguing new interfacial phenomenon. A considerably larger body of experimental data, both as a function of the alkane and the surfactant length are required to determine the behaviour of the IF effect for the water/Cn/alkyl-TAB system, and the physics underlying this behaviour. These measurements are now in progress. Moreover, X-ray reflectivity, and, if feasible, grazing incidence diffraction measurements on the liquid and frozen interface are required to determine the structure of the IF monolayer in the interface-normal and interface-parallel directions. Such X-ray measurements are extremely demanding, as demonstrated by the liquid–liquid interface measurements of Schlossman and coworkers[20] due to two factors. First, there is the need to diffract through a condensed phase (the upper alkane phase) which greatly increases the scattered background. Second, the extremely low γ_i values of these systems, result in large capillary-wave-induced interfacial roughnesses. These, in turn, reduce the reflected and diffracted signals, thereby restricting severely the measurable angular range, and, consequently, the achievable resolution. Nevertheless, even these limited-range measurements may be worthwhile, and are now in progress.

Acknowledgements

We thank Henning Kraack and Zvi Sapir (Bar-Ilan) for important discussions and assistance with the measurements.

References

1 (*a*) J. G. Dash, *Contemp. Phys.*, 1989, **30**, 89; (*b*) R. Lipowsky, *J. Appl. Phys.*, 1984, **55**, 2485; (*c*) G. An and M. Schick, *Phys. Rev. B*, 1988, **37**, 7534.
2 (*a*) J. W. M. Frenken and J. F. van der Veen, *Phys. Rev. Lett.*, 1985, **54**, 134; (*b*) H. Dosch, T. Höfer, J. Peisl and R. L. Johnson, *Europhys. Lett.*, 1991, **15**, 527; (*c*) D. M. Zhu and J. G. Dash, *Phys. Rev. Lett.*, 1986, **57**, 2959; (*d*) M. Elbaum and M. Schick, *Phys. Rev. Lett.*, 1991, **66**, 1713; (*e*) S. Chandavarkar, R. M. Geertman and W. H. de Jeu, *Phys. Rev. Lett.*, 1992, **69**, 2384.
3 (*a*) J. C. Earnshaw and C. J. Hughes, *Phys. Rev. A*, 1992, **46**, R4494; (*b*) X. Z. Wu, E. B. Sirota, S. K. Sinha, B. M. Ocko and M. Deutsch, *Phys. Rev. Lett.*, 1993, **70**, 958.
4 (*a*) B. M. Ocko, A. Braslau, P. S. Pershan, J. Als-Nielsen and M. Deutsch, *Phys. Rev. Lett.*, 1986, **57**, 94; (*b*) J. Als-Nielsen, F. Christensen and P. S. Pershan, *Phys. Rev. Lett.*, 1982, **48**, 1107; (*c*) P. S. Pershan and J. Als-Nielsen, *Phys. Rev. Lett.*, 1984, **52**, 759; (*d*) G. J. Kellogg, P. S. Pershan, E. H. Kawamoto, W. Foster, M. Deutsch and B. M. Ocko, *Phys. Rev. E*, 1995, **51**, 4709.
5 B. M. Ocko, X. Z. Wu, E. B. Sirota, S. K. Sinha, O. Gang and M. Deutsch, *Phys. Rev. E*, 1997, **55**, 3164.
6 O. Gang, *Surface Ordering in Chain Molecules*, PhD Thesis, Bar-Ilan University, 1999, unpublished.
7 O. Gang, X. Z. Wu, B. M. Ocko, E. B. Sirota and M. Deutsch, *Phys. Rev. E*, 1998, **58**, 6086.
8 O. Gang, B. M. Ocko, X. Z. Wu, E. B. Sirota and M. Deutsch, *Phys. Rev. Lett.*, 1998, **80**, 1264.
9 O. Gang, B. M. Ocko, X. Z. Wu, E. B. Sirota and M. Deutsch, *Phys. Rev. Lett.*, 1999, **82**, 588.
10 (*a*) T. K. Xia and U. Landman, *Phys. Rev. B*, 1993, **48**, 11313; (*b*) J. G. Harris, *J. Phys. Chem.*, 1992, **96**, 5077; (*c*) F. A. M. Leermakers and M. A. Cohen Stuart, *Phys. Rev. Lett.*, 1996, **76**, 82; (*d*) G. A. Sefler, G. Du, P. B. Miranda and Y. R. Shen, *Chem. Phys. Lett.*, 1995, **235**, 347.
11 (*a*) A. Weinstein and S. A. Safran, *Phys. Rev. E*, 1996, **53**, R45; (*b*) A. ten Bosch, *J. Chem. Phys.*, 1998, **108**, 2228; (*c*) A. ten Bosch, *Phys. Rev. E*, 2001, **63**, 61808.

12 (*a*) P. Smith P, R. M. Lynden-Bell, J. C. Earnshaw and W. Smith, *Mol. Phys.*, 1999, **96**, 249; (*b*) H. Z. Li and T. Yamamoto, *J. Phys. Soc. Jpn.*, 2002, **71**, 1083; (*c*) T. Shimizu and T. Yamamoto, *J. Chem. Phys.*, 2000, **113**, 3359.

13 (*a*) A. V. Tkachenko and Y. Rabin, *Phys. Rev. Lett.*, 1996, **76**, 2527; (*b*) A. V. Tkachenko and Y. Rabin, *Phys. Rev. Lett.*, 1997, **79**, 532; (*c*) Y. Rabin, *Phys. Rev. E.*, 1998, **55**, 778.

14 A. J. Colussi, M. R. Hoffmann and Y. Tang, *Langmuir*, 2000, **16**, 5213.

15 E. B. Sirota, X. Z. Wu, B. M. Ocko and M. Deutsch, *Phys. Rev. Lett.*, 1997, **79**, 531.

16 (*a*) N. Maeda and V. V. Yaminsky, *Phys. Rev. Lett.*, 2000, **84**, 698; (*b*) H. Gang, J. Patel, X. Z. Wu, M. Deutsch, O. Gang, B. M. Ocko and E. B. Sirota, *Europhys. Lett.*, 1998, **43**, 314.

17 (*a*) C. Merkl, T. Pfohl and H. Riegler, *Phys. Rev. Lett.*, 1997, **79**, 4625; (*b*) N. Maeda, N. M. Kohonen and H. K. Christenson, *Phys. Rev. E*, 2000, **61**, 7239; (*c*) H. Schollmeyer, B. Struth and H. Riegler, *Langmuir*, 2003, **19**, 5042; (*d*) A. Holzwarth, S. Leporatti and H. Riegler, *Europhys. Lett.*, 2000, **52**, 653; (*e*) Y. Yamamoto, H. Ohara, K. Kajikawa, H. Ishii, N. Ueno, K. Seki and Y. Ouchi, *Chem. Phys. Lett.*, 1999, **304**, 231.

18 (*a*) M. Aratono, T. Takiue, N. Ikeda, A. Nakamura and K. Motomura, *J. Phys. Chem.*, 1992, **96**, 9422; (*b*) M. Aratono, T. Takiue, N. Ikeda, A. Nakamura and K. Motomura, *J. Phys. Chem.*, 1993, **97**, 5141.

19 (*a*) J. Glinski, G. Chavapeyer, J. K. Platten and C. De Saedeleer, *J. Colloid Interface Sci.*, 1993, **158**, 382; (*b*) M. Salajan, J. Glinski, G. Chavapeyer, J. K. Platten and C. De. Saedeleer, *J. Colloid Interface Sci.*, 1994, **164**, 387.

20 (*a*) A. M. Tikhonov, S. V. Pingali and M. L. Schlossman, *J. Chem. Phys.*, 2004, **120**, 11822; (*b*) M. L. Schlossman and A. M. Tikhonov, in *Mesoscale Phenomena in Fluid Systems*, ed. F. Case and P. Alexandridis, ACS Symposium Series 861, OUP, 2003, p. 81; (*c*) A. M. Tikhonov and M. L. Schlossman, *J. Phys. Chem. B*, 2003, **107**, 3344; (*d*) M. L. Schlossman, *Curr. Opin. Coll. Int. Sci.*, 2002, **7**, 235.

21 Q. Lei and C. D. Bain, *Phys. Rev. Lett.*, 2004, **94**, 176103.

22 (*a*) E. Sloutskin, X. Z. Wu, T. B. Peterson, O. Gang, B. M. Ocko, E. B. Sirota and M. Deutsch, *Phys. Rev. E*, 2003, **68**, 31605–31606; (*b*) E. Sloutskin, E. B. Sirota, O. Gang, X. Z. Wu, B. M. Ocko and M. Deutsch, *Eur. Phys. J. E*, 2004, **13**, 109.

23 X. Z. Wu, B. M. Ocko, E. B. Sirota, S. K. Sinha, M. Deutsch, G. H. Cao and M. W. Kim, *Science*, 1993, **261**, 1018.

24 E. Sloutskin, O. Gang, H. Kraack, B. M. Ocko and M. Deutsch, *Phys. Rev. Lett.*, 2002, **89**, 65501.

25 E. Sloutskin, H. Kraack, O. Gang, B. M. Ocko, E. B. Sirota and M. Deutsch, *J. Chem. Phys.*, 2003 **118**, 10729.

26 (*a*) J. H. Hildebrand and R. L. Scott, *The Solubility of Nonelectrolytes*, Reinhold, New York, 1950; (*b*) E. A. Guggenheim, *Mixtures*, OUP, Oxford, 1952; (*c*) R. Defay, I. Prigogine, A. Bellemans and D. H. Everett, *Surface Tension and Adsorption Wiley*, NY, 1966.

27 X. Z. Wu, B. M. Ocko, H. Tang, E. B. Sirota, S. K. Sinha and M. Deutsch, *Phys. Rev. Lett.*, 1995, **75**, 1332.

28 R. R. Matheson Jr. and P. Smith, *Polymer*, 1985, **26**, 288.

29 G. A. Sefler, Q. Du, P. B. Miranda and Y. R. Shen, *Chem. Phys. Lett.*, 1995, **235**, 347.

30 M. M. Knock, G. R. Bell, E. K. Hill, H. J. Turner and C. D. Bain, *J. Phys. Chem. B*, 2003, **107**, 10801.

31 (*a*) D. Turnbull and R. L. Cormia, *J. Chem. Phys.*, 1961, **34**, 3, 820; (*b*) K. F. Kelton, *Solid State Phys.*, 1991, **45**, 75.

32 H. Kraack, E. B. Sirota and M. Deutsch, *J. Chem. Phys.*, 2000, **112**, 6873 and references therein.

33 (*a*) E. B. Sirota and A. B. Herhold, *Science*, 1999, **283**, 529; (*b*) A. B. Herhold, H. E. King and E. B. Sirota, *J Chem. Phys.*, 2002, **116**, 9036.

34 J. Frenkel, *Kinetic Theory of Liquids*, Clarendon, Oxford, 1946.

35 (*a*) A cmc = 0.31 mM at $T = 34\ ^\circ$C is reported by J. R. Lu, E. A. Simister, R. K. Thomas and J. Penfold, *J. Phys. Chem.*, 1993, **97**, 6024; (*b*) Jaeger *et al.*,[36] however, report a *lower* value of cmc = 0.28 mM at a higher temperature, $T = 65\ ^\circ$C.; (*c*) J. F. Paddy, *J. Phys. Chem.*, 1967, **71**, 3488 reports yet another value: cmc = 0.15 mM. This, however, was measured at $T = 25\ ^\circ$C, well below T_K and may not represent an equilibrium value. In view of this uncertainty we adopted a value of cmc = 0.21 mM extrapolated from CnTAB ($n = 10$–16) measurements quoted in J. Israelachvili, *Intermolecular and Surface Forces*, 2nd Edn., Academic, London, 1992.

36 D. A. Jaeger, G. Li, W. Subotkowski and K. T. Carron, *Langmuir*, 1997, **13**, 5563.

37 (*a*) K. Shinoda and P. Becher, *Principles of Solution and Solubility*, Dekker, New York, 1974; (*b*) T. Lyklema, *Fundamentals of Interface and Colloid Science*, Academic, London, 1991, vol. **1**: *Fundamentals*.

38 C. D. Bain, Q.F. Lei and K. Wilkinson, unpublished.

39 D. M. Small, *Physical Chemistry of Lipids*, Plenum, New York, 1986.

General discussion

Prof. Scholz opened the discussion of Prof. Eisenthal's paper: The temperature should affect your results. Did you study this effect?

Prof. Eisenthal answered: Certainly temperature affects the kinetics of molecular transport across a bilayer. It does so in at least two ways. One is the effect of temperature on the fluidity of the membrane, which reflects the frictional force that a molecule experiences in traversing the bilayer. Another temperature factor is the free energy barrier that separates a molecule located in the hydrophilic water phase at the membrane interface, from its location in the hydrophobic phospholipid chain region of the membrane interior. The height of the barrier determines the thermal energy that a molecule must have in order to pass over the barrier, enter the membrane and subsequently cross the bilayer. Clearly temperature is an important variable, which we plan to exploit in our studies of molecular transport across membrane bilayer structures.

Prof. Scholz added: Did you study other phospholipids, *i.e.* those with higher phase transition temperature?

Prof. Eisenthal replied: In addition to our studies of the liposome made up of DOPG lipids we have carried out studies using the phospholipid dipalmitoylphosphotidylglycerol (DPPG), which has a higher gel to liquid crystalline phase transition temperature. The aim as you have surmised was to determine the effect of the bilayer phase on the dynamics of molecular transport across a bilayer. The liquid crystalline phase is a more fluid phase, reflecting the greater disorder in the packing of the phospholipids than in the more tightly packed gel phase. With DOPG, whose gel to liquid transition temperature is -18 °C, we found that the organic cation, malachite green (MG), penetrates the membrane at the temperature of the experiment, 20 °C, and crosses with a time constant of 90 s. We selected DPPG because its transition temperature of 41 °C is above the temperature of the experiment, and therefore the bilayer would be in the gel phase. It is of interest to note that DPPG has the same headgroup, the same charge, -1, as DOPG, but differs only in that DOPG has one double bond in one of its two chains. Because DPPG does not have this one double bond it can achieve the tight packing of an *all-trans* configuration at a higher temperature than is possible with DOPG. To measure the transit time of MG across the DOPG bilayer we rapidly mixed the DPPG liposome solution with an aqueous MG solution. From the SHG measurements we found that MG rapidly adsorbs to the outer surface of the liposome, but does not cross the bilayer in a ten minute time period. Clearly the phase of the bilayer is important in the transport of molecules across the bilayer.[1]

1 A. Srivistava and K. B. Eisenthal, *Chem. Phys. Lett.*, 1998, **292**, 345.

Prof. Scholz asked: How did you ensure that you have only unilamellar liposomes?

Prof. Eisenthal responded: We demonstrated that the liposomes in our experiments were unilamellar, *i.e.* consisted of only a single bilayer, with the aid of an established ^{31}P NMR method. First we measured the magnitude and position of the ^{31}P line, noting that every phospholipid contains phosphorus.

A chemical shift reagent is then added to the liposome solution. The length scale of the shift reagent's activity is very small compared with the bilayer dimension, 4–5 nm. Thus only the ^{31}P in the outerlayer undergo a chemical shift, not the ^{31}P at the surface of the inner layer. If the liposome

DOI: 10.1039/b416303f

is unilamellar the magnitude of the ^{31}P NMR signal should be reduced by about 50%. If there is more than one bilayer then the signal would be reduced by more than 50%. We found that the ^{31}P signal was reduced to a value close to 50%. Therefore the liposomes were unilamellar.

Dr Bain asked: Presumably, you could do similar SHG measurements on a planar black lipid membrane and then simultaneously measure the ion current across the membrane. Have you done SHG measurements on black membranes? Would there be particular problems with such measurements?

Prof. Eisenthal replied: It would be of interest to carry out SHG measurements of the transport kinetics of a charged chemical species, such as MG, simultaneously with ion current measurements of the charged species crossing the bilayer. As you note this would be possible using the configuration of a black membrane bilayer separating two bulk solutions. Unlike the liposome experiments, a very large number of MG would cross the black membrane into a very large volume, compared with the liposome interior of $\sim 10^{-15}$ cm^3. The decay of the SHG signal is due to the MG that cross the bilayer, adsorb on the bilayer surface, and thereby cancel the contribution of the oppositely oriented MG on the other side of the bilayer. Therefore for the black membrane the competition of MG adsorption with MG entering the bulk solution, suggests that it would take longer to achieve the inner surface population needed to effect a decrease in the SHG signal than is the case with the liposomes. Note that the ion current measures the number of MG that cross the black membrane per unit time, whereas SHG measures the number of MG that cross the black membrane and adsorb to the inner membrane surface per unit time. The number of MG that cross per unit time is not necessarily, and probably not equal to the number that cross and adsorb to the inner surface per unit time. This leads us to surmise that the SHG kinetics would not correspond to the ion current measurements of MG crossing the black membrane bilayer.

Prof. Rusling asked: Would it be possible to use your technique to study electron transfer reactions, such as with a donor on one side and an acceptor on the other side of the membrane?

Prof. Eisenthal replied: Yes it is possible to study electron transfer at interfaces with SHG. One example is electron transfer, at the 1,2 dichloroethane/water interface, from photoexcited Ru(bpy)$_3$$^{2+}$ present only in the water phase, to a ferrocene derivative present only in the organic phase. The risetime kinetics in the SHG signal was determined by the diffusion of the photoexcited Ru(bpy)$_3$$^{2+}$ to the interface.[1] To circumvent the diffusion controlled kinetics, and thereby obtain the kinetics of the electron transfer reaction, we are studying ultrafast electron transfer at an organic liquid/water interface, where the organic liquid serves as the electron donor. The acceptor molecules in the water phase are sufficiently dilute such that electron transfer will only take place between the photoexcited acceptor molecules, which are at the interface at the time of the femtosecond excitation pulse. Because the donor organic molecules constitute one side of the interface and the acceptor molecules that can participate in an electron transfer are at the interface, translational diffusion is not required. Using SHG as the probe we observe multiexponential dynamics in the femtosecond to picosecond time scale which involve the electron transfer kinetics, and the solvation and rotation dynamics of the photoexcited acceptor molecules at the interface. The results will be published shortly.

1   K. L. Kott, D. A. Higgins, R. J. McMahon and R. M. Corn, *J. Am. Chem. Soc.*, 1993, **115**, 5342.

Prof. Lynden-Bell asked: Could you tell us how your elegant technique compares with the work of Harden McConnell using spin labels?

Prof. Eisenthal responded: Harden McConnell's introduction of spin labels to probe membranes was a major advance by providing a new and powerful method to probe the environment and chain motions at specific positions along the phospholipid chains. The SHG and SFG spectroscopies are interface selective when bounded by time averaged centrosymmetric bulk media whereas the McConnell method can be used in any environment to which a spin label can be attached. All molecules that are not centrosymmetric can be SHG and SFG active, whereas a spin label is a specially designed molecule for attachment to something else. One other difference of note is the time scale of physical and chemical events that the ESR and SHG/SFG can probe. SHG and SFG permit real time measurements that extend from femtoseconds to whatever, whereas ESR provides dynamic information in the form of a time constant, but not the time dependent curves that SHG/SFG yield. Overall the information from spin label spectroscopy and that from SHG/SFG spectroscopies are complementary, each having its own advantages.

Dr Leermakers asked: Why doesn't the SH signal go to zero in the absence of valinomycin? In principle, Na^+ or K^+ has a finite mobility across the bilayer and the signal should keep going down. If you plot the SH signal as a function of $\ln(t)$ can you see the time constant for unfacilitated Na^+ or K^+ transport?

Prof. Eisenthal answered: The reason that the SHG signal does not go to zero in the absence of the valinomycin ion transporter, even though Na^+ has a finite mobility for crossing a bilayer on its own, is the time scale of our measurement *versus* the time scale for unassisted Na^+ crossing of a bilayer. The duration of our experiments is minutes, whereas the Na^+ crossing time is on the order of ten hours. If we were to extend the duration of our experiments to hours, then the SHG signal should approach zero without the assistance of valinomycin. Recall that the reason that we added valinomycin to the liposome solution was to determine if an electrostatic barrier developed as MG crossed the bilayer, which in turn limited the MG population at the inner layer. Our results showed that an electrostatic barrier was responsible for the SHG signal not going to zero and was eliminated by the addition of valinomycin to the system.

Prof. Rathman said: This is a very interesting, useful technique. Mixtures of lipids may be especially interesting if the lipid composition on the outer layer differs from the inner-layer composition. In this case, the equilibrium distribution of probe may not be 0.

Differences in composition may arise spontaneously, especially if curvature is high. Another approach would be to change some property (pH, ionic strength) of the external solution—this may promote flip–flop of lipids and give rise to differences in composition between the inner and outer layers.

Prof. Eisenthal responded: It certainly would be of value to have bilayers that have different lipid compositions in the inner and outer layers. Living cells have such different compositions. It is not clear how to achieve this using the present method of liposome preparation. Currently the bilayers are formed by a self assembly process, which thereby results in both layers of the bilayer having the same average composition. The idea of enhancing the flip–flop rate by perturbing a particular type of phospholipid in the outer layer by the introduction of an acid, or a salt, or some other species, is an interesting one. The bilayer would of course have to be made up of two or more different phospholipids. The time scale for changing the inner and outer compositions would probably be rather long, considering the typical flip–flop rate is many hours. However, we can vary the composition of a bilayer, without differentiating the inner and outer layers, by mixing different phospholipids prior to bilayer formation. We have done this using mixtures of neutral palmitoyloleoylphosphotidylcholine (POPC) and the negatively charged palmitoyloleoylphosphotidylglycerol (POPG), which mix ideally at the temperature of the experiment, 20 °C. The aim for preparing known mixtures of the two lipids was to investigate the effect of charge on the adsorption of MG and on the kinetics of MG crossing the bilayer. Indeed, we found a substantial effect.[1]

1   Y. Liu, E. C. Y. Yan and K. B. Eisenthal, *Biophys. J.*, 2001, **80**, 1004.

Dr Walker asked: All of the systems you discussed had excess negative charge at the vesicle surface and malachite green as a cationic probe.

Have you tried using neutral probes? If so, what resulted from the study and if not, could you speculate whether the observations (magnitude and kinetics) would be qualitatively different from the MG cation case.

Prof. Eisenthal replied: Using SHG to measure the kinetics of a neutral molecule crossing a bilayer is certainly of interest. Here are a few thoughts on what might happen. For neutral molecules we expect the crossing dynamics and the number that cross to be markedly different from that of a charged molecule such as malachite green. First, the molecular charge affects the free energy of adsorption and thus the interface population. With a negatively charged liposome the interface population of the positively charged MG would be increased relative to a neutral liposome. The larger MG population would accelerate the kinetics of crossing. Clearly for a positively charged liposome the MG kinetics of crossing the bilayer would be slowed down compared with the negatively charged liposome. Second, the free energy barrier to a neutral molecule penetrating the membrane should be smaller, considering electrostatic effects only. In the bulk water outer membrane side of the interface the neutral molecule is stabilized less than a charged molecule. In addition the free energy of a neutral molecule in the low polarity membrane interior should be less than that of a charged molecule. These two effects reduce the free energy barrier, and would thereby result in faster kinetics. As a last point, the number of neutral molecules that cross the bilayer should be larger than for charged molecules. The reason for this is that neutral molecules on crossing to the inner layer do not generate a repulsive potential, as we observed took place with the charged MG.

Thus, for neutral molecules we expect the time dependent SHG signal should approach zero without the need of any transporter molecules such as valinomycin.

Dr Cicuta opened the discussion of Prof. Samec's paper: I would like to make some remarks on the surface light scattering technique used in your measurements.

(a) In the theoretical section of your paper, you mention that the surface tension is a complex quantity entering the dispersion relation of surface waves. While this has been a controversial point of discussion, I believe that it is now widely recognized that the correct parametrization of the surface is in terms of a purely real surface tension and a complex compression modulus.[1]

(b) Is the dispersion relation in eqn. (2) of your paper equivalent to the form more often used in the literature?[2,3]

(c) Are the data describing the widening of the scattered light power spectrum as a function of wavenumber (Fig. 6 of your paper) consistent with the expression of eqn. (8) in your paper, which is the more widely used in the literature on capillary waves?

These are technical issues that do not cast doubt over the use of the technique to measure the interface tension of liquid/liquid interfaces. I also do not doubt the conclusion that for liquid/liquid interfaces this technique is not sensitive to the dilational elasticity or viscosity. The latter point has often been made in the literature, in particular by Earnshaw *et al.*[4] In contrast, it is possible to use this technique to measure the complex dilational for monolayers at liquid/air interfaces.[2,3]

1   D. M. A. Buzza, *Langmuir*, 2002, **18**, 8418.
2   D. Langevin, *Light Scattering by Liquid Surfaces and Complementary Techniques*, Dekker, New York, 1992.
3   P. Cicuta and I. Hopkinson, *Colloids Surf., A*, 2004, **233**, 97.
4   J. C. Earnshaw, R. C. McGivern, A. C. McLaughlin and P. J. Winch, *Langmuir*, 1990, **6**, 649.

Prof. Samec responded: Concerning your first point, I would agree. Actually, we used the theoretical expression derived by Laudon[1] to interpret the measured quasi-elastic light scattering spectrum. He argued that an imaginary component in the dilatational modulus ε leads to an imaginary part in the interfacial tension through the equation

$$\gamma = \gamma_{\text{free}} - \int_{0}^{\gamma_0} (\varepsilon/\gamma)\,\mathrm{d}\gamma \tag{1}$$

which follows by integrating the equation defining ε,[1] *cf.* eqn. (1) of our paper. However, we have not been able to trace neither the effect of the dilatational modulus, nor the effect of the complex part of the interfacial tension in the measured spectra.

The dispersion relation obtained by setting the denominator in eqn. (2) equal to zero reproduces an earlier theoretical result.[1,2] Besides, eqn. (2) correctly reduces to the expressions for the case of a single liquid covered by a film, which upon setting $\varepsilon = 0$ becomes equivalent to the expression used in the literature for a liquid–air interface.[1] On the other hand, eqns. (6) and (7) for the linewidth $\Delta\omega$ in the absence and presence of the very stiff film, respectively, do not reduce to the those derived for the single-liquid case. Neither of these expressions reduces to eqn. (8) of our paper that was often used in the literature, which thus seems to represent a very specific low-viscosity limit. Therefore, our conclusion has been that eqn. (2) should always be used as a basis for interpretation of the QELS spectra.

Regarding the third point in your comment, changes in the linewidth with the wavenumber (Fig. 6) and the potential (Fig. 11) are not consistent with eqn. (8), but they can be described by eqn. (6) or eqn. (7), though a direct calculation from the theoretical equation, eqn. (2), leads to a better agreement between theory and experiment (Table 2).

1   *Surface Excitations*, ed. V. M. Agranovich and R. Laudon, North Holland, Amsterdam, 1984, pp. 591–638.
2   E. H. Lucassen-Reynders and J. Lucassen, *Adv. Colloid Interface Sci.*, 1969, **2**, 347.

Prof. Schlossman asked: Studies on phospholipid adsorption at the alkane/water interface, as opposed to the DCE/water interface you studied, have observed the formation of very thick multilayers (including budding and other structures that are visible by optical microscopy) at the interface. Have you observed effects of this nature at the DCE/water interface?

Prof. Samec replied: Except for the sudden decrease in the surface excess concentration of DPPC when its bulk concentration rises above *ca.* 2.5 μmol L^{-1} (Fig. 10 of our paper), we do not have any other evidence pointing to the formation of a multi-layer. The observed collapse of the adsorbed layer is likely to represent the very first stage of the multi-layer formation or emulsification.

Prof. Schlossman added: Is the interface stable under conditions of higher concentration at which multilayers may form?

Prof. Samec answered: The water–1,2-dichloroethane interface is rather stable even under these conditions, no chaotic oscillations and other anomalies in voltammetric or interfacial tension behaviour are observed. Such phenomena have been known to accompany the spontaneous emulsification of the polarised liquid–liquid interface.[1]

1   T. Kakiuchi, N. Nishi, T. Kasahara and M. Chiba, *ChemPhysChem*, 2003, **4**, 179.

Prof. Scholz asked: What studies do you envisage to understand the collapse of the monolayer? Is the collapse associated with liposome formation?

Prof. Samec replied: Neutron reflectivity as reported here by Webster *et al.* could perhaps be a convenient experimental approach to study the structural aspects of the collapse of the monolayer, though it is not clear whether the method can be adapted for measurements at polarised liquid–liquid interfaces. Unfortunately, we have been unable to identify the emerging structures.

Prof. Kornyshev asked: Have you measured the amplitude of fluctuations and how they depended on electric field?

Prof. Samec responded: Unlike X-ray studies by Schlossman *et al.*, the analysis of the power spectrum of the capillary wave as measured by the quasi-elastic light scattering technique does not allow one to evaluate the amplitude of thermally excited fluctuations. What we can infer from measurements is the capillary wave frequency and damping factor for a selected wavenumber (wavelength).

Prof. Kornyshev added: There is also another method to evaluate the interfacial area and the potential effect on it. These are Direct Energy Transfer (DET) experiments.

The theory of this method has been published[1] but no experiments of the field effect have yet been studied.

In DET experiments one observes the time-dependent survival probability of the excited donor in the presence of acceptors; both localised at the interface.

$$\Phi(t) = \exp\{-t/\tau - \int dr \rho(r)[1 - \exp(-tW(r))]\}$$

Where W the rate of energy transfer $W(r) = \frac{1}{\tau}\left(\frac{a}{r}\right)^6$

At ITIES we have $\Phi(t) = \Phi_0(t)\exp\left[-\alpha\theta\Delta S/S_0\left(\frac{t}{\tau}\right)^{1/3}\right]$

Where $\Phi_0(t)$ is the DET signal at pzc and $\Delta S/S_0$ represents a relative increase of the interfacial area with a potential.

Thus studies of potential dependence of DET can give direct information on the effect of potential on the area of ITIES.

1 A. A. Kornyshev and M. I. Urbakh, *Electrochem. Commun.*, 2004, **6**, 703.

Prof. Eisenthal commented: I found it quite interesting to learn how interfacial energy transfer experiments could be used to measure interfacial areas, and thereby provide a method to study the effects of an interfacial electric field on the area of an interface. Prof. Kornyshev gives an expression that relates the energy transfer to the interfacial area and a time dependence that varies as $\exp(-at^{1/3})$. Theory predicts this time dependence for energy transfer in two dimensions, when the donor–acceptor transfer probability per unit time varies as the inverse 6th power of their separation. I would like to know if Prof. Kornyshev has observed this $t^{1/3}$ time dependence experimentally. I note that we have used SHG to measure interfacial electronic energy transfer from rhodamine 6G to a cyanine dye at the air/water interface. The predicted $t^{1/3}$ time dependence was observed.[1]

1 E. V. Sitzmann and K. B. Eisenthal, *J. Chem. Phys.*, 1989, **90**, 2831.

Prof. Samec said: A problem associated with the capacitance measurements at the interface between two immiscible electrolyte solutions (ITIES) is that ac polarization should lead to periodic changes in the interfacial tension, which is related to the width of the interface, *i.e.* to the mean square amplitude of the surface fluctuations (capillary waves). In this way, the inner-layer thickness is periodically perturbed, inducing a periodic change in the inner-layer potential difference. As a result, the Randles equivalent circuit is not quite an adequate description of ITIES, but an internal source of ac voltage in parallel to capacitance should be added.[1,2] Simulations based on the revised equivalent circuit suggest that the observed sharp enhancement of the Randles capacitance that is observed close to the potential window could be an artefact.[2]

1 Z. Samec, J. Langmaier and A. Trojánek, *J. Electroanal. Chem.*, 1998, **444**, 1.
2 Z. Samec, J. Langmaier and A. Trojánek, *J. Electroanal. Chem.*, 1999, **463**, 232.

Prof. Kornyshev replied: The large potentials will first effect the amplitude of corrugations and through it the "Gouy–Chapman" capacitance of the back-to-back double layers. I expect this effect to be stronger and show up first, before the "inner-layer" capacitance was influenced by the electric field.

Prof. Urbakh asked: It has been noted in the talk by Schlossman on X-ray studies that liquid/liquid interface can have some elasticity. Can you estimate the interfacial elasticity from a spectrum of capillary waves that you measure in quasi-elastic light scattering experiments?

Prof. Samec responded: The order of magnitude of the dilatational modulus ε can be estimated from the changes in the interfacial tension and the relative surface excess, *cf.* Figs. 9 and 10 of our paper. However, the simulation of the fluctuation spectra using this estimate indicated that its effect is too weak to be safely evaluated from the measured power spectrum, *e.g.* by its fitting to eqn. (2).

Prof. Kornyshev commented: I think I have to step back a bit from the propaganda of this method, since in order to localise the fluorescence probes we will need to anchor them at the interface, and their presence as "surfactants" may affect surface tension. At low concentrations this will not be a problem, but if we wanted to test surface coverage dependence of this signal we may face this complication in the treatment of the data. But in principle the corresponding correction may be taken into account. Hopefully, we might be even able to retrieve the surface tension dependence on surface coverage, if all other parameters in the equations that I have mentioned are considered to be known.

Prof. Scholz opened the discussion of Dr Leermaker's paper: Why do you make the assumption of symmetric interactions A–C and B–C?

Dr Leermakers replied: We deliberately focussed on the system in which the minority component C would not have a finite flux in the system when the two bulk phases A and B were symmetrically put in an off-equilibrium state, *i.e.* away for coexistence. In this case the flux of A is exactly opposite of that of B. The steady state accumulation of C at the interface between A and B was one of the main targets for this highly idealized example. In order to achieve this, we not only had to take symmetric interactions between A–C and B–C, but we also had to assume that all the molecules in the system have the same mobility and the same degree of polymerisation. Even in this simple system we were not able to solve the equations analytically and had to do this numerically. Results from this idealized system can now relatively easily be explained because by having a choice of parameters we removed various complications that would otherwise have obstructed this.

Prof. Kornyshev opened the discussion of Prof. Wennerström's paper: How would you correspond your work to the previous work of, say, Kozlov and Chernomordik?

Prof. Wennerström replied: One of the aims of our study was to arrive at quantitative tests of theoretical models ascribed to Kozlov and Markin,[1] Chernomordik and coworkers[2] and Siegel and coworkers.[3] Previously we have used these concepts to derive an expression for the barrier of emulsion droplet coalescence in systems with only short range forces.[4] This model describes emulsion stability for emulsion systems with an unprecedented accuracy.[5] This experimental study illustrates the virtue of working with a well characterized system so that the parameters entering the expression for the barrier are previously determined avoiding adjustable parameters in the description.

1   M. M. Kozlov and V. S. Markin, *Biofizika*, 1983, **28**, 242.
2   L. V. Chernomordik, M. M. Kozlov and J. Zimmerberg, *J. Membr. Biol.*, 1995, **146**, 1.
3   D. P. Siegel, *Biophys. J.*, 1993, **76**, 291.
4   A. Kabalnov and H. Wennerström, *Langmuir*, 1996, **12**, 276.
5   A. Kabalnov and J. Weers, *Langmuir*, 1996, **12**, 1931.

Prof. Scholz asked: What is the nucleation site for the sponge-to-lamellae transition? Is it the wall?

Prof. Wennerström answered: The data are consistent with the model based on random homogeneous nucleation. However, as expressed in the paper, we cannot rule out that the nucleation in fact occurs at the walls of the sample tube.

Dr Leermakers asked: Why do you discuss the formation of the L_3 phase by focusing on H_0 and not on $\bar{k}$ (Gaussian bending modulus)? It is possible to show that for $C_{12}E_5$ $\bar{k}$ is negative, which stabilizes bilayers (L_α-phase). With increasing temperature $\bar{k}$ becomes less negative. In addition the uptake of decane makes $\bar{k}$ less negative. Both trends are well known and consistent with a transition from the L_α-phase to the L_3 phase. H_0 can not be measured directly and is in my opinion not straightforwardly defined if the uptake of alkane is finite (*i.e.* if the decane is sub-saturated, *i.e.*when there is no alkane–water interface onto which a monolayer of $C_{12}E_5$ is adsorbed).

Prof. Wennerström replied: This is an issue that has been extensively discussed in the literature. Our conclusion that it is much more fruitful to focus on the monolayer curvature properties is based on the following circumstances:

(i) Our ambition is to describe all phases with monolayer and bilayer structures occurring in the global phase diagram, using the same basic concepts.[1]

(ii) A consequence is that we by temperature extrapolation have a prediction for the monolayer H_0 in both the lamellar and sponge phases.[1]

(iii) The theoretical description based on the Helfrich second order expansion of the bilayer curvature fails in predicting the phase equilibria of the sponge phase. It is then necessary to include fourth order terms.[2]

Whether one uses the monolayer or the bilayer as a basis for the description of curvature energies it is necessary to include the Gaussian curvature term. The added alkane influences the spontaneous curvature as well as both κ and $\bar{\kappa}$ in the monolayer description and κ and $\bar{\kappa}$ in case of the bilayer description. Manifestations of the alkane effect are discussed for example by Anderson *et al.*[3]

We do not agree on the measurability of the monolayer H_0. It is obviously measurable at lower temperatures of the present system, and the "measurements" at higher temperatures simply requires more sophisticated theoretical arguments.[1]

1   T. D. Le, U. Olsson, H. Wennerström and P. Schurtenberger, *Phys. Rev. E.*, 1999, **60**, 4300.
2   H. Wennerström and U. Olsson, *Langmuir*, 1993, **9**, 365.
3   D. Anderson, H. Wennerström and U. Olsson, *J. Phys. Chem.*, 1989, **93**, 4243.

Dr Leermakers asked: Do you assume $\bar{k}$ to be not too negative when $L_\alpha \rightarrow L_3$ takes place? Or can $\bar{k}$ be very negative in the L_3 phase?

Prof. Wennerström replied: It is important to be clear whether we evaluate the curvature elastic properties at the bilayer midplane or at the polar/apolar interface of the monolayers. We base our theoretical description of the lamellar–sponge equilibrium basically on a free energy density which expanded in the bilayer volume fraction can be written as $G/V = a_3\phi^3 + a_5\phi^5$.[1,2] For the sponge phase the a_3 term contains contributions from the monolayer H_0, κ and $\bar{\kappa}$. These can, within a parallel surface approximation, be combined to a bilayer $\bar{\kappa}$ according to $\bar{\kappa}_{\text{bilayer}} = 2\bar{\kappa} - 2\kappa H_0\delta$,[3] where δ is the bilayer thickness. Additionally there is a contribution from the entropy of the disordered network. For large negative H_0, κ_{bilayer} can be strongly positive but we expect a breakdown of the description when $|H_0 l| > 1$.

1   H. Wennerström and U. Olsson *Langmuir*, 1993, **9**, 365.
2   T. D. Le, U. Olsson, H. Wennerström and P. Schurtenberger, *Phys. Rev. E.*, 1999, **60**, 4300.
3   D. Anderson, H. Wennerström and U. Olsson, *J. Phys. Chem.*, 1989, **93**, 4243.

Prof. Scholz asked: How is the fusion/fission kinetics affected by the ionic strength in the aqueous phase?

Prof. Wennerström replied: For the nonionic systems low electrolyte concentrations (≈ 1 mM) have a negligible effect on the aggregation properties and this can be utilized to study connectivity through conductance measurements. At higher electrolyte levels approaching molar concentrations the weak but selective interaction between ions and the polyoxyethylene groups has an effect on the spontaneous curvature H_0. This can be quantitatively determined and it provides an accurate characterization of the Hofmeister series.[1]

1   A. Kabalnov, U. Olsson and H. Wennerström, *J. Phys. Chem.*, 1995, **99**, 6220.

Dr Bain asked: Is there any difference at a molecular level in the structures/conformations of the surfactant and alkane in the L_3 and L_α phases?

Prof. Wennerström responded: We have previously made extensive studies of the molecular characteristics in the respective phases of the system studied in this report. Methods used are NMR diffusometry,[1] NMR quadrupolar relaxation,[2] NMR quadrupolar splittings,[3] small angle neutron scattering[4] and light scattering.[1] One result of these studies is the observation that the small changes that certainly occur locally with respect to properties as area per molecule, local diffusivity, local mean orientation and local reorientation are all within experimental accuracy unaffected by phase changes. In the literature one can find scattered but numerous reports of the contrary but this typically boils down to a failure of separating local effects from observations due to changes in the global structure.

1   U. Olsson and P. Schurtenberger, *Langmuir*, 1993, **9**, 3389.
2   M. S. Leaver, U. Olsson, H. Wennerström and R. Strey, *J. Phys. II* , 1994, **4**, 515.
3   M. S. Leaver, U. Olsson, H. Wennerström, R. Strey and U. Würz, *J. Chem. Soc., Faraday Trans.*, 1995, **91**, 4269.
4   H. B. Jörgensen, U. Olsson and K. Mortensen, *Langmuir*, 1997, **13**, 1413.

Dr Dagastine opened the discussion of Prof. Deutsch's paper: The interfacial tensions scale on Fig. 5 of your paper seems unrealistic. The interfacial tension of the alkane–water interface is on the order of 50 mN m^{-1}. The scale on Fig. 5 is comparable to the air–alkane interface, not the water–alkane interface. This appears to be a typographical error or the inclusion of the wrong plot.

Prof. Deutsch responded: This is indeed a typographical error, which we will correct in the final version of the paper. We measure surface tensions of a few mN m^{-1} below 50 mN m^{-1}, using the Wilhelmy plate method. Earlier measurements, using the pendant drop method,[1] obtain a few mN m^{-1} above 50 mN m^{-1}. Both measurements used the same material cleaning methods, and the source of the different results is not clear.

1(*a*) D. M. Small, *Physical Chemistry of Lipids*, Plenum, NY, 1986; (*b*) R. Aveyard, B. J. Briscoe and J. Chapman, *J. Chem. Soc., Faraday Trans.*, 1972, **68**, 10.

Dr Dagastine asked: Have the authors considered using pendant drop to look at the interfacial freezing instead of Wilhelmy plate. This would then allow the authors to probe the interfacial rheological response of the surface frozen layer through dilation and compression measurements. Since this layer is a frozen film, one would expect the layers to exhibit elasticity, but a visco-elastic response may be present as well. Furthermore, is it possible to induce surface freezing through compression or dilation using a pendant drop method?

Prof. Deutsch replied: To the best of our knowledge the rheology of the surface-frozen films has not been investigated to date. The suggestion of Dr Dagastine is a good approach to the problem, though not the only one. Light scattering can also be used, and the method presented in this meeting by Dr Mitani seems also to be a very interesting way to probe the mechanical properties of the surface-frozen monolayers. In fact, the different methods may probe the rheological properties on different length scales, and thus a comparison of their respective results will be very interesting. Rheological studies of the surface- and interface-frozen monolayers are high on our list of priorities. For now, we just comment that the appearance of a frozen layer at the surface or interface should, in principle, result in a higher bending rigidity for the surface, which will tend to damp the capillary waves. This damping will yield a lower apparent surface roughness in the fits of X-ray reflectivity data. As shown in ref. 1 we do not observe this effect when the surface–frozen layer has a hexagonal rotator structure, *i.e.* for C$_{40}$ and shorter alkanes. However, for C$_{44}$ and longer alkanes, where the surface-frozen layer is a herringbone-packed crystal, a reduction of about 0.5–0.6 Å is observed in the effective roughness, as derived from the X-ray reflectivity fits, when the surface freezes. The values of the effective surface tensions, and the reductions upon appearance of the herringbone structure, are rather similar to those in Table 1 in Prof. Schlossman's paper. These reductions result most probably from the same effect suggested by Prof. Schlossman to explain his results: the appearance of a non-zero bending rigidity (although in our case it is not due to dipole interactions). At the time we published ref. 1, we had not attempted to extract the bending rigidity from the effective roughness values because of the uncertainties in the best-fit parameters (due to small, but non-negligible, cross-correlations between the fit parameters). However, in conjunction with the planned rheological measurements mentioned above, we will also take a more careful look at the bending rigidity issue.

1 B. M. Ocko, X. Z. Wu, E. B. Sirota, S. K. Sinha, O. Gang and M. Deutsch, *Phys. Rev. E*, 1997, **55**, 3164.

Dr Dagastine asked: Have the authors considered using pendant drop methods to probe the time scale of surface freezing? It is conceivable to equilibrate the system at a temperature where surface freezing is observed and form a droplet using pendant drop. Most commercial pendant drop instruments can capture drop profiles at 50 frames a second, easily capturing interfacial tension dynamics that change on the order of seconds. Could this be a method to better understand the nucleation kinetics of surface freezing?

Prof. Deutsch replied: This is, again, a very interesting suggestion. No, we have not attempted following the nucleation of the surface-frozen layer by this method. We did try to follow the appearance of the surface-frozen layer by time-resolved, fixed-incidence-angle X-ray reflectivity measurements, which had a time resolution of about 100 ms. On this time scale the surface-freezing appears abruptly. I certainly would like to see higher-time-resolution measurements. However, I suspect that temperature uniformity and equilibration issues may become increasingly more important with increasing time resolution.

Prof. Lynden-Bell asked: Can you speculate on the mechanism by which the surfactant causes surface freezing?

Prof. Deutsch answered: As in the famous story of the blind men and the elephant,[1] speculations from incomplete data may be misleading. With this cautionary note in mind, we observe that the amphiphilic nature of a surfactant molecule at the water/alkane interface requires its hydrophilic headgroup to reside in the water while its hydrophobic tail prefers to reside in the alkane. To maximize the average distance of the methylenes from the water, at least the part of the surfactant's alkyl tail closest to the headgroup (and thus closest to the interface) will be aligned normal to the surface. Since the persistence length of an alkyl chain is about 4–5 methylenes, one would expect the surface-normal part to be about that length at least. These interface-normal "rigid" rods should impart, through an attractive van der Waals interaction, a preferential interface-normal alignment to the alkane chains close to the interface. This partial orientation allows the near-interface layer to freeze slightly above the temperature at which the chains freeze anyway in the bulk. Some support for this speculation is provided by a recent study by Prof. Schlossman and coworkers on the conformation of long-chain alcohols adsorbed at the water/hexane interface.[2] For a dense adsorbed layer their X-ray reflectivity measurements are consistent with interface-normal aligned alcohol molecules, the methylene groups of which deviate increasingly from an extended, *all-trans* conformation with distance from the interface.

Interestingly, our measurements show that a coverage of just 10–20% of the water/alkane interface area by the surfactant "support poles" are enough to induce interfacial freezing in the alkane. An interesting question remaining is whether these "poles" are ordered, vineyard-like, or disordered, in-plane. Unfortunately, grazing-incidence X-ray diffraction methods, which are able in principle to answer this question, are most probably not practical at the liquid/liquid interface due to the very high diffuse background scattering from the bulk liquid alkane.

1   http://www.kheper.net/topics/blind_men_and_elephant/.
2   A. M. Tikhonov, S. V. Pingali and M. L. Schlossman, *J. Chem. Phys.*, 2004, **120**, 11822.

Prof. Penfold said: The point about the surface freezing being templated by the cationic surfactant is well made, as we know from our neutron reflection studies, a $C_{16}TAB$ monolayers at the air–water interface where the mean conformation was inferred.[1] The alkyl chain region closest to the head group is orientated almost perpendicular to the plane with progressive disorder further up the chain. That initial aligned region could indeed template the ordering associated with the surface freezing. In contrast, our results[2] show that the non-ionic surfactants C_nEO_m show a different orientational order at the interface with the initial part of the alkyl chain ordered in the vertical plane. Hence I would expect that the non-ionic surfactants would not induce surface freezing at the hexadecane and water interface.

1   J. R. Lu, Z. X. Li, J. Smallwood, R. K. Thomas and J. Penfold, *J. Phys. Chem.*, 1995, **99**, 8233.
2   J. R. Lu, R. K. Thomas and J. Penfold, *Adv. Colloid Interface Sci.*, 2000, **84**, 143.

Prof. Deutsch responded: Prof. Penfold makes a very good point. A few years ago we studied the nucleation of bulk and of emulsions of alkanes (C_n, $n = 15$–60) using calorimetry.[1] We found that the bulk could not be supercooled, which was interpreted as being due to the nucleation of the bulk by the surface-frozen layer. However, the emulsion could be supercooled considerably (by 13–27 °C, depending on the chain length). This was interpreted as being due to the suppression of interfacial freezing by the combination of a water matrix and the surfactant used in the emulsification, the commercial non-ionic surfactant Igepal. We have also recently started similar measurements using CTAB (and its different-length homologues) as a surfactant. On the few samples measured so far, the supercooling in the emulsion is significantly reduced as compared to that observed with Igepal. This is, again, assigned tentatively to interfacial freezing at the droplet's alkane/water interface, induced by the surfactant. Moreover, preliminary microscopy studies (of the same type that was presented in my talk) of emulsions made with Igepal rather than STAB do not show the same droplet elongation effects observed with STAB. We have also tried the anionic surfactant SDS, which was also found not to show the effects observed with STAB, although this may be due to the fact that the chain length of SDS is just $n = 12$, while the alkane used had a longer chain, $n = 16$. As shown in Fig. 7 in our paper, and in ref. 2, when the surfactant's chain is shorter than that of the alkane, the interfacial freezing effect vanishes. The subject of the influence of the type of the surfactant, and its length, on the interfacial freezing is now under study in our laboratory.

1(*a*) H. Kraack, E. B. Sirota and M. Deutsch, *J. Chem. Phys.*, 2000, **112**, 6873; (*b*) H. Kraack, E. B. Sirota and M. Deutsch, *Macromolecules*, 2000, **33**, 6174; (*c*) H. Kraack, E. B. Sirota and M. Deutsch, *Polymer*, 2001, **42**, 8225.
2   Q. Lei and C. D. Bain, *Phys. Rev. Lett.*, 2004, **92**, 176103.

Prof. Deutsch commented: As stated in the paper, a situation where the surface tension between the alkane and the water phases vanishes due to surface freezing at $T < T_{mix}$ should lead to a number of interesting effects. One example is the appearance of spontaneous emulsification. Another effect predicted in the paper is that of the deformation of alkane-in-water emulsion droplets from spherical to elongated shapes in this temperature range. This effect has now been observed and is presented below for an $H_2O/C_{16}/STAB$ system. The width of the photographs in Figs. A and B of our paper is 250 μm. The shape of all droplets observable in Fig. A is spherical, as this minimizes surface energy. In Fig. B, however, the surface tension is zero, and elongated shapes can, and do, form, as this entails no increase in surface energy. Many droplets (excluding the larger ones) are elongated rods, although triangular and non-regular shapes can also be observed in the lower part of the figure.

The time evolution of the shape of the droplets was shown in a short video clip, screened at the meeting. In that clip the evolution of a droplet from a spherical to a faceted to a triangular shape was shown. A tail of "beads on a string" was then observed to form at one of the vertexes of the triangle. The tail, drawing material from the triangular "droplet", grew into a very long coiled tail, which eventually thinned out to below the resolution of the microscope. A snapshot of the triangular "droplet" and its tail at an early stage of this process is shown in Fig. C, the width of which is about 50 μm. The beaded tail (though not the string which connects the beads) can be observed emanating from the upper vertex.

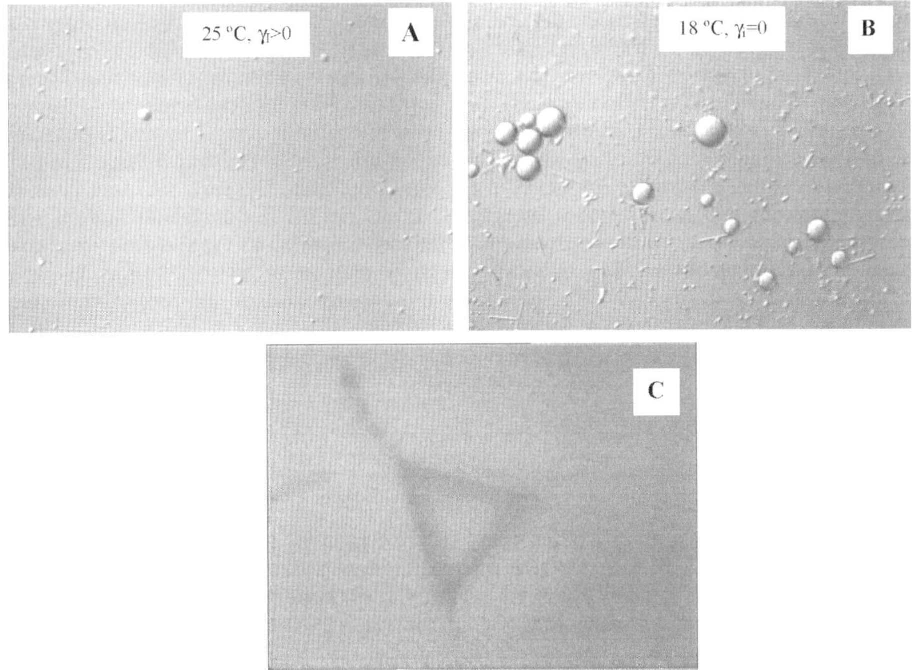

Prof. Scholz asked: Is the frozen surface layer the nucleation site for bulk freezing?

Prof. Deutsch replied: While this seems very likely, no direct X-ray evidence for that is available. In surface-specific X-ray measurements, the abrupt bulk freezing breaks up the surface, so that the reflected X-ray signal vanishes, and no preferred orientation of the bulk-frozen sample can be observed. However, there is indirect evidence that the surface frozen layer serves as a template which nucleates bulk freezing. For example, a recent experimental study of the crystallization of levitated microdroplets of C_{15} and C_{17} concludes from the long induction times for bulk crystallization, that it is initiated by the surface-frozen layer.[1] The relatively large undercoolings observed in calorimetry of alkane-in-water emulsion droplets (where surface freezing does not occur), and the virtually zero undercoolings observed in calorimetry of non-emulsified bulk alkane melts (where surface freezing does occur),[2] also supports the assumption that bulk nucleation of alkanes from the melt is templated on the surface-frozen layer, where it occurs.[3] Finally, it is interesting to note that measurements made by Sirota and Herhold[4] indicate that the critical nucleus does not always have the same structure as that of the macroscopic bulk that is formed from that nucleus. The same holds for the structure of the surface-frozen layer *vs.* that of the bulk. In these cases the bulk nucleation barrier may not vanish upon the appearance of the surface frozen layer, but only diminish by a small or a large factor, depending on the mismatch in structure.

1   I. M. Weidinger, J. Klein, P. Stöckel, E. Biller, H. Baumgärtel and T. Leisner, *Z. Phys. Chem.*, 2003, **217**, 1597.
2   H. Kraack, E. B. Sirota and M. Deutsch, *J. Chem. Phys.*, 2000, **112**, 6873.
3   E. Sloutskin, E. B. Sirota, H. Kraack, B. M. Ocka and M. Deutsch, *Phys. Rev. E*, 2001, **64**, 31708.
4   E. B. Sirota and A. B. Herhold, *Science*, 1999, **283**, 529.

Prof. Lynden-Bell asked: Do you have any information about the dynamics in the frozen surface layer? Our simulations (ref. 12*a* of your paper) suggested significant lateral diffusion in spite of the hexagonal order.

Prof. Deutsch responded: We do not have any concrete information on the lateral movement of molecules within the frozen surface layer. I find it difficult to believe in a significant diffusion within the layer, considering the fact that the layer shows experimental characteristics of an ordered solid layer, *i.e.* long-range in-plane order, and is densely packed. Nevertheless, if we take the existence of a significant lateral diffusion as a given, it is more likely to occur in alkanes comprising 40 carbons or less, where the surface-frozen monolayer is in the rotator phase. For molecules of 44 carbons or more, the surface frozen layer is in a herringbone-ordered crystalline phase, which is less likely to allow diffusion. Finally, regardless of the existence, or otherwise, of *lateral* diffusion within the layer, *vertical* exchange of molecules between the liquid bulk and the surface-frozen layer (both comprising the same molecules) should be easy, as it does not involve molecular motion *within* a closely-packed solid phase. We have observed a demixing transition in the surface-frozen layer of a binary mixture of alcohols.[1] This transition certainly requires molecular motion. However, a molecular exchange with the bulk is more likely in this case than lateral diffusion of molecules.

1   E. Sloutskin, O. Gang, H. Kraack, B. M. Ocka, E. B. Sirota and M. Deutsch, *Phys. Rev. Lett.*, 2002, **89**, 065501.

Dr Oldfield asked: Have you or anyone else determined the ternary phase diagram for this system at the relevant temperature? If so, what are its principal features? Are there any microemulsion phases?

Prof. Deutsch replied: I am not aware of any such experimental determination for the present system, which employs STAB, and we certainly have not measured it. I was unable to find a ternary phase diagram for similar systems comprising even CTAB, which is a much more widely used surfactant. Note that in our case, a single phase diagram is not enough: we would need to determine the phase diagram for a range of chain lengths of both the surfactant (DTAB, CTAB, STAB) and the alkane (C_8–C_{20}). If Dr Oldfield knows of any such measurements I would be very grateful for the reference.

Prof. Kornyshev asked: Have you measured the thickness of the frozen layer as a function of proximity of T to the critical temperature of freezing temperature on the bulk, T^*? Is there some logarithmic growth of this thickness with $(T - T^*) \to 0$, like in the "wetting" theories?

Prof. Deutsch answered: Yes, we have measured the dependence of the layer thickness on T. In alkanes, the surface-frozen layer is only a single molecule thick from its formation temperature, T_s, down to the bulk freezing temperature T_b.[1] In terms of a "wetting" transition, the formation of the surface frozen layer constitutes an extreme case of partial wetting, *i.e.* a wetting transition where the wetting layer's thickness grows to a *finite* thickness at T_b. I say "extreme" since in our case the growth is zero, and the final, finite thickness at T_b is the same as the initial one, at T_s.

In surface freezing of alcohols a bilayer, rather than a monolayer, is formed upon surface freezing. However, even in that case the single bilayer persists down to T_b and no additional layers are formed upon approaching T_b.[2] Only a single case was found, in a mixture of two different-length alcohols within a narrow concentration range, where a sequence: liquid surface $\rightarrow$ single surface-frozen bilayer $\rightarrow$ double surface-frozen bilayer $\rightarrow$ bulk freezing could be observed upon cooling from above T_s to below T_b.[3] Thus, in principle, we cannot exclude the possibility that other partial wetting scenarios, or even a complete wetting one, may be hiding out somewhere in phase space. How can we find an actual system which can be tuned to that region in phase space so that we can observe these wetting behaviours experimentally, is, of course, a different matter altogether.

1 B. M. Ocko, X. Z. Wu, E. B. Sirota, S. K. Sinha, O. Gang and M. Deutsch, *Phys. Rev. E*, 1997, **55**, 3164.
2 O. Gang, X. Z. Wu, B. M. Ocka, E. B. Sirota and M. Deutsch, *Phys. Rev. E*, 1998, **58**, 6086.
3 E. Sloutskin, H. Kraack, O. Gang, B. M. Ocka, E. B. Sirota and M. Deutsch, *J. Chem. Phys.*, 2003, **118**, 10729.

Prof. Scholz asked: Is it possible to freeze the bulk using the frozen surface layer as template when freezing is performed in a temperature gradient?

Prof. Deutsch responded: This is an interesting idea, which we toyed with a number of years ago. The idea was that with a tunable gradient we may be able to observe perhaps a layer by layer freezing from the surface into the bulk. At that time we wanted to have a large enough sample, say a few cm long by a few mm wide, to allow carrying out X-ray measurements. We were unsuccessful in creating a temperature field which will have simultaneously a millidegree-level lateral uniformity, a uniform vertical gradient tunable from a few tens of millidegrees mm^{-1} to a few degrees mm^{-1}, and a stability of a few hours, required for measuring an X-ray reflectivity curve. Since we had so many other interesting issues related to this effect to study, we have abandoned this particular line of research. I agree, however, that it is an interesting approach and well worth exploring.

Dr Leermakers said: Surface freezing induced by surfactants at the water alkane interface is interesting from the following point of view.

If you consider a slab of alkanes sitting in between two water–alkane interfaces onto which surfactants are adsorbed you expect attraction between the interfaces if the oil is simply polymer-like (depletion interaction). However, if the alkane is orientated by the surfactants you will see repulsion. Again the CH_3–CH_2 disparity may help to orientate the alkane as it does in the classical system of alkane uptake.

Prof. Deutsch responded: This is an interesting point. The attraction is, presumably, related to the Casimir effect. What is the source of the repulsion? Is it the electric repulsion between the like-charged headgroups at the two interfaces? Or is the repulsion a result of the molecules changing from a flexible random-walk shape to a rigid rod aligned normal to the surfaces, thus pushing away each other? At any rate, it seems to me that the difference in the interaction between the interfaces will be significant only for interfaces not too far from each other, perhaps a few hundred Ångströms apart for the Casimir effect and even less for the repulsion, which is likely to be significant only for interfaces separated by a few molecular lengths. The realization of a system where these effects could be observed would require some thought.

Dr Clarke asked: Are the droplets birefringent (on the video)?

Prof. Deutsch replied: We have not checked the birefringence of these droplets. However, I would not expect any *observable* birefringence effects since at the temperature these photographs were taken (25 °C and 18 °C) the alkane in the droplets is liquid.

I would like to point out in that respect of the work of Prof. Baumgärtel and coworkers at the Freie Universität, Berlin.[1] They have investigated surface freezing in electrically levitated droplets of alkanes, detecting the freezing by intensity variation of polarized light in transmission.

1   I. M. Weidinger, J. Klein, P. Stöckel, E. Biller, H. Baumgärtel and T. Leisner, *Z. Phys. Chem.*, 2003, **217**, 1597 and references within.

Dr Clegg asked: Are they alkane or water droplets?

Prof. Deutsch answered: The droplets I have shown in the stills and in the video clip are all C_{16} alkane droplets, stabilized by STAB, in a water matrix.

Prof. Scholz asked: Are the small droplets in the tails shown in the video interconnected?

Prof. Deutsch replied: Yes. They are connected by a thin "string" which may be a continuous tube or a string of smaller droplets, too small to resolve by the microscope. As time progresses, the "string" thins out (as do the droplets on the string), until they diminish below the resolution limit of the microscope.

Dr Bain said: The spontaneous creation of fresh interface in Prof. Deutsch's fascinating video requires that the surface tension vanishes. The surface tension only tends to zero in the surface frozen phase, which requires the presence of surfactant. The dynamics that are observed may therefore be associated with mass transport of STAB to the growing oil–water interface, as well as with the fluid dynamical instabilities of a cylindrical interface.

Prof. Deutsch replied: This seems reasonable, although there may be other, equally plausible, mechanisms. See for example the review paper of deGennes and Taupin on spontaneous emulsification.[1] I prefer to reserve judgement until we had more time to think about, and further study, these shape changes.

1   P. G. deGennes and C. Taupin, *J. Phys. Chem.*, 1982, **86**, 2294.

Dr Oldfield asked: The natural consequence of surface tension lowering to nearly zero for an oil–water system is the formation of a microemulsion, that is the stable absorption of one phase by the other, in this case of oil by water forming and oil-in-water microemulsion. So it seems that you are monitoring the early stages of microemulsion formation in a static (*i.e.* diffusion limited) system. Would you agree? Do you know if the compositions in your study are microemulsions at equilibrium?

Prof. Deutsch responded: Although I am not an expert on microemulsions, I agree with Dr Oldfield's analysis. The formation of a microemulsion is a consequence of the vanishing of the surface tension of the droplets in an emulsion. What is new in our case is the mechanism by which this occurs. In a conventional microemulsion this occurs *via* an increase in the bulk concentration of the surfactant, which increases the surface coverage of the droplets by the surfactant. With the right combination, this increased coverage leads to a decrease in the surface tension to zero. In our case, the reduction in surface tension is a consequence of the freezing of the surface, and thus the surface tension vanishes as the system's temperature is lowered to a point T within the range $T_b < T < T_{mix} < T_f$. The microemulsion phase should be stable, in principle, as long as it is kept below T_{mix}. Although we have no proof positive of this (the measurements are only a few weeks old), we did not observe aggregation within observation times of many hours. It is important to note that this system can be tuned into, and out of, the stable microemulsion phase easily by lowering and raising the temperature over a rather small range of a degree or two straddling T_{mix}. This is considerably more convenient and certainly faster, than doing so by changing the bulk surfactant concentration by adding or removing surfactant molecules.

Concluding remarks

Hubert H. Girault

Laboratoire d'Electrochimie Physique et Analytique, Ecole Polytechnique Fédérale de Lausanne, CH-1015, Lausanne, Switzerland

Received 11th October 2004, Accepted 11th October 2004
First published as an Advance Article on the web 15th December 2004

Dynamics and structure of the liquid–liquid interface

Compared to our understanding of the behaviour of molecules in the gas phase, or our knowledge of the solid state, it is regretful to say that our comprehension of the liquid phase suffers from a major gap between our perception at the molecular level and that at the hydrodynamic level. This inability to span from the microscopic to the macroscopic does not allow us to have a comprehensive vision of the structure of liquids and consequently of chemical reactivity in solution.

Most of the knowledge available at the end of the 20th century on the structure and dynamics of liquid–liquid interfaces stemmed mainly from surface tension measurements and thermodynamic analysis. Classical textbooks such as "The Physics and Chemistry of Surfaces" by N. K. Adam,[1] "Interfacial Phenomena" by J. T. Davies and E. K. Rideal[2] or more recently "Physical Chemistry of Surfaces" by A. W. Adamson[3] should be consulted to be reminded how rather simple experimental approaches could yield a rather good global image of the interface. What was missing at the time was direct experimental evidence of molecular properties that can now be provided as has been clearly shown in this *Faraday Discussion*.

We should thank the organising committee for having gathered scientists from different backgrounds in order to provide a rather comprehensive overview of the state-of-the-art about the most modern techniques to investigate liquid–liquid interfaces. We have seen during these three days that the recent experimental and theoretical developments converge to give a picture of the interface that depends strongly on the nature of the intermolecular forces. The purpose of these remarks is not to provide a summary of the conference, but to discuss some of the issues addressed during the meeting.

Probing the interface

Non-linear optical methods, developed over the last couple of decades, have the specific properties to enable the study of regions where a breakdown of centrosymmetry occurs, and in this respect, are ideally suitable to study liquid–liquid interfaces.

In the introductory Spiers Memorial Lecture, Prof. Geraldine Richmond[4] has shown how Sum Frequency Generation (SFG) coupled with molecular dynamics can unravel the structure of a series of interfaces such as alkane–water, carbon tetrachloride–water or even 1,2-dichloroethane–water. These experiments are difficult to perform, as we have been reminded, and are also difficult to analyse. However, when coupled to molecular dynamics, the data presented at this conference, in particular regarding the free OH stretch observed when water is in contact with hydrophobic solvents such as alkanes, provide direct information on the orientation of the interfacial water molecules. SFG data suggest that alkane–water and carbon tetrachloride–water are rather sharp

DOI: 10.1039/b415755a

interfaces whereas the 1,2-dichloroethane–water interface seems to be more "mixed" or diffuse as no free OH stretch band has been observed. It is clear that SFG is becoming an essential technique to probe the orientation and interactions of interfacial molecules.

X-Ray reflectivity measurements probe the electron density profile normal to the interface. Schlossman *et al.*[5] who have pioneered the use of this technique to study liquid–liquid interfaces, have clearly shown in this meeting that X-ray surface scattering measurements as a function of the wave vector transfer Q can provide useful and reliable information on the interfacial widths found to be equal to 450 pm and 710 pm, at 25 °C, for the water–nitrobenzene and water–2-heptanone interfaces, respectively. Variation of the temperature of the water–nitrobenzene interface from 25 °C to 55 °C also indicates that the role of the bending rigidity decreases with increasing temperature. In the case of polarised interfaces, preliminary results are very encouraging showing the possibility in the near future to study charge distribution at polarized interfaces. Of course, the analysis of X-ray reflectivity data is model dependent, but coupled with other techniques, this methodology has the major advantage to be able to study neat interfaces without recourse to the use of molecular probes.

Neutron reflectivity measurements at bare interfaces are extremely challenging, the key difficulty being to prepare large planar interfaces with a thin liquid film on a silicon or silica block. Also attenuation of the neutron transmission through the liquids is a critical constraint. To circumvent this inherent problem, contrast enhancers such as surfactants can be used to probe the interfacial region. As shown again during these discussions, the analysis of neutron reflectivity measurements are strongly model dependent, and this experimental approach is not mature enough yet to provide useful information. Neutron scattering on emulsions (as presented in the poster session) seems to be able to circumvent many experimental difficulties and may open the way to a wider use of neutron scattering for the study of liquid–liquid interfaces.

Surface second harmonic generation (SSHG) is also a non-linear optical technique that has been successfully used over the last decades. If it is also possible to probe neat interfaces, where quadrupoles or octupoles play a non-negligible role, this technique has been mainly used with molecular probes (such as push–pull molecules) where the data can be analysed within the simple dipolar approximation. In the meeting, the use of "molecular rulers" to probe the surface charge effects at the water–cyclohexane has been presented; in particular, the influence of the charged moiety on the electronic properties of the chromophore has been discussed. During the oral presentation of these concluding remarks, we have debated the philosophy on the use of molecular probes for SSHG. Often, the data provide more information on the probe itself than on the region it is supposed to probe. May be this issue would disappear if we had more specifically tailored and better-characterised probe molecules.

It has been shown in recent years that SHG could also be observed on spherical objects. During the meeting, it has been elegantly shown that this specific experimental approach could be used to monitor the transfer of hydrophobic ions across liposome membranes so as to gain access to ion transfer kinetics.

Time resolved spectroscopy at liquid–liquid interfaces can provide not only information on the solvation dynamics but also the surface roughness. Unfortunately, no papers in this area were submitted. The availability of ultrafast lasers and the fast response detectors make it possible to investigate directly electronic processes and the molecular motion at the interfaces. Time resolved fluorescence studies at liquid–liquid interfaces coupled with molecular dynamics calculations should provide new structural and dynamic information about the interfacial region.

If the topological structure of solid interfaces can be studied by scanning probe techniques (STM, AFM, *etc.*) or electron spectroscopy (LEED, Auger, *etc.*), probing liquid–liquid interfaces with scanning probes remains a challenge for the experimentalists. Nevertheless, preliminary STM data and a more detailed AFM study of interfacial molecular forces have been presented, and I would not be surprised to see in the future major developments obtained from scanning probe measurements.

On the experimental side, I really enjoyed the presentation of a novel laser technique to measure surface tensions. There are many ways to measure surface tension, but the method presented based on momentum conservation when a light beam crosses an interface is particularly elegant and rather efficient. Because this approach does not require any specific equipment, it is likely to be adopted by many research groups.

Molecular dynamics

To a certain extent, molecular dynamics fill part of the gap mentioned above between a molecular description of liquids and their hydrodynamic behaviour. The time domain remains limited to the nanosecond timescale and the number of molecules considered to the performance of the computer! As far as simulation is concerned, liquid–liquid interfaces being molecular by nature are rather easier to treat than metal–solution interfaces.

Over the three days of the meeting, different papers reported computer simulation results. Monte-Carlo calculations have been performed to calculate the average water dipole orientation at the water–benzene interface (Note in passing that benzene was discovered by Michael Faraday). Interfacial water molecules have been found to have dual orientation preferences: the molecules located nearest to the organic phase prefer to stay perpendicular to the interface, pointing flatly toward the non-polar phase by their dipole vectors, whereas the waters located somewhat farther from the organic phase prefer the parallel alignment with the interface.

Simulation can also be used to predict reactivity at the interface as long as the time scale of the reaction is compatible. The photo-dissociation of I_2^- has been studied *in silico*. When the photo-excited ion is located at the interface, a separation of the photo-products can be observed on some trajectories.

Room temperature molten salts or ionic liquids have received a lot of interest from the synthetic chemistry community, and computer simulation has been used to study their interface. Interestingly, the presence and orientation of the different ions can be distinguished according to their charge.

Personally, I have some reservations about molecular dynamics results that are not supported by experimental verification. As discussed a couple of times during this meeting, the interplay between theory and experiment is absolutely necessary to be able to draw any reliable conclusions.

ITIES

From the end of the 19th century, electrochemists such as Walter Nernst used the interface between two immiscible liquids to assemble concentration cells to measure transport numbers. The concept of polarising such an interface as a metal–solution interface was pioneered by Claude Gavach[6] in the seventies. Since then, the Interface between Two Immiscible Electrolyte Solutions (ITIES, an acronym coined by Jiri Koryta[7]) is a recognised field of interfacial electrochemistry. A major issue in the field is the Galvani potential difference scale. Indeed, a Galvani potential difference can be considered as the sum of an ionic term associated to the space charge regions and a dipolar term associated to the orientation of the interfacial dipoles. By definition, at the potential of zero charge (pzc) the former term is zero and the key question is to know if the latter term is significant or not. If it is not, the pzc can be easily measured by streaming a solution in the other and by measuring the resulting open circuit voltage, which would represent the zero of the Galvani potential difference scale.

The subject of ion transfer kinetics has been discussed briefly in these discussions. The major issue is to know if the timescale associated to the transfer of an ion from one phase to the other is slower than the timescale of diffuse layer relaxation. Unfortunately, the field of electron transfer reactions was not part of this meeting, as the measurement of the solvent re-organisation energy can yield indirect information on the interfacial structure.

In 1997, the group of Richard Compton in Oxford proposed a novel type of system whereby an interface is formed on a solid electrode.[8] The resulting so-called three-phase system is now developing as a field on its own as it provides new routes to combine ion transfer and electron transfer reactions. Two presentations were given to report how this technique can yield Gibbs energy of transfer values, but it is clear that the fundamentals of the three-phase systems remain a subject of discussion. The key issue to be resolved is that of the potential distribution especially in the vicinity of the triple point. Then, understanding the coupling of the different mass transfer processes will require major investigation.

Organised molecular assemblies at the liquid–liquid interface

Having studied and characterised the liquid–liquid interface, it is clear that the next step is to use the interface to construct molecular and supramolecular assemblies.

In this spirit, the reported interfacial freezing phenomenon induced by the surfactant template shows how one can tailor the properties of the interface. In the longer term, I would not be surprised to see molecular engineering developing on soft interfaces. Perhaps, at the next *Faraday Discussion* on liquid–liquid interfaces, some will report "molecular boats cruising on the interface" powered, for example, by light.

"If science is expensive, dreaming does not cost much..."

References

1   N. K. Adam, *The Physics and Chemistry of Surfaces*, Clarendon Press, Oxford, 1930.
2   J. T. Davies and E. K. Rideal, *Interfacial Phenomena*, Academic Press, New York, 1963.
3   A. W. Adamson, *Physical Chemistry of Surfaces*, John Wiley & Sons, New York, 1997.
4   M. A. Leich and G. L. Richmond, *Faraday Discuss.*, 2005, **129**, 1.
5   G. Luo, S. Malkova, S. V. Pingali, D. G. Schulz, B. Lin, M. Meron, T. G. Graber, J. Gebhardt, P. Vanysek and M. Schlossman, *Faraday Discuss.*, 2005, **129**, 23.
6   C. Gavach and F. Henry, *J. Electroanal. Chem.*, 1974, **54**(2), 361–370.
7   J. Koryta, P. Vanysek and M. Brezina, *J. Electroanal. Chem.*, 1976, **67**(2), 263–266.
8   F. Marken, R. G. Compton, C. H. Goeting, J. S. Foord, S. D. Bull and S. G. Davies, *Electroanalysis*, 1998, **10**(12), 821–826.

List of Posters

Bending rigidity of aqueous biopolymer mixtures **E. Scholten, L. M. C. Sagis** and **E. van der Linden**, *Wageningen University, The Netherlands*

Cyclic voltammetry with partially blocked electrodes: A similar probe for liquid–liquid reactions **T. J. Davies** and **R. G. Compton**, *Oxford University, UK*

Biphasic aqueous electrochemistry of redox oils: A novel biomimetic methodology? **R. G. Compton, A. J. Wain, C Amatore** and **J. D. Wadhawan**, *Oxford University, UK and Ecole Normale Supérieure, France*

Electrochemically driven ion transfer across liquid–liquid interfaces supported by composite ceramic carbon materials **M. Opallo, G. Shul** and **F. Marken**, *Polish Academy of Sciences, Poland*

Polyion adsorption and transfer at polarised liquid–liquid interfaces **E. Samcová, J. Langmaier, A. Trojánek** and **Z. Samec**, *Charles University, Czech Republic*

Study on liquid surface physics of surfactant solutions **A. Ozawa** and **A. Minamisawa**, *Japan Women's University, Japan*

X-Ray studies of surfactant ordering at water–hexane interfaces **A. M. Tikhonov, S. V. Pingali** and **M. L. Schlossman**, *University of Illinois at Chicago, USA*

The pristine oil/water interface **J. K. Beattie** and **A. M. Djerdjev**, *University of Sydney, Australia*

Controlled deposition at the liquid/liquid interface, growth of metal and binary alloy nanoparticles **M. Platt** and **R. A. W. Dryfe**, *University of Manchester (UMIST), UK*

Measurement of surface tension and elasticity in gel by laser manipulation technique **Y. Yoshitake, S. Mitani, K. Sakai** and **K. Takagi**, *University of Tokyo, Japan*

Wetting transitions of alkanes on cationic surfactant solutions **K. M. Wilkinson, H. Matsubara** and **C. D. Bain**, *Oxford University, UK*

Adsorbed films of alkanol and fluoroalkanol mixtures at the hexane/water interface studied by synchrotron X-ray reflectivity **T. Takiue, S. Venkatesh Pingali, G. Luo, A. Tikhonov, N. Ikeda, M. Schlossman** and **M. Aratono**, *Kyushu University, Japan*

A novel automated dispersion stability analyser **T. S. Horozov** and **B. P. Binks**, *University of Hull, UK*

Electrolytes at aqueous interfaces with room temperature ionic liquids: A MD investigation **A. Chaumont, R. Schurhammer** and **G. Wipff**, *Université Louis Pasteur, France*

A molecular dynamics study of a liquid–liquid-interface: Structure and dynamics **J. B. Buhn, P. A. Bopp** and **M. J. Hampe**, *Technische Universität Darmstadt, Germany*

Lactone cleavage reaction of rhodamine dye at liquid/liquid interface studied by micro-two-phase sheath flow/two-photon excitation fluorescence microscopy **T. Tokimoto, S. Tsukahara** and **H. Watarai**, *Osaka University, Japan*

Interfacial kinetics of synergistic extraction of samarium(III) studied by micro-two-phase sheath flow/fluorescence microscopy **T. Tokimoto, S. Tsukahara** and **H. Watarai**, *Osaka University, Japan*

Mass transfer of an alkyl sulfate anion with a ferrocene derivative cation across single microdroplet/water interface **T. Negishi** and **K. Nakatani**, *University of Tsukuba, Japan*

Soft structures formed by colloids due to liquid–liquid interfaces **P. S. Clegg, N. M. Phillips, J. Cleaver, W. C. K. Poon** and **M. E. Cates**, *University of Edinburgh, UK*

Experimental and molecular dynamic study on extraction kinetics of lithium with a mixture
of a fluorine-containing ß-diketone and a neutral organophosphorus compound
A. Sakoguchi, Y. Yanagiya and **F. Nakashio**, *Sojo University, Japan*

Correlation of the polarized potential window of the interfaces between water and
room-temperature molten salts with their solubility in water **N. Nishi** and **T. Kakiuchi**,
Kyoto University, Japan

A new method of surface rheology applied to viscoelasticity of a protein monolayer **P. Cicuta**
and **E. M. Terentjev**, *Cambridge University, UK*

Numerical modelling of a liquid–liquid interface under cycloidic motion **C. S. König** and
I. A. Sutherland, *Brunel University, UK*

Electrochemical synthesis of polymer coated gold nanoparticles at a liquid–liquid interface
R. Knake, A. W. Fahmi and **V. J. Cunnane**, *University of Limerick, Ireland*

Small angle neutron scattering to investigate the liquid–liquid interface **K. M. Knight,**
A. S. Wills and **D. J. Caruana**, *University College London, UK*

Modeling vapor–liquid and liquid–liquid aqueous interfaces with the gradient theory in
combination with the CPA EOS **A. J. Queimada, C. Miqueu, I. M. Marrucho,**
G. M. Kontogeorgis, and **J. A. P. Coutinho**, *Aveiro University, Portugal and*
Technical University of Denmark, Denmark

A flexible interfacial rheometer **C. A. Hickey, F. Le Moal, K. B. Peters** and **S. J. Roser,**
Camtel Ltd., UK

Thermodynamics of liquid/liquid interface at solid electrode shielded with the thin layer
of organic solvent: Towards electroactivity of redox-inactive proteins **A. A. Karyakin** and
M. Yu. Vagin, *M. V. Lomonosov Moscow State University, Russia*

Rapid emulsifier selection and evaluation of emulsion stability by analytical centrifugation
D. Lerche and **T. Sobisch**, *L.U.M. GmbH, Germany*

The formation of thin films from colloidal dispersions containing vermiculite to form an
electrical fuel cell: An illustration of various forces **S. G. Grayeff**, *Israel Institute of Technology,*
The Technicon, Haifa, Israel

List of Participants

Dr S. Appleyard, *Royal Society of Chemistry, UK*
Dr C. Bain, *University of Oxford, UK*
Professor J. Beattie, *University of Sydney, Australia*
Professor I. Benjamin, *University of California, USA*
Professor P. Bopp, *Université Bordeaux 1, France*
Mr J. Buhn, *TU Darmstadt, Germany*
Dr D. Caruana, *University College London, UK*
Miss C. Charlton, *Royal Society of Chemistry, UK*
Dr P. Cicuta, *University of Cambridge, UK*
Dr S. Clarke, *University of Cambridge, UK*
Dr P. Clegg, *University of Edinburgh, UK*
Professor R. Compton, *University of Oxford, UK*
Dr H. Crichton, *Royal Society of Chemistry, UK*
Dr R. Dagastine, *University of Melbourne, Australia*
Mr T. Davies, *University of Oxford, UK*
Professor M. Deutsch, *Bar-Ilan University, Israel*
Dr R. Dryfe, *University of Manchester (UMIST), UK*
Professor K. Eisenthal, *Columbia University, USA*
Professor D. Fermín, *Universität Bern, Switzerland*
Dr A. Fisher, *University of Cambridge, UK*
Mr A. Graham, *Scientific & Medical Products Ltd, UK*
Professor H. Girault, *EPFL, Switzerland*
Dr S. Grayeff, *The Technicon, Haifa, Israel*
Miss C. Hall, *Royal Society of Chemistry, UK*
Professor M. Hampe, *TU Darmstadt, Germany*
Dr T. Horozov, *University of Hull, UK*
Professor J. Janata, *Georgia Tech, USA*
Dr P. Jedlovszky, *ELTE University, Hungary*
Professor A. Karyakin, *M. V. Lomonosov Moscow State University, Russia*
Dr R. Knake, *University of Limerick, Ireland*
Miss K. Knight, *University College London, UK*
Dr C. König, *Brunel Institute for Bio-engineering, UK*
Professor A. Kornyshev, *Imperial College London, UK*
Mr R. Le Roux, *University of Cambridge, UK*
Dr F. Leermakers, *Wageningen University, The Netherlands*
Mr G. Lian, *Unilever R&D Colworth, UK*
Dr L. Liggieri, *Istituto der l'Energetica e le Interfasi, Italy*
Dr H. Lunn, *Royal Society of Chemistry, UK*
Professor R. Lynden-Bell, *University of Cambridge, UK*
Dr F. Marken, *Loughborough University, UK*
Miss S. Matthews, *University of Cambridge, UK*
Dr S. Mitani, *University of Tokyo, Japan*
Mr T. Negishi, *University of Tsukuba, Japan*
Dr N. Nishi, *Kyoto University, Japan*
Dr C. Oldfield, *MicroScience Technologies Group, UK*
Dr M. Opallo, *Polish Academy of Sciences, Poland*
Dr A. Ozawa, *Japan Women's University, Japan*
Ms S. Patel, *Bayer CropScience AG, Germany*
Professor J. Penfold, *CCLRC Rutherford Appleton Laboratory, UK*
Mr K. Peters, *Camtel Ltd., UK*
Professor M. Pilling, *University of Leeds, UK*
Mr M. Platt, *University of Manchester (UMIST), UK*
Mr A. Queimada, *Universidade De Aveiro, Portugal*
Miss A. Querol Suquia, *Queen Mary, University of London, UK*
Professor J. Rathman, *Ohio State University, USA*
Professor G. Richmond, *University of Oregon, USA*

Dr D. Rossetti, *Unilever, UK*
Professor J. Rusling, *University of Connecticut, USA*
Dr A. Sakoguchi, *Sojo University, Japan*
Dr E. Samcova, *Charles University, Czech Republic*
Professor Z. Samec, *J. Heyrovsky Institute of Physical Chemistry, Czech Republic*
Dr M. Sastry, *National Chemical Laboratory, Pune, India*
Professor M. Schlossman, *University of Illinois at Chicago, USA*
Miss E. Scholten, *Wageningen University, The Netherlands*
Professor F. Scholz, *Universitat Greifswald, Germany*
Miss R. Schurhammer, *Université Louis Pasteur, France*
Dr M. Stephenson, *University of Manchester (UMIST), UK*
Professor K. Takagi, *University of Tokyo, Japan*
Dr T. Takiue, *Kyushu University, Japan*
Dr T. Tokimoto, *Osaka University, Japan*
Professor T. Tominaga, *Okayama University of Science, Japan*
Professor P. Unwin, *University of Warwick, UK*
Professor M. Urbakh, *Tel Aviv University, Israel*
Professor B. Van der Veken, *University of Antwerp, Belgium*
Dr J. Wadhawan, *Ecole Normale Superieure, France*
Dr A. Wakisaka, *National Institute of Advanced Industrial Science and Technology, Japan*
Dr R. Walker, *University of Maryland, USA*
Dr J. Webster, *CCLRC, UK*
Professor H. Wennerström, *Lund University, Sweden*
Mr T. Wharton, *Christison Particle Technologies, UK*
Dr C. Whitby, *University of Hull, UK*
Miss K. Wilkinson, *University of Oxford, UK*
Miss Y. Yoshitake, *University of Tokyo, Japan*
Mr T. Youngs, *Queens University Belfast, UK*
Dr A. Zarbakhsh, *Queen Mary, University of London, UK*

Index of Contributors*

* The page numbers in **bold** type indicate papers submitted for discussions.